THE HUMAN MICROBIOTA

THE HUMAN MICROBIOTA

How Microbial Communities Affect Health and Disease

Edited by
David N. Fredricks

WILEY Blackwell

Cover Design: Michael Rutkowski
Cover Illustrations: top five panels © courtesy of David N. Fredricks; right side art © Mads Abildgaard/iStockphoto

Published by John Wiley & Sons, Inc., Hoboken, New Jersey
Published simultaneously in Canada

Library of Congress Cataloging-in-Publication Data:

The human microbiota: how microbial communities affect health and disease / edited by David N. Fredricks.
p. cm.
Includes bibliographical references and index.
ISBN 978-0-470-47989-6 (cloth)
1. Human body–Microbiology. 2. Microorganisms. I. Fredricks, David N.
QR46.H86 2013
579–dc23

2012015251

Printed and bound in Singapore by Markono Print Media Pte Ltd

10 9 8 7 6 5 4 3 2 1

CONTENTS

PREFACE vii

CONTRIBUTORS xi

1 **THE NIH HUMAN MICROBIOME PROJECT** 1
Lita M. Proctor, Shaila Chhibba, Jean McEwen, Jane Peterson, Chris Wellington, Carl Baker, Maria Giovanni, Pamela McInnes, and R. Dwayne Lunsford

2 **METHODS FOR CHARACTERIZING MICROBIAL COMMUNITIES ASSOCIATED WITH THE HUMAN BODY** 51
Christine Bassis, Vincent Young, and Thomas Schmidt

3 **PHYLOARRAYS** 75
Eoin L. Brodie and Susan V. Lynch

4 **MATHEMATICAL APPROACHES FOR DESCRIBING MICROBIAL POPULATIONS: PRACTICE AND THEORY FOR EXTRAPOLATION OF RICH ENVIRONMENTS** 85
Manuel E. Lladser and Rob Knight

5 **TENSION AT THE BORDER: HOW HOST GENETICS AND THE ENTERIC MICROBIOTA CONSPIRE TO PROMOTE CROHN'S DISEASE** 105
Daniel N. Frank and Ellen Li

6 **THE HUMAN AIRWAY MICROBIOME** 119
Edith T. Zemanick and J. Kirk Harris

7 **MICROBIOTA OF THE MOUTH: A BLESSING OR A CURSE?** 135
Angela H. Nobbs, David Dymock, and Howard F. Jenkinson

8 **MICROBIOTA OF THE GENITOURINARY TRACT** 167
Laura K. Sycuro and David N. Fredricks

9 **FUNCTIONAL STRUCTURE OF INTESTINAL MICROBIOTA IN HEALTH AND DISEASE** 211
Alexander Swidsinski and Vera Loening-Baucke

10 **FROM FLY TO HUMAN: UNDERSTANDING HOW COMMENSAL MICROORGANISMS INFLUENCE HOST IMMUNITY AND HEALTH** **255**
June L. Round

11 **INSIGHTS INTO THE HUMAN MICROBIOME FROM ANIMAL MODELS** **273**
Bethany A. Rader and Karen Guillemin

12 **TO GROW OR NOT TO GROW: ISOLATION AND CULTIVATION PROCEDURES IN THE GENOMIC AGE** **289**
Karsten Zengler

13 **NEW APPROACHES TO CULTIVATION OF HUMAN MICROBIOTA** **303**
Slava S. Epstein, Maria Sizova, and Amanda Hazen

14 **MANIPULATING THE INDIGENOUS MICROBIOTA IN HUMANS: PREBIOTICS, PROBIOTICS, AND SYNBIOTICS** **315**
George T. Macfarlane and Sandra Macfarlane

INDEX **339**

PREFACE

The human body is a marvelously intricate machine, and the parts list includes trillions of microbial cells that colonize epithelial surfaces such as those found in the mouth and gut. There is increasing evidence that these microbes do more than just reside on tissues—they play key roles in human physiology and organ function. Indeed, there are 100 times more genes in our microbiome compared to our human genome, and these microbial genes code for proteins that impact diverse processes such as digestion, immunity, and development. The goal of this book is to provide an overview of the microbial diversity found in humans and to describe efforts linking microbial communities to human health. Attempts to understand human-associated microbial communities were given a boost by Human Microbiome Project (HMP) initiatives in the United States, Canada, Europe, and Asia. The National Institutes of Health in the United States has devoted more than $150 million to support these studies that are now maturing with release of data and a flurry of publications. This is a time of unprecedented discovery, and although still young, the field is sufficiently advanced to warrant a book summarizing progress. Answers to many questions are now emerging. How do microbial communities differ across body sites? What is the variability in microbial composition across healthy and diseased humans at the same body site? How do certain microbial communities foster healthy tissues? What are the microbial community profiles associated with disease states, and are these communities markers of disease or causes of disease? How can microbial communities be manipulated to optimize health and minimize disease risk? How do microbial communities change over the course of human development? What are the internal factors (genetic, anatomic, hormonal, physiologic) and external environmental factors (diet, sexual activity, hygiene) that shape human-associated microbial communities? With partial answers to these questions come many additional questions about the intimate relationships between human and microbial cells in our bodies.

Our excursion into the human microbiome begins with an introduction to the Human Microbiome Project by Lita Proctor and colleagues from the NIH (Chapter 1). This chapter provides an excellent description of the HMP with its many research initiatives and early progress. It also provides some historical context and a vision for future research. We then shift to chapters focused on tools for studying the human microbiome, including a methodological overview chapter (Chapter 2) by Christine Bassis, Vincent Young, and Tom Schmidt that lays the groundwork for later chapters. Bassis and colleagues compare different genomic cultivation-independent methods for characterizing microbial communities, highlighting the advantages and limitations of commonly used techniques. They also consider the role of cultivation methods in the genomic era, and provide advice about designing microbiome studies. In their chapter, Susan Lynch and Eoin Brodie delve more deeply into the use of phyloarrays for microbial community analysis (Chapter 3), a technique that has some distinct advantages for microbial community profiling. Manuel Lladser and Rob Knight round out our methods section by providing a mathematical perspective on interpreting microbial community structure and diversity. This chapter (Chapter 4) is highly relevant in the era of high-throughput sequencing of phylogenetically informative microbial gene sequences [generated by polymerase chain reaction (PCR) or from metagenomic methods] for describing microbial populations.

We then begin a tour of various microbial niches of the human body, such as the gut, respiratory tract, mouth, and genital tract. These chapters highlight the different microbial populations found in different human tissues, and describe how microbial communities change with conditions such as gingivitis, inflammatory bowel disease, and bacterial vaginosis. Dan Frank and Ellen Li start this tour with a description of how host genetics (immune response) and the gut microbiota may interact to facilitate Crohn's disease (Chapter 5). They also introduce the concept of dysbiosis that will be used in other chapters. Edith Zemanick and J. Kirk Harris then describe the microbiota of the human respiratory tract (Chapter 6), focusing on the normal microbiota and alterations in conditions such as cystic fibrosis and ventilator-associated pneumonia. The oral microbiota is described in a chapter by Angela Nobbs, David Dymock, and Howard Jenkinson (Chapter 7). Here they note some of the physical and metabolic interactions among the 600 different bacteria species that live in the mouth. They also describe the connections between some oral microbial communities and local conditions such as caries, and systemic diseases such as endocarditis. Laura Sycuro and myself review the genital tract microbiota of women and men, with a particular focus on the condition bacterial vaginosis that is associated with numerous adverse health outcomes in women and neonates (Chapter 8). Alexander Swidsinski and Vera Loening-Baucke end this section with a chapter on use of *in situ* hybridization methods combined with fluorescence microscopy for describing the spatial relationships of microbes and human cells in the gut. This chapter (Chapter 9) is notable for moving beyond the description of "who's there" to a description of the structural and functional features of the gut microbiota. Note that not every human body niche is covered in these chapters.

Two chapters focus on the use of animal models to manipulate the microbiota and understand how changes impact health. June Round tackles the use of a variety of animal models to study host immunity, including the fruit fly, zebrafish, and mouse (Chapter 10). She highlights key lessons that can be learned from these models regarding human immune responses to our indigenous microbiota. The team of Bethany Rader and Karen Guillemin describe new insights that have been produced by animal models, including fish (Chapter 11). Several important questions are more easily answered with animal models, including how microbial communities assemble in space and time, and identifying the relative contributions of host genetics, environmental factors (such as diet), and stochastic sampling. Microbes and animals can be studied in the laboratory in ways that are not possible in human studies.

The use of cultivation-independent methods suggests that many human-associated microbes still resist cultivation in the laboratory. Karsten Zengler tackles this issue in his chapter on cultivation procedures in the genomic age (Chapter 12), reinforcing the challenges and rewards of cultivating microbes from the human body. So, how can we cultivate fastidious members of the human microbiota? The chapter by Slava Epstein, Maria Sizova, and Amanda Hazen (Chapter 13) provides many novel approaches for cultivating human microbes using cutting-edge techniques in order to resolve the "great plate count anomaly."

Finally, George and Sandra Macfarlane (Chapter 14) address a key issue in the microbiome field: how one can manipulate the human microbiota. These investigators provide a thoughtful and balanced review on the use of prebiotics, probiotics, and synbiotics to alter the human indigenous microbiota for the purpose of enhancing health.

What does it mean to be human? The authors of these chapters provide a compelling argument that we are far from alone, and that our microbiota helps mold the human form. Please enjoy the insights from this outstanding collection of investigators who are unraveling the mysteries of the human microbiome.

Special thanks to Sue Bartlett for helping to bring these chapters together and assisting the editors with organization of the project.

David N. Fredricks, MD
Fred Hutchinson Cancer Research Center
Seattle, Washington

CONTRIBUTORS

Carl Baker, MD, PhD National Institute of Arthritis and Musculoskeletal and Skin Diseases (NIAMS), National Institutes of Health (NIH), Bethesda, Maryland 20892

Christine Bassis, PhD Department of Internal Medicine, University of Michigan, Ann Arbor, Michigan 48109

Eoin L. Brodie, PhD Ecology Department, Earth Sciences Division, 1 Cyclotron Road, Lawrence Berkeley National Laboratory, Berkeley, California 94720

Shaila Chhibba National Human Genome Research Institute (NHGRI), National Institutes of Health (NIH), Bethesda, Maryland 20892

David Dymock, PhD School of Oral and Dental Sciences, University of Bristol, Lower Maudlin Street, Bristol BS1 2LY, United Kingdom

Slava S. Epstein, PhD Department of Biology, Northeastern University, Boston, Massachusetts 02115

Daniel N. Frank, PhD Division of Infectious Diseases, School of Medicine, University of Colorado, Mucosal and Vaccine Research Program Colorado (MAVRC), and UC-Denver Microbiome Research Consortium (MiRC), Denver, Colorado 80045

David N. Fredricks, MD Vaccine and Infectious Disease Division, Fred Hutchinson Cancer Research Center, 1100 Fairview Avenue North, Seattle, Washington 98109

Maria Giovanni, PhD National Institute of Allergy and Infectious Diseases (NIAID), National Institutes of Health (NIH), Bethesda, Maryland 20892

Karen Guillemin, PhD Institute of Molecular Biology, University of Oregon, Eugene, Oregon 97403

J. Kirk Harris, PhD Department of Pediatrics, University of Colorado Anschutz Medical Campus, Aurora, Colorado 80045

Amanda Hazen, MS Department of Biology, Northeastern University, Boston, Massachusetts 02115

Howard F. Jenkinson, PhD School of Oral and Dental Sciences, University of Bristol, Lower Maudlin Street, Bristol BS1 2LY, United Kingdom

Rob Knight, PhD Howard Hughes Medical Institute and Department of Chemistry and Biochemistry, University of Colorado, Boulder, Colorado 80309

Ellen Li, MD, PhD Department of Medicine, Stony Brook University, Stony Brook, New York 11790

Manuel E. Lladser, PhD Department of Applied Mathematics, University of Colorado, Boulder, Colorado 80309

Vera Loening-Baucke, MD The University Hospital Charité of the Humboldt University at Berlin, Chariteplatz 1, 10117 Berlin, Germany

R. Dwayne Lunsford, PhD Program Director, Microbiology Program, Integrative Biology and Infectious Disease Branch, Division of Extramural Research, National Institute of Dental and Craniofacial Research (NIDCR), National Institutes of Health (NIH), Bethesda, Maryland 20892

Susan V. Lynch, PhD Division of Gastroenterology, Department of Medicine, University of California, 513 Parnassus Avenue, San Francisco, California 94143

George T. Macfarlane, PhD The University of Dundee, Microbiology and Gut Biology Group, Ninewells Hospital Medical School, Dundee DD1 9SY, United Kingdom

Sandra Macfarlane, PhD The University of Dundee, Microbiology and Gut Biology Group, Ninewells Hospital Medical School, Dundee DD1 9SY, United Kingdom

Jean McEwen, JD, PhD National Human Genome Research Institute (NHGRI), National Institutes of Health (NIH), Bethesda, Maryland 20892

Pamela McInnes, DDS, MSc National Institute of Dental and Craniofacial Research (NIDCR), National Institutes of Health (NIH), Bethesda, Maryland 20892

Angela H. Nobbs, PhD School of Oral and Dental Sciences, University of Bristol, Lower Maudlin Street, Bristol BS1 2LY, United Kingdom

Jane Peterson, PhD National Human Genome Research Institute (NHGRI), National Institutes of Health (NIH), Bethesda, Maryland 20892

Lita M. Proctor, PhD National Human Genome Research Institute (NHGRI), National Institutes of Health (NIH), Bethesda, Maryland 20892

Bethany A. Rader, PhD Department of Molecular and Cell Biology, University of Connecticut, Storrs, Connecticut 06269

June L. Round, PhD Department of Pathology, Division of Microbiology and Immunology, University of Utah, Salt Lake City, Utah 84112

Thomas Schmidt, PhD Department of Microbiology and Molecular Genetics, Michigan State University, East Lansing, Michigan 48824

Maria Sizova, PhD Department of Biology, Northeastern University, Boston, Massachusetts 02115

Alexander Swidsinski, MD, PhD The University Hospital Charité of the Humboldt University at Berlin, Chariteplatz 1, 10117 Berlin, Germany

Laura K. Sycuro, PhD, MSc Vaccine and Infectious Disease Division, Fred Hutchinson Cancer Research Center, 1100 Fairview Avenue North, Seattle, Washington 98109

Chris Wellington National Human Genome Research Institute (NHGRI), National Institutes of Health (NIH), Bethesda, Maryland 20892

Vincent Young, MD, PhD Department of Internal Medicine, University of Michigan, Ann Arbor, Michigan 48109

Edith T. Zemanick, MD Department of Pediatrics, University of Colorado Anschutz Medical Campus, Aurora, Colorado 80045

Karsten Zengler, PhD Department of Bioengineering, University of California, San Diego, 9500 Gilman Drive, La Jolla, California 92093

1

THE NIH HUMAN MICROBIOME PROJECT

LITA M. PROCTOR, SHAILA CHHIBBA, JEAN McEWEN, JANE PETERSON, and CHRIS WELLINGTON

NHGRI/NIH, Bethesda, Maryland

CARL BAKER

NIAMS/NIH, Bethesda, Maryland

MARIA GIOVANNI

NIAID/NIH, Bethesda, Maryland

PAMELA McINNES and R. DWAYNE LUNSFORD

NIDCR/NIH, Bethesda, Maryland

1.1. INTRODUCTION

The human microbiome is the full complement of microbial species and their genes and genomes that inhabit the human body. The National Institutes of Health (NIH) Human Microbiome Project (HMP) is a community resource project designed to promote the study of complex microbial communities involved in human health and

The Human Microbiota: How Microbial Communities Affect Health and Disease, First Edition. Edited by David N. Fredricks.

disease. The HMP has increased the appreciation for the features of the human microbiome that all people share as well as the features that are highly personalized. Host genetics, the environment, diet, the immune system, and many other factors all interact with the human microbiota to regulate the composition and function of the microbiome. As a scientific resource, the HMP has publically deposited to date or made available over 800 reference microbial genome sequences, hundreds of microbial isolates from the human microbiome, over 3 terabases (Tbp) of metagenomic microbial sequence, over 70 million 16S rRNA reads, close to 700 microbiome metagenome assemblies, over 5 million unique predicted genes, and a comprehensive bodywide survey of the human microbiome in hundreds of individuals from a healthy adult cohort. A number of demonstration projects are contributing a wealth of knowledge about the association of the microbiome with specific gut, skin, and urogenital diseases. Other key resources include the development of new computational tools, technologies, and scientific approaches to investigate the microbiome, and studies of the ethical, legal, and social implications of human microbiome research. This chapter captures the historical context of the HMP and other international research endeavors in the human microbiome, highlights the multiple initiatives of the HMP program and the products from this activity, and closes with some suggestions for future research needs in this emerging field.

1.2. GENESIS OF HUMAN MICROBIOME RESEARCH AND THE HUMAN MICROBIOME PROJECT (HMP)

It sometimes seems that research on the human microbiome blossomed overnight. However, the conceptual and technological foundations for the study of the human microbiome began to emerge before the 1990s and can be found within many disciplines. Microbial ecologists who studied microorganisms and microbial communities in the environment recognized early on that most microorganisms in nature were not culturable and so developed alternate approaches to the study of microbial communities. An early and broadly adopted approach for investigating microorganisms in the environment, based on the three-domain system for biological classification [1], was the use of the 16S ribosomal RNA gene as a taxonomic marker for interrogating microbial diversity in nature [2]. With the growth of non-culture-based, molecular techniques in the 1980s and 1990s for study of environmental microorganisms and communities, some medical microbiologists turned these tools to the human body and found far greater microbial diversity than expected, even in well-studied sites such as the oral cavity [3–5].

In the infectious disease field, recognition was growing that many diseases could not satisfy Koch's postulates as the pathogenesis of many of these diseases appeared to involve multiple microorganisms. The term *polymicrobial diseases* was coined to describe those diseases with multiple infectious agents [6]. We now recognize that many of these formerly classified polymicrobial diseases, such as abscesses, AIDS-related opportunistic infections, conjunctivitis, gastroenteritis, hepatitis, multiple sclerosis, otitis media, periodontal diseases, respiratory diseases, and genital infections, are associated with multiple microbial factors, that is, with the entire microbiome. In an essay on the history of microbiology and infectious disease, Lederberg [7], who coined the term *microbiome*, called for "a more ecologically informed metaphor" to understand the relationship between humans and microbes.

The field of immunology was also undergoing its own revolution with the recognition that the innate and adaptive immune systems not only evolved to eliminate specific pathogens but are also intimately involved in shaping the composition of the commensal intestinal microbiota [8–10]. Recognition was also growing in this field that the microbiota is involved in regulating gut development and function [11,12].

Another key catalyst for discussions about the inclusion of the microbiome in the study of human health and disease was the publication of the first drafts of the human genome sequence. Relman and Falkow [13] noted on this occasion that a "second human genome project" should be undertaken to produce a comprehensive inventory of microbial genes and genomes associated with the human body. Lead by Davies [14], they renewed a call for considering the role of the human-associated microorganisms in development and in health and disease. Also, by 2005 or so, as sequencing costs began to drop, sequencing technology offered the opportunity to consider extensive surveys of the microbial communities associated with the human. Early human studies focusing on the most complex of human microbiomes, the digestive tract [15,16], demonstrated the tremendous complexity as well as the functional potential of the human microbiome.

The time appeared right to undertake a comprehensive study of the human microbiome—the full complement of microbial species and their genes and genomes that inhabit the body. A meeting, organized by the French National Institute for Agricultural Research (INRA), of European, North American and Asian scientists and government agency and private-sector representatives was convened in Paris in 2005 to discuss how to approach such a comprehensive study. This 2-day meeting covered a broad range of topics, including sequencing all of the bacteria in the human microbiome, the impact of the human microbiome on the study of health, and the possible structure of a human digestive tract microbiome program. Recommendations from this first international meeting included the formation of an International Human Microbiome Consortium and an agreement to release data rapidly, share data standards, and develop reference datasets (`http://www.human-microbiome.org/fileadmin/user_upload/Paris-recommendations.pdf`). Around this same time, the National Academy of Sciences published a report on metagenomics [17] (`http://books.nap.edu/catalog.php?record_id=11902`), which highlighted this new discipline with its focus on the combination of genomics, bioinformatics, and systems biology to study microbial communities in nature; this report also informed the scientific community of the potential of this new discipline. The Paris meeting was followed by several other international meetings in 2007 and 2008.

These discussions led to the formation of the European Commission's call for studies on human metagenomics. The NIH also invited community comment during this incubation period. A number of white papers identified specific needs for the field that included a reference microbial genome sequence catalog, animal models for microbiome studies, benchmarking studies for the analysis of 16*S* rRNA and microbiome metagenome sequencing, computational tools for the field, and considerations of the ethical aspects of human microbiome research. Pilot projects to develop protocols for sequencing the human microbiome were begun by the NIH National Human Genome Research Institute (NHGRI) in mid-2007. The

NIH Common Fund–supported Human Microbiome Project (HMP) was formally launched in late 2007 with the intent to produce a number of major community resources: a reference catalog of microbial genome sequences, a large cohort study to survey microbiomes across the human body in healthy adults, a suite of demonstration projects to examine correlations of changes in the microbiome with disease, and the computational tools to analyzing microbiome metagenomic sequence data (http://commonfund.nih.gov/hmp/). Funding of the Metagenomics of the Human Intestinal Tract (MetaHIT) program began in 2008, which included scientific partnerships across eight European countries (http://www.metahit.eu/). Other large-scale efforts in human microbiome research emerged in close order around the world and include, among others, the NIH HIV Lung Microbiome Project, the Gambian Gut Microbiome Project, the INRA French/China program MicroObes, the Canadian Human Microbiome Initiative, the Australian Jumpstart Human Microbiome Project, and the Korean Twin Cohort Microbiome Diversity project.

1.3. GUIDING PRINCIPLES, STRUCTURE, AND INITIATIVES OF THE HMP PROGRAM

1.3.1. HMP Guiding Principles and Creation of a Community Resource Project

The Human Microbiome Project was envisioned as a community resource program. A community resource program is defined as a research project "specifically devised and implemented to create a set of data, reagents or other material whose primary utility will be as a resource for the broad scientific community" (http://www.genome.gov/10506537). It was recognized that the metagenomic and associated metadata from human microbiome research are unique research resources. In order to establish and serve as a community resource, the guiding principles for the HMP included rapid data release into public databases. These follow the guiding principles that were created for the Human Genome Project and have been used for all large genome projects at NIH (https://commonfund.nih.gov/hmp/datareleaseguidelines.aspx).

At the same time, it was expected that users of the prepublication data would acknowledge the scientific contribution of the HMP data producers by following normal standards of scientific etiquette and fair use of unpublished data. These standards were outlined in the 2003 Fort Lauderdale agreement (http://www.genome.gov/10506537) and further elaborated in the 2009 Toronto meeting agreement (Toronto International Data Release Workshop Authors, [18]; doi: 10.1038/461168a). An HMP Research Network Consortium was established to enhance collaborative activities and to support large-scale analyses of the HMP data, the products of which would contribute to the overall community resource. A consortium agreement, signed by all members outlined the request to acknowledge the data producers' contributions. New consortium members, nominated by existing consortium members, are asked to agree to the consortium statement. A marker paper that described the HMP and its data release policy was published (NIH HMP Working Group, [19]; doi 10:1101/gr096651.109) and serves as an outline of the large-scale analyses that the HMP Consortium is undertaking.

In addition, a data use agreement was drafted to provide guidance for users of the prepublication data from the larger community. The data use agreement, posted on the DACC website (http://hmpdacc.org/resources/data_browser.php), reiterated the Fort Lauderdale and Toronto meeting guidelines and also provided guidance on how publications that use HMP data should acknowledge and cite the HMP Consortium and the NIH as a source of the data. Finally, an agreement was made that all reagents, such as the reference microbial strains to be sequenced, should be deposited in appropriate repositories.

For the healthy cohort study, it was recognized that whole-genome shotgun sequencing (WGS) of nucleic acid extracts would capture various amounts of the human subject genome sequence, depending on the amount of human tissue collected during the microbiome sampling procedure. It was decided that the human genome sequence would not be made publically available but that the research community, with appropriate authorization, should have access to human subject data for research on the human microbiome. The NIH National Center for Biotechnology Information (NCBI) database of genotypes and phenotypes (dbGaP: http://www.ncbi.nlm.nih.gov/projects/gap/cgi-bin/about.html) was adopted at the public database for the HMP clinical metadata and sequence data (http://www.ncbi.nlm.nih.gov/gap?term=Human%20Microbiome%20Project). The dbGaP has two levels of access—open and controlled—in order to regulate the distribution of the sequence and health information of the study volunteers. Open access contains publically accessible data. Controlled access requires approval by a NIH Data Access Committee (DAC) for legitimate microbiome research purposes.

The WGS sequence data were computationally filtered to remove the human subject sequence before these data were deposited in the open access portion of the sequence read archive (SRA) in dbGaP. The criteria and procedure for removing human sequence is described later in this chapter. Clinical patient metadata were deposited in the controlled access portion of dbGaP. The procedures for requesting access to the controlled data can be found at the following website: https://dbgap.ncbi.nlm.nih.gov/aa/wga.cgi?login=&page=login. The HMP-targeted 16*S* ribosomal RNA gene sequence data were deposited in the open access SRA in dbGaP as there is no human sequence associated with these data.

Whereas other national and international programs focused on the microbiome of a specific body site, the HMP decided to survey the microbiomes of multiple body sites in healthy adults to produce baseline data for healthy microbiomes, develop a catalog of microbial genome sequences of microbiome reference strains, and evaluate the associations of microbial communities with specific diseases. A Data Analysis and Coordination Center (DACC) was created to manage the data from the sequencing activities, process the sequence data to consortium agreed-on standards for further analysis, coordinate the data analysis activities in the consortium, and serve as a portal for the scientific community to access the datasets, tools, and other resources generated by the program. In addition, initiatives in technology development; computational tools; and the ethical, legal, and societal implications of microbiome research were created to support the field. There are three sources of information about the HMP program. The NIH Common Fund website provides an overview of the main initiatives in the program (https://commonfund.nih.gov/hmp/). The NCBI Bioprojects pages describe the data types produced in the program (http://www.ncbi.nlm.nih.gov/bioproject/43021). There are four projects listed by NCBI under the HMP umbrella based on the four data types produced: (1) the 16S rRNA gene and (2) whole-genome shotgun metagenome datasets produced from the healthy adult cohort study, (3) the reference strain microbial genome

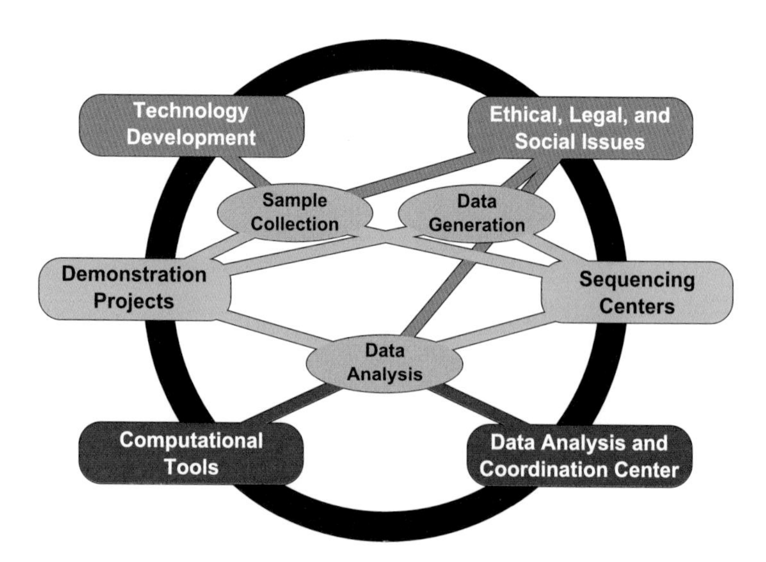

Figure 1.1. Conceptual diagram of the NIH Human Microbiome Project. The HMP program is comprises of six formal Initiatives, shown around the circle and include technology development, ethical, legal, and social issues; sequencing centers, the data analysis and coordination center; computational tools; and the demonstration projects. These initiatives interact through the activities of the ≥200-member HMP research network consortium, which also includes members of the larger scientific community and NIH program staff. The consortium activities, shown in the three interior bubbles, include (1) sample collection, which includes the clinical protocols development and collection of microbiome specimens and nucleic acid extract sample preparation from the specimens in the healthy cohort study and in the demonstration projects; (2) data generation, which includes the sequencing activities for the healthy cohort, demonstration projects, and the reference strain microbial genomes; and (3) data analysis, which includes the extensive data processing, benchmarking, and quality control steps needed to produce data for public release and for the analysis of microbiome sequence data by the consortium. The connecting lines graphically depict the major interactions between the initiatives.

sequence dataset, and (4) the datasets produced in the individual demonstration project activities. Finally, the DACC provides an extensive web resource that describes the datasets produced by the program, the derivative datasets developed by the HMP Working Groups, the suite of computational tools developed for the analyses, and other contextual information about the HMP (`www.hmpdacc.org`). A conceptual diagram of the initiatives within the HMP program and their interrelationships and how the initiative research teams and the research consortium interacts provides another view of this program (Figure 1.1). Using this figure, the HMP program is described below.

1.3.2. HMP Large-Scale Sequencing Centers

In order to establish scientific approaches and protocols for the Human Microbiome Project and to be able to sequence very large numbers of HMP samples, the first initiative in the HMP included the support of four large-scale sequencing centers:

Baylor College of Medicine (http://www.hgsc.bcm.tmc.edu/), the Broad Institute (http://www.broadinstitute.org/), the J. Craig Venter Institute (http://www.jcvi.org/), and Washington University at St. Louis (http://genome.wustl.edu/). These sequencing centers are responsible for (1) developing the protocols, (2) sequencing microbiome samples from a baseline adult population of healthy human subjects and reference strain microbial genomes, (3) analyzing the microbiome sequence data, (4) providing computational approaches, and (5) contributing to the analysis of the healthy subject microbiome data. These centers are also responsible for supporting the sequencing activities in several of the demonstration projects (discussed in further detail below). Further, the sequencing center project investigators provided oversight for data production objectives and goals.

1.3.3. Data Coordination and Analysis

Data Analysis and Coordination Center (DACC)

The Data Analysis and Coordination Center (DACC) was established in order to facilitate data deposition and to coordinate processing and analysis of the very large datasets produced by the HMP (www.hmpdacc.org). In order to support HMP activities, the DACC established a human microbiome database and developed a comprehensive analysis pipeline. The DACC plays a major role in the establishment, coordination, and support of an HMP Research Network Consortium, which was made up of members of the microbiome community interested in participating in the analysis of the large HMP dataset as well as the various workgroups, which focus on specific tasks. The DACC hosts an electronic collaboration site where data analyses, workgroup discussions, and publication drafts can be shared within the consortium. The DACC supports extensive community outreach and training activities. For example, the DACC website includes the project catalog of the reference genome sequences, a browser that includes links to the datasets from the benchmarking activities, the healthy cohort study, the demonstration projects, and many of the bioinformatics and computational tools that are used in the project.

HMP Workgroups

It was recognized that large-scale analyses of these new and complex datasets, particularly of the healthy adult cohort data (discussed below) would add value to the resources emerging from the program. This would require the efforts of a large group of scientists. Thus, the Data Analysis Working Group (DAWG) was formed and consisted of a combination of HMP grantees as well as individuals in the scientific community with specific expertise in the analysis of metagenomic data, all who joined the Research Network Consortium. During the 2 years of active data processing, analysis, and interpretation of the healthy cohort dataset, this group met weekly on conference calls, held biannual research network consortium meetings; held a virtual jamboree, which was a 1-day online meeting to discuss the healthy cohort data analyses with experts in the microbiology and diseases of the body sites; and exchanged computational tools, analyses and draft manuscripts through a consortium-managed electronic resource.

At one time or another, there were over 200 members of the 20 workgroups tackling specific tasks; several of these workgroups also work together toward larger goals or provide oversight and guidance toward major program objectives. For example, the Strains Working Group works with the Annotation and the Finishing Working Groups to coordinate the selection, sequencing, and annotation of the reference strains for the project. The Data Generation and Processing Working Groups works with the Data Release Working Group to agree on common processed datasets for downstream analysis. As the consortium is working together to analyze these datasets for major publications and for companion papers, each member of the consortium agrees to guiding principles on data use and HMP consortium acknowledgment in publications.

1.3.4. Reference Strain Microbial Genome Sequences

The HMP sought to create a public reference dataset of microbial (primarily from bacteria but also from some archaea, viruses, bacteriophages, and eukaryotic microbes) genome sequences of microorganisms collected from the major body sites. The goal was to create a catalog of genome sequences from 3000 bacterial strains and as many viral/phage and eukaryotic microbial strains as possible. The microbial genome sequence dataset is intended to provide a reference for the interpretation of 16S rRNA sequences and to serve as scaffolding for assemblies of the metagenomic sequences derived from microbiome samples. As an extension of this public resource, cultures of sequenced strains that were donated from personal laboratory collections were deposited at the HMP Repository with the NIAID Biodefense and Emerging Infections Research Resource Repository (BEI: `http://www.beiresources.org/`). Approximately 100 of these cultures that are expected to be in high demand will be in a "shelf-ready" state and will be available immediately to the scientific community. Another several hundred cultures are archived and can be prepared once requests for specific cultures are received by BEI.

At project inception, guidelines for inclusion of strains in the microbial reference genome dataset were established and focused on aspects of each nominated organism including (1) its phylogeny and uniqueness, (2) its established clinical significance, (3) its abundance or dominance in a body site, (4) whether identical species were found in different body sites, and (5) whether there was an opportunity to explore pangenomes (pangenome, the core genome containing genes present in all strains of a microbial species plus other genes present in one or more strains of the species) (`http://www.hmpdacc.org/doc/sops/reference_genomes/strains/StrainSelection.pdf`). Microbiologists and clinicians with body-site-specific expertise were consulted to identify and provide, when possible, strains for sequencing based on these guidelines. In addition, the HMP has continued to solicit feedback and strain nominations from the global community and hosts a web portal for this purpose (`http://www.hmpdacc.org/outreach/feedback.php`). All nominations are discussed and decided on by the Strains Working Group, representing all sequencing centers, DACC and NIH.

Microbiome strains were contributed by investigators in the field from their personal laboratory collections or were identified from public culture collections, including the American Type Culture Collection (ATCC), the German Collection of Microorganisms and Cell Cultures (DSMZ), the UK National Collection of Type Cultures (NCTC), the Belgian Co-ordinated Collections of Microorganisms (BCCM) as well as the Culture Collection from University of Goteborg (CCUG)

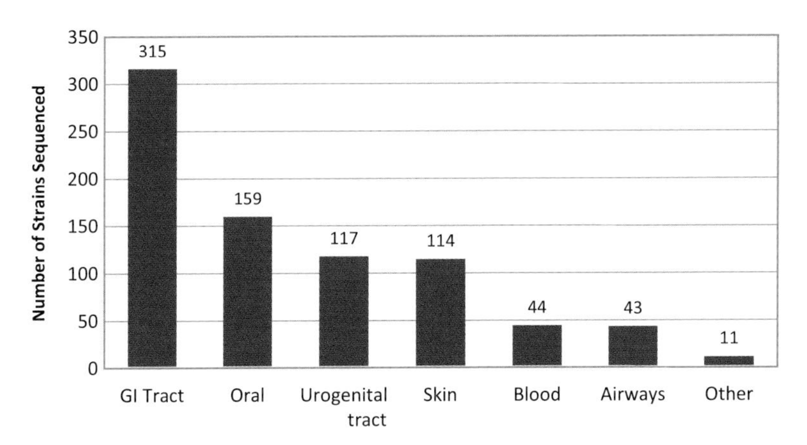

Figure 1.2. Distribution of HMP reference sequence bacterial strains by major body site. Note that additional body sites (blood) outside of the typical HMP major body sites served as sources of the isolates. "Other" refers to isolates collected from other, miscellaneous body sites. (Data and figure courtesy of Drs. Heather Huot-Creasy, DACC and Ashlee Earl, Broad Institute. Additional details are available at http://www.hmpdacc.org/refernce_genomes/statistics_specific.php.)

and the Biological Resource Center of Institut Pasteur (CIP). Workgroups of different body sites experts were convened to identify the sources of strains to be sequenced. These microbial strains came from a wide variety of body sites, with GI tract samples contributing about a third of the strains and oral, skin, and urogenital samples contributing approximately equal numbers of strains. The airway, blood, and additional body site samples make up the remaining sources for these strains (Figure 1.2). A publication documenting the analysis of the first 178 microbial isolates was published (viz., the Human Microbiome Jumpstart Reference Strains Consortium [20]). This analysis described 550,000 predicted genes, 30,000 of which are novel.

As of this writing, over 1300 strains have been sequenced (~800) or targeted for sequencing (~500) by the four sequencing centers (http://www.hmpdacc-resources.org/hmp_catalog/main.cgi?section=HmpSummary&page=showSummary). This list comprises primarily bacterial strains, although some bacteriophages, eukaryotic microbes, and methanogenic archaea have been included. The sequences are available in GenBank. The Strains Working Group made a decision to finish the completed sequences to various levels; approximately 30 are finished genome sequences, and most are at the high quality draft level of finishing [21].

Because only a fraction (current estimate ~60%) of the human-associated microbes are in culture and available for sequencing, a technology development initiative aimed at isolating uncultivable microorganisms was created. This program included support for innovative cultivation techniques to isolate new strains from the body sites and the application of single cell genomics methodologies to reach this project goal.

In order to guide this effort, the Strains Working Group has conducted an analysis of the healthy cohort 16S data to develop a priority list of the top 100 most desirable bacterial strains to target for sequencing. The approach used to identify new or novel taxa that have not yet been sequenced was to select 16*S* sequences for all of the body sites that had less than 90% identity to already sequenced strains and were found in at least 30% of all samples from a particular body site. Then, using the 16*S* data, the body sites were identified that contained most of the strains that had not yet been sequenced; this analysis resulted in a little over 100 targeted strains. This analysis showed that 73 of the 100 desired strains were located in the oral cavity and 30 were located in the gut; the remainder were evenly distributed across the other three major body sites. These data are being used to guide the technology development teams in their sample sorting efforts and in their searches for novel strains. In addition, collaborations between the demonstration project teams and the technology development teams are endeavoring out to identify tissue types and samples that could serve as material for isolating new strains for cultivation or cells for further analysis.

1.3.5. Healthy Adult Cohort Study of Multiple Microbiomes

The third initiative of the HMP represents the largest cohort study to date of the microbiomes of the multiple body habitats of healthy adults. There have been differences in the terms used to describe the microbiome body habitats sampled for this study. In this chapter, we will consistently refer to the oral, skin, nares, gut, and vagina areas as the major body *sites*. Specific areas within each major body site will be called body *subsites*. As these volunteers were clinically evaluated and determined to be healthy, this study is typically called the *healthy adult cohort study*, and the goal of the study was to collect and analyze minimally disturbed microbiomes. The study can be broken down into three components: the clinical phase, the sequencing phase, and the data analysis phase.

Clinical Phase

Experts in clinical research and ethical issues advised on the inclusion and exclusion criteria and on the consent forms developed for the study. Extensive exclusion criteria for the selection of healthy volunteers were developed and were based on a combination of health history (particularly systemic disorders such as hypertension, cancer, autoimmune disorders), use of antibiotics, probiotics or immunomodulators, and body mass index, as well as physical examination of the volunteers such as presence of skin lesions and oral and dental health status. It was common to find that these apparently healthy volunteers were not always "healthy" in all body sites. An example of this dichotomy was with the oral cavity, where otherwise healthy volunteers had dental caries that resulted because they were not eligible for enrollment until the dental disease was treated and the mouth determined to be healthy. Women were required to have a history of regular menstrual cycles.

The subjects were informed that their microbiome samples and microbiome sequence data would be coded to anonymize study participants, that controlled access databases would be used to store the clinical metadata and human genome sequence data, and that permission to use these data for microbiome research purposes would be regulated by the NIH Data Access Committee (DAC) to ensure that the data were being used properly. The volunteers consented to allow researchers to use their human sequence data for microbiome research but were assured that their identities would not be revealed to the researchers or to the public. The volunteers were also assured that all reasonable effort would be expended to separate their human sequence from the microbial sequence data before the microbial data were deposited in open access databases, which is open to all users of the database and does not require a DAC review.

A comprehensive clinical protocol was developed to ensure that minimally disturbed microbiomes were sampled. All of the body sites were directly sampled except for the digestive tract, in which stool served as a proxy for all distal gut regions. Saliva was collected from each subject at each visit. Blood and serum were also collected from each subject at the first visit, DNA was extracted from one aliquot of the blood for future whole-genome sequencing, and lymphocytes were harvested from a second aliquot and stored at –80°C for future preparation of cell lines. The human subject genome sequences, the bulk DNA, and the cell lines will be made into additional community resources. The blood, DNA extracts and serum is stored at the NHGRI Sample Repository for Human Genetic Research (Coriell Institute for Medical Research, Camden, NJ). The two clinical laboratories (Baylor College of Medicine and Washington University in St. Louis) extracted the DNA from the body site samples using the same commercial kit and standard operational procedures and distributed the DNA to the four institutions (Broad Institute, Baylor College of Medicine, J. Craig Venter Institute, and Washington University in St. Louis) carrying out the sequencing activities. The MoBio Powersoil DNA extraction kit (www.mobio.com) was selected after pilot studies to test different commercial extraction kits.

In this study, 300 adult volunteers were selected from a total of approximately 550 screened individuals. Approximately 20% self-identified as a racial minority and about 11% self-identified as Hispanic. The total pool of volunteers was split between two clinical sites: one in the southwestern United States (Houston, TX) and the other in the midwestern United States (St. Louis, MO). An equal number of adult men and women in the 18–40 year-old range were recruited for the study. The body mass index (BMI) range for the volunteers was 18–35. The mean blood pressure of the volunteers was 120/70, and the vast majority did not smoke. In addition, the majority of the volunteers self-reported as generally meat eaters and that they had been breastfed during infancy.

Enrollment and sampling of the volunteers commenced in December 2008 and were completed in October 2010. Of the 300 study participants, 279 were sampled twice and 100 were sampled a third time; the interval between the first and third samplings averaged approximately 10 months. A number of subsites within each body site were sampled, so there were 18 total subsites in five major body sites (oral, skin, nares, gut, and vagina for women) sampled; the oral body site had the largest number of subsites sampled (9) (Figure 1.3). Deposition of the full clinical metadata set in dbGaP was completed in February 2011, approximately 4 months after the last sampling was completed (http://www.ncbi.nlm.nih.gov/projects/gap/cgi-bin/study.cgi?study_id=phs000228.v3.p1). These metadata were released in editions because the clinical teams conducted continuous in-house analysis of the metadata to verify that there were no "identifiable" traits or combinations of traits in the metadata that could reveal a specific clinical subject. A manual of procedures detailing the clinical sampling protocol and criteria for sampling can be found at the dbGaP website (http://www.ncbi.nlm.nih.gov/projects/gap/cgi-bin/study.cgi?study_id=phs000228.v2.p1).

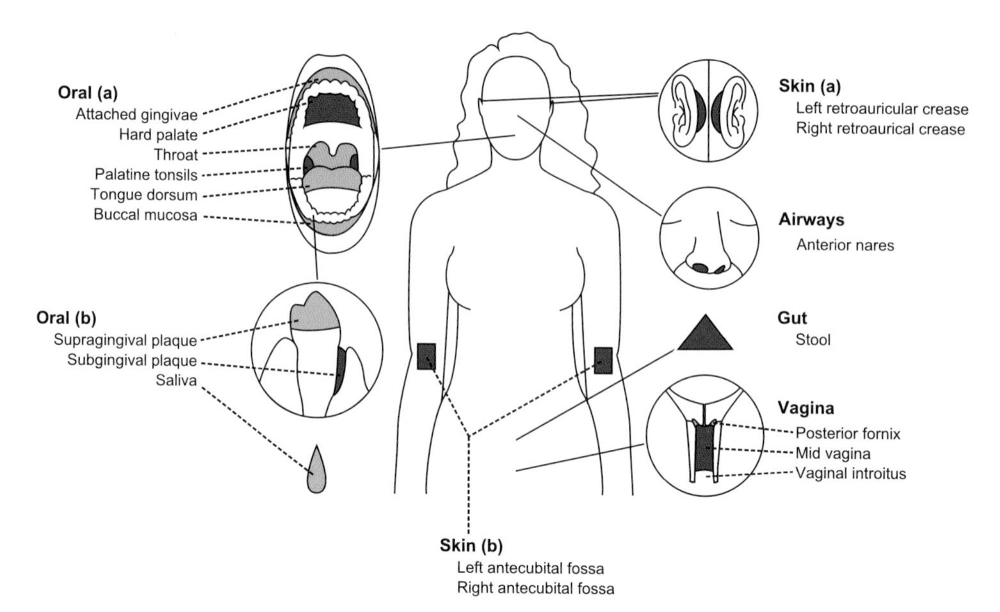

Figure 1.3. Schematic of the body sites sampled for the HMP healthy adult cohort study. Three hundred individuals were sampled across a total of 18 body subsites in five major body sites to collect tissue or body fluids for nucleic acid extraction and subsequent sequence analysis. The oral cavity, skin, airway, and gut sites were sampled in males, and the vagina was additionally sampled in females as the fifth major body site for the study. Eight distinct soft and hard surface subsites were sampled in the oral cavity with saliva representing the ninth oral subsite, four subsites were sampled on the skin, and three subsites were sampled in the vagina. The airway was represented by a pooled sample of the anterior nares, and the distal gut tract region was represented by one sample of stool. (This figure was adapted from the Sitepainter visualization tool figure, courtesy of R. Knight, M. Perrung, and A. Gonzalez, University of Colorado. Tool available at `www.hmpdacc/sp`.)

Sequencing Phase

As a part of the pilot project for this initiative, the four sequencing centers undertook a series of benchmarking exercises to determine appropriate protocols for sequencing the healthy human microbiome DNA and to compare consistency of results across the sequencing facilities. The group developed a mock microbiome community of a 22 bacterial species assemblage as a test specimen to evaluate DNA extraction, primer selection for library construction, and sequencing protocols. On the basis of these data, the group decided that primers for the variable region V3–V5 of the 16*S* rRNA gene would be used for the targeted 16*S* sequencing of all of the samples and, as needed, the V1–V2, V1–V3, and or V6–V9 regions would be targeted to amplify specific bacterial groups that do not amplify well with the V3–V5 primers. A manuscript describing the benchmarking exercise is in review.

As might be expected, DNA yield varied greatly across the body site samples (Table 1.1). As an example, stool yielded the greatest amount of total DNA (~9.5–21.0 ng/μL) whereas skin samples yielded the lowest, at 0.001 ng/μL. There were over 12,000 unique primary samples collected from the 300 subjects. Primary samples included samples collected in order to sequence the 16*S* rRNA gene or the metagenome of the microbiota as well as urine, blood, and saliva; 11,000 of those samples

TABLE 1.1 Range in DNA Yield (ng/μL) of Samples Collected from the Five Major Body Sites in the HMP Healthy Adult Cohort Study[a]

Body Site	DNA Yield (ng/μL)
GI tract (1)	9.49–21.08
Oral (9)	0.16–4.72
Nares (1)	1.05–2.10
Skin (4)	0.001–0.156
Vagina (3)	4.02–8.57

[a]Values in parentheses indicate the number of subsites sampled within each body site. Skin is reported to three places because overall yield was lower than that for other body site samples. Single swab (nares, vagina, skin, soft oral subsites) and curette (hard oral sites) samples and single stool subsamples (50–800 μL) were directly extracted using the MoBio PowerSoil kit and DNA extract eluted in 10 μL. DNA concentrations measured by fluorometric assay by the Baylor College of Medicine and Washington University clinical labs. DNA concentrations for each body site derived from three replicate extracts.
Source: Data and table courtesy of Dr. Joe Petrosino, Baylor College of Medicine.

were used for nucleic acid extraction. A majority of the samples were analyzed by targeted sequencing of 16S clone libraries with the Roche 454 sequencing technology. In addition, a fraction of the samples were analyzed by metagenomic whole genome shotgun sequencing using both the 454 and the Illumina GAII technologies.

The targeted 16S sequences and WGS sequences were deposited at NCBI databases by the participating sequencing centers. The 16S sequences were deposited in the open access sequence read archive (SRA) of dbGaP. The metagenomic sequences as well as the clinical metadata were deposited in the controlled access portion of dbGaP since they included information about the human subjects. Clinical metadata collected from these volunteers included elements such as gender, age, BMI, vital signs, vaginal pH, medical history, and other key information about the subjects. Since these WGS sequences contained human subject sequence, NCBI developed a computational tool, Bestmatch Tagger (BMTagger), to computationally filter the human sequence from the total sequence. The algorithm discriminates between human reads and microbial reads by comparing consecutive sequences of 18mer-length nucleotides found in the total sequence with those found in the human genome sequence and then includes an alignment procedure that finds all matches for any missing alignments. The human genome reference sequence used was the Genome Reference Consortium's most current refinement of the human genome sequence (GRCh36, `http://www.ncbi.nlm.nih.gov/projects/genome/assembly/grc/human/index.html`) (S. Sherry, K. Rotmistrovsky, R. Agarwala, and NCBI, personal communication, 08/01/11). The filtered WGS sequence was deposited in the open access SRA as microbiome metagenomic sequence data.

Data Processing and Analysis Phase

In preparation for the data analysis phase, a group of scientists from the microbiome community, the sequencing centers, and the DACC as well as NIH staff were brought together to form a HMP Data Analysis Working Group (DAWG). As there was continuous sequence data production, the DAWG declared a data freeze on May 1, 2010 on a subset of the 16*S* rRNA sequence data and on July 1, 2010 on a subset of the WGS metagenomic sequence data in order to define a common, master dataset for the follow-on global analysis activities to be undertaken by the research consortium.

Of the >11,000 primary samples collected for the full study, the May 1 freeze targeted 16*S* rRNA data and included 5300 samples from 18 body subsites of 5 major body sites from 242 subjects (113 females, 129 males), and the July 1 freeze WGS data included 736 samples from 16 body subsites of 5 major body sites from 102 subjects. No third visit samples had been sequenced by the data freeze, but of the 242 subjects, a subset of 131 (~54%) included samples from two visits generally spaced by 6 months and up to a year between visits. These datasets included a total of ~74 million 16*S* rRNA reads. Once the contaminating human sequence was removed (which represented on average ~60% of the total sequence), a total of 3.5 terabases (Tbp) of metagenomic WGS sequence was generated for subsequent analysis.

For each major body site, the typical sequence generated from a sample ranged between 10.8 and 12.8 gigabases (Gbp) (average 12 Gbp). However, the ratio of microbial sequence reads to total sequence reads (i.e., the percent of human DNA sequence and sequence from other contaminating DNA) varied greatly across the body sites (Figure 1.4). The largest fraction of microbial reads to total reads was found in the gut samples (stool, ~98%). Nares, skin, and vaginal samples yielded about 10–25% microbial sequence reads to total reads.

Two kinds of metagenome assemblies were produced from the processed whole-genome shotgun data. The processed metagenome sequences were assembled using SOAPdenovo. Hybrid metagenome assemblies from processed Illumina and Roche 454 sequence reads were also produced using Newbler. These two kinds of metagenome assemblies were prepared in order to support different types of analyses. For example, the de novo assemblies were used for comparisons against the reference microbial genome sequences to determine microbiome community composition, and the hybrid assemblies were used for the reconstruction of metabolic modules and pathways inferred from the whole-genome shotgun data.

The DAWG and its various workgroups developed processed datasets in 2010–2011 that the DAWG agreed would serve as the common, master processed datasets for downstream data analyses. These finalized datasets include (1) 16*S* data that had been quality-controlled and processed to remove errors at agreed-on stringency levels, (2) metagenomic data mapped to a global list of microbial reference genome sequences from both the HMP sequencing efforts and microbiome reference strain data available in GenBank, (3) metagenomic assemblies produced either de novo or as hybrid assemblies, and (4) other such data products for use by the DAWG (Table 1.2). The approximate sizes of each data type are also shown.

The results from the global analysis of the healthy cohort study describe the range of normal microbial variation among healthy adults in a Western population. The microbial composition differed among individuals when these communities were analyzed at several taxonomic levels (genera, species, strains). Further, previous observations about community structure seem to be true for all of the major body sites examined in this study: the microbial communities grouped by body site and not by individual. In addition, there was great variability in microbial

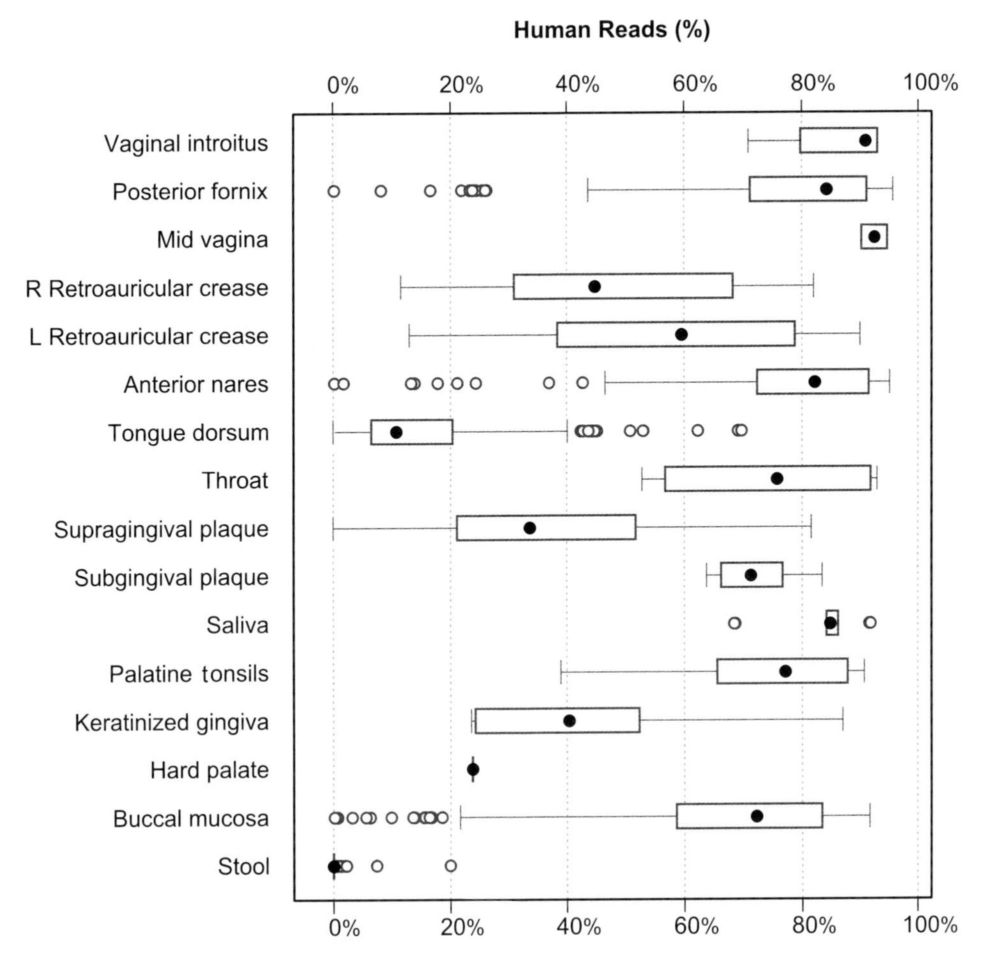

Figure 1.4. Percent human sequence reads in total sequences of whole-genome shotgun reads from HMP healthy cohort microbiome nucleic acid extracts. Boxplots represent the range in percent of human reads per body site (*x* axis) with black dot representing the mean. Body sites are listed on *y* axis. Note that the majority of samples had significant human contamination, at levels of ≥60% of total sequence. (Analysis and graph courtesy of Drs. Dirk Gevers and Katherine Huang of the Broad Institute.)

composition between subsites within a body site. As one example, even adjacent surfaces of the oral cavity separated by only millimeters or in closer proximity within the same subject exhibited strikingly different community structures.

Even though community structure varied greatly between body sites, the potential metabolic capabilities encoded in these metagenomes were much more constant, both among body sites and between individuals. Over 5 million unique genes were cataloged from the healthy cohort analysis. However, although the microbial community composition in the healthy microbiome varied among individuals, the predicted core functions that the microbiota are equipped to carry out remain remarkably stable within each body site, particularly for major metabolic pathways.

TABLE 1.2. Finalized Datasets[a] Used by HMP DAWG for Analysis of Healthy Cohort Data (May 1, 2010 and July 1, 2010 Data Freezes)

Name	Description	Approixmate Size
16S high quality (V1–3, V3–5, V6–9)	Aggressively filtered and trimmed reads (low error rates, short reads) (454)	1.5-terabyte data
16S low quality (V1–3, V3–5, V6–9)	Less aggressively filtered and trimmed reads (higher error rates, longer reads) (454)	1.5-terabyte data
WGS mappings to reference genomes	Alignments between WGS data (both reads and ORFs) and reference genomes	2-terabyte data
MetaHIT mappings to reference genomes	Alignments between metaHIT WGS data and reference genomes	0.1-terabyte data
WGS pretty good assemblies (PGAs)	De novo assemblies from the WGS data	0.3-terabyte data
WGS hybrid assemblies	Mixed Illumina and 454 WGS metagenomic data, resulting in long, high-quality assemblies	0.1-terabyte data
WGS read annotations	Gene predictions and functional assignments from assemblies	3.6-terabyte data
Orthologous gene family abundances	Relative abundances of KOs from read-level blastx results	741 samples, 13,328 KO families
Functional/metabolic pathway coverage	Presence/absence of KEGG modules and pathways	741 samples, 246 small modules, 290 large pathways
Functional/metabolic pathway abundance	Relative abundances of KEGG modules and pathways	741 samples, 246 small modules, 290 large pathways

[a]These datasets are available on the HMP DACC website: www.hmpdacc.org.

These results also suggest that a careful examination of specialized metabolic functions, such as vitamin, toxin, or antimicrobial production or the production of signaling molecules or novel metabolites, will be key to deciphering the signature characteristics of each microbiome of the body.

Although major metabolic pathways appear to be common across all microbiomes, in fact we still know little about most of the predicted genes or proteins in the human microbiome. In analysis of the healthy cohort data, a large fraction (43%) of the metagenome sequence from the five major body sites could not be aligned to the reference genome sequences and the majority of the annotated genes (80–90% or over 4 million genes) and predicted proteins (75–85%) could not be assigned a function. Clearly, a next key step is to characterize the functional properties of the microbiome at both the strain and total community levels.

Further, most (although by no means all) communities are colonized predominantly by one specific group of bacteria. Most signature groups, in turn, consist of predominantly one specific microbial taxon, with subtypes present in lower abundance. This likely reflects niche specialization within these communities. Further, localized environmental factors such as vaginal pH were important in some communities. A very interesting future question will be what the "most important" factors are influencing lifelong microbiome composition, whether they are genetics, diet, birth environment, geography, or combinations of these factors.

1.3.4. Demonstration Projects of Microbiome–Disease Associations

The fourth resource of the HMP included a group of projects that were designed to determine whether correlations between microbiome community composition and specific diseases can be detected. It was recognized at the inception of the initiative that studies could not yet be conducted to determine whether there are causal relationships between specific diseases and changes in the microbiome. There was, however, sufficient evidence for a number of diseases that appeared to include a role for microbial communities in the disease processes. The "demonstration projects" program has this question as its goal in a number of different putative microbiome-associated diseases. The demonstration projects began with a 1-year pilot phase during which 15 projects recruited subjects and tested sampling protocols. Following an administrative review, 11 projects from the initial pool of 15 were funded to continue their work for 3 additional years.

Of these 11 studies, six projects study the microbiome associated with gut diseases, three study the microbiome and urogenital conditions or diseases, and two study the microbiome and skin diseases (Table 1.3). Depending on the study, the age groups recruited ranged from birth to over 50 years old, and the number of subjects recruited ranged from 19 to 489. Most are case–control studies. Almost all of the studies included targeted 16*S* rRNA gene sequencing, and some included WGS metagenome sequencing of the microbiomes inhabiting unaffected body sites and the diseased tissue of interest. Some of these studies also included the analysis of functional markers of the microbiome such as gene expression or gene products of the microbial communities or metabolomic studies of the microbiome.

These projects are a diverse set of carefully controlled case studies with large cohort sizes that support the correlation of microbiome changes with development of specific diseases. These studies will contribute valuable datasets for further study as they include detailed clinical metadata such as the disease phenotype along with phylogenetic and total community analysis of the microbiomes from controls and disease-associated tissues. Many of these studies also include microbial genome sequences from reference strains isolated from the diseased tissue of interest. The data are rapidly released into the public domain. Many of these studies also include characterization of the microbiomes prior to disease development, in response to the presence of disease or, in some cases, in response to standard-of-care interventions and so include additional dimensions of analysis to the study of the associations of microbial communities with specific diseases.

Early results from some of these demonstration project studies are beginning to suggest that a characteristic microbiome community appears to be associated with the specific disease under study. For example, neonatal enterocolitis, esophageal adenocarcinoma, ulcerative colitis, Crohn's disease, and eczema all appear to have a characteristic microbial community associated with the disease state, which is different from the microbial composition of control tissues. Further, the microbial signatures associated with the some of these disease states include both structural markers, such as the community composition, as well as functional markers, such as

TABLE 1.3. Summary of HMP Demonstration Projects[a]

Principal Investigator(s)	Short Title	Number of Subjects	Age and Other Criteria	Study Type	Sequence Data Type, Technology	WHO ICD General Disease	WHO ICD Specific Disease
Phillip Tarr, Washington Univ. School of Medicine, St. Louis, MO	The Neonatal Microbiome and NEC	489	<1500 g at birth, other criteria	Case–control	Metagenome, Roche 454 GS FLX titanium	Certain conditions originating in perinatal period	Necrotizing enterocolitis
Gary Wu, James Lewis, Frederic D. Bushman, Univ. Pennsylvania School of Medicine, Philadelphia	Diet, Genetic Factors, and the Gut Microbiome in Crohn's Disease	128	4 studies: FSM (>18 yo); CAFE(18–40 yo); COMBO (2–50 yo); PLEASE (<22 yo); other criteria	Cross-sectional, controlled trial, longitudinal cohort	16S rRNA and metagenome, Roche 454 GS FLX titanium	Diseases of digestive system	Crohn's disease
James Versalovic, Baylor College of Medicine, Texas Children's Hospital, Houston	The Human Gut Microbiome and Recurrent Abdominal Pain in Children	44	7–12 yo; other criteria	Case–control	16S rRNA, Roche 454 GS FLX titanium	Diseases of digestive system	Irritable bowel syndrome

Vincent Young, Univ. Michigan , Ann Arbor; Eugene Change, Univ. Chicago; Folker Meyer, Argonne National Lab, Argonne, IL; Tom Schmidt and James Tiedje, Michigan State Univ., East Lansing, MI; Mitchell Sogin, Marine Bio Lab, Woods Hole, MA	Ulcerative Colitis Human Microbiome Project	23	>18 yo, other criteria	Longitudinal	16S rRNA and metagenome, Roche 454 GS FLX titanium	Diseases of digestive system	Ulcerative colitis
Claire Fraser-Liggett, Univ. Maryland School of Medicine, Baltimore	Human Gut Microbiome in Crohn's Disease	19	Five twin pairs, other criteria	Twin	Metagenome, Roche 454 GS FLX titanium	Diseases of digestive system	Crohn's disease
Zhiheng Pei, New York Univ. Langone Medical Center, New York; Karen Nelson, J. Craig Venter Institute, Rockville, MD	Foregut Microbiome in Development of Esophageal Adenocarcinoma	42	>50 yo, other criteria	Case–control	Metagenome, Roche 454 GS FLX titanium	Neoplasms	Malignant neoplasm of esophagus
J. Dennis Fortenberry, Indiana Univ. School of Medicine, Indianapolis	Urethral Microbiome of Adolescent Males	55	14–17 yo, other criteria	Longitudinal, observational, cohort	Metagenome, Roche 454 GS FLX titanium	Certain infectious and parasitic diseases	Infections with a predominantly sexual mode of transmission

(*Continued*)

TABLE 1.3. *(Continued)*

Principal Investigator(s)	Short Title	Number of Subjects	Age and Other Criteria	Study Type	Sequence Data Type, Technology	WHO ICD General Disease	WHO ICD Specific Disease
Gregory Buck, Virginia Commonwealth Univ., Richmond	The Vaginal Microbiome: Disease, Genetics, and the Environment	460	>18 yo, other criteria	Twin, clinical cohort	16S rRNA, Roche 454 GS FLX titanium	Diseases of genitourinary system	Bacterial vaginosis
Jacques Ravel, Univ. Maryland School of Medicine, Baltimore; Larry Forney, Univ. Idaho	The Microbial Ecology of Bacterial Vaginosis	200	>18 yo, reproductive age, other criteria	Longitudinal, prospective	16S rRNA, Roche 454 FLX titanium	Diseases of genitourinary system	Bacterial vaginosis
Martin Blaser, New York Univ., Langone Medical Center, New York	Cutaneous Microbiome in Psoriasis	200	Not specified	Longitudinal, case–control	16S rRNA and metagenome, Roche 454 GS titanium, RNAseq metatranscriptomics	Diseases of skin and subcutaneous tissue	Psoriasis vulgaris
Julie Segre, National Human Genome Research Institute, NIH, Bethesda, MD; Heidi Kong, National Cancer Institute, NIH, Bethesda, MD	Skin Microbiome, Atopic Dermatitis, and Immunodeficiency	33	3–40 yo, other criteria	Longitudinal, case–control	Whole-genome genotyping and ABI de novo sequencing	Diseases of skin and subcutaneous tissue	Atopic dermatitis

[a]This tabulation includes project investigator names and affiliations (PIs), short titles, number of subjects, age range (yo = years old) and main inclusion criteria, study type, sequence data type, and sequencing technology used for the project. Please refer to the Nature Preceding marker papers for more detail on each study. Categorization of the general and specific disease(s) under study are indicated according to the WHO International Code of Diseases 2010 (ICD10) classification system. The WHO ICD10 general and specific disease categories are linked to the ICD website.

the metaproteome of the disease-associated microbial community, providing a suite of markers for possible diagnostic or prognostic applications.

In addition, it appears in some cases that microbial markers may be emerging that appear to precede the disease state. For example, gastric esophageal reflux disease (GERD) is characterized by a series of diseases, starting with reflux esophagitis, continuing to Barrett's esophagus in about 20% of cases, and in rare cases of Barrett's esophagus, proceeding to the development of esophageal adenocarcinoma. In the Pei/Nelson esophageal adenocarcinoma demonstration project study, it was found that the microbiome of the intermediate stage of the disease (Barrett's esophagus) appears to be very similar to the microbiome of those patients who go on to develop adenocarcinoma, suggesting that the microbiome in Barrett's esophagus is a potential precursor state to the cancer. In this case, it may be possible to develop diagnostic biomarkers for adenocarcinoma far before the cancer develops. A summary (Table 1.3) includes links to the Bioprojects pages at NCBI that describes the projects and leads the user to the data. This table also lists the general and specific disease(s) under study in each project as they are categorized according to the WHO International Disease Classification 2010 (ICD10) system (`http://www.who.int/classifications/icd/en/`). Summaries of the hypotheses, aims, and data types to be produced are documented as marker papers in Nature Proceedings (`http://precedings.nature.com/search?query=human+microbiome+project`). Highlights of the early results from these projects are provided below and are categorized by body site.

Gut Diseases and the Microbiome

Three of the gut disease projects include the microbiomes of younger populations, such as neonates [Tarr, necrotizing enterocolitis (NEC)] or children [Versalovic, irritable bowel syndrome (IBS)]. Wu, Lewis and Bushman, in their multifaceted project on Crohn's disease (CD), also included a study of the effects of an elemental diet on pediatric patients with inflammatory bowel disease (IBD). Four of the gut projects focus on gut diseases in adults [Fraser–Liggett, CD; three of the four studies by Wu et al. involve CD in adults; Young, ulcerative colitis (UC)]. The fourth digestive tract project was described earlier and is somewhat unusual in that it is the one study of the association of the microbiome with a cancer (Pei/Nelson, esophageal adenocarcinoma).

The gut disease studies with young patients are showing some promising early results. The Tarr study on NEC found that antibiotic treatment, the standard of care for premature infants, decreased gut microbial diversity and that this was associated with the development of NEC. Further, they found that key host immune system markers increased before the appearance of NEC, although the specificity and timing of these host signals need further study. In the Versalovic study, there appear to be IBS-specific microbial signatures in those children with IBS and further, assemblages of specific gut microbial taxa that may distinguish between the occurrence of subtypes of pediatric IBS. In the pediatric study of the Wu et al. project, elemental dietary interventions appeared to change gut microbiome composition within 24 h of intervention, suggesting that elemental diet therapy can have a major impact on the composition of the gut microbiome in pediatric patients with CD and possibly the disease itself.

Early results from the studies of adult microbiomes and gut diseases are promising as well. The Fraser–Liggett CD study compared twins with either ileal CD (iCD) or colonic CD (cCD). This was a multifaceted study, and many aspects of the microbiome (microbial composition, gene content, and gene products) were correlated with patient clinical metadata. Early results suggest that, although the picture is not yet clear for cCD, a combination of specific microbial assemblages and their genes and products appear to correlate with iCD and are consistent with the increased inflammation seen in the iCD gut. These markers may lead to diagnostic tools for assessing the development of iCD.

The Young et al. study is also a multifaceted study of the association of a gut disease, UC, with the microbiome and includes an interesting experimental model component. IBD consists of iCD, cCD, and UC, a disease of the colon. For some UC patients, the colon must be removed (colectomy) and a pseudorectum ("pouch") is formed from a segment of their small intestines. In over 50% of these patients, the pseudorectum may itself become inflamed, a condition known as "pouchitis." The Young et al. study follows patients who have undergone the pseudorectum surgery and has found that the microbiome composition of patients with pouchitis shifts to a microbial community more similar to the colon microbiome composition of UC patients, even though the pouchitis occurs in a structure formed from the small intestine, not the colon. This pouchitis study appears to be a good experimental model for UC and provides insights for isolating the role of the gut microbiome in the cause or contribution to the development of UC.

Urogenital Diseases and the Microbiome

The microbiome–urogenital disease association studies include bacterial vaginosis and sexually transmitted infections (STIs) and the vaginal microbiome (Buck) and included both longitudinal studies and twin studies. The Fortenberry study included the relationship of circumcision, sexual history, and STIs with the penile microbiome of adolescent males. Early results from the Buck study suggest that there may be a genetic component to the composition of the vaginal microbiome. Further, the vaginal microbiome composition appears to respond to the hormonal cycle as microbial diversity appears to be lowest at midmenstrual cycle or, in other words, during ovulation.

The Fortenberry study included monthly sampling of both the urethra and the coronal sulcus of the penis of adolescent males to characterize changes in the microbiome over time, in response to sexual activity and between circumcised and uncircumcised males. Early results show that, although there are differences in the penile microbiome between, for example, circumcised and uncircumcised males, the microbial composition appeared to be fairly stable over time. Further, the urethral microbiome composition differed between those males with and without STIs. These results may be applicable to the treatment of sexually transmitted diseases.

Skin Diseases and the Microbiome

One microbiome-associated skin disease project is a study of atopic dermatitis in children (eczema, Segre). Eczema is characterized by periodic exacerbations (known as *flares*) that result in highly inflamed skin. Until recently, eczema has been studied as a single pathogen disease. Segre's research team has examined the role of the skin microbiome in modulating the extent and duration of the disease and includes both pediatric and adult patients. In a longitudinal study of pediatric patients with eczema, the Segre lab found that total skin microbial diversity was reduced during flares with a concomitant increase in *Staphylococcus aureus*. Whether this dominant organism is a consequence or cause of the shifts in microbial diversity and of the disease is currently under study.

Because of the nature of the work, the demonstration projects are more akin to individual investigator projects, so no formal HMP analysis workgroup was formed. However, an informal workgroup has come together to discuss strategies for the analysis of each project's sequence data, to participate in tutorials created for the projects on the various computational tools being developed for human microbiome studies and to discuss common data standards in order to make the results from each study comparable. Publications from the demonstration projects are described in Section 1.4.

1.3.7. Technology Development

Two additional HMP initiatives were designed to provide resources for the HMP effort and for the field in general, technology development to isolate novel microorganisms and computational tools development. Because early estimates indicated that a large fraction (~40%) of the microorganisms associated with the human microbiome were not yet in culture, it was recognized that there was a critical need for new approaches that could isolate or enrich for new and novel microorganisms from the microbiome. In order to address this need, the technology development program supported 10 projects that are working on a wide variety of methodologies to enrich for and isolate specific populations of cells for downstream applications (Table 1.4). In many cases, these projects were intended as an investment in the long-term development of new technologies that could be applied in a 5–10-year time horizon. A few details on each project are provided below.

Five of the projects focus on enrichment and isolation of specific populations of cells by a variety of flow cytometric or microfluidic approaches (Han/Bradbury, Podar, Singh, Worthen, Relman, Chang). The Relman project includes the use of optical tweezers for isolating specific cells and *in situ* gene expression measurements of individual cells. The Chang lab is developing sorting and enriching techniques for specific populations from the colonic mucosa-associated microbial communities. Two projects are applying novel cultivation methodologies, one to isolate microaerophilic bacteria, those cells that grow best under low oxygen tension (Young/Schmidt); and a second to sort and encapsulate single cells into a gel matrix for microcolony cultivation (Doktycz). One project is developing a pipeline to process cells from flow sorting to single-cell genome sequence analysis (Zhang/Lo). The Marzaili project is focusing on a DNA purification methodology to enrich for low-abundance microbial sequences from a mixed assemblage.

The intended products from these activities include pure cultures, DNA isolated from a culture, whole-genome amplification products from single cells, or enrichments of specific strains within a mixture of cells. Some of these investigators whose methodologies had sufficiently matured will also collaborate with some of the

TABLE 1.4. HMP Technology Development Projects Listed by Project Title and Investigator

Project Titles	Investigator
FACS-MABE: a method for sorting and enriching the as-yet uncultured bacterial species from the human distal gut	Emma Allen-Vercoe (Univ. Guelph)
Species-by-species dissection of microbiomes using phage display and flow sorting	Andrew Bradbury (LANL)
Isolation, selection, and polony amplification of single cells in a gel matrix	Ronald Davis (Stanford)
Functional sorting of microbial cells from complex microbiota	Mitchel Doktycz (UT-Battelle, ORNL)
Novel cultivation methods for the domestication of vaginal bacteria	David Fredricks (Fred Hutchinson Cancer Research Center)
Confining single cells to enhance and target cultivation of human microbiome	Rustem Ismagilov (Univ. Chicago)
Culturing uncultivatable gut microorganisms	Kim Lewis (Northeastern Univ.)
Metagenomic dissestion of the gut microbiota	Xiaoxia Lin (Univ. Mich)
Tools for human microbiome studies	John Nelson (GE Global Research)
Targeted genomic characterization of uncultured bacteria from human microbiota	Mircea Podar (UT-Battelle, ORNL)
Optimization of a microfluidic device for single bacterial cell genomics	David Relman (Stanford)
Cultivation and characterization of microaerobes from the human microbiome	Thomas Schmidt (MSU)
FISH "N" Chips: A microfluidic processor for isolating and analyzing microbes	Anup Singh (Sandia National Laboratories)
Multidimensional separation of bacteria	Scott Worthen (CHOP)
An integrated lab-on-chip system for genome sequencing of single microbial cells	Kun Zhang (UCSD)

demonstration project research teams to apply their methodologies to enrich for and isolate specific cells from control and diseased tissue microbiome samples.

1.3.8. Computational Tools

The sixth Initiative of the HMP includes the development of computational and bioinformatic tools to support and advance human microbiome sequence data analysis, particularly of metagenomic sequence data. The computational tools program supports 10 projects for a wide variety of tools for this purpose (Table 1.5). The common theme across all of these projects was the recognition that next-generation sequencing technologies are able to produce orders of magnitude more sequence than the traditional Sanger chemistry sequence technologies. In fact, terabase-range datasets of metagenomes from complex microbial communities are increasing in frequency, yet computational tools are not routinely available that can accommodate these massive datasets.

It is perhaps valuable to take a moment to note the fundamental differences between sequence analysis of a metagenome and sequence analysis of the genome

TABLE 1.5. HMP Computational Tools Projects Listed by Project Title and Investigator

Project Title or Description	Investigator
Algorithmically Tuned Protein Families, Rule-Base and Characterized Proteins	Daniel Haft (J. Craig Venter Institute)
Novel Computational Tools for Studying the Human Microbiome	David Fredricks (Fred Hutchinson Cancer Research Center)
Functional activity and interorganismal interactions in the human microbiome	Curtis Huttenhower (Harvard Univ.)
New Tools for Understanding the Composition and Dynamics of Microbial Communities	Robin Knight (Univ. Colorado-Boulder)
Novel Methods for Effective Analysis Assembly and Comparison of HMP Sequences	Weizhong Li (UC San Diego)
High Performance Validation and Classification of Metagenomic Ribosomal-RNA Sequences	Dan Franks (Univ. Denver)
Assembly and analysis software for exploring the human microbiome	Mihai Pop (Univ. Maryland)
Identifying population-level variation in cross-sectional and longitudinal HMP studies	Patrick Schloss (Univ. Michigan)
Exploiting Microbiome Sequences for Improved Models of Protein-DNA Interactions	Gary Stormo (Washington Univ. – St. Louis)
Fragment assembly and metabolic/species diversity analysis for HMP data	Yuzhen Ye (Indiana Univ.)

of a single microorganism. The sequence reads derived from the DNA of a single microbial species can be assembled, aligned, and annotated because (1) one is trying to reconstruct the genome of a single microorganism and all of the sequence come from a single organism, (2) there are databases with the genome sequences of similar or related microorganisms that can be used for comparison and as a guide to reconstructing the new genome, and (3) most microbial genomes are closed, circular structures, which increases the probability that a full genome sequence can be completed.

A robust microbial genome sequence database is needed in order to identify organisms and their close relatives in WGS metagenome datasets. To address this need, the HMP set a goal to add the sequence of at least 3000 new bacterial genomes to the current database. As of this writing, 800 microbial genome sequences had been completed and deposited in NCBI while another 500 microbial genome sequences are in progress (`http://www.ncbi.nlm.nih.gov/genomes/lproks.cgi, accessed 02/13/12`).

The four traits of massive size, fragmentary nature of the data, data complexity, and rapidly changing sequence technologies for metagenomic data have demanded the development of a wide range of new and novel analytical tools that can operate efficiently, dependably handle large datasets, and operate at sufficient speeds for routine data analysis of metagenomic data. The computational tools program projects can generally be grouped into two categories: (1) those that focus on the diversity of the microbial community and (2) those that focus on the metabolic potential of the microbial community. Some projects are contributing to the development of tools for the upstream aspects of metagenomic sequence analysis. Franks and Eddy are developing tools to improve the quality of the primary sequence data and to guide 16S rRNA gene sequence alignments using this gene's primary and secondary structure characteristics. The individual sequence reads need to be systematically pieced together, but assembling metagenomic data is very different from assembling a genome sequence as all of the sequences in the community have not been sampled or species information about the community is incomplete so that few reference genomes are available for piecing together a metagenome sequence. Pop is developing metagenomic assembly validation/quality control tools and alignment tools to address these issues. Li is creating a metagenomic meta-assembler from existing assembly tools in order to select the best features of each tool that will increase speed, performance, and capability of an assembly tool. See Tables 1.5 and 1.7.

Several investigators are developing a suite or pipeline of tools that can process sequence data and produce analytical results at different endpoints of completion. Some pipelines address the annotation component of metagenomic analysis; for example, Ye has developed a suite of tools that can take the metagenomic sequence data from the assembled state through to annotation and metabolic pathway reconstruction. In order to accommodate the current inability to name all of the species in a microbial community, Schloss has developed a pipeline of tools that employs both a taxon-free approach (i.e., does not require the identification of the microorganism) and a phylogenetic approach (i.e., evolutionary relatedness of microorganisms) to characterize microbial community diversity and visualization tools to describe this diversity. Fredricks is developing novel methods for refining placement of microbial sequences in phylogenetic trees. Knight has produced a suite of tools that include particularly strong visualization tools for conducting time series analyses of microbial composition. See Tables 1.5 and 1.7.

Three projects have focused on the functional properties of the microbiome. Huttenhower is developing tools for analysis of the metabolic potential in microbiomes from metagenomic data and to identify the interspecies regulatory networks in the microbiome. Stormo is developing tools to analyse the regulatory properties of the microbiome, particularly the DNA-protein interactions that play a role in regulating the transcription of genes across the microbiome. Haft is developing tools for the analysis of protein families predicted from the metagenome data. Publications and links to tools that are available on the grantee's website, the DACC or on `www.sourceforge.net` are listed in Section 1.4. See also Tables 1.5 and 1.7.

1.3.9. Ethical, Legal, and Societal Implications of Microbiome Research

A unique feature of the HMP among the many human microbiome programs around the world is the inclusion of a program in the ethical, legal, and social implications (ELSI) of human microbiome research. ELSI studies have become a legacy of the Human Genome Project and of the extramural research program at the National Human Genome Research Institute (NHGRI), which is congressionally mandated to allocate 5% of its research budget to the support of studies in this area. A number of interesting ethical issues have arisen in the course of HMP. Other issues could potentially arise and are worthy of study. Studies funded under HMP include issues related to the equitable selection of research participants, identifiability of individuals through microbiome profiles, informed consent to participate

TABLE 1.6. The HMP Ethical, Legal, and Societal Implications (ELSI) Projects Listed by Project Investigator and Title or Description

Investigator	Institution	Title or Description
Mildred Cho and Pamela Sankar	Stanford Univ.	Toward a Framework for Policy Analysis of Microbiome Research
Paul Spicer	Univ. Oklahoma, Norman	Indigenous Communities and Human Microbiome Research
Diane Hoffmann	Univ. Maryland, Baltimore	Federal Regulation of Probiotics: An Analysis of Existing Regulatory Framework
Amy Lynn Mcguire	Baylor College of Medicine	Ethical, Legal, and Social Dimensions of Human Microbiome Research
Rosamond Rhodes	Mount Sinai School of Medicine	Human Microbiome Research and the Social Fabric
Ruth Farrell and Richard Sharp	Cleveland Clinic	Patient perceptions of bioengineered probiotics and clinical metagenomics

in human microbiome research, data sharing practices and protection of privacy, invasiveness of sampling protocols, and the return of research results and incidental findings to the research participants. Further, the research results may have a significant impact on the nature and direction of clinical medicine and on potential products, such as probiotics, that could be developed on the basis of the research findings, and all of these raise broader societal implications.

The HMP supports six investigative teams to conduct research on a broad range of ELSI topics (Table 1.6). The Cho–Sankar team is conducting an analysis of risk/benefit concepts in microbiome research and of how experimental design may reflect perceptions of ethical issues in human microbiome research. Spicer is conducting an analysis of the potential impact of human microbiome research results on the concepts of social and ancestral identity in indigenous peoples. Hoffmann is conducting an analysis of federal regulation of probiotics, in order to assess whether the current regulatory framework ensures probiotic safety and the accuracy of health-related claims. McGuire is studying the perceptions and attitudes of current research participants in the "healthy cohort" study of the HMP, as well as of HMP researchers, regarding a wide range of ethical issues in human microbiome research, with the goal of making recommendations for guidelines for the management of ethical issues in future microbiome research. Rhodes is developing guidelines for educating the lay and scientific public about ethical issues with respect to human subject research, biobanking, public health, and commercialization of products for treating the human microbiome. The Sharp–Farell team is conducting an analysis of patient perceptions of probiotics for the treatment of medical conditions.

1.4. PRODUCTS FROM THE HUMAN MICROBIOME PROJECT

There are a number of products from the research conducted within the HMP. Several have already been outlined in previous sections such as the reference strains sequencing activities and cultures at BEI. Research efforts are now focusing on expanding the collection to include eukaryotic microbes, eukaryotic viruses, and bacteriophages, and outreach activities are underway to communicate with those groups that specialize in these microbial taxa to contribute strains to the HMP sequencing efforts.

1.4.1. Derivative Datasets from the Healthy Adult Cohort Study

Sequence data prepared by the participating sequencing centers were deposited in the sequence read archive (SRA) as the data were being produced. As noted earlier, the DACC, DAWG, and their workgroups processed the data according to agreed-on parameters and also developed a number of derivative datasets of the 16*S* and metagenomic WGS sequence data. The group carried out these data processing steps in order to create master, common sets of data for downstream analyses. In 2011, the DAWG also decided that two specific derivative datasets from the suite of datasets should also be released to provide a community resource for microbiome researchers. This set included deconvoluted, trimmed 16*S* data that included 74 million 16*S* reads from over 6000 of the healthy adult cohort study samples; the 16*S* rRNA variable region V3–V5 was sequenced for all of these, while the V1–V3 and V6–V9 regions were also sequenced for a subset of these 6000 samples. De novo metagenomic assemblies from an initial group of 690 of these samples were also included in this release. In addition, the DACC is preparing a gene index of all proteins predicted from these assemblies, which will also be released as a community resource. In keeping with rapid release of data to the public, these data derivative releases were posted on the DACC website (`http://www.hmpdacc.org/doc/PGA_16SData_post.pdf`), distributed to Newswire (`http://www.newswire.com`) and posted on ASM's MicrobcWorld (`http://www.microbeworld.org/index.php?option=com_jlibrary&view=article&id=6682`).

1.4.2. Computational Tools for Human Microbiome Research

Many computational and bioinformatic tools have been developed by the HMP Computational Tools grantees and were either refined under HMP support or were already developed and adapted to HMP data types. The list of tools is provided (Table 1.7), along with links to their lab websites, the DACC or `www.sourceforge.net`. Other tools were developed by the sequencing centers and other members of the HMP Research Network Consortium, or existing tools were adapted for use in human microbiome analysis. These are included on the DACC website filed under "Get Tools."

1.4.3. Publications from the HMP

As of this writing 194 publications cite the NIH Human Microbiome Project for support. Online supplemental material for this chapter provided by the publisher includes the HMP PubMed publications list (see Table 1.8).

TABLE 1.7. Computational Tools Developed or Modified for HMP

Yuzhen Ye, Indiana Univ.: fragment assembly and metabolic/species diversity analysis for HMP (Ye lab website: http://omics.informatics.indiana.edu/hmp/index.php)
- FragGeneScan: a tool for fragmental gene prediction in short reads (http://omics.informatics.indiana.edu/FragGeneScan/)
- AbundanceBin: abundance-based tool for binning metagenomic sequences (http://omics.informatics.indiana.edu/AbundanceBin/)
- MinPath: a parsimony approach for biological pathway reconstructions using protein family predictions (http://omics.informatics.indiana.edu/MinPath/)
- AbundantOTU (and AbundantOTU+): a tool for fast and accurate identification and quantification of abundant species from pyrosequences of 16S rRNA by using consensus alignment (http://omics.informatics.indiana.edu/AbundantOTU/)
- PHYLOSHOP: a tool for extracting ribosomal RNA fragments from WGS and simple analysis of ribosomal RNAs (http://omics.informatics.indiana.edu/mg/phyloshop/)
- RAPSearch: a fast tool for protein similarity search (http://omics.informatics.indiana.edu/mg/RAPSearch/)
- RAPSearch2: an even faster RAPSearch that supports multithreading (http://omics.informatics.indiana.edu/mg/RAPSearch2/)
- SWIFT: a fast protein similarity search tool that utilizes a reduced amino acid alphabet and suffix arrays to detect seeds of flexible length; in development
- PathRecruit: an online resource for computing and visualizing both the functional diversity and species diversity for metagenomic samples, in development

Mihai Pop, Univ. Maryland, College Park: assembly and analysis software for exploring the human microbiome (Pop lab software website: http://www.cbcb.umd.edu/~mpop/Software.shtml)
- Minimus-SR: short-read version of our assembler Minimus (available open-source at amos.sourceforge.net in short-read assembly package http://sourceforge.net/apps/mediawiki/amos/index.php?title=Minimus)
- Bambus 2: extensions to AMOS package for analysis of assembly graphs (http://sourceforge.net/apps/mediawiki/amos/index.php?title=Bambus2)
- Crossbow: cloud-computing-enabled sequence aligner and variant caller (http://bowtie-bio.sourceforge.net/crossbow/index.shtml)
- Metastats: statistical package for comparing metagenomic samples (http://metastats.cbcb.umd.edu/)
- Metapath: statistical package for comparing metagenomic samples at the pathway level (http://cbcb.umd.edu/~boliu/metapath/)
- Phymm: statistical binning for metagenomic data (http://cbcb.umd.edu/software/phymm/)
- Contrail: cloud-enabled assembler (http://sourceforge.net/apps/mediawiki/contrail-bio/index.php?title=Contrail)
- DNAclust: fast and accurate clustering of DNA sequences (http://dnaclust.sourceforge.net/)

Rob Knight, University of Colorado, Boulder: New tools for understanding the composition and dynamics of microbial communities
- QIIME (Quantitative Insights Into Microbial Ecology): software package for comparison and analysis of microbial communities, primarily based on high-throughput amplicon sequencing data generated on a variety of platforms, but also supporting analysis of other types of data (such as shotgun metagenomic data) (http://qiime.sourceforge.net/)
- SitePainter: allows users to visualize the different HMP body sites based on gradients of colors to represent available datasets (http://www.hmpdacc.org/sp/)

Daniel Haft, J. Craig Venter Institute: algorithmically-tuned protein families, also rule_base and characterized proteins
- TIGRFAMs—resource consisting of curated multiple sequence alignments, hidden Markov models (HMMs) for protein sequence classification, and associated information designed to support automated annotation of (mostly prokaryotic) proteins (http://www.jcvi.org/cgi-bin/tigrfams/index.cgi)

(*Continued*)

TABLE 1.7. *(Continued)*

- Partial phylogenetic profiling (PPP) software and comparative genomics database: (ftp://ftp.icvi.org/pub/data/ppp)
- CHAR (Database of Experimentally Characterized Proteins): new annotation rules created and included in distributions of the JCVI-produced tool AutoAnnotate (http://www.jcvi.org/cms/research/projects/annotation-service/)
- TIGRFAMs: added collections of HMM-based protein family definitions for automated annotation pipelines (ftp://ftp.icvi.Qrg/pub/data/TIGRFAMs/)
- CRISPR—Cas system classification (http://www.nature.com/nrmicro/journal/v9/n6/full/nrmicro2577.html)

Gary Stormo, Washington University at St Louis: exploiting microbiome sequences for improved models of protein-DNA interactions (Stormo lab page: http://ural.wustl.edu/resources.html#Software)

Daniel Frank, University of Colorado, Denver: high performance validation and classification of metagenomic ribosomal-RNA sequences (Software that is at least in the beta testing stage is provided with no restrictions at: http://www.phyloware.com/Phyloware/Home.html; XplorSeq—Mac OSX software for sequence analysis: http://www.phyloware.com/Phyloware/XplorSeq.html)

Patrick Schloss, University of Michigan: identifying population-level variation in cross-sectional and longitudinal HMP studies. Mothur: single resource that incorporates functionality of several tools to analyze microbial ecology data (http://www.mothur.org/)

Weizhong Li, University of California, San Diego: novel methods for effective analysis assembly and comparison of HMP sequences

- Meta-assembler for 454 reads (http://camera.calit2.net/)
- FR-HIT: A new fragment recruitment method called FR-HIT has been implemented. FR-HIT has similar sensitivity as BLASTN but is about 2 orders of magnitude faster in recruiting raw reads. FR-HIT is slower than some mapping programs, but it can recruit several times more reads (http://weizhong-lab.ucsd.edu/frhit/). Source code (http://code.google.com/p/frhit/)
- Meta-RNA(H3): The rRNA prediction method Meta-RNA was improved using Hmmer3 (http://weizhong-lab.ucsd.edu/meta_rna/)
- WebMGA: A collection of web servers, WebMGA, has been created. In addition to the new Meta-rRNA and cd-hit-454, WebMGA includes ~20 other commonly used tools such as ORF calling, sequence clustering, quality control of raw reads, removal of contaminations and functional annotation (http://weizhong-lab.ucsd.edu/metagenomic-analysis/)

Curtis Huttenhower, Harvard University School of Public Health: functional activity and inter-organismal interactions in the human microbiome (http://huttenhower.org/galaxy/)

- The LEfSe algorithm has been developed to discover and explain microbial and functional biomarkers in the human microbiota and other microbiomes. Its accuracy in comparison to existing methods using both synthetic data and published metagenomic functional gene family catalogs has been validated. The method is freely available online and has already, before publication, received nearly 200 unique nonrobot visitors.
- HUMAnN, an end-to end system for reconstructing gene families and functional and metabolic pathways from metagenomic (or metratranscriptomic) data, has been developed. HUMAnN has been validated as a significant improvement over state-of-the-art using a synthetic metagenomes and was used for metabolic reconstruction of 649 samples (>2.5Tbp sequence) from 7 body sites on 102 individuals as part of the HMP (http://huttenhower.sph.harvard.edu/humann)

David Fredricks, Fred Hutchinson Cancer Research Center: novel computational tools for studying the human microbiome

- Bioconductor software package for processing of high throughput sequence reads (http://bioconductor.org/packages/devel/html/microbiome454.html)
- Reference package standard (http://github.com/fhcrc/taxtastic/wiki/refpkg)

TABLE 1.8. HMP Publications in PubMed (Updated 02/15/12)

1. Gonzalez A, Stombaugh J, Lauber CL, Fierer N, Knight R. SitePainter: A tool for exploring biogeographical patterns. *Bioinformatics* **28**(3):436–468 (2012).
2. Gonzalez A, Knight R. Advancing analytical algorithms and pipelines for billions of microbial sequences. *Curr Opin Biotechnol* **23**(1):64–71 (2012).
3. Mercer M, Brinich MA, Geller G, Harrison K, Highland J, James K, Marshall P, McCormick JB, Tilburt J, Achkar JP, et al. How patients view probiotics: Findings from a multicenter study of patients with inflammatory bowel disease and irritable bowel syndrome. *J Clin Gastroenterol* **46**(2):138–144 (2012).
4. Bik HM, Porazinska DL, Creer S, Caporaso JG, Knight R, Thomas WK. Sequencing our way towards understanding global eukaryotic biodiversity. *Trends Ecol Evol* **27**(4):233–243 (2012).
5. Cuellar-Partida G, Buske FA, McLeay RC, Whitington T, Noble WS, Bailey TL. Epigenetic priors for identifying active transcription factor binding sites. *Bioinformatics* **28**(1):56–62 (2012).
6. Zhao Y, Tang H, Ye Y. RAPSearch2: A fast and memory-efficient protein similarity search tool for next-generation sequencing data. *Bioinformatics* **28**(1):125–126 (2012).
7. Yao G, Ye L, Gao H, Minx P, Warren WC, Weinstock GM. Graph accordance of next-generation sequence assemblies. *Bioinformatics* **28**(1):13–16 (2012).
8. Sun Y, Cai Y, Huse SM, Knight R, Farmerie WG, Wang X, Mai V. A large-scale benchmark study of existing algorithms for taxonomy-independent microbial community analysis. *Brief Bioinform* **13**(1):107–121 (2012).
9. Werner JJ, Koren O, Hugenholtz P, DeSantis TZ, Walters WA, Caporaso JG, Angenent LT, Knight R, Ley RE. Impact of training sets on classification of high-throughput bacterial 16s rRNA gene surveys. *ISME (International Society for Microbial Ecology) J* **6**(1):94–103 (2012).
10. Markowitz VM, Chen IM, Palaniappan K, Chu K, Szeto E, Grechkin Y, Ratner A, Jacob B, Huang J, Williams P, et al. IMG: The Integrated Microbial Genomes database and comparative analysis system. *Nucleic Acids Res* **40**(database issue):D115–D122 (2012).
11. Madupu R, Richter A, Dodson RJ, Brinkac L, Harkins D, Durkin S, Shrivastava S, Sutton G, Haft D. CharProtDB: A database of experimentally characterized protein annotations. *Nucleic Acids Res* **40**(database issue):D237–D241 (2012).
12. Pagani I, Liolios K, Jansson J, Chen IM, Smirnova T, Nosrat B, Markowitz VM, Kyrpides NC. The Genomes OnLine Database (GOLD) v.4: Status of genomic and metagenomic projects and their associated metadata. *Nucleic Acids Res* **40**(database issue):D571–D579 (2012).
13. Markowitz VM, Chen IM, Chu K, Szeto E, Palaniappan K, Grechkin Y, Ratner A, Jacob B, Pati A, Huntemann M, et al. IMG/M: The integrated metagenome data management and comparative analysis system. *Nucleic Acids Res* **40**(database issue):D123–D129 (2012).
14. Ji X, Pushalkar S, Li Y, Glickman R, Fleisher K, Saxena D. Antibiotic effects on bacterial profile in osteonecrosis of the jaw. *Oral Dis* **18**(1):85–95 (2012).
15. Lewis CM Jr, Obregón-Tito A, Tito RY, Foster MW, Spicer PG. The Human Microbiome Project: Lessons from human genomics. *Trends Microbiol* **20**(1):1–4 (2012).
16. Kuczynski J, Lauber CL, Walters WA, Parfrey LW, Clemente JC, Gevers D, Knight R. Experimental and analytical tools for studying the human microbiome. *Nat Rev Genet* **13**(1):47–58 (2011).
17. Gonzalez A, King A, Robeson Ii MS, Song S, Shade A, Metcalf JL, Knight R. Characterizing microbial communities through space and time. *Curr Opin Biotechnol* **23**(3):431–436 (2011).

(*Continued*)

TABLE 1.8. (*Continued*)

18. Liu Z, Hsiao W, Cantarel BL, Drábek EF, Fraser-Liggett C. Sparse distance-based learning for simultaneous multiclass classification and feature selection of metagenomic data. *Bioinformatics* **27**(23):3242–3249 (2011).
19. Kuczynski J, Stombaugh J, Walters WA, González A, Caporaso JG, Knight R. Using QIIME to analyze 16S rRNA gene sequences from microbial communities. In *Current Protocols in Bioinformatics*, AD Baxevanis et al., eds., Wiley, 2011, Chap. 10, Unit 10.7.
20. McDonald D, Price MN, Goodrich J, Nawrocki EP, Desantis TZ, Probst A, Andersen GL, Knight R, Hugenholtz P. An improved Greengenes taxonomy with explicit ranks for ecological and evolutionary analyses of bacteria and archaea. *ISME J* **6**(3):610–628 (2011).
21. Fierer N, Lauber CL, Ramirez KS, Zaneveld J, Bradford MA, Knight R. Comparative metagenomic, phylogenetic and physiological analyses of soil microbial communities across nitrogen gradients. *ISME J* **6**(5):1007–1017 (2011).
22. Basu MK, Selengut JD, Haft DH. ProPhylo: Partial phylogenetic profiling to guide protein family construction and assignment of biological process. *BMC Bioinformatics* **12**:434 (2011).
23. Schwab AP, Frank L, Gligorov N. Saying privacy, meaning confidentiality. *Am J Bioeth* **11**(11):44–45 (2011).
24. Rhodes R, Azzouni J, Baumrin SB, Benkov K, Blaser MJ, Brenner B, Dauben JW, Earle WJ, Frank L, Gligorov N, et al. De minimis risk: A proposal for a new category of research risk. *Am J Bioeth* **11**(11):1–7 (2011).
25. Youssef NH, Blainey PC, Quake SR, Elshahed MS. Partial genome assembly for a candidate division OP11 single cell from an anoxic spring (Zodletone Spring, Oklahoma). *Appl Environ Microbiol* **77**(21):7804–7814 (2011).
26. Koren S, Treangen TJ, Pop M. Bambus 2: Scaffolding metagenomes. *Bioinformatics* **27**(21):2964–2971 (2011).
27. Pirrung M, Kennedy R, Caporaso JG, Stombaugh J, Wendel D, Knight R. TopiaryExplorer: Visualizing large phylogenetic trees with environmental metadata. *Bioinformatics* **27**(21):3067–3069 (2011).
28. Holtz LR, Wylie KM, Sodergren E, Jiang Y, Franz CJ, Weinstock GM, Storch GA, Wang D. Astrovirus MLB2 viremia in febrile child. *Emerg Infect Dis* **17**(11):2050–2052 (2011).
29. Saulnier DM, Riehle K, Mistretta TA, Diaz MA, Mandal D, Raza S, Weidler EM, Qin X, Coarfa C, Milosavljevic A, et al. Gastrointestinal microbiome signatures of pediatric patients with irritable bowel syndrome. *Gastroenterology* **141**(5):1782–1791 (2011).
30. Plottel CS, Blaser MJ. Microbiome and malignancy. *Cell Host Microbe* **10**(4):324–335 (2011) (review).
31. Knights D, Parfrey LW, Zaneveld J, Lozupone C, Knight R. Human-associated microbial signatures: Examining their predictive value. *Cell Host Microbe* **10**(4):292–296 (2011).
32. Proctor LM. The Human Microbiome Project in 2011 and beyond. *Cell Host Microbe* **10**(4):287–391 (2011).
33. Wu GD, Chen J, Hoffmann C, Bittinger K, Chen YY, Keilbaugh SA, Bewtra M, Knights D, Walters WA, Knight R, et al. Linking long-term dietary patterns with gut microbial enterotypes. *Science* **334**(6052):105–108 (2011).
34. Chen CH, Cho SH, Chiang HI, Tsai F, Zhang K, Lo YH. Specific sorting of single bacterial cells with microfabricated fluorescence-activated cell sorting and tyramide signal amplification fluorescence in situ hybridization. *Anal Chem* **83**(19):7269–7275 (2011).

TABLE 1.8. (*Continued*)

35. Zaneveld JR, Parfrey LW, Van Treuren W, Lozupone C, Clemente JC, Knights D, Stombaugh J, Kuczynski J, Knight R. Combined phylogenetic and genomic approaches for the high-throughput study of microbial habitat adaptation. *Trends Microbiol* **19**(10):472–482 (2011) (review).
36. McKenzie VJ, Bowers RM, Fierer N, Knight R, Lauber CL. Co-habiting amphibian species harbor unique skin bacterial communities in wild populations. *ISME J* **6**(3):588–596 (2011).
37. Sczesnak A, Segata N, Qin X, Gevers D, Petrosino JF, Huttenhower C, Littman DR, Ivanov II. The genome of th17 cell-inducing segmented filamentous bacteria reveals extensive auxotrophy and adaptations to the intestinal environment. *Cell Host Microbe* **10**(3):260–272 (2011).
38. Arias CA, Panesso D, McGrath DM, Qin X, Mojica MF, Miller C, Diaz L, Tran TT, Rincon S, Barbu EM, et al. Genetic basis for in vivo daptomycin resistance in enterococci. *N Engl J Med* **365**(10):892–900 (2011).
39. Wu S, Zhu Z, Fu L, Niu B, Li W. WebMGA: A customizable web server for fast metagenomic sequence analysis. *BMC Genomics* **12**:444 (2011).
40. Bowers RM, Sullivan AP, Costello EK, Collett JL Jr, Knight R, Fierer N. Sources of bacteria in outdoor air across cities in the midwestern United States. *Appl Environ Microbiol* **77**(18):6350–6356 (2011).
41. Mei Z, Wu TF, Pion-Tonachini L, Qiao W, Zhao C, Liu Z, Lo YH. Applying an optical space-time coding method to enhance light scattering signals in microfluidic devices. *Biomicrofluidics* **5**(3):34116–341166. (2011).
42. Lane MM, Czyzewski DI, Chumpitazi BP, Shulman RJ. Reliability and validity of a modified Bristol Stool Form Scale for children. *J Pediatr* **159**(3):437–441 (2011).
43. Frank DN, Zhu W, Sartor RB, Li E. Investigating the biological and clinical significance of human dysbioses. *Trends Microbiol* **19**(9):427–434 (2011).
44. Liu P, Meagher RJ, Light YK, Yilmaz S, Chakraborty R, Arkin AP, Hazen TC, Singh AK. Microfluidic fluorescence in situ hybridization and flow cytometry (μFlowFISH). *Lab Chip* **11**(16):2673–2679 (2011).
45. Edgar RC, Haas BJ, Clemente JC, Quince C, Knight R. UCHIME improves sensitivity and speed of chimera detection. *Bioinformatics* **27**(16):2194–2200 (2011).
46. Tito RY, Belknap SL 3rd, Sobolik KD, Ingraham RC, Cleeland LM, Lewis CM Jr. Brief communication: DNA from early Holocene American dog. *Am J Phys Anthropol* **145**(4):653–657 (2011).
47. González-Rivera R, Culverhouse RC, Hamvas A, Tarr PI, Warner BB. The age of necrotizing enterocolitis onset: an application of Sartwell's incubation period model. *J Perinatol* **31**(8):519–523 (2011).
48. Brunicardi FC, Gibbs RA, Wheeler DA, Nemunaitis J, Fisher W, Goss J, Chen C. Overview of the development of personalized genomic medicine and surgery. *World J Surg* **35**(8):1693–1699 2011 (review).
49. Harring TR, Guiteau JJ, Nguyen NT, Cotton RT, Gingras MC, Wheeler DA, O'Mahony CA, Gibbs RA, Brunicardi FC, Goss JA. Building a comprehensive genomic program for hepatocellular carcinoma. *World J Surg* **35**(8):1746–1750 (2011).
50. Nguyen NT, Cotton RT, Harring TR, Guiteau JJ, Gingras MC, Wheeler DA, O'Mahony CA, Gibbs RA, Brunicardi FC, Goss JA. A primer on a hepatocellular carcinoma bioresource bank using the cancer genome atlas guidelines: Practical issues and pitfalls. *World J Surg* **35**(8):1732–1737 (2011).
51. Knights D, Kuczynski J, Charlson ES, Zaneveld J, Mozer MC, Collman RG, Bushman FD, Knight R, Kelley ST. Bayesian community-wide culture-independent microbial source tracking. *Nat Methods* **8**(9):761–763 (2011).

(*Continued*)

TABLE 1.8. (*Continued*)

52. Bergmann GT, Bates ST, Eilers KG, Lauber CL, Caporaso JG, Walters WA, Knight R, Fierer N. The under-recognized dominance of Verrucomicrobia in soil bacterial communities. *Soil Biol Biochem* **43**(7):1450–1455 (2011).
53. Ghodsi M, Liu B, Pop M. DNACLUST: Accurate and efficient clustering of phylogenetic marker genes. *BMC Bioinformatics* **12**:271 (2011).
54. Segata N, Izard J, Waldron L, Gevers D, Miropolsky L, Garrett WS, Huttenhower C. Metagenomic biomarker discovery and explanation. *Genome Biol* **12**(6):R60 (2011).
55. Stombaugh J, Widmann J, McDonald D, Knight R. Boulder ALignment Editor (ALE): A web-based RNA alignment tool. *Bioinformatics* **27**(12):1706–1707 (2011).
56. Niu B, Zhu Z, Fu L, Wu S, Li W. FR-HIT, a very fast program to recruit metagenomic reads to homologous reference genomes. *Bioinformatics* **27**(12):1704–1705 (2011).
57. Haft DH, Basu MK. Biological systems discovery in silico: Radical S-adenosylmethionine protein families and their target peptides for posttranslational modification. *J Bacteriol* **193**(11):2745–2755 (2011).
58. Bucher BT, McDuffie LA, Shaikh N, Tarr PI, Warner BB, Hamvas A, White FV, Erwin CR, Warner BW. Bacterial DNA content in the intestinal wall from infants with necrotizing enterocolitis. *J Pediatr Surg* **46**(6):1029–1033 (2011).
59. Makarova KS, Haft DH, Barrangou R, Brouns SJ, Charpentier E, Horvath P, Moineau S, Mojica FJ, Wolf YI, Yakunin AF, et al. Evolution and classification of the CRISPR-Cas systems. *Nat Rev Microbiol* **9**(6):467–477 (2011).
60. Kong HH. Skin microbiome: Genomics-based insights into the diversity and role of skin microbes. *Trends Mol Med* **17**(6):320–328 (2011).
61. Muegge BD, Kuczynski J, Knights D, Clemente JC, González A, Fontana L, Henrissat B, Knight R, Gordon JI. Diet drives convergence in gut microbiome functions across mammalian phylogeny and within humans. *Science* **332**(6032):970–974 (2011).
62. Ye Y, Choi JH, Tang H. RAPSearch: A fast protein similarity search tool for short reads. *BMC Bioinform* **12**:159 (2011).
63. Kellermayer R, Dowd SE, Harris RA, Balasa A, Schaible TD, Wolcott RD, Tatevian N, Szigeti R, Li Z, Versalovic J, Smith CW. Colonic mucosal DNA methylation, immune response, and microbiome patterns in Toll-like receptor 2-knockout mice. *FASEB J* **25**(5):1449–1460 (2011).
64. Dominguez-Bello MG, Blaser MJ, Ley RE, Knight R. Development of the human gastrointestinal microbiota and insights from high-throughput sequencing. *Gastroenterology* **140**(6):1713–1719 (2011) (review).
65. Reeves AE, Theriot CM, Bergin IL, Huffnagle GB, Schloss PD, Young VB. The interplay between microbiome dynamics and pathogen dynamics in a murine model of Clostridium difficile Infection. *Gut Microbes* **2**(3):145–158 (2011).
66. Bates ST, Berg-Lyons D, Caporaso JG, Walters WA, Knight R, Fierer N. Examining the global distribution of dominant archaeal populations in soil. *ISME J* **5**(5):908–917 (2011).
67. DeSantis TZ, Keller K, Karaoz U, Alekseyenko AV, Singh NN, Brodie EL, Pei Z, Andersen GL, Larsen N. Simrank: Rapid and sensitive general-purpose k-mer search tool. *BMC Ecol* **11**:11 (2011).
68. Walters WA, Caporaso JG, Lauber CL, Berg-Lyons D, Fierer N, Knight R. PrimerProspector: De novo design and taxonomic analysis of barcoded polymerase chain reaction primers. *Bioinformatics* **27**(8):1159–1161 (2011).
69. Ballal SA, Gallini CA, Segata N, Huttenhower C, Garrett WS. Host and gut microbiota symbiotic factors: Lessons from inflammatory bowel disease and successful symbionts. *Cell Microbiol* **13**(4):508–517 (2011).
70. Fierer N, McCain CM, Meir P, Zimmermann M, Rapp JM, Silman MR, Knight R. Microbes do not follow the elevational diversity patterns of plants and animals. *Ecology* **92**(4):797–804 (2011).

TABLE 1.8. (*Continued*)

71. Pushalkar S, Mane SP, Ji X, Li Y, Evans C, Crasta OR, Morse D, Meagher R, Singh A, Saxena D. Microbial diversity in saliva of oral squamous cell carcinoma. *FEMS Immunol Med Microbiol* 61(3):269–277 (2011).
72. Knights D, Kuczynski J, Koren O, Ley RE, Field D, Knight R, DeSantis TZ, Kelley ST. Supervised classification of microbiota mitigates mislabeling errors. *ISME J* **5**(4):570–573 (2011).
73. Spor A, Koren O, Ley R. Unravelling the effects of the environment and host genotype on the gut microbiome. *Nat Rev Microbiol* **9**(4):279–290 (2011) (review).
74. Grice EA, Segre JA. The skin microbiome. *Nat Rev Microbiol* **9**(4):244–253 (2011) (review). [Erratum in *Nat Rev Microbiol* **9**(8):626 (2011).]
75. Hansen EE, Lozupone CA, Rey FE, Wu M, Guruge JL, Narra A, Goodfellow J, Zaneveld JR, McDonald DT, Goodrich JA, et al. Pan-genome of the dominant human gut-associated archaeon, Methanobrevibacter smithii, studied in twins. *Proc Natl Acad Sci USA*. **108(**Suppl 1):4599–4606 (2011).
76. Koren O, Spor A, Felin J, Fåk F, Stombaugh J, Tremaroli V, Behre CJ, Knight R, Fagerberg B, Ley RE, et al. Human oral, gut, and plaque microbiota in patients with atherosclerosis. *Proc Natl Acad Sci USA* 108(Suppl 1):4592–4598 (2011).
77. Ravel J, Gajer P, Abdo Z, Schneider GM, Koenig SS, McCulle SL, Karlebach S, Gorle R, Russell J, Tacket CO, et al. Vaginal microbiome of reproductive-age women. *Proc Natl Acad Sci USA* **108**(Suppl 1):4680–4687 (2011).
78. Caporaso JG, Lauber CL, Walters WA, Berg-Lyons D, Lozupone CA, Turnbaugh PJ, Fierer N, Knight R. Global patterns of 16S rRNA diversity at a depth of millions of sequences per sample. *Proc Natl Acad Sci USA* **108**(Suppl 1):4516–4522 (2011).
79. Treangen TJ, Sommer DD, Angly FE, Koren S, Pop M. Next generation sequence assembly with AMOS. In *Current Protocols in Bioinformatics*, AD Baxevanis et al., eds., Wiley, 2011, Chap. 11, Unit 11.8.
80. Haas BJ, Gevers D, Earl AM, Feldgarden M, Ward DV, Giannoukos G, Ciulla D, Tabbaa D, Highlander SK, Sodergren E, et al. Human Microbiome Consortium, (with Petrosino JF, Knight R, Birren BW). Chimeric 16S rRNA sequence formation and detection in Sanger and 454-pyrosequenced PCR amplicons. *Genome Res* **21**(3):494–504 (2011).
81. Frank DN. Growth and Development Symposium: Promoting healthier humans through healthier livestock: Animal agriculture enters the metagenomics era. *J Anim Sci* **89**(3):835–844 (2011).
82. Wu YW, Ye Y. A novel abundance-based algorithm for binning metagenomic sequences using l-tuples. *J Comput Biol* **18**(3):523–534 (2011).
83. Bradbury AR, Sidhu S, Dübel S, McCafferty J. Beyond natural antibodies: The power of in vitro display technologies. *Nat Biotechnol* **29**(3):245–254 (2011).
84. Blainey PC, Quake SR. Digital MDA for enumeration of total nucleic acid contamination. *Nucleic Acids Res* **39**(4):e19 (2011).
85. Park J, Kerner A, Burns MA, Lin XN. Microdroplet-enabled highly parallel co-cultivation of microbial communities. *PLoS ONE* **6**(2):e17019 (2011).
86. Caporaso JG, Knight R, Kelley ST. Host-associated and free-living phage communities differ profoundly in phylogenetic composition. *PLoS ONE*. **6**(2):e16900 (2011).
87. Blainey PC, Mosier AC, Potanina A, Francis CA, Quake SR. Genome of a low-salinity ammonia-oxidizing archaeon determined by single-cell and metagenomic analysis. *PLoS ONE* **6**(2):e16626 (2011).
88. Chuang HS, Raizen DM, Lamb A, Dabbish N, Bau HH. Dielectrophoresis of Caenorhabditis elegans. *Lab Chip* **11**(4):599–604 (2011).
89. Ye Y. Identification and quantification of abundant species from pyrosequences of 16S rRNA by consensus alignment. *Proc IEEE Int Conf Bioinformatics Biomedicine*, 2/4/10, IEEE, 2011, pp. 153–157.

(*Continued*)

TABLE 1.8. (*Continued*)

90. Bates ST, Cropsey GW, Caporaso JG, Knight R, Fierer N. Bacterial communities associated with the lichen symbiosis. *Appl Environ Microbiol* **77**(4):1309–1314 (2011).
91. Wardwell LH, Huttenhower C, Garrett WS. Current concepts of the intestinal microbiota and the pathogenesis of infection. *Curr Infect Dis Rep* **13**(1):28–34 (2011).
92. Hu S, Dong TS, Dalal SR, Wu F, Bissonnette M, Kwon JH, Chang EB. The microbe-derived short chain fatty acid butyrate targets miRNA-dependent p21 gene expression in human colon cancer. *PLoS ONE* **6**(1):e16221 (2011).
93. Haft DH. Bioinformatic evidence for a widely distributed, ribosomally produced electron carrier precursor, its maturation proteins, and its nicotinoprotein redox partners. *BMC Genom* **12**:21 (2011).
94. Kalisky T, Blainey P, Quake SR. Genomic analysis at the single-cell level. *Annu Rev Genet* **45**:431–445 (2011).
95. Liu B, Gibbons T, Ghodsi M, Treangen T, Pop M. Accurate and fast estimation of taxonomic profiles from metagenomic shotgun sequences. *BMC Genom* **12**(Suppl 2):S4 (2011).
96. Gonzalez A, Stombaugh J, Lozupone C, Turnbaugh PJ, Gordon JI, Knight R. The mind-body-microbial continuum. *Dialog Clin Neurosci* **13**(1):55–62 (2011).
97. Parfrey LW, Walters WA, Knight R. Microbial eukaryotes in the human microbiome: ecology, evolution, and future directions. *Front Microbiol* **2**:153 (2011).
98. Young VB, Kahn SA, Schmidt TM, Chang EB. Studying the enteric microbiome in inflammatory bowel diseases: Getting through the growing pains and moving forward. *Front Microbiol* **2**:144 (2011).
99. Preidis GA, Hill C, Guerrant RL, Ramakrishna BS, Tannock GW, Versalovic J. Probiotics, enteric and diarrheal diseases, and global health. *Gastroenterology* **140**(1):8–14 (2011).
100. Caporaso JG, Lauber CL, Costello EK, Berg-Lyons D, Gonzalez A, Stombaugh J, Knights D, Gajer P, Ravel J, Fierer N, et al. Moving pictures of the human microbiome. *Genome Biol* **12**(5):R50 (2011).
101. Frank DN, Robertson CE, Hamm CM, Kpadeh Z, Zhang T, Chen H, Zhu W, Sartor RB, Boedeker EC, Harpaz N, et al. Disease phenotype and genotype are associated with shifts in intestinal-associated microbiota in inflammatory bowel diseases. *Inflamm Bowel Dis* **17**(1):179–184 (2011).
102. Haft DH, Varghese N. GlyGly-CTERM and rhombosortase: A C-terminal protein processing signal in a many-to-one pairing with a rhomboid family intramembrane serine protease. *PLoS ONE* **6**(12):e28886 (2011).
103. Schloss PD, Gevers D, Westcott SL. Reducing the effects of PCR amplification and sequencing artifacts on 16S rRNA-based studies. *PLoS ONE* **6**(12):e27310 (2011).
104. Flores GE, Bates ST, Knights D, Lauber CL, Stombaugh J, Knight R, Fierer N. Microbial biogeography of public restroom surfaces. *PLoS ONE* **6**(11):e28132 (2011).
105. Cantarel BL, Erickson AR, VerBerkmoes NC, Erickson BK, Carey PA, Pan C, Shah M, Mongodin EF, Jansson JK, Fraser-Liggett CM, et al. Strategies for metagenomic-guided whole-community proteomics of complex microbial environments. *PLoS ONE* **6**(11):e27173 (2011).
106. Ferrara F, Listwan P, Waldo GS, Bradbury AR. Fluorescent labeling of antibody fragments using split GFP. *PLoS ONE* **6**(10):e25727 (2011).
107. Segata N, Huttenhower C. Toward an efficient method of identifying core genes for evolutionary and functional microbial phylogenies. *PLoS ONE* **6**(9):e24704 (2011).
108. Ahn J, Yang L, Paster BJ, Ganly I, Morris L, Pei Z, Hayes RB. Oral microbiome profiles: 16S rRNA pyrosequencing and microarray assay comparison. *PLoS ONE* **6**(7):e22788 (2011).
109. Lladser ME, Gouet R, Reeder J. Extrapolation of urn models via poissonization: Accurate measurements of the microbial unknown. *PLoS ONE* **6**(6):e21105 (2011).

TABLE 1.8. (*Continued*)

110. Dong Q, Nelson DE, Toh E, Diao L, Gao X, Fortenberry JD, Van der Pol B. The microbial communities in male first catch urine are highly similar to those in paired urethral swab specimens. *PLoS ONE* **6**(5):e19709 (2011).
111. Cho SH, Godin JM, Chen CH, Qiao W, Lee H, Lo YH. Recent advancements in optofluidic flow cytometer. *Biomicrofluidics* **4**(4):43001 (2010) (review).
112. Hu S, Wang Y, Lichtenstein L, Tao Y, Musch MW, Jabri B, Antonopoulos D, Claud EC, Chang EB. Regional differences in colonic mucosa-associated microbiota determine the physiological expression of host heat shock proteins. *Am J Physiol Gastrointest Liver Physiol* **299**(6):G1266–G1275 (2010).
113. Wang Y, Antonopoulos DA, Zhu X, Harrell L, Hanan I, Alverdy JC, Meyer F, Musch MW, Young VB, Chang EB. Laser capture microdissection and metagenomic analysis of intact mucosa-associated microbial communities of human colon. *Appl Microbiol Biotechnol* **88**(6):1333–1342 (2010).
114. Harrington ED, Arumugam M, Raes J, Bork P, Relman DA. SmashCell: A software framework for the analysis of single-cell amplified genome sequences. *Bioinformatics* **26**(23):2979–1980 (2010).
115. Krentz BD, Mulheron HJ, Semrau JD, Dispirito AA, Bandow NL, Haft DH, Vuilleumier S, Murrell JC, McEllistrem MT, Hartsel SC, et al. A comparison of methanobactins from Methylosinus trichosporium OB3b and Methylocystis strain Sb2 predicts methanobactins are synthesized from diverse peptide precursors modified to create a common core for binding and reducing copper ions. *Biochemistry* **49**(47):10117–10130 (2010).
116. Nelson DE, Van Der Pol B, Dong Q, Revanna KV, Fan B, Easwaran S, Sodergren E, Weinstock GM, Diao L, Fortenberry JD. Characteristic male urine microbiomes associate with asymptomatic sexually transmitted infection. *PLoS ONE* **5**(11):e14116 (2010).
117. Redford AJ, Bowers RM, Knight R, Linhart Y, Fierer N. The ecology of the phyllosphere: Geographic and phylogenetic variability in the distribution of bacteria on tree leaves. *Environ Microbiol* **12**(11):2885–2893 (2010).
118. Costello EK, Gordon JI, Secor SM, Knight R. Postprandial remodeling of the gut microbiota in Burmese pythons. *ISME J* **4**(11):1375–1385 (2010).
119. Selengut JD, Haft DH. Unexpected abundance of coenzyme F(420)-dependent enzymes in Mycobacterium tuberculosis and other actinobacteria. *J Bacteriol* **192**(21):5788–5798 (2010).
120. Rho M, Tang H, Ye Y. FragGeneScan: Predicting genes in short and error-prone reads. *Nucleic Acids Res* **38**(20):e191 (2010).
121. Wang Y, Devkota S, Musch MW, Jabri B, Nagler C, Antonopoulos DA, Chervonsky A, Chang EB. Regional mucosa-associated microbiota determine physiological expression of TLR2 and TLR4 in murine colon. *PLoS ONE*. **5**(10):e13607 (2010).
122. Goll J, Rusch DB, Tanenbaum DM, Thiagarajan M, Li K, Methé BA, Yooseph S. METAREP: JCVI metagenomics reports—an open source tool for high-performance comparative metagenomics. *Bioinformatics* **26**(20):2631–2632 (2010).
123. So A, Pel J, Rajan S, Marziali A. *Efficient Genomic DNA Extraction from Low Target Concentration Bacterial Cultures Using SCODA DNA Extraction Technology*, Cold Spring Harbor Protocol, 2010(10/1/10).
124. Dougan G, Weinstock GM. A new era in the genomics of bacteria. *Curr Opin Microbiol* **13**(5):616–618 (2010).
125. Manichanh C, Reeder J, Gibert P, Varela E, Llopis M, Antolin M, Guigo R, Knight R, Guarner F. Reshaping the gut microbiome with bacterial transplantation and antibiotic intake. *Genome Res* **20**(10):1411–1419 (2010).

(*Continued*)

TABLE 1.8. *(Continued)*

126. Gao Z, Perez-Perez GI, Chen Y, Blaser MJ. Quantitation of major human cutaneous bacterial and fungal populations. *J Clin Microbiol* **48**(10):3575–3581 (2010).
127. Chumpitazi BP, Lane MM, Czyzewski DI, Weidler EM, Swank PR, Shulman RJ. Creation and initial evaluation of a stool form scale for children. *J Pediatr* **157**(4):594–597 (2010).
128. Kuczynski J, Liu Z, Lozupone C, McDonald D, Fierer N, Knight R. Microbial community resemblance methods differ in their ability to detect biologically relevant patterns. *Nat Meth* **7**(10):813–819 (2010).
129. McGuire AL, Beskow LM. Informed consent in genomics and genetic research. *Annu Rev Genom Hum Genet* **11**:361–381 (2010).
130. Koren S, Miller JR, Walenz BP, Sutton G. An algorithm for automated closure during assembly. *BMC Bioinformatics* **11**:457 (2010).
131. Nossa CW, Oberdorf WE, Yang L, Aas JA, Paster BJ, Desantis TZ, Brodie EL, Malamud D, Poles MA, Pei Z. Design of 16S rRNA gene primers for 454 pyrosequencing of the human foregut microbiome. *World J Gastroenterol* **16**(33):4135–4144 (2010).
132. Venkatesh M, Flores A, Luna RA, Versalovic J. Molecular microbiological methods in the diagnosis of neonatal sepsis. *Expert Rev Anti Infect Ther* **8**(9):1037–1048 (2010).
133. Reeder J, Knight R. Rapidly denoising pyrosequencing amplicon reads by exploiting rank-abundance distributions. *Nat Methods* **7**(9):668–669 (2010).
134. Cho SH, Qiao W, Tsai FS, Yamashita K, Lo YH. Lab-on-a-chip flow cytometer employing color-space-time coding. *Appl Phys Lett* **97**(9):093704 (2010).
135. Yeoman CJ, Yildirim S, Thomas SM, Durkin AS, Torralba M, Sutton G, Buhay CJ, Ding Y, Dugan-Rocha SP, Muzny DM, et al. Comparative genomics of Gardnerella vaginalis strains reveals substantial differences in metabolic and virulence potential. *PLoS ONE* **5**(8):e12411 (2010).
136. Brotman RM, Ravel J, Cone RA, Zenilman JM. Rapid fluctuation of the vaginal microbiota measured by Gram stain analysis. *Sex Transm Infect* **86**(4):297–302 (2010).
137. Wu GD, Lewis JD, Hoffmann C, Chen YY, Knight R, Bittinger K, Hwang J, Chen J, Berkowsky R, Nessel L, et al. Sampling and pyrosequencing methods for characterizing bacterial communities in the human gut using 16S sequence tags. *BMC Microbiol* **10**:206 (2010).
138. Marri PR, Paniscus M, Weyand NJ, Rendón MA, Calton CM, Hernández DR, Higashi DL, Sodergren E, Weinstock GM, Rounsley SD, et al. Genome sequencing reveals widespread virulence gene exchange among human Neisseria species. *PLoS ONE* **5**(7):e11835 (2010).
139. Erez A, Plunkett K, Sutton VR, McGuire AL. The right to ignore genetic status of late onset genetic disease in the genomic era; prenatal testing for Huntington disease as a paradigm. *Am J Med Genet A* **152A**(7):1774–1780 (2010).
140. Huse SM, Welch DM, Morrison HG, Sogin ML. Ironing out the wrinkles in the rare biosphere through improved OTU clustering. *Environ Microbiol* **12**(7):1889–1898 (2010).
141. Bennett WE Jr, González-Rivera R, Puente BN, Shaikh N, Stevens HJ, Mooney JC, Klein EJ, Denno DM, Draghi A II, Sylvester FA, et al. Proinflammatory fecal mRNA and childhood bacterial enteric infections. *Gut Microbes* **1**(4):209–212 (2010).
142. Zaneveld JR, Lozupone C, Gordon JI, Knight R. Ribosomal RNA diversity predicts genome diversity in gut bacteria and their relatives. *Nucleic Acids Res* **38**(12):3869–3879 (2010).
143. Dominguez-Bello MG, Costello EK, Contreras M, Magris M, Hidalgo G, Fierer N, Knight R. Delivery mode shapes the acquisition and structure of the initial microbiota across multiple body habitats in newborns. *Proc Natl Acad Sci USA* **107**(26):11971–11975 (2010).

TABLE 1.8. (*Continued*)

144. Cho SH, Chen CH, Tsai FS, Godin JM, Lo YH. Human mammalian cell sorting using a highly integrated micro-fabricated fluorescence-activated cell sorter (microFACS). *Lab Chip* **10**(12):1567–1573 (2010).
145. Harwich MD Jr, Alves JM, Buck GA, Strauss JF 3rd, Patterson JL, Oki AT, Girerd PH, Jefferson KK. Drawing the line between commensal and pathogenic Gardnerella vaginalis through genome analysis and virulence studies. *BMC Genomi* **11**:375 (2010).
146. Pei AY, Oberdorf WE, Nossa CW, Agarwal A, Chokshi P, Gerz EA, Jin Z, Lee P, Yang L, Poles M, et al. Diversity of 16S rRNA genes within individual prokaryotic genomes. *Appl Environ Microbiol* **76**(12):3886–3897. [erratum in *Appl Environ Microbiol* **76**(15):5333 (2010)].
147. Lauber CL, Zhou N, Gordon JI, Knight R, Fierer N. Effect of storage conditions on the assessment of bacterial community structure in soil and human-associated samples. *FEMS Microbiol Lett* **307**(1):80–86 (2010).
148. Kong HH, Segre JA. Bridging the translational research gap: A successful partnership involving a physician and a basic scientist. *J Invest Dermatol* **130**(6):1478–1480 (2010).
149. Wang GP, Sherrill-Mix SA, Chang KM, Quince C, Bushman FD. Hepatitis C virus transmission bottlenecks analyzed by deep sequencing. *J Virol* **84**(12):6218–6228 (2010).
150. Pati A, Ivanova NN, Mikhailova N, Ovchinnikova G, Hooper SD, Lykidis A, Kyrpides NC. GenePRIMP: A gene prediction improvement pipeline for prokaryotic genomes. *Nat Meth* **7**(6):455–457 (2010).
151. Haft DH, Basu MK, Mitchell DA. Expansion of ribosomally produced natural products: a nitrile hydratase- and Nif11-related precursor family. *BMC Biol* **8**:70 (2010).
152. Human Microbiome Jumpstart Reference Strains Consortium, Nelson KE, Weinstock GM, Highlander SK, Worley KC, Creasy HH, Wortman JR, Rusch DB, Mitreva M, Sodergren E, Chinwalla AT, et al. A catalog of reference genomes from the human microbiome. *Science* **328**(5981):994–999 (2010).
153. Cho SH, Chen CH, Tsai FS, Godin J, Lo YH. Mammalian cell sorting using µFACS, *Proc Conf Lasers Electro Optics*, **2010**:CTuD1 (2010).
154. Thomas CM, Versalovic J. Probiotics-host communication: Modulation of signaling pathways in the intestine. *Gut Microbes* **1**(3):148–163 (2010) (review).
155. Lewis T, Loman NJ, Bingle L, Jumaa P, Weinstock GM, Mortiboy D, Pallen MJ. High-throughput whole-genome sequencing to dissect the epidemiology of Acinetobacter baumannii isolates from a hospital outbreak. *J Hosp Infect* **75**(1):37–41 (2010).
156. Caporaso JG, Kuczynski J, Stombaugh J, Bittinger K, Bushman FD, Costello EK, Fierer N, Peña AG, Goodrich JK, Gordon JI, et al. QIIME allows analysis of high-throughput community sequencing data. *Nat Methods* **7**(5):335–336 (2010).
157. McGuire AL, Lupski JR. Personal genome research: What should the participant be told? *Trends Genet* **26**(5):199–201 (2010).
158. Blaser MJ. Harnessing the power of the human microbiome. *Proc Natl Acad Sci USA* **107**(14):6125–6126 (2010).
159. Fierer N, Lauber CL, Zhou N, McDonald D, Costello EK, Knight R. Forensic identification using skin bacterial communities. *Proc Natl Acad Sci USA* **107**(14):6477–6481 (2010).
160. Fujimura KE, Slusher NA, Cabana MD, Lynch SV. Role of the gut microbiota in defining human health. *Expert Rev Anti Infect Ther* **8**(4):435–454 (2010) (review).
161. White JR, Navlakha S, Nagarajan N, Ghodsi MR, Kingsford C, Pop M. Alignment and clustering of phylogenetic markers—implications for microbial diversity studies. *BMC Bioinformatics* **11**:152 (2010).
162. Jones RT, Knight R, Martin AP. Bacterial communities of disease vectors sampled across time, space, and species. *ISME J* **4**(2):223–231 (2010).

(*Continued*)

TABLE 1.8. (*Continued*)

163. Selengut JD, Rusch DB, Haft DH. Sites inferred by metabolic background assertion labeling (SIMBAL): Adapting the partial phylogenetic profiling algorithm to scan sequences for signatures that predict protein function. *BMC Bioinformatics* **11**:52 (2010).
164. Caporaso JG, Bittinger K, Bushman FD, DeSantis TZ, Andersen GL, Knight R. PyNAST: A flexible tool for aligning sequences to a template alignment. *Bioinformatics* **26**(2):266–267 (2010).
165. Kuczynski J, Costello EK, Nemergut DR, Zaneveld J, Lauber CL, Knights D, Koren O, Fierer N, Kelley ST, Ley RE, et al. Direct sequencing of the human microbiome readily reveals community differences. *Genome Biol* **11**(5):210 (2010) (review).
166. Zhou X, Brotman RM, Gajer P, Abdo Z, Schüette U, Ma S, Ravel J, Forney LJ. Recent advances in understanding the microbiology of the female reproductive tract and the causes of premature birth. *Infect Dis Obstet Gynecol* **2010**:737425 (2010) (review).
167. Hamady M, Lozupone C, Knight R. Fast UniFrac: Facilitating high-throughput phylogenetic analyses of microbial communities including analysis of pyrosequencing and PhyloChip data. *ISME J* **4**(1):17–27 (2010).
168. Salzman NH, Hung K, Haribhai D, Chu H, Karlsson-Sjöberg J, Amir E, Teggatz P, Barman M, Hayward M, Eastwood D, et al. Enteric defensins are essential regulators of intestinal microbial ecology. *Nat Immunol* **11**(1):76–83 (2010).
169. Chen CH, Cho SH, Tsai F, Erten A, Lo YH. Microfluidic cell sorter with integrated piezoelectric actuator. *Biomed Microdevices* **11**(6):1223–1231 (2009).
170. NIH HMP Working Group, Peterson J, Garges S, Giovanni M, McInnes P, Wang L, Schloss JA, Bonazzi V, McEwen JE, Wetterstrand KA, Deal C, et al. The NIH Human Microbiome Project. *Genome Res* **19**(12):2317–2323 Epub (2009).
171. Lazarevic V, Whiteson K, Huse S, Hernandez D, Farinelli L, Osterås M, Schrenzel J, François P. Metagenomic study of the oral microbiota by Illumina high-throughput sequencing. *J Microbiol Meth* **79**(3):266–271 (2009).
172. Blaser MJ, Falkow S. What are the consequences of the disappearing human microbiota? *Nat Rev Microbiol* **7**(12):887–894 (2009).
173. Hildebrandt MA, Hoffmann C, Sherrill-Mix SA, Keilbaugh SA, Hamady M, Chen YY, Knight R, Ahima RS, Bushman F, Wu GD. High-fat diet determines the composition of the murine gut microbiome independently of obesity. *Gastroenterology* **137**(5):1716–1724, e1–2 (2009).
174. Sillanpää J, Nallapareddy SR, Qin X, Singh KV, Muzny DM, Kovar CL, Nazareth LV, Gibbs RA, Ferraro MJ, Steckelberg JM, et al. A collagen-binding adhesin, Acb, and ten other putative MSCRAMM and pilus family proteins of Streptococcus gallolyticus subsp. gallolyticus (Streptococcus bovis group, biotype I). *J Bacteriol* **191**(21):6643–6653 (2009).
175. Ghodsi M, Pop M. Inexact local alignment search over suffix arrays. *Proc (IEEE Int Conf Bioinformatics Biomed)*, 11/1/9, IEEE, 2009, pp. 83–97.
176. Frank DN. BARCRAWL and BARTAB: Software tools for the design and implementation of barcoded primers for highly multiplexed DNA sequencing. *BMC Bioinformatics* **10**:362 (2009).
177. Chain PS, Grafham DV, Fulton RS, Fitzgerald MG, Hostetler J, Muzny D, Ali J, Birren B, Bruce DC, Buhay C, et al. Genomic Standards Consortium Human Microbiome Project Jumpstart Consortium, Detter JC. Genomics. Genome project standards in a new era of sequencing. *Science* **326**(5950):236–237 (2009).
178. Bennett WE Jr, González-Rivera R, Shaikh N, Magrini V, Boykin M, Warner BB, Hamvas A, Tarr PI. A method for isolating and analyzing human mRNA from newborn stool. *J Immunol Meth* **349**(1–2):56–60 (2009).

TABLE 1.8. (*Continued*)

179. Loman NJ, Snyder LA, Linton JD, Langdon R, Lawson AJ, Weinstock GM, Wren BW, Pallen MJ. Genome sequence of the emerging pathogen Helicobacter canadensis. *J Bacteriol* **191**(17):5566–5567 (2009).
180. Brady A, Salzberg SL. Phymm and PhymmBL: Metagenomic phylogenetic classification with interpolated Markov models. *Nat Methods* **6**(9):673–676 (2009).
181. Pel J, Broemeling D, Mai L, Poon HL, Tropini G, Warren RL, Holt RA, Marziali A. Nonlinear electrophoretic response yields a unique parameter for separation of biomolecules. *Proc Natl Acad Sci USA* **106**(35):14796–14801 (2009).
182. Yang L, Lu X, Nossa CW, Francois F, Peek RM, Pei Z. Inflammation and intestinal metaplasia of the distal esophagus are associated with alterations in the microbiome. *Gastroenterology* **137**(2):588–597 (2009).
183. Wang Y, Hoenig JD, Malin KJ, Qamar S, Petrof EO, Sun J, Antonopoulos DA, Chang EB, Claud EC. 16S rRNA gene-based analysis of fecal microbiota from preterm infants with and without necrotizing enterocolitis. *ISME J* **3**(8):944–954 (2009).
184. Ye Y, Doak TG. A parsimony approach to biological pathway reconstruction/inference for genomes and metagenomes. *PLoS Comput Biol* **5**(8):e1000465 (2009).
185. Pop M. Genome assembly reborn: recent computational challenges. *Brief Bioinform* **10**(4):354–366 (2009).
186. Ye Y, Tang H. An ORFome assembly approach to metagenomics sequences analysis. *J Bioinform Comput Biol* **7**(3):455–471 (2009).
187. Petrosino JF, Highlander S, Luna RA, Gibbs RA, Versalovic J. Metagenomic pyrosequencing and microbial identification. *Clin Chem* **55**(5):856–866 (2009) (review).
188. Mahowald MA, Rey FE, Seedorf H, Turnbaugh PJ, Fulton RS, Wollam A, Shah N, Wang C, Magrini V, Wilson RK, et al. Characterizing a model human gut microbiota composed of members of its two dominant bacterial phyla. *Proc Natl Acad Sci USA* **106**(14):5859–5864 (2009).
189. Sharp RR, Achkar JP, Brinich MA, Farrell RM. Helping patients make informed choices about probiotics: A need for research. *Am J Gastroenterol* **104**(4):809–813 (2009) (review).
190. White JR, Nagarajan N, Pop M. Statistical methods for detecting differentially abundant features in clinical metagenomic samples. *PLoS Comput Biol* **5**(4):e1000352 (2009).
191. Cho SH, Chen CH, Tsai FS, Lo YH. Micro-fabricated fluorescence-activated cell sorter. *Proc Conf IEEE Eng Med Biol Soc IEEE*, 2009, pp. 1075–1078.
192. Langmead B, Schatz MC, Lin J, Pop M, Salzberg SL. Searching for SNPs with cloud computing. *Genome Biol* **10**(11):R134 (2009).
193. Langmead B, Trapnell C, Pop M, Salzberg SL. Ultrafast and memory-efficient alignment of short DNA sequences to the human genome. *Genome Biol* **10**(3):R25 (2009).
194. Wooley JC, Ye Y. Metagenomics: Facts and artifacts, and computational challenges. *J Comput Sci Technol* **25**(1):71–81 (2009).
195. Pei A, Nossa CW, Chokshi P, Blaser MJ, Yang L, Rosmarin DM, Pei Z. Diversity of 23S rRNA genes within individual prokaryotic genomes. *PLoS ONE* **4**(5):e5437 (2009).

1.5. OTHER NIH-SUPPORTED HUMAN MICROBIOME RESEARCH ACTIVITIES

One of the long-term goals of the Human Microbiome Project was to provide data for the biomedical research community regarding the role of the microbiome in human health and in disease. It was hoped that the results from the HMP would encourage additional investments in the field.

Support for human microbiome research has clearly grown at the NIH. Support from 9 microbiome-specific RFAs/PAs (Table 1.9) included 34 projects funded for a total of $36M over 5 years. This analysis does not include 8 microbiome-specific RFAs/PAs that are active as of this writing or three other active RFAs/PAs in which the awards are yet to be made, so this is a conservative estimate of support for microbiome research across the NIH.

Further, growth of research support across the NIH institutes and centers (ICs) has been somewhat organic as most of the microbiome grants have been individual investigator-initiated projects and not written in response to specific RFAs or PAs. Starting from basal levels and only one or two ICs over 2005–2006, the levels of support for human microbiome research and the number of ICs engaged in support of this research notably increased over the next 5-year period to 2009–2010.

In order to classify the diseases under study in those projects investigating microbiome and disease associations, the WHO International Classification of Diseases 2010 (ICD10) system (`http://www.who.int/classifications/icd/en`) was used to categorize the diseases. Ten NIH ICs focused their microbiome association studies on 9 of the 22 ICD major categories of disease and related health problems (Table 1.10) and included 38 different kinds of specific diseases (Table 1.11). Together, NIDCR, NHLBI, and NIDDK supported microbiome association studies of >60% of these 38 specific diseases, which fell into 6 of the 9 ICD major disease categories, including diseases of the digestive system; infectious and parasitic diseases; neoplasms; diseases of the respiratory system; endocrine, nutritional and

TABLE 1.9. Previously Funded NIH Microbiome-Related RFAs and PAs Not Part of the HMP[a]

RFA/PA Title	RFA/PA Number	Total Awarded ($ M)	Number of Projects
Metagenomic Analyses of the Oral Microbiome (R01)	PA04-131	3.6	1
Partnerships to Develop Tools to Evaluate Women's Health	AI05-029	4	1
New Approaches for the Prevention and Treatment of Necrotizing Enterocolitis (R01)	HD07-018	5.6	8
Microbicide Innovation Program (MIP III) (R21/R23)	AI07-034	0.96	1
Metagenomic Analyses of the Oral Microbiome (R01)	PA08-090	0.75	1
Microbiome of the Lung and Respiratory Tract in HIV-Infected Individuals and HIV-Uninfected Controls (U01)	HL09-006	10.7	7
Enterics Research Investigational Network Cooperative Research Centers (U19)	AI09-023	4.3	4
Metagenomic Evaluation of Oral Polymicrobial Disease (R01)	DE10-003	4.6	8
Gut–Liver–Brain Interactions in Alcohol-Induced Pathogenesis (R01)	AA10-007	1.3	3
Total		35.81	34

[a]As of fiscal year 2010.

TABLE 1.10. General Disease Categories for Non-HMP Microbiome and Disease Association Projects[a]

Category	NIAAA	NIAID	NIAMS	NCCAM	NCI	NIDCR	NIDDK	NICHD	NHLBI	NINR	Total
Diseases of digestive system	2	—	—	1	—	4	2	—	—	—	9
Diseases of genitourinary system	—	2	—	—	—	—	—	—	—	—	2
Certain infectious and parasitic diseases	—	2	—	—	—	2	—	—	5	—	9
Injury, poisoning, and certain other consequences of external causes	—	1	—	—	—	—	—	—	—	—	1
Diseases of musculoskeletal system and connective tissue	—	—	1	—	—	—	—	—	—	—	1
Neoplasms	—	—	—	—	1	2	—	—	—	—	3
Diseases of respiratory system	—	—	—	—	—	1	—	—	1	—	2
Endocrine, nutritional and metabolic diseases	—	—	—	—	—	—	5	—	1	—	6
Certain conditions originating in the perinatal period	—	—	—	—	—	—	—	3	1	1	5
Total	2	5	1	1	1	9	7	3	8	1	38

[a]Indicated in figure 5 of the WHO International Disease Classification (ICD) system.

TABLE 1.11. Specific Disease Categories for Non-HMP Microbiome and Disease Association Projects[a]

Disease Category	NIAAA	NIAID	NIAMS	NCCAM	NCI	NIDCR	NIDDK	NICHD	NHLBI	NINR
Diseases of digestive system	Alcoholic liver disease	—	—	Inflammatory bowel disease	—	Periodontitis; dental caries; oral mucositis	Inflammatory bowel disease; nonalcholic fatty liver disease	—	—	—
Diseases of the genitourinary system	—	Bacterial vaginosis; pelvic inflammatory disease	—	—	—	—	—	—	—	—
Certain infectious and parasitic diseases	—	Enteric disease	—	—	—	HIV, SIV	—	—	HIV	—
Injury, poisoning, and certain other consequences of external causes	—	Food allergies	—	—	—		—	—	—	—
Diseases of musculoskeletal system and connective tissue	—	—	Rheumatoid arthritis	—	—		—	—	—	—
Neoplasms	—	—	—	—	Colon cancer	Head and neck cancer	—	—	—	—
Diseases of respiratory system	—	—	—	—	—	Chronic rhinitis	—	—	Chronic obstructive pulmonary disease	—
Endocrine, nutritional and metabolic diseases	—	—	—	—	—		Obesity	—	Cystic fibrosis	—
Certain conditions originating in the perinatal period	—	—	—	—	—		—	Necrotizing enterocolitis	Bronchopulmonary dysplasia	Necrotizing enterocolitis

[a]Indicated in figure 5 of the WHO International Disease Classification (ICD) system for fiscal year 2010.

metabolic diseases; and conditions originating in the perinatal period (Table 1.10). As a point of comparison, in the HMP, necrotizing enterocolitits, inflammatory bowel disease, and bacterial vaginosis are being studied in the HMP demonstration projects (Table 1.3).

1.6. FUTURE DIRECTIONS FOR HUMAN MICROBIOME RESEARCH

The field of human microbiome research is at a pivotal point. The NIH Human Microbiome Project has provided an extensive resource of datasets, tools, and protocols for the study of both the healthy microbiome and the microbiomes of a diversity of diseases. It has considered some of the ethical issues that microbiome research may raise and in some cases has provided guidance for handling those issues. Here, we mention some key foundational studies and resources needed to move the field forward. Although institutes and agencies with disease-specific interests have already initiated mission-focused research programs, some suggestions are made for well-placed investments to accelerate progress in the field. Some of these ideas have also been outlined in a commentary [22].

We now know that the microbiome is acquired anew from the environment at birth and that the maturing immune system [23] and successional stages in the assembling microbial community interact to establish the microbiome in the first 2–3 years of life [24–26]. But why and how does the microbiome mature over these 2–3 years? And are there other fundamental changes that continue into adulthood? For example, what is the effect of hormonal changes, if any, at puberty or at menopause on the microbiome? We do not yet have a mechanistic understanding of the roles of the source inoculum in the maturing microbiome, the host immune system in regulating colonization by specific members of the microbiome, the successional events that result in a mature microbiome, the roles of the microbiota in resisting colonization by new microbes, or other host genetic factors in the selection of microbial composition of the microbiome.

Further, studies are suggesting that contemporary practices such as delivery by Caesarean section versus vaginal birth [27] and formula use versus breastfeeding [28,29] may affect the communities that assemble in the infant and therefore have an impact on the growth of beneficial microbes, particularly in the gut microbiome. In addition, the current practice of antibiotic use in mothers giving Caesarean birth and in infants and children appears to impact the microbiota, and this impact seems to last for months to years after antibiotic use [30]. Also, we do not yet understand the role of the early microbiome in the composition and function of the microbiome throughout life and in the development of later disorders or disease. For example, some studies suggest that a disturbed microbiome at infancy, through antibiotic use, may predispose one to allergies later in childhood [31]. In fact, a number of disorders (e.g. Crohn's disease, asthma, hay fever, type 1 diabetes, inflammatory bowel disease, multiple sclerosis, autism, celiac disease) may be associated with a disturbed, altered, or impoverished microbiome at infancy [32–36]. A working hypothesis for these observations of an association between a disturbed microbiome and subsequent disease is that induction of immune system maturation in these infants may be delayed, thereby rendering them more susceptible to diseases later in life.

A foundational study of microbiome development from birth through early childhood is needed. These studies should include the mother's microbiome and those of the child's immediate family. This study will need to include parameters beyond microbial community composition. In fact, what seems to be emerging from early human microbiome studies is that there is less diversity in the major metabolic pathways of the microbiome than in the diversity of the microbiome community that carry out these activities [37]. Just as advances in sequencing technologies paved the way for the characterization of microbiome composition, new technologies are now needed to study microbiome function and its interactions with the host. Technologies are now becoming available for the study of microbiome function, such as metabolomics, metatranscriptomics, and metaproteomics, all of which capture different aspects of the activities of the microbiota. Development of these large-scale methodologies to become high-throughput technologies will be an important resource for the field. Along with these technologies, additional approaches are needed to measure strain-level functional properties as well as the host immune system responses to microbial signals.

It appears that the microbiome retains much of its dynamic quality throughout life [38] and suggests that we may not yet understand what constitutes a healthy microbiome, particularly over the lifetime of an individual [39], even in Western populations. Some of the dynamic quality of the microbiome may be due to the factors in play during establishment of the early microbiome. Other studies are showing that diet is key to microbiome dynamics throughout life. In fact, few studies have broadly sampled the human population, particularly populations of non-European ancestry, to capture the breadth of these factors, and more work is needed to define the factors that regulate microbiome stability in life.

Further, no major microbiome study has yet included genetic analysis of the host. It is imperative that we begin to include host genomics in our efforts to understand what factors control and affect the microbiome. Human microbiome research raises its own unique questions about ethical issues such as return of research results, intellectual property, and ownership of the microbiome materials and confidentiality [40–42]. Efforts should be directed toward educating the public about the role of the microbiome in health and the need for volunteers in clinical studies of the microbiome to be broadly accepted so that both the host factors and microbiome factors can be included in these studies.

Although the microbiome of each body region is important to the health of that region, the gut microbiome could arguably be considered the "cardinal microbiome" as this is the microbial community that contributes to food digestion, directly supplies energy for host cell metabolism, and directly interacts with the host immune system [43]. Studies of diet and microbiome composition verify that microbiome composition appears closely associated with long term dietary practices [44,45] but not short-term diet changes [46]. The gut microbiome also directly and indirectly communicates with the microbiomes of other body regions through signaling molecules of microbial origin that circulate throughout the body. A concerted effort to study, the relationship between diet and the gut microbiome would be an important foundational study as would an effort to understand the systemic role of the gut microbiome and how it interacts with the organ systems and with the microbiomes across the human body.

Perhaps one of the most effective means of addressing these key areas would be through large cohort studies that include racially and ethnically diverse populations. These studies would serve as the foundation from which numerous studies could address the properties during microbiome assembly, variability of the microbiome across populations, and through time. Studies would follow the successional development of microbiome composition and also the changing functional properties of the microbiome as it matures. Opportunities to integrate microbiome studies with large cohort studies may now become available. There are a number of longitudinal birth cohort studies in place or planned in the near future that may serve as models in this endeavor. For example, the National Children's Study (NCS) (http://www.nationalchildrensstudy.gov/Pages/default.aspx) is a multidisciplinary US agency study designed to examine the environmental factors (broadly defined as diet, genetics, and other factors) that affect growth, development, and health of children in the United States in a prospective study from birth to adulthood. The NCS is designed as a platform to enable research and birth cohorts are being recruited across the country as the basis of the study. There are plans to collect samples and data from pregnant mothers, their subsequent newborns, and immediate family members, including siblings, family pets, and the immediate environment of the child. The goal of the program is to recruit up to 100,000 children over the course of the full study and follow them to 21 years of age. The study is sampling a representative distribution of the US population, so it can expect to include many racial and ethnic groups in the full study. The NCS full study is scheduled to begin in 2014. Large international birth cohort studies are also underway, such as the French Longitudinal Study of Children, is a cohort study of 20,000 children followed from birth to adulthood (ELFE; http://www.cls.ioe.ac.uk/text.asp?section=00010001000500090016; http://www.biomedcentral.com/1471-2431/9/58). An international consortium of scientists is proposing to collect stool samples for microbiome analysis of these children. Young Lives, a British-led international study of childhood poverty (http://www.younglives.org.uk/), is following 12,000 children in four developing countries, Ethiopia, India, Peru, and Vietnam. Proposals have been made to include microbiome sampling and analysis of the children in this study. Partnerships with other large birth cohort studies would be invaluable, as would partnerships with other large cohort studies where the subjects have consented broadly and where genomic and phenotype data are available. With appropriate coordination and consents from the study participants, such studies could provide the ideal framework from which to analyze the microbiome and its functional properties from time of birth across diverse populations.

In order for the results from these studies to support the research interests of the broadest community, these activities will require a flexible and user-friendly infrastructure that links all of the different microbiome datasets, which include microbial composition and microbial function and the host phenotype and genotype data with appropriate, ready-to use computational tools for analysis that are accessible to all regardless of the bioinformatic resources or computational expertise at one's home institution. Massive microbiome datasets of terabase and petabase sizes will be the order of the day in the very near future. New approaches and tools are needed that can accommodate these large datasets and support data transfer, analysis, and interpretation. In fact, it will be the routine access and use of this network of data and tools by a broader community that will move this field into the clinical realm as microbiome and related data are applied to questions in the treatment of disease and in the support of health. A broadly available resource that will support the needs of both the research community and the clinical community will be crucial to the full integration of the microbiome into scientific and clinical studies and is a fundamental community resource.

The microbiome coevolved with its human host over millennia. Studies from the field of human anthropology are suggesting that, over the last 2 million years, there were several waves of early hominid migrations from the African continent across the globe. More recent comparative genomics studies of the Denisovans [47], who started migrating ~1,000,000 years ago, the Neanderthals [48], who started migrating ~600,000 years ago, and, *Homo sapiens*, who started migrating ~100,000–200,000 years ago, suggest there was interbreeding between these early hominid species and *H. sapiens*. Perhaps more importantly, a more recent study provided evidence that this interbreeding may have conferred increased fitness to our species, particularly in the *H. sapiens* immune system. Analysis of a key group of immune system genes, the human leukocyte antigen (HLA) class I genes, shows these genes are particularly variable with thousands of alleles across different populations. Analyses suggest that the HLA gene alleles spread rapidly from early hominids to *H. sapiens* as a large fraction (~50–95%) of these alleles appear to be derived from early hominids [49]. But it may not only be heritable traits that were acquired during this interbreeding. The earlier hominid microbiomes may have been very different from the microbiomes of migrating *H. sapiens* as diets, time elapsed since migration across many biomes, and environmental exposures would have likely been quite different between the early hominid groups and *H. sapiens*. It is quite conceivable that interbreeding also led to the acquisition of new beneficial microbes, leading to a more robust microbiome for *H. sapiens*, conferring the ability to defend against new opportunistic pathogens and to diversify the diet. These studies suggest that human microbiome studies should be conducted with this evolutionary context in mind.

More recent reviews of human microbiome studies argue that the field will need to move beyond an understanding of the fundamental properties of microbiome composition to an understanding of the fundamental properties of microbiome function if the microbiome is to be integrated into the study of human health and disease. Future microbiome function studies should include diverse populations in order to circumscribe and associate the functional properties of the microbiome with other features of these populations. Collaboration with large cohort studies, particularly birth cohorts of diverse populations, may be one means of focusing such an effort in order to develop the resources needed to study the role of the microbiome in health and in disease. Development of high-through put methodologies to measure microbiome function in conjunction with large cohort studies, all of which are supported by a well-designed, user-friendly infrastructure, will establish the needed resources and data for future research of the microbiome in health and in disease. Finally, it is important to ground human microbiome studies in the appropriate evolutionary and ecological context if we are to understand the drivers behind microbiome assembly, homeostasis, and its role in human health maintenance.

ACKNOWLEDGMENTS

We would like to acknowledge numerous individuals outside the NIH who provided data, figures, suggestions, and edits on the chapter. The HMP research network consortium members who contributed willingly to data and figures include Drs. Ashlee Earl (Broad Institute) and Heather Huot-Creasy (DACC/Univ MD School of Medicine), who provided the data for Figure 1.2. Dr. Earl also provided extensive comments on an earlier version of the chapter. Drs. Dirk Gevers and Katherine Huang of the Broad Institute provided the data and figures for Figure 1.4. Dr. Gevers and also Drs. Rob Knight of the University of Colorado, Curtis Huttenhower of the Harvard School of Public Health, and Owen White of the University of Maryland School of Medicine also provided valuable comments on the chapter draft. Any errors are solely ours to claim.

REFERENCES

1. Woese CR, Fox GE. *Proc Natl Acad Sci USA* **74**:5088–5090 (1977).
2. Lane DJ, Pace B, Olsen GJ, Stahl DA, Sogin ML, Pace NR. *Proc Natl Acad Sci USA* **82**(20):6955–6959 (1985).
3. Choi BK, Paster BJ, Dewhirst FE, Gōbel UB. *Infect Immun* **62**:1889–1895 (1994).
4. Ashimoto A, Chen C, Bakker I, Slots J. *Oral Microbiol. Immunol* **11**:266–273 (1996).
5. Kroes I, Lepp PW, Relman DA. *Proc Natl Acad Sci USA* **96**:14547–14552 (1999).
6. Brogden KA, Guthmiller JM. *Polymicrobial Diseases*, ASM Press, 2002.
7. Lederberg J. *Science* **288**:287–293 (2000).
8. Falk PG, Hooper LV, Midtvedt T, Gordon JI. *Microbiol Molec Biol Rev* **62**:1157–1170 (1998).
9. Hooper LV, Gordon JI. *Science* **292**:1115–1118 (2001).
10. Macpherson AJ, Harris NL. *Nat Rev Immunol* **4**:478–485 (2004).
11. Macpherson AJ, Gatto D, Sainsbury E, Harriman GR, Hengartner H, Zinkemagel RM. *Science* **288**:2222–2226 (2000).
12. Hooper LV. *Trends Microbiol* **12**:129–134 (2004).
13. Relman A, Falkow S. *Trends Microbiol* **9**:206–208 (2001).
14. Davies J. *Science* **291**:2316 (2001).
15. Eckburg PB, Bik EM, Bernstein CN, Purdom E, Dethlefsen L, Sargent M, Gill SR, Nelson KE, Relman DA. *Science* **308**:1635–1638 (2005).
16. Gill SR, Pop M, Deboy RT, Eckburg PB, Turnbaugh PJ, Samuel BS, Gordon JI, Relman DA, Fraser-Liggett CM, Nelson KE. *Science* **312**:1355–1359 (2006).
17. National Research Council. *The New Science of Metagenomics*, (2007). ISBN-10 0-309-10676-1.
18. Toronto International Data Release Workshop Authors. *Nature* **461**:168–170 (2009).
19. NIH HMP Working Group. *Genome Res* **19**(12):2317–2323 (2009).
20. Human Microbiome Jumpstart Reference Strains Consortium. *Science* **328**(5981):994–999 (2010).
21. Chain PS, Grafham DV, Fulton RS, Fitzgerald MG, Hostetler J, Muzny D, Ali J, Birren B, Bruce DC, Buhay C et al. *Science* **326**:236–237 (2009).
22. Proctor LM. *Cell Host Microbe* **10**:287–291 (2011).
23. Mazmanian SK, Liu CH, Tzianabos AO, Kasper DL. *Cell* **122**:107–118 (2005).
24. Dethlefsen L. *Trends Ecol Evol* **21**:517–523 (2006).
25. Palmer C, Bik EM, DiGiulio DB, Relman DA, Brown PO. *PLoS Biol* **5**(7):e177 (2007).
26. Koenig GM Jr, Lin IH, Abbott NL. *Proc Natl Acad Sci USA*. **107**:3998–4003 (2010).
27. Dominguez-Bello MG, Costello EK, Contreras M, Magris M, Hidalgo G, Fierer N, Knight R. *Proc Natl Acad Sci USA*. **107**:11971–11975 (2010).
28. Poroyko V, White JR, Wang M, Donovan S, Alverdy J, Liu DC, Morowitz MJ. *PLoS One*. **5**:e12459 (2010).

29. Sánchez E, De Palma G, Capilla A, Nova E, Pozo T, Castillejo G, Varea V, Marcos A, Garrote JA, Polanco I, López A, Ribes-Koninckx C, García-Novo MD, Calvo C, Ortigosa L, Palau F, Sanz Y. *Appl Environ Microbiol* **77**(15):5316–5323 (2011).
30. Jernberg C, Löfmark S, Edlund C, Jansson JK. *Microbiology* **156**(Pt 11):3216–3223 (2010).
31. Bisgaard H, Li N, Bonnelykke K, Chawes BL, Skov T, Paludan-Müller G, Stokholm J, Smith B, Krogfelt KA. *J Allergy Clin Immunol.* **128**:646–652 (2011).
32. Finegold SM, Molitoris D, Song Y, Liu C, Vaisanen ML, Bolte E, McTeague M, Sandler R, Wexler H, Marlowe EM, Collins MD, Lawson PA, Summanen P, Baysallar M, Tomzynski TJ, Read E, Johnson E, Rolfe R, Nasir P, Shah H, Haake DA, Manning P, Kaul A. *Clin Infect Dis* **35**(Suppl 1):S6–S16 (2002).
33. Kummeling I, Stelma FF, Dagnelie PC, Snijders BE, Penders J, Huber M, van Ree R, van den Brandt PA, Thijs C. *Pediatrics* **119**(1):e225–e231 (2007).
34. Ponsonby AL, Catto-Smith AG, Pezic A, Dupuis S, Halliday J, Cameron D, Morley R, Carlin J, Dwyer T. *Inflamm Bowel Dis* **15**(6):858–866 (2009).
35. Ochoa-Reparaz J, Mielcarz DW, Ditrio LE, Burroughs AR, Foureau DM, Haque-Begum S, Kasper LH. *J Immunol* **183**(10):6041–6050 (2009).
36. Decker E, Engelmann G, Findeisen A, Gerner P, Laass M, Ney D, Posovszky C, Hoy L, Hornef MW. *Pediatrics* **125**(6):e1433–e1440 (2010).
37. Qin J, Li R, Raes J, Arumugam M, Burgdorf KS, Manichanh C, Nielsen T, Pons N, Levenez F, Yamada T et al. *Nature* **464**:59–65 (2010).
38. Costello EK, Lauber CL, Hamady M, Fierer N, Gordon JI, Knight R. *Science* **326**:1694–1697 (2009).
39. Claesson MJ, Cusack S, O'Sullivan O, Greene-Diniz R, de Weerd H, Flannery E, Marchesi JR, Falush D, Dinan T, Fitzgerald G et al. *Proc Natl Acad Sci USA*. **108**:4586–4591 (2011).
40. Hawkins AK, O'Doherty KC. *BMC Medical Genomics* **4**:72.
41. Blaser MJ. *Proc Natl Acad Sci USA*. **107**:6125–6126 (2010).
42. McGuire AL, Colgrove J, Whitney SN, Diaz CM, Bustillos D, Versalovic J. *Genome Res* **18**(12):1861–1864 (2008).
43. Mazmanian SK, Round JL, Kasper DL. *Nature*. **453**:620–625 (2008).
44. Turnbaugh PJ, Ridaura VK, Faith JJ, Rey FE, Knight R, Gordon JI. *Sci Transl Med* **1**:6ra14 (2009).
45. De Filippo C, Cavalieri D, Di Paola M, Ramazzotti M, Poullet JB, Massart S, Collini S, Pieraccini G, Lionetti P. *Proc Natl Acad Sci USA*. **107**:14691–14695 (2010).
46. Wu GD, Chen J, Hoffmann C, Bittinger K, Chen YY, Keilbaugh SA, Bewtra M, Knights D, Walters WA, Knight R et al. *Science.* **334**:105–108 (2011).
47. Reich D, Green RE, Kircher M, Krause J, Patterson N, Durand EY, Viola B, Briggs AW, Stenzel U, Johnson PL, Maricic T, Good JM, Marques-Bonet T, Alkan C, Fu Q, Mallick S, Li H, Meyer M, Eichler EE, Stoneking M, Richards M, Talamo S, Shunkov MV, Derevianko AP, Hublin JJ, Kelso J, Slatkin M, Pääbo S. *Nature* **468**(7327):1053–1060 (2010).
48. Green RE, Krause J, Briggs AW, Maricic T, Stenzel U, Kircher M, Patterson N, Li H, Zhai W, Fritz MH, Hansen NF, Durand EY, Malaspinas AS, Jensen JD, Marques-Bonet T, Alkan C, Prüfer K, Meyer M, Burbano HA, Good JM, Schultz R, Aximu-Petri A, Butthof A, Höber B, Höffner B, Siegemund M, Weihmann A, Nusbaum C, Lander ES, Russ C, Novod N, Affourtit J, Egholm M, Verna C, Rudan P, Brajkovic D, Kucan Z, Gusic I, Doronichev VB, Golovanova LV, Lalueza-Fox C, de la Rasilla M, Fortea J, Rosas A, Schmitz RW, Johnson PL, Eichler EE, Falush D, Birney E, Mullikin JC, Slatkin M, Nielsen R, Kelso J, Lachmann M, Reich D, Pääbo S. *Science* **328**(5979):710–722 (2010).
49. Abi-Rached L, Jobin MJ, Kulkarni S, McWhinnie A, Dalva K, Gragert L, Babrzadeh F, Gharizadeh B, Luo M, Plummer FA et al. *Science.* **334**:89–94 (2011).

2

METHODS FOR CHARACTERIZING MICROBIAL COMMUNITIES ASSOCIATED WITH THE HUMAN BODY

CHRISTINE BASSIS and VINCENT YOUNG

Department of Internal Medicine, University of Michigan, Ann Arbor, Michigan

THOMAS SCHMIDT

Department of Microbiology and Molecular Genetics, Michigan State University, East Lansing, Michigan

2.1. INTRODUCTION

In the healthy human body, microbial cells famously outnumber human cells 10 to 1 and the bacterial communities, which make up the vast majority of the human-associated microbial cells, vary greatly between individuals [28,115]. It is widely recognized that the human microbiota play important roles in health and disease, but much about these complex communities, including interactions with their hosts, remains unknown. In this chapter we will introduce techniques used to study the structure, function and dynamics of the human microbiome (Figure 2.1).

Learning what roles the human microbiota play in health and disease is a key goal driving the study of human microbiota and is the focus of efforts such as the NIH Human Microbiome Project (`http://nihroadmap.nih.gov/hmp/`) [51,138], the international MetaHIT project (`www.metahit.eu`) [106], and the International Human Microbiome Consortium (`www.human-microbiome.org`). Our ability to develop predictive models of the microbiome and manipulate its activity will be enhanced by understanding the relationship between the structure of the microbial community and its functions. Addressing this complex area of research requires

The Human Microbiota: How Microbial Communities Affect Health and Disease, First Edition. Edited by David N. Fredricks.

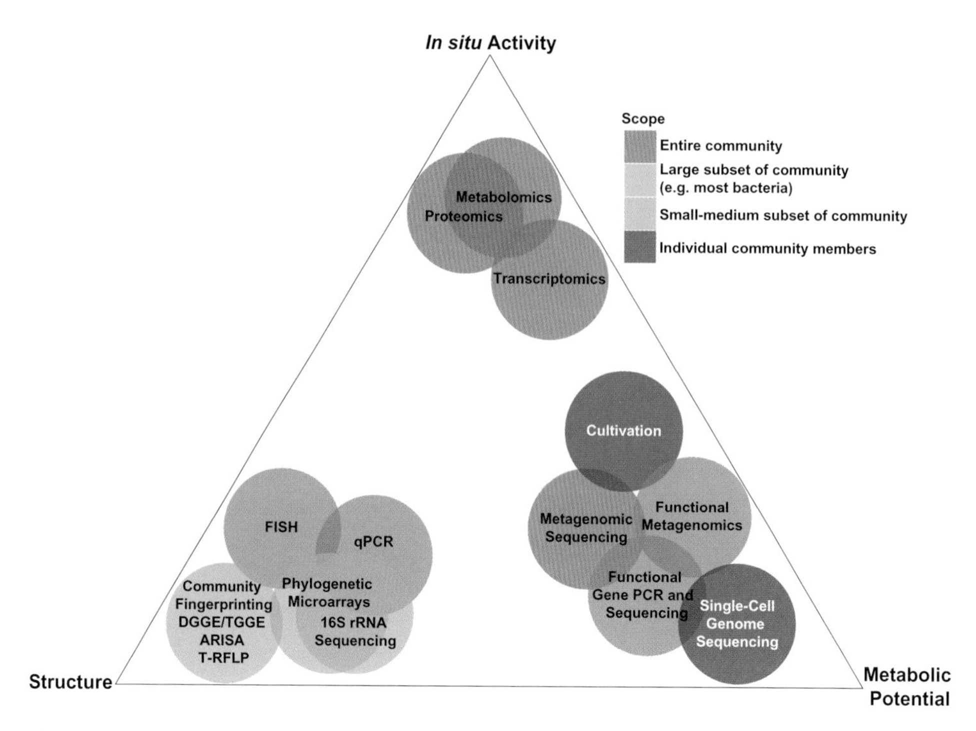

Figure 2.1. Methods for Characterizing Human-Associated Microbial Communities. The type of information provided by each method is indicated by its position within the triangle and its color. Each vertex represents a broad type of information: structure, metabolic potential, and *in situ* activity. Methods are positioned closest to the vertex that best describes the type of information provided by the method. If a method provides more than one type of information, it is positioned between the appropriate vertices. The method color indicates the scope encompassed by the method from the entire community (dark teal) to individual community members (dark purple).

studies of community structure, dynamics, and function that lay the groundwork for answering such basic questions as

- Which microbes are present at specific sites on the human body?
- Do the microbial communities change (with time, diet, disease, etc.)?
- What are the human-associated microbes doing?

Significant progress has been made, particularly in cataloging the microbial communities at various sites on the human body [28,106,136], but much remains to be learned.

2.2. CULTIVATION-INDEPENDENT TECHNIQUES

Relating microbial community structure, dynamics, and function to human health requires the ability to compare these aspects of microbial communities between different individuals and over time. The methods for characterizing and comparing human-associated microbial communities are the focus of this chapter. Huge advances in the ability to compare complex microbial communities, such as those associated with humans, have accompanied the development of culture-independent techniques. Until the 1980s, microbial identification depended on isolating microbes in culture and classifying according to various phenotypic characteristics, including morphology, motility, metabolism, and biochemistry. Distinguishing different types of bacteria or determining that two isolates were, indeed, the same strain was not a trivial task. The ability to compare bacterial cultivars reliably improved with the development of DNA-DNA hybridization assays [50]. However, determining evolutionary relationships among isolates remained difficult [150]. Furthermore, the vast majority of microbes don't grow under standard cultivation conditions. Therefore, most uncultivated microbes were overlooked or observed only microscopically. This began to change with the development of nucleic acid sequencing technology and the recognition that sequences, particularly ribosomal RNA sequences, were useful phylogenetic markers and microbial identifiers [98,99,150]. These developments, along with those of PCR and *in situ* hybridization, threw open the door for identifying and comparing microbes, both cultivated and uncultivated, from multiple environments and have led to a shift toward cultivation-independent studies of microbial communities. Current studies of the human microbiota rely largely on cultivation-independent molecular techniques that have been adapted from successful studies of microbial communities inhabiting other environments. Although these techniques do not require cultivation, interpretation of the results from molecular surveys is typically enhanced by making comparisons to cultivars. Thus, despite the ease with which massive quantities of culture-independent data can be gathered, cultivation remains a valuable complementary approach [48], and will be discussed later.

2.2.1. Community Structure and Composition Using Phylogenetic Markers

In cultivation-independent studies of microbial community composition, the microbes are characterized by specific phylogenetic marker genes. The techniques presented in this section use these genes to characterize different aspects of microbial communities. The genes encoding for the RNA component of the ribosome's small subunit (SSU rRNA), which has a sedimentation coefficient of 16*S* in bacteria and archaea, and 18*S* in eukaryotes, are by far the most widely used phylogenetic markers for microbes. The SSU rRNA genes contain both highly conserved regions, useful for targeting broad-range PCR primers, and variable regions, useful for microbial identification and establishing phylogenetic relationships [5]. The number of 16*S* rRNA sequences available for comparison in public databases is large and constantly growing [98]. As of February 2, 2011 there were 2,827,757 16*S* rRNA sequences in GenBank [8]. To make meaningful comparisons of these genes, they must first be aligned to accommodate internal variation in length of the stem and loop structures. There are several databases that provide extensive numbers of aligned SSU rRNA genes, including (as of February 2, 2011) 1,498,677 16*S* rRNA sequences in RDP [24], 1,471,257 SSU rRNA (16*S* and 18*S*) sequences in the SILVA web database [105], and 752,661 16*S* rRNA sequences (>1250 bases) in Greengenes [35]. Overall the SSU rRNA genes perform very well as microbial identifiers and phylogenetic markers. Furthermore, when microbes were chosen for genome sequencing based on SSU phylogeny, the discovery rate of new protein families was significantly higher than if microbes were chosen randomly, demonstrating the usefulness of SSU gene sequences as predictors of genomic novelty [153]. However, one complication of using SSU rRNA genes to study microbial communities is that the number of copies of the SSU gene varies from 1 to 15 in the genomes of bacteria and archaea [78]. Therefore community analyses based on SSU rRNA genes can be misleading because microbes with many SSU rRNA genes are likely to be overrepresented, which, for example, means that a microbe with many SSU rRNA gene copies may appear to be more abundant than an equally abundant microbe with few SSU rRNA gene copies [29,78]. The Ribosomal RNA Database (rrnDB; `http://ribosome.mmg.msu.edu/rrndb/index.php`) catalogs rRNA copy numbers and can be used to improve the interpretation of SSU rRNA-based microbial community studies [78]. Finally, there are cases where the SSU rRNA gene is not a useful marker such as in distinguishing between or determining the relationships of closely related strains [43]. In these cases, several protein coding genes, including *rpoB* [1], *gyrA* [22,89], and *recA* [70,101], have been used as phylogenetic markers.

Community Fingerprinting

Community fingerprinting techniques provide a relatively fast and inexpensive overview of microbial diversity and allow the structures of different communities to be compared. They can be used to determine whether different communities contain common members and, to some extent, compare the abundances of community members. The most common fingerprinting techniques start with PCR amplification to obtain a pool of amplicons that represents members of the microbial community in the sample. The community fingerprinting techniques target different genomic regions and use different approaches to assess the composition of the pool of PCR amplicons. However, specific community members are not identified by these techniques. Therefore, as DNA sequencing has become less expensive, these techniques are being used more frequently as a "first pass" to determine which samples to analyze further.

1. *DGGE/TGGE.* For denaturing gradient gel electrophoresis (DGGE), the gene of interest, typically the 16*S* rRNA gene for studies of overall bacterial diversity, is amplified by PCR using primers with a guanine cytosine (GC) clamp that holds both strands of an amplicon together during electrophoresis [90]. The diversity of the amplicon pool is assessed by running the PCR products on a denaturing gradient gel, which results in a series of bands on the gel. The approach is similar for temperature gradient gel electrophoresis (TGGE), except that there is a temperature gradient along the gel [91]. In both cases, the degree to which the double stranded PCR products denature depends on their sequence and GC DNA content and determines the mobility of the specific products in the pool on the gel. Ideally, each band on the gel represents a unique amplicon sequence and thus, a single type of microbe. By comparing the DGGE or TGGE patterns of different microbial communities, it is possible to observe common and distinct members. Bands of interest can be cut out of the gel and sequenced to determine specific identities. One drawback of this technique is that bands from different samples that have the

same mobility may not actually represent the same sequences. Additionally, a single band may consist of multiple amplicon sequences. DGGE has been used to provide an overview of community dynamics, such as in a study following the development of the microbial community in the guts of infants over the first year of life [40].

2. *T-RFLP.* For terminal restriction fragment length polymorphism (T-RFLP), one or both of the primers used to amplify the community DNA is fluorescently labeled [80]. This results in a pool of amplicons with one or both ends fluorescently labeled. The amplicons are then digested with at least one restriction enzyme. The sizes and abundances of the fragments containing the fluorescently labeled amplicon ends (terminal restriction fragments) are measured by gel electrophoresis or by capillary electrophoresis with a DNA sequencer. Since the amplicon sequence determines the size of the terminal restriction fragment, different-sized terminal restriction fragments represent different microbes in the community. So, different microbes that yield terminal restriction fragments of the same size cannot be distinguished by this method. Typically restriction enzymes that recognize 4 bp sequences are used for T-RFLP [123]. The specific restriction enzyme used greatly affects the ability to resolve different sequences [38]. In communities with more than 50 different types of bacteria, at best it is possible to resolve 70% of the bacteria present [38]. However, actual T-RFLP results agree very well with predicted T-RFLP results based on cloning and sequencing 16*S* rRNA-encoding genes from the same community [71,75]. Applications and advances in T-RFLP have been reviewed by Schutte et al. [123].

3. *ARISA.* In automated rRNA intergenic spacer analysis (ARISA) each type of microbe is represented by the highly variable length of the region between the 16*S* rRNA gene and the 23*S* rRNA gene (16*S*–23*S* ITS). The 16*S*–23*S* ITS is amplified by PCR with primers targeting the conserved regions at the ends of the 16*S* and 23*S* genes. One of the primers is fluorescently labeled, allowing for the automated determination of the 16*S*–23*S* ITS size using a capillary genetic analyzer/sequencing machine [41]. ARISA is better able to distinguish subspecies than 16*S* rRNA sequencing and is useful for monitoring communities in low-diversity systems [102].

DNA Sequencing of Phylogenetic Markers

A convenient and powerful way to determine the microbial composition of a sample is by directly sequencing a phylogenetic marker gene, such as the SSU rRNA gene. DNA sequence data allow the microbes in a sample to be identified and the microbial community structures of different samples to be compared. Similar to the community fingerprinting techniques, the first step in microbial community analysis by DNA sequencing of phylogenetic markers is PCR amplification of microbial community DNA, typically using primers that target a broad group of organisms such as primers targeting conserved regions of the bacterial 16S rRNA gene. This yields a pool of amplicons that represent the community. These amplicons are then sequenced by one of the methods described below. DNA sequencing costs have dramatically decreased since the technology was introduced, allowing this approach to become routine.

1. *Clone Libraries and Sanger Sequencing.* In a community made up of multiple types of organisms, PCR with broad-range primers results in a pool of amplicons with different sequences. In order to obtain the sequences that represent a community, PCR products can be cloned into standard sequencing vectors and transformed into *Escherichia coli* cells. The *E. coli* cells are plated on selective media, and each colony that grows contains many copies of one particular amplicon-containing vector. The vector is purified from transformants, and traditional dideoxy Sanger sequencing of the cloned amplicons from many transformants allows the community members to be identified and the community structure to be determined. Sanger sequencing of cloned amplicons typically results in relatively long read length sequences. Single sequencing reads are typically ~700 nucleotides; however, multiple reads can be obtained and assembled for a single clone, allowing the acquisition of nearly full-length 16S rRNA gene sequences (~1500 bp). Studies vary in how many clones are sequenced, but it is currently typical to sequence on the order of tens to hundreds of sequences per sample [49,58]. For example, in a study of the structure of bacterial communities on human skin at different body sites a total of 112,283 near-full-length 16*S* rRNA gene sequences were obtained from clones with a mean of 277 sequenced clones per sample [49]. Therefore, in communities consisting of 1000s of different types of organisms, the less abundant community members are likely to be missed by this method.

2. *Next-Generation Sequencing.* In order to get a more complete census of the microbes present in an environment, many researchers are shifting to next-generation sequencing technologies, including Roche 454, Illumina/Solexa, and ABI SOLiD. Next-generation sequencing yields a massive number of sequences per run, from hundreds of thousands to over 100 million depending on the method [83]. However, one disadvantage is the relatively short read lengths, with 35–500 bases, although this is increasing [4,83,125]. Of the next-generation sequencing methods, method 454 was the first to be exploited for microbial community structure analysis [129], and most studies target the 16S rRNA gene. Because the sequence read lengths are shorter than those obtained with Sanger sequencing, specific variable regions, rather than the full 16*S* rRNA gene, are typically amplified in the primary PCR for 454 [36,104,112,156]. The variable region chosen for sequencing can impact the overall results of the study. For example, the number of observed operational taxonomic units (OTUs), groups of organisms defined by sequence similarity, is likely to depend on the variable region sequenced [117,154]. In order to distinguish sequences from different samples within the same 454 run, each reverse primer includes a unique barcode. The Human Microbiome Project (HMP) has developed and vetted 96 of these barcoded primers for the V5–V3 and V3–V1 regions of the 16*S* rRNA gene (for current HMP SOPs, see `http://www.hmpdacc.org/tools_protocols/tools_protocols.php`). There is a serious concern about the abundance of sequencing errors generated in 454 and that these errors may be inflating estimates of diversity [64,76]. Detection of homopolymers (such as a run of guanines) is particularly problematic [76,108]. However, bioinformatic tools for quality control can greatly improve error rates by removing suspect sequences and clustering sequences that likely originate from identical templates [65,107]. Furthermore, continued development of software for the raw data analysis means that saving image files from 454 runs may allow additional data to be extracted in the future.

3. *Analytic Methods for 16S rRNA Community Sequence Data.* The analytic methods for using 16*S* rRNA gene sequences to compare microbial communities can be applied to sequences resulting from a variety of methods ranging from Sanger sequencing of clone libraries to next-generation sequencing. The main difference between the two is scale; the vast amount of data from next-generation sequencing underscores the need for efficient bioinformatic tools. In order to use 16*S* rRNA sequences to assess community structure and compare the structures of different communities accurately, it is necessary to trim the primer and low-quality bases from the ends of each sequence. For many analyses it is also critical to align each sequence with a reference alignment [118]. The removal of chimeric (artifactual hybrid) sequences is also critical for accurate analyses [52]. After trimming, alignment, and chimera removal, sequences can be grouped into OTUs on the basis of sequence similarity cutoffs [118,119], typically 90–97% similarity, or classified to a particular taxonomic level (phylum, class, order, family, genera) [149]. Once the sequences are grouped, estimates of diversity (richness and evenness) can be calculated and taxon-based community structures can be compared. Chao1 [20,21,100] is commonly used to estimate community richness. The Simpson index [127] and the Shannon index [124] incorporate richness and evenness to quantify the diversity of a community. There are several possibilities for comparing communities and determining how similar or different they are: Measurements that factor in presence or absence only include the classic Jaccard index [63,66] and the classic Sφrensen index [63,131], while other measurements, including Morisita–Horn [61], Bray–Curtis [13], and θ [155], factor in community structure (both membership and abundance). Accounting for relative abundance in community analyses can provide insights on the functional importance of different microbial groups. However, relative abundance is not always reliably measured using SSU rRNA-based methods and may not be indicative of functional importance, so in some cases it may be more appropriate to compare communities on the basis of their membership only. These taxon-based community diversity measures and comparisons of communities can be based on OTUs or sequence classifications. Since OTUs are defined by sequence similarity cutoffs, all sequences can be assigned to an OTU. In contrast, classification relies on similarities to previously classified sequences in the databases, so analyses that move away from higher taxonomic levels (phylum, class) towards lower levels (family, genus) are likely to have more unclassified sequences and, thus, missing information for analysis, and while the sequences in a particular group are all similar to a reference sequence, they may not be that similar to each other. However, classification is still a useful tool, in part because it can promote understanding and facilitate discussion (e.g., discussing differences in the relative abundance of proteobacteria vs. OTU 35). Alternatively, rather than comparing communities according to specific OTUs or taxonomic classifications, it is possible to compare the overall phylogenetic differences of different communities using Unifrac [53,82]. Two examples of comprehensive software packages where many tools for calculating richness and diversity and comparing communities can be implemented are mothur (`www.mothur.org`) [120] and QIIME (`http://qiime.sourceforge.net/`) [19]. These programs allow large numbers of sequences to be trimmed, aligned, checked for chimeras, grouped into OTUs or taxonomic groups, and analyzed efficiently [19,120].

Phylogenetic Microarrays

Phylogenetic microarrays are used to determine the structure of microbial communities by identifying the microbes in a sample without sequencing. Instead of sequencing, DNA or cDNA from a sample of interest is PCR amplified, labeled, and applied to a microarray, which typically contain 100s–1000s of different oligonucleotide probes in known positions. So far, most phylogenetic microarrays have been designed to identify the bacteria in a sample by targeting the 16*S* rRNA (or 16*S* rRNA gene). The versatile PhyloChip was designed to detect a broad range of bacteria (>30,000 bacterial 16*S* rRNA gene sequences were taken into account in designing the probes) and has been used in a variety of environments from uranium contaminated soil [14,111] to the lungs of intubated patients [42]. Other arrays, such as the human intestinal tract chip (HITChip) [109], target the bacteria in a specific environment while others, such as the *Burkholderia* PhyloChip [122], target smaller subsets of bacteria. Viral identification is another promising application of microarrays. Since there is no conserved genetic element shared by all viruses (i.e., no equivalent to the 16*S* rRNA gene for identification by PCR and sequencing), identification of viruses in complex environmental or clinical samples is extremely challenging. However, viruses have been identified from a variety of samples using carefully designed microarrays [73,87,147,148]. There are also a handful of studies using phylogenetic microarrays to identify fungi [60]. The phylogenetic microarray is well suited for identifying the less abundant members of a community. Comparisons between phylogenetic microarrays and clone libraries have shown that additional bacterial taxa were detected from the same sample using phylogenetic microarrays that were missed by analyzing 141–429 clone sequences per sample [14,34,42]. While the data from phylogenetic microarrays and from pyrosequencing were shown to be generally concordant in one study [23], exactly how phylogenetic microarray data compare with pyrosequencing data for detecting less abundant types of bacteria remains to be determined. Phylogenetic microarrays can be faster and less expensive than sequencing. However, only taxa represented on the microarray can be detected, so unexpected or novel taxa may be missed [34]. Changes in the abundances of particular bacterial strains can be determined, but the relative abundances of different bacteria can't be determined. Nonspecific hybridization is another concern, but mismatched control probes can be included on the array to demonstrate specificity [34].

qPCR

To compare the abundance of a particular microbe or group of microbes in different samples, the abundance of a proxy gene (such as the SSU rRNA-encoding gene) can be measured using quantitative PCR (qPCR), also called *real-time PCR* (reviewed by Smith and Osborn [128]). As a qPCR proceeds through its cycles and more amplicons are produced, more fluorescence is also produced and measured in real time [12]. The cycle at which fluorescence reaches a threshold level (C_T) can be used to calculate the relative abundance of template in the sample compared to a reference gene. Alternatively, the number of copies of the target in the sample can be calculated if a standard curve is generated.

There are two main ways of detecting amplicons in qPCR: using a dye that is fluorescent only when bound to dsDNA (e.g., SYBR Green) or using a specific probe with a fluorescent tag. There are a few variations on the qPCR assays that use specific probes, including (1) hydrolysis probe assays (TaqMan) in which the probe has a fluorescent tag at the 5′ end and a quencher at the 3′ end such that emission of fluorescence occurs only after the probe binds the target and is degraded by the nuclease activity of Taq as it proceeds along the template during polymerization, releasing the fluorescent tag [57], and (2) molecular beacons that form hairpins at low temperatures, bringing the fluorescent tag and quencher together until the denaturing step of PCR when the hairpin binding is released and the probe can bind its target as the temperature is lowered for the annealing step; only probes bound to their targets will emit fluorescence at this step—the rest will convert back to hairpins, and their fluorescence will be quenched [141]. Using a dsDNA-binding dye can be more convenient and economical because a different probe is not required for each different target. However, the dsDNA-binding dyes are not specific for a particular target, so primer dimers and other nonspecific products are also detected. Like standard endpoint PCR, the qPCR target can range from broad groups of bacteria to specific strains but the length of the amplicon is typically limited to 200–300 bp. The qPCR technique is also commonly used to measure gene expression levels by targeting cDNA (reviewed in Ref. 128).

FISH

Fluorescence *in situ* hybridization (FISH) uses fluorescently tagged oligonucleotide probes to hybridize with complementary RNA inside microbial cells (reviewed in 2008 by Amann and Fuchs [2]). When combined with fluorescence microscopy, the abundance, morphology and location of labeled cells can be determined. This can provide insights into interactions with host organisms, other microbes, and environmental surfaces. Another option for determining the abundance of labeled cells, rather than by microscopy, is to count the labeled cells with a flow cytometer [30]. 16*S* rRNA is a common FISH target because, in addition to providing a convenient method for microbial identification, ribosomal RNA is often a very abundant target, which yields a strong fluorescent signal.

As with PCR primers, FISH probes can be designed to target broad groups of microbes (e.g., Eub338 [3] hybridizes to most bacteria), specific subsets of microbes (e.g., ALF968 [92] hybridizes to α-proteobacteria) or even specific strains. In addition to choosing an appropriate sequence for a 16*S* rRNA FISH probe, it is important to consider that some regions of the 16*S* rRNA are more accessible to probes than other regions because of the structure of the molecule [6]. However, the addition of unlabeled helper oligonucleotides can improve accessibility of hard-to-label regions of the 16*S* rRNA molecule [45].

The fluorescent signal can be increased by using catalyzed reporter deposition–fluorescence *in situ* hybridization (CARD-FISH). For CARD-FISH the oligonucleotide probe has a horseradish peroxidase (HRP) tag, which results in the deposition of multiple fluorescent molecules in the vicinity of the target bound by the HRP-labeled oligonucleotide when fluorescently labeled tyramides are applied [121]. This method was developed to enhance the detection of slow–growing bacteria with relatively few rRNA molecules per cell and is also useful for detecting mRNA, which is typically less abundant than rRNA [27]. It has also been useful for visualizing bacteria in environments with high background autofluorescence, such as on the surface of algae [137]. Other ways of increasing the fluorescent signal include using locked nucleic acid (LNA) probes, which bind to their targets with high affinity [74], and using double-labeled probes [135]. Another way to increase the signal of labeled cells is to use *in situ* PCR [77]. Not actually a version of FISH, fluorescently labeled nucleotides are incorporated into *in situ* PCR products inside cells, and since the cells are sufficiently permeable to allow PCR reagents in and out but not PCR products, the labeled PCR products are retained inside the cell [59,67].

It is possible to attribute certain metabolic activities to specific community members by combining FISH with detection of radioisotope incorporation using microautoradiography (MAR)-FISH (reviewed in Ref. 146) or stable isotope incorporation (reviewed in Ref. 96) using the Raman-FISH [62] or FISH-SIMS (secondary–ion mass spectrometry) method [7,79,97].

2.2.2. Diversity of Genetic Potential

The previous sections have focused on using phylogenetic markers to examine community structure. Although these methods tell us what types of microbes are present, they don't tell us about the metabolism of those microbes. While phylogeny can give hints about metabolic capabilities, even organisms that have highly similar 16*S* rRNA sequences may vary remarkably in the rest of their genomes and the functions thus imparted.

Functional Gene PCR, Cloning, and Sequencing

Similar to cloning and sequencing phylogenetic marker genes (e.g., the 16*S* rRNA gene), it is possible to amplify other genes from microbial communities by PCR with degenerate primers targeting conserved regions of specific genes of interest. These PCR products can then be cloned and sequenced to get a picture of the diversity of a particular set of genes, which can provide insights about a particular function or type of metabolism in the community. The diversity of butyrate-producing bacteria in human feces was examined by amplifying the butyryl-CoA:acetate CoA-transferase gene with degenerate primers, cloning, and sequencing [81]. Methanogen diversity in human feces has also been studied by amplifying methyl-coenzyme M reductase (*mcrA*) genes with degenerate primers, cloning, and sequencing [116]. One disadvantage is that less conserved versions of genes may escape detection by this method.

Shotgun Metagenomics

In shotgun metagenomics, the total DNA from a particular environment is sequenced without a selective PCR step. Ideally this will provide a more complete view of the genetic content of the organisms in a sample. Sequences can be obtained by Sanger sequencing of a metagenomic library constructed by directly cloning the DNA from a sample, without PCR amplification (e.g., see Ref. 47) or by pyrosequencing without cloning (e.g., see Ref. 106). Since specific genes are not targeted, one advantage of metagenomic sequencing is the potential to discover completely novel or unexpected genes or sequences. Although metagenomic sequencing studies avoid the bias of PCR-based studies targeting specific genes, other biases, including DNA extraction biases, cloning biases and sequencing biases, may remain [88]. A more recent metagenomic sequencing study of an artificial community, consisting of a known mixture of microbes, illustrated that the abundance of each organism in the sample was not reflected in the sequence data [88].

Metagenomic sequencing of communities with limited diversity (e.g., biofilms in acid mine drainage and symbionts of a gutless marine worm) has allowed the assembly of nearly complete genomes from uncultured organisms, providing insights to the metabolism and interactions of the community members [142,152]. In more complex communities assembling genomes is currently less feasible, but it is still possible to gain insights into overall community function and metabolism. The MetaHIT consortium reported sequencing of 576.7 Gbp from the human gut microbiome [106]. Metagenomic sequencing has been particularly illuminating in the area of viral diversity since there is not a universal gene that all viruses contain for targeting in a selective PCR approach [37].

With so much data, metagenomic sequence analysis is a major computational challenge [151]. Part of the challenge of making sense of the genetic data is the lack of similar reference genome sequences in the databases. Partly to address this problem, the HMP aims to sequence 900 genomes from the human microbiome [136]. By May 2010 the genomes of 356 phylogenetically diverse isolates from the human microbiome had been sequenced by the HMP, and these reference genomes greatly improved their ability to analyze metagenomic data [136].

Functional Gene Arrays

Functional gene arrays are used to determine the diversity of functional genes present in a microbial community without sequencing. Fluorescently labeled DNA is produced from the microbial community DNA of interest with random primers and the Klenow fragment of DNA polymerase [55,56]. The labeled DNA is then applied to a microarray with oligonucleotide probes for functional genes, and the hybridization pattern is used to determine which genes are present in the sample. The latest version of the GeoChip has ~28,000 probes designed to recognize a variety of genes, including those involved in nutrient cycling and antibiotic resistance [54]. So far functional gene arrays have been used to investigate a range of environmental microbial communities (e.g., marine basalts [86], coral [72], soil [9]), but not the human microbiota. However, the extensive cataloging of the human gut microbiome metagenome [47,106] could be used to design a functional gene microarray for the human gut communities. As with the phylogenetic microarrays, one limitation is that functional gene microarrays are unable to recognize anything not included on the array.

Functional Metagenomics

Functional metagenomics is a way to screen microbial communities for specific functions and to identify the genes that encode those functions. Similar to metagenomic sequencing, there is no PCR step; DNA from the sample is cloned directly into an expression vector. The cloned DNA is then expressed in a heterologous host (usually a strain of *E. coli*) and screened for a particular activity such as antibiotic resistance or bile salt hydrolysis, two examples from the human microbiome [68,130]. This technique is particularly well suited for discovering novel or poorly conserved genes that encode a particular function that would be missed by targeted PCR approaches. It also has the advantage of linking a function to a particular gene, something lacking from studies based solely on sequencing. However, many genes are not functional when expressed in a heterologous host and are likely to be overlooked by this method [143].

Single-Cell Genome Amplification and Sequencing

Whole-genome sequencing is no longer limited to cultured microbes. It is possible to sequence the genome of a single cell, obtained by one of a variety of methods, including microfluidics [85], fluorescence-activated cell sorting (FACS) [114], or micromanipulation [114]. Multiple displacement amplification (MDA) can produce enough DNA from a single cell for genome sequencing [31,32] by either Sanger sequencing or next-generation methods. This allows the genomic content of individual cells to be determined and variations between cells to be compared, something that is lost in shotgun metagenomics. When applied to an uncultured TM7 cell from the human mouth, this method has yielded a partial genome sequence in 1825 scaffolds [85]. There are still several technical challenges to obtaining a complete, assembled genome sequence from a single cell with this technology. In MDA some regions of the genome are amplified to much higher levels (>1000-fold) than other regions, which results in very high sequence coverage of some regions at the expense of less amplified regions [114]. Another complication with MDA is the production of chimeras, which complicates genome assembly [114]. Double-stranded breaks in the chromosome during cell lysis also prevent complete assembly if there is only a single copy of the chromosome in each cell. Avoiding contamination is also critical; since random primers are used for MDA, any contaminating DNA will be amplified with the target DNA, resulting in large amounts of sequence contamination. Improvements in the combination of technologies that make single-cell genome sequencing possible, from cell capture to sequencing, are making this exciting approach more feasible [114].

2.2.3. Beyond Metabolic Potential: Transcriptomics, Proteomics, Metabolomics

This section serves as a brief overview of the techniques available for looking beyond the metabolic potential of a community to what the microbial community is actually doing in its natural environment. Metatranscriptomics is used to determine which genes are expressed in a microbial community. Total RNA is isolated, rRNA is removed to enrich for mRNA [134], and the remaining RNA is reverse-transcribed into cDNA for sequencing. Metatranscriptomics has been applied to many marine environments [33,44,46,103,126], and this technique is now being used for the first time on the human microbiota [139,140]. Metaproteomics gets a step closer to actual community function by using mass spectrometry to determine which proteins are present (and presumably functioning) in a sample (for a review, see Ref. 145). Finally, in metabolomics metabolites are analyzed to determine the effect of the microbiota on the metabolism of drugs and dietary components by methods including NMR spectroscopy, high-performance liquid chromatography–mass spectrometry (HPLC-MS), and gas chromatography–mass spectrometry (GC-MS) [95,144].

2.3. CULTIVATION

Although the majority of microbes have not been cultivated, cultivation is the foundation of microbiology. Most of what we know about the capabilities of microbes relies on our ability to culture them. From understanding the roles of specific genes to discovering microbes with novel metabolic activities, cultivation is key to our understanding of the microbial world. Even uncultivated microbes are better understood, owing to the genes that they share with their cultivated relatives.

Microbiologists have long realized that conventional cultivation methods typically yield <1% of the microbes present in a sample, a phenomenon dubbed the "great plate count anomaly" [132]. This anomaly is not due solely to the plating efficiency of individual populations of bacteria; of the >70 Bacterial phyla, less than half include a cultured representative [98]. Given the value of having genomes of characterized strains to interpret metagenomic data from microbial communities, there has been a renewed effort to culture microbes that are known only by their SSU rRNA gene sequences.

The main strategies are to mimic the natural environment of the microbes and to allow extended periods of growth. Some of the specific culture conditions that have been used to mimic the environment include low nutrient levels [26,157], low oxygen, and high CO_2 [133]. Additionally, using components of the natural environment in cultivation has been successful. In one study novel marine bacteria were cultured from intertidal sediments; the microbial community was embedded in agar, placed in a diffusion chamber that allowed molecules but not bacteria through, and placed in an aquarium set up with sediment and seawater [69]. There is also a focus on separating microbes so that slow growers are not overrun by rapidly growing microbial weeds, either by dilution to extinction methods [18], in gel microdroplets (GMDs) [157], or in an isolation chip (ichip) [93]. Some cultivation approaches, including GMDs and the ichip, also attempt to accommodate microbes that grow better with other microbes by allowing diffusion of molecules between microbes while maintaining physical separation between different microbes. Similarly, the "bacterial lobster trap," which can trap a single bacterium, allows the exchange of nutrients, waste, and small molecules during growth [25]. There are various possible reasons for the dependence of some microbes on others for growth, including the supply of a particular nutrient or signal by a microbial partner. Indeed, adding microbial signaling molecules, including cyclic AMP, homoserine lactones, and putative peptide signals, to culture medium has improved the ability to cultivate some bacteria [15–17,94,133]. Similarly, the failure to culture many obligate symbiotic bacteria is attributed to specific unknown (or irreproducible) growth requirements provided by the host organism. For example, *Capnocytophaga canimorsus*, a normal member of the mouth microbiota of cats and dogs, exhibited robust growth when cocultured with a variety of cultured animal cells [mouse macrophages, human epithelial cells (HeLa), and canine epithelial cells] but only meager growth in isolation on conventional media [84]. The addition of catalase or pyruvate to combat reactive oxygen species also seems to ease the transition into culture [10,133]. In some intriguing cases it has been shown that microbes can be domesticated, growing well in pure culture without special treatment after several transfers [11,39,93,94]. In other cases, pure cultures have been achieved but cell densities have not exceeded their environmental densities, which are much lower than concentrations seen in conventional cultures [26,48].

Detecting the growth of target microbes is a key part of novel cultivation techniques, and it often requires special effort because the microbes are slow growers, achieve lower cell concentrations than conventional cultures, or are relatively rare among isolates. Flow cytometry has been used to detect and sort GMDs containing microcolonies for transfer to a liquid medium [157]. In other cases cultures were screened by microscopy [110]. These methods have contributed to the cultivation of several novel microbes; however, screening for specific microbes by microscopy can be quite labor-intensive. Plate wash PCR (PWPCR) is an alternative method to detect target organisms growing on a plate in which group-specific primers are used with DNA isolated from the mixture of colonies on the plate [133]. If the target organism is detected by PWPCR, then colonies or groups of colonies, including very small colonies visible only with a dissecting microscope, from the replicate plates are transferred to individual wells containing a liquid version of the plate medium. Portions of the liquid cultures are pooled and screened by PCR until the culture containing the target organism is identified.

2.4. CHOOSING AN APPROACH

Important insights into the roles of microbial communities in human health and disease have resulted from collaborations between environmental microbial ecologists and clinical researchers [113]. However, the range of experimental approaches available for studying the complex microbial communities of humans can be overwhelming, and collaborations between environmental microbial ecologists and clinical researchers are complicated by a lack of familiarity across the disciplines. The ability to find some common ground with which to start discussions between collaborators is an essential aspect of undertaking this type of research. Therefore, it may be useful to conclude this discussion of techniques with some suggestions for applying studies of microbial diversity to clinical research issues.

When considering which of the experimental methods presented in this review might be an appropriate entry point into studies of microbial communities, it would be quite useful to consider which type of information might be most informative for the question at hand. As such, in Figure 2.1 the techniques that are discussed in this review are presented in terms of the type of information that can be gained as well as what portion of the community of interest is being interrogated. We have divided the information that is generated into three large categories. The first is *structure*, which we define as a census of the community in terms of the diversity of organisms and some information with regards to their relative abundance. The next category is *metabolic potential*, where information about the possible functions of a microbial community is gained. In the final category, which we refer to as *in situ activity*, specific functions of the microbial consortium are assessed directly. When deciding which of these categories to address, considerations of cost, availability, and the specific research question all play a role.

Using the complementary approaches described in this chapter, it is possible to study the structure, function, and dynamics of human-associated microbial communities and to make connections between microbial communities and human health. Investigations that determine whether there are differences in microbial community structures between health and disease are the first step. Once such differences are established, studies of a more functional nature such as those provided by techniques that provide information about metabolic potential or *in situ* activity are appropriate to gain further insight into the roles of microbial communities in human health and disease.

ACKNOWLEDGMENTS

Funding from R01 HG004906 and UH3 DK083933 was granted to TS and VY (both HMP initiatives) and R01 DK070875, to VY. CB was support by T32 HL007749.

REFERENCES

1. Adékambi T, Drancourt M, Raoult D. The rpoB gene as a tool for clinical microbiologists. *Trends Microbiolo* **17**:37–45 (2009).
2. Amann R, Fuchs BM. Single-cell identification in microbial communities by improved fluorescence *in situ* hybridization techniques. *Nat Rev Microbiol* **6**:339–348 (2008).
3. Amann RI, Binder BJ, Olson RJ, Chisholm SW, Devereux R, Stahl DA. Combination of 16S rRNA-targeted oligonucleotide probes with flow cytometry for analyzing mixed microbial populations. *Appl Environ Microbiol* **56**:1919–1925 (1990).
4. Ansorge WJ. Next-generation DNA sequencing techniques. *New Biotechnol* **25**:195–203 (2009).
5. Ashelford KE, Chuzhanova NA, Fry JC, Jones AJ, Weightman AJ. At Least 1 in 20 16S rRNA sequence records currently held in public repositories is estimated to contain substantial anomalies. *Appl Environ Microbiol* **71**:7724–7736 (2005).
6. Behrens S, Fuchs BM, Mueller F, Amann R. Is the in situ accessibility of the 16S rRNA of Escherichia coli for Cy3-labeled oligonucleotide probes predicted by a three-dimensional structure model of the 30S ribosomal subunit? *Appl Environ Microbiol* **69**:4935–4941 (2003).
7. Behrens S, Losekann T, Pett-Ridge J, Weber PK, Ng W-O, Stevenson BS, Hutcheon ID, Relman DA, Spormann AM. Linking microbial phylogeny to metabolic activity at the single-cell level by using enhanced element labeling-catalyzed reporter deposition fluorescence in situ hybridization (EL-FISH) and nanoSIMS. *Appl Environ Microbiol* **74**:3143–3150 (2008).
8. Benson DA, Karsch-Mizrachi I, Lipman DJ, Ostell J, Wheeler DL. GenBank. *Nucleic Acids Res* **36**:D25–D30 (2008).

9. Berthrong ST, Schadt CW, Pineiro G, Jackson RB. Afforestation alters the composition of functional genes in soil and biogeochemical processes in South American grasslands. *Appl Environ Microbiol* **75**:6240–6248 (2009).

10. Bogosian G, Aardema ND, Bourneuf EV, Morris PJL, O'Neil JP. Recovery of hydrogen peroxide-sensitive culturable cells of Vibrio vulnificus gives the appearance of resuscitation from a viable but nonculturable state. *J Bacteriol* **182**:5070–5075 (2000).

11. Bollmann A, Lewis K, Epstein SS. Incubation of environmental samples in a diffusion chamber increases the diversity of recovered isolates. *Appl Environ Microbiol* **73**:6386–6390 (2007).

12. Bonetta L. Prime time for real-time PCR. *Nat Meth* **2**:305–312 (2005).

13. Bray JR, Curtis JT. An ordination of the upland forest communities of southern Wisconsin. *Ecol Monogr* **27**:325–349 (1957).

14. Brodie EL, DeSantis TZ, Joyner DC, Baek SM, Larsen JT, Andersen GL, Hazen TC, Richardson PM, Herman DJ, Tokunaga TK, et al. Application of a high-density oligonucleotide microarray approach to study bacterial population dynamics during uranium reduction and reoxidation. *Appl Environ Microbiol* **72**:6288–6298 (2006).

15. Bruns A, Cypionka H, Overmann J. Cyclic AMP and acyl homoserine lactones increase the cultivation efficiency of heterotrophic bacteria from the central Baltic Sea. *Appl Environ Microbiol* **68**:3978–3987 (2002).

16. Bruns A, Nubel U, Cypionka H, Overmann J. Effect of signal compounds and incubation conditions on the culturability of freshwater bacterioplankton. *Appl Environ Microbiol* **69**:1980–1989 (2003).

17. Bussmann I, Philipp B, Schink B. Factors influencing the cultivability of lake water bacteria. *J Microbiol Meth* **47**:41–50 (2001).

18. Button DK, Schut F, Quang P, Martin R, Robertson BR. Viability and isolation of marine bacteria by dilution culture: Theory, procedures, and initial results. *Appl Environ Microbiol* **59**:881–891 (1993).

19. Caporaso JG, Kuczynski J, Stombaugh J, Bittinger K, Bushman FD, Costello EK, Fierer N, Pena AG, Goodrich JK, Gordon JI, et al. QIIME allows analysis of high-throughput community sequencing data. *Nat Meth* **7**:335–336 (2010).

20. Chao A. Estimating population size for sparse data in capture-recapture experiments. *Biometrics* **45**:427–438 (1989).

21. Chao A. Nonparametric estimation of the number of classes in a population. *Scand J Stat* **11**:265–270 (1984).

22. Chun J, Bae KS. Phylogenetic analysis of Bacillus subtilis and related taxa based on partial gyrA gene sequences. *A Van Leeuwenhoek* **78**:123–127 (2000).

23. Claesson MJ, O'Sullivan O, Wang Q, Nikkilä J, Marchesi JR, Smidt H, de Vos WM, Ross RP, O'Toole PW. Comparative analysis of pyrosequencing and a phylogenetic microarray for exploring microbial community structures in the human distal intestine. *PLoS ONE* **4**:e6669 (2009).

24. Cole JR, Wang Q, Cardenas E, Fish J, Chai B, Farris RJ, Kulam-Syed-Mohideen AS, McGarrell DM, Marsh T, Garrity GM, et al. The Ribosomal Database Project: Improved alignments and new tools for rRNA analysis. *Nucleic Acids Res* **37**:D141–D145 (2009).

25. Connell JL, Wessel AK, Parsek MR, Ellington AD, Whiteley M, Shear JB. Probing prokaryotic social behaviors with bacterial "lobster traps." *mBio* **1** (2010).

26. Connon SA, Giovannoni SJ. High-throughput methods for culturing microorganisms in very-low-nutrient media yield diverse new marine isolates. *Appl Environ Microbiol* **68**:3878–3885 (2002).

27. Constant P, Chowdhury SP, Pratscher J, Conrad R. Streptomycetes contributing to atmospheric molecular hydrogen soil uptake are widespread and encode a putative high-affinity [NiFe]-hydrogenase. *Environ Microbiol* **12**:821–829 (2010).
28. Costello EK, Lauber CL, Hamady M, Fierer N, Gordon JI, Knight R. Bacterial community variation in human body habitats across space and time. *Science* **326**:1694–1697 (2009).
29. Crosby LD, Criddle CS. Understanding bias in microbial community analysis techniques due to rrn operon copy number heterogeneity. *Biotechniques* **34**:790–802 (2003).
30. Czechowska K, Johnson DR, van der Meer JR. Use of flow cytometric methods for single-cell analysis in environmental microbiology. *Curr Opin Microbiol* **11**:205–212 (2008).
31. Dean FB, Hosono S, Fang L, Wu X, Faruqi AF, Bray-Ward P, Sun Z, Zong Q, Du Y, Du J, et al. Comprehensive human genome amplification using multiple displacement amplification. *Proc Natl Acad Sci USA* **99**:5261–5266 (2002).
32. Dean FB, Nelson JR, Giesler TL, Lasken RS. Rapid amplification of plasmid and phage DNA using Phi29 DNA polymerase and multiply-primed rolling circle amplification. *Genome Res* **11**:1095–1099 (2011).
33. DeLong EF. The microbial ocean from genomes to biomes. *Nature* **459**:200–206 (2009).
34. DeSantis T, Brodie E, Moberg J, Zubieta I, Piceno Y, Andersen G. High-density universal 16S rRNA microarray analysis reveals broader diversity than typical clone library when sampling the environment. *Microbial Ecol* **53**:371–383 (2007).
35. DeSantis TZ, Hugenholtz P, Larsen N, Rojas M, Brodie EL, Keller K, Huber T, Dalevi D, Hu P, Andersen GL. Greengenes, a chimera-checked 16S rRNA gene database and workbench compatible with ARB. *Appl Environ Microbiol* **72**:5069–5072 (2006).
36. Dominguez-Bello MG, Costello EK, Contreras M, Magris M, Hidalgo G, Fierer N, Knight R. Delivery mode shapes the acquisition and structure of the initial microbiota across multiple body habitats in newborns. *Proc Natl Acad Sci* **107**:11971–11975 (2010).
37. Edwards RA, Rohwer F. Viral metagenomics. *Nat Rev Microbiol* **3**:504–510 (2005).
38. Engebretson JJ, Moyer CL. Fidelity of select restriction endonucleases in determining microbial diversity by terminal-restriction fragment length polymorphism. *Appl Environ Microbiol* **69**:4823–4829 (2003).
39. Epstein SS. General model of microbial uncultivability. In Epstein SS, ed., *Uncultivated Microorganisms*, Springer-Verlag, Berlin, 2009, pp. 131–159.
40. Favier CF, Vaughan EE, De Vos WM, Akkermans ADL. Molecular monitoring of succession of bacterial communities in human neonates. *Appl Environ Microbiol* **68**:219–226 (2002).
41. Fisher MM, Triplett EW. Automated approach for ribosomal intergenic spacer analysis of microbial diversity and its application to freshwater bacterial communities. *Appl Environ Microbiol* **65**:4630–4636 (1999).
42. Flanagan JL, Brodie EL, Weng L, Lynch SV, Garcia O, Brown R, Hugenholtz P, DeSantis TZ, Andersen GL, Wiener-Kronish JP, et al. Loss of bacterial diversity during antibiotic treatment of intubated patients colonized with *Pseudomonas aeruginosa*. *J Clin Microbiol* **45**:1954–1962 (2007).
43. Fox GE, Wisotzkey JD, Jurtshuk PJ. How close is close: 16S rRNA sequence identity may not be sufficient to guarantee species identity. *Int J Syst Bacteriol* **42**:166–170 (1992).
44. Frias-Lopez J, Shi Y, Tyson GW, Coleman ML, Schuster SC, Chisholm SW, DeLong EF. Microbial community gene expression in ocean surface waters. *Proc Natl Acad Sci USA* **105**:3805–3810 (2008).

45. Fuchs BM, Glockner FO, Wulf J, Amann R. Unlabeled helper oligonucleotides increase the in situ accessibility to 16S rRNA of fluorescently labeled oligonucleotide probes. *Appl Environ Microbiol* **66**:3603–3607 (2000).
46. Gilbert JA, Field D, Huang Y, Edwards R, Li W, Gilna P, Joint I. Detection of large numbers of novel sequences in the metatranscriptomes of complex marine microbial communities. *PLoS ONE* **3**:e3042 (2008).
47. Gill SR, Pop M, DeBoy RT, Eckburg PB, Turnbaugh PJ, Samuel BS, Gordon JI, Relman DA, Fraser-Liggett CM, Nelson KE. Metagenomic analysis of the human distal gut microbiome. *Science* **312**:1355–1359 (2006).
48. Giovannoni S, Stingl U. The importance of culturing bacterioplankton in the "omics" age. *Nat Rev Microbiol* **5**:820–826 (2007).
49. Grice EA, Kong HH, Conlan S, Deming CB, Davis J, Young AC, NISC Comparative Sequencing Program, Bouffard GG, Blakesley RW, Murray PR, Green ED, et al. Topographical and temporal diversity of the human skin microbiome. *Science* **324**:1190–1192 (2009).
50. Grimont PA. Use of DNA reassociation in bacterial classification. *Can J Microbiol* **34**:541–546 (1988).
51. Group TNHW, Peterson J, Garges S, Giovanni M, McInnes P, Wang L, Schloss JA, Bonazzi V, McEwen JE, Wetterstrand KA, et al. The NIH Human Microbiome Project. *Genome Res* **19**:2317–2323 (2009).
52. Haas BJ, Gevers D, Earl AM, Feldgarden M, Ward DV, Giannoukos G, Ciulla D, Tabbaa D, Highlander SK, Sodergren E, et al. Chimeric 16S rRNA sequence formation and detection in Sanger and 454-pyrosequenced PCR amplicons. *Genome Res* **21**:494–504 (2011).
53. Hamady M, Lozupone C, Knight R. Fast UniFrac: Facilitating high-throughput phylogenetic analyses of microbial communities including analysis of pyrosequencing and PhyloChip data. *ISME J* **4**:17–27 (2009).
54. He Z, Deng Y, Van Nostrand JD, Tu Q, Xu M, Hemme CL, Li X, Wu L, Gentry TJ, Yin Y, et al. GeoChip 3.0 as a high-throughput tool for analyzing microbial community composition, structure and functional activity. *ISME J* **4**:1167–1179 (2010).
55. He Z, Gentry TJ, Schadt CW, Wu L, Liebich J, Chong SC, Huang Z, Wu W, Gu B, Jardine P, et al. GeoChip: A comprehensive microarray for investigating biogeochemical, ecological and environmental processes. *ISME J* **1**:67–77 (2007).
56. He Z, Wu L, Li X, Fields MW, Zhou J. Empirical establishment of oligonucleotide probe design criteria. *Appl Environ Microbiol* **71**:3753–3760 (2005).
57. Heid CA, Stevens J, Livak KJ, Williams PM. Real time quantitative PCR. *Genome Res* **6**:986–994 (1996).
58. Hilty M, Burke C, Pedro H, Cardenas P, Bush A, Bossley C, Davies J, Ervine A, Poulter L, Pachter L, et al. Disordered microbial communities in asthmatic airways. *PLoS ONE* **5**:e8578 (2010).
59. Hodson R, Dustman W, Garg R, Moran M. In situ PCR for visualization of microscale distribution of specific genes and gene products in prokaryotic communities. *Appl Environ Microbiol* **61**:4074–4082 (1995).
60. Hong JW, Park JY, Fomina M, Gadd GM. Development and optimization of an 18S rRNA-based oligonucleotide microarray for the fungal order Eurotiales. *J Appl Microbiol* **108**:985–997 (2010).
61. Horn HS. Measurement of "overlap" in comparative ecological studies. *Am Naturalist* **100**:419–424 (1966).
62. Huang WE, Stoecker K, Griffiths R, Newbold L, Daims H, Whiteley AS, Wagner M. Raman-FISH: Combining stable-isotope Raman spectroscopy and fluorescence *in situ*

hybridization for the single cell analysis of identity and function. *Environ Microbiol* **9**:1878–1889 (2007).

63. Hubálek Z. Coefficients of association and similarity, based on binary (presence-absence) data: An evaluation. *Biolo Rev* **57**:669–689 (1982).
64. Huse S, Hubery J, Morrison H, Sogin M, Welch D. Accuracy and quality of massively parallel DNA pyrosequencing. *Genome Biol* **8**:R143 (2007).
65. Huse SM, Welch DM, Morrison HG, Sogin ML. Ironing out the wrinkles in the rare biosphere through improved OTU clustering. *Environ Microbiol* **12**:1889–1898 (2010).
66. Jaccard P. Distribution de la flore alpine dans le Bassin des Dranses et dans quelques régions voisines. *Bull Soc Vaudoise Sci Nat* **37**:241–272 (1901).
67. Jacobs D, Angles ML, Goodman AE, Neilan BA. Improved methods for in situ enzymatic amplification and detection of low copy number genes in bacteria. *FEMS Microbiol Lett* **152**:65–73 (1997).
68. Jones BV, Begley MI, Hill C, Gahan CGM, Marchesi JR. Functional and comparative metagenomic analysis of bile salt hydrolase activity in the human gut microbiome. *Proc Natl Acad Sci USA* **105**:13580–13585 (2008).
69. Kaeberlein T, Lewis K, Epstein SS. Isolating "uncultivable" microorganisms in pure culture in a simulated natural environment. *Science* **296**:1127–1129 (2002).
70. Karlin S, Weinstock G, Brendel V. Bacterial classifications derived from recA protein sequence comparisons. *J Bacteriol* **177**:6881–6893 (1995).
71. Kibe R, Sakamoto M, Hayashi H, Yokota H, Benno Y. Maturation of the murine cecal microbiota as revealed by terminal restriction fragment length polymorphism and 16S rRNA gene clone libraries. *FEMS Microbiol Lett* **235**:139–146 (2004).
72. Kimes NE, Nostrand JDV, Weil E, Zhou J, Morris PJ. Microbial functional structure of *Montastraea faveolata*, an important Caribbean reef-building coral, differs between healthy and yellow-band diseased colonies. *Environ Microbiol* **12**:541–556 (2010).
73. Kistler A, Avila PC, Rouskin S, Wang D, Ward T, Yagi S, Schnurr D, Ganem D, DeRisi JL, Boushey HA. Pan-viral screening of respiratory tract infections in adults with and without asthma reveals unexpected human coronavirus and human rhinovirus diversity. *J Infect Dis* **196**:817–825 (2007).
74. Kubota K, Ohashi A, Imachi H, Harada H. Improved in situ hybridization efficiency with locked-nucleic-acid-incorporated DNA probes. *Appl Environ Microbiol* **72**:5311–5317 (2006).
75. Kuehl CJ, Wood HD, Marsh TL, Schmidt TM, Young VB. Colonization of the cecal mucosa by Helicobacter hepaticus impacts the diversity of the indigenous microbiota. *Infect Immun* **73**:6952–6961 (2005).
76. Kunin V, Engelbrektson A, Ochman H, Hugenholtz P. Wrinkles in the rare biosphere: Pyrosequencing errors can lead to artificial inflation of diversity estimates. *Environ Microbiol* **12**:118–123 (2010).
77. Laflamme C, Gendron L, Turgeon N, Filion G, Ho J, Duchaine C. In situ detection of antibiotic-resistance elements in single *Bacillus cereus* spores. *Syst Appl Microbiol* **32**:323–333 (2009).
78. Lee ZM-P, Bussema C, Schmidt TM. rrnDB: Documenting the number of rRNA and tRNA genes in bacteria and archaea. *Nucleic Acids Res* **37**:D489-D493 (2009).
79. Li T, Wu TD, Mazeas L, Toffin L, Guerquin-Kern JL, Leblon G, Bouchez T. Simultaneous analysis of microbial identity and function using NanoSIMS. *Environ Microbiol* **10**:580–588 (2008).
80. Liu WT, Marsh TL, Cheng H, Forney LJ. Characterization of microbial diversity by determining terminal restriction fragment length polymorphisms of genes encoding 16S rRNA. *Appl Environ Microbiol* **63**:4516–4522 (1997).

81. Louis P, Young P, Holtrop G, Flint HJ. Diversity of human colonic butyrate-producing bacteria revealed by analysis of the butyryl-CoA:acetate CoA-transferase gene. *Environ Microbiol* **12**:304–314 (2010).

82. Lozupone C, Knight R. UniFrac: A new phylogenetic method for comparing microbial communities. *Appl Environ Microbiol* **71**:8228–8235 (2005).

83. MacLean D, Jones JDG, Studholme DJ. Application of "next-generation" sequencing technologies to microbial genetics. *Nat Rev Microbiol* **7**:287–296 (2009).

84. Mally M, Shin H, Paroz C, Landmann R, Cornelis GR. *Capnocytophaga canimorsus*: A human pathogen feeding at the surface of epithelial cells and phagocytes. *PLoS Pathogens* **4**:e1000164 (2008).

85. Marcy Y, Ouverney C, Bik EM, Losekann T, Ivanova N, Martin HG, Szeto E, Platt D, Hugenholtz P, Relman DA, et al. Dissecting biological "dark matter" with single-cell genetic analysis of rare and uncultivated TM7 microbes from the human mouth. *Proc Natl Acad Sci USA* **104**:11889–11894 (2007).

86. Mason OU, Di Meo-Savoie CA, Van Nostrand JD, Zhou J, Fisk MR, Giovannoni SJ. Prokaryotic diversity, distribution, and insights into their role in biogeochemical cycling in marine basalts. *ISME J* **3**:231–242 (2008).

87. Mihindukulasuriya KA, Wu G, St. Leger J, Nordhausen RW, Wang D. Identification of a novel coronavirus from a beluga whale by using a panviral microarray. *J Virol* **82**:5084–5088 (2008).

88. Morgan JL, Darling AE, Eisen JA. Metagenomic sequencing of an *in vitro*-simulated microbial community. *PLoS ONE* **5**:e10209 (2010).

89. Mun Huang W. Bacterial diversity based on type II DNA topoisomerase genes. *Annu Rev Genet* **30**:79–107 (1996).

90. Muyzer G, de Waal EC, Uitterlinden AG. Profiling of complex microbial populations by denaturing gradient gel electrophoresis analysis of polymerase chain reaction-amplified genes coding for 16S rRNA. *Appl Environ Microbiol* **59**:695–700 (1993).

91. Muyzer G, Smalla K. Application of denaturing gradient gel electrophoresis (DGGE) and temperature gradient gel electrophoresis (TGGE) in microbial ecology. *A Van Leeuwenhoek* **73**:127–141 (1998).

92. Neef A. *Anwendung der in situ Einzelzell-Identifizierung von Bakterien zur Populationsanalyse in Komplexen Mikrobiellen Biozönosen*. PhD thesis, Technische Univ. München, Munich, Germany, 1997.

93. Nichols D, Cahoon N, Trakhtenberg EM, Pham L, Mehta A, Belanger A, Kanigan T, Lewis K, Epstein SS. Use of Ichip for high-throughput in situ cultivation of "uncultivable" microbial species. *Appl Environ Microbiol* **76**:2445–2450 (2010).

94. Nichols D, Lewis K, Orjala J, Mo S, Ortenberg R, O'Connor P, Zhao C, Vouros P, Kaeberlein T, Epstein SS. Short peptide induces an "uncultivable" microorganism to grow in vitro. *Appl Environ Microbiol* **74**:4889–4897 (2008).

95. Nicholson JK, Holmes E, Wilson ID. Gut microorganisms, mammalian metabolism and personalized health care. *Nat Rev Microbiol* **3**:431–438 (2005).

96. Orphan VJ. Methods for unveiling cryptic microbial partnerships in nature. *Curr Opin Microbiol* **12**:231–237 (2009).

97. Orphan VJ, House CH, Hinrichs KU, McKeegan KD, DeLong EF. Methane-consuming archaea revealed by directly coupled isotopic and phylogenetic analysis. *Science* **293**:484–487 (2001).

98. Pace NR. Mapping the tree of life: Progress and prospects. *Microbiol Mol Biol Rev* **73**:565–576 (2009).

99. Pace NR. A molecular view of microbial diversity and the biosphere. *Science* **276**:734–740 (1997).

100. Pan HY, Chao A, Foissner W. A nonparametric lower bound for the number of species shared by multiple communities. *J Agric Biol Environ Stat* **14**:452–468 (2009).

101. Payne GW, Vandamme P, Morgan SH, LiPuma JJ, Coenye T, Weightman AJ, Jones TH, Mahenthiralingam E. Development of a recA gene-based identification approach for the entire burkholderia genus. *Appl Environ Microbiol* **71**:3917–3927 (2005).

102. Popa R, Mashall MJ, Nguyen H, Tebo BM, Brauer S. Limitations and benefits of ARISA intra-genomic diversity fingerprinting. *J Microbiol Meth* **78**:111–118 (2009).

103. Poretsky RS, Hewson I, Sun S, Allen AE, Zehr JP, Moran MA. Comparative day/night metatranscriptomic analysis of microbial communities in the North Pacific subtropical gyre. *Environ Microbiol* **11**:1358–1375 (2009).

104. Price LB, Liu CM, Johnson KE, Aziz M, Lau MK, Bowers J, Ravel J, Keim PS, Serwadda D, Wawer MJ, et al. The effects of circumcision on the penis microbiome. *PLoS ONE* **5**:e8422 (2010).

105. Pruesse E, Quast C, Knittel K, Fuchs BM, Ludwig W, Peplies J, Glockner FO. SILVA: A comprehensive online resource for quality checked and aligned ribosomal RNA sequence data compatible with ARB. *Nucleic Acids Res* **35**:7188–7196 (2007).

106. Qin J, Li R, Raes J, Arumugam M, Burgdorf KS, Manichanh C, Nielsen T, Pons N, Levenez F, Yamada T, et al. A human gut microbial gene catalogue established by metagenomic sequencing. *Nature* **464**:59–65 (2010).

107. Quince C, Lanzen A, Curtis TP, Davenport RJ, Hall N, Head IM, Read LF, Sloan WT. Accurate determination of microbial diversity from 454 pyrosequencing data. *Nat Meth* **6**:639–641 (2009).

108. Quinlan AR, Stewart DA, Stromberg MP, Marth GT. Pyrobayes: An improved base caller for SNP discovery in pyrosequences. *Nat Meth* **5**:179–181 (2008).

109. Rajili-Stojanovi M, Heilig HGHJ, Molenaar D, Kajander K, Surakka A, Smidt H, de Vos WM. Development and application of the human intestinal tract chip, a phylogenetic microarray: Analysis of universally conserved phylotypes in the abundant microbiota of young and elderly adults. *Environ Microbiol* **11**:1736–1751 (2009).

110. Rappe MS, Connon SA, Vergin KL, Giovannoni SJ. Cultivation of the ubiquitous SAR11 marine bacterioplankton clade. *Nature* **418**:630–633 (2002).

111. Rastogi G, Osman S, Vaishampayan PA, Andersen GL, Stetler LD, Sani RK. Microbial diversity in uranium mining-impacted soils as revealed by high-density 16S microarray and clone library. *Microbial Ecol* **59**:94–108 (2010).

112. Ravel J, Gajer P, Abdo Z, Schneider GM, Koenig SSK, McCulle SL, Karlebach S, Gorle R, Russell J, Tacket CO, et al. Vaginal microbiome of reproductive-age women. *Proc Natl Acad Sci USA* **108**:4680–4687 (2011).

113. Robinson CJ, Bohannan BJM, Young VB. From structure to function: The ecology of host-associated microbial communities. *Microbiol Mol Biol Rev* **74**:453–476 (2010).

114. Rodrigue S, Malmstrom RR, Berlin AM, Birren BW, Henn MR, Chisholm SW. Whole genome amplification and *de novo* assembly of single bacterial cells. *PLoS ONE* **4**:e6864 (2009).

115. Savage DC. Microbial ecology of the gastrointestinal tract. *Annu Rev Microbiol* **31**:107–133 (1977).

116. Scanlan P, Shanahan F, Marchesi J. Human methanogen diversity and incidence in healthy and diseased colonic groups using mcrA gene analysis. *BMC Microbiol* **8**:79 (2008).

117. Schloss PD. The effects of alignment quality, distance calculation method, sequence filtering, and region on the analysis of 16S rRNA gene-based studies. *PLoS Comput Biol* **6**:e1000844 (2010).
118. Schloss PD. A high-throughput DNA sequence aligner for microbial ecology studies. *PLoS ONE* **4**:e8230 (2009).
119. Schloss PD, Handelsman J. Introducing DOTUR, a computer program for defining operational taxonomic units and estimating species richness. *Appl Environ Microbiol* **71**:1501–1506 (2005).
120. Schloss, PD, Westcott SL, Ryabin T, Hall JR, Hartmann M, Hollister EB, Lesniewski RA, Oakley BB, Parks DH, Robinson CJ, et al. Introducing mothur: Open-source, platform-independent, community-supported software for describing and comparing microbial communities. *Appl Environ Microbiol* **75**:7537–7541 (2009).
121. Schonhuber W, Fuchs B, Juretschko S, Amann R. Improved sensitivity of whole-cell hybridization by the combination of horseradish peroxidase-labeled oligonucleotides and tyramide signal amplification. *Appl Environ Microbiol* **63**:3268–3273 (1997).
122. Schonmann S, Loy A, Wimmersberger C, Sobek J, Aquino C, Vandamme P, Frey B, Rehrauer H, Eberl L. 16S rRNA gene-based phylogenetic microarray for simultaneous identification of members of the genus *Burkholderia*. *Environ Microbiol* **11**:779–800 (2009).
123. Schutte UM, Abdo Z, Bent SJ, Shyu C, Williams CJ, Pierson JD, Forney LJ. Advances in the use of terminal restriction fragment length polymorphism (T-RFLP) analysis of 16S rRNA genes to characterize microbial communities. *Appl Microbiol Biotechnol* **80**:365–380 (2008).
124. Shannon CE. A mathematical theory of communication. *Bell Syst Technical J* **27**:379–423 (1948).
125. Shendure J, Ji H. Next-generation DNA sequencing. *Nat Biotechnol* **26**:1135–1145 (2008).
126. Shi Y, Tyson GW, DeLong EF. Metatranscriptomics reveals unique microbial small RNAs in the ocean's water column. *Nature* **459**:266–269 (2009).
127. Simpson EH. Measurement of diversity. *Nature* **163**:688 (1949).
128. Smith CJ, Osborn AM. Advantages and limitations of quantitative PCR (Q-PCR)-based approaches in microbial ecology. *FEMS Microbiol Ecol* **67**:6–20 (2009).
129. Sogin ML, Morrison HG, Huber JA, Welch DM, Huse SM, Neal PR, Arrieta JM, Herndl GJ. Microbial diversity in the deep sea and the underexplored "rare biosphere." *Proc Natl Acad Sci* **103**:12115–12120 (2006).
130. Sommer MOA, Dantas G, Church GM. Functional Characterization of the antibiotic resistance reservoir in the human microflora. *Science* **325**:1128–1131 (2009).
131. Sørensen T. A method of establishing groups of equal amplitude in plant sociology based on species content and its application to analyses of the vegetation on Danish commons. *Kongelige Danske Videnskabernes Selskab Biol Skrifter* **5**:1–34 (1948).
132. Staley JT, Konopka A. Measurement of in situ activities of nonphotosynthetic microorganisms in aquatic and terrestrial habitats. *Annu Rev Microbiol* **39**:321–346 (1985).
133. Stevenson BS, Eichorst SA, Wertz JT, Schmidt TM, Breznak JA. New strategies for cultivation and detection of previously uncultured microbes. *Appl Environ Microbiol* **70**:4748–4755 (2004).
134. Stewart FJ, Ottesen EA, DeLong EF. Development and quantitative analyses of a universal rRNA-subtraction protocol for microbial metatranscriptomics. *ISME J* **4**:896–907 (2010).

135. Stoecker K, Dorninger C, Daims H, Wagner M. Double labeling of oligonucleotide probes for fluorescence *in situ* hybridization (dope-fish) improves signal intensity and increases rRNA accessibility. *Appl Environ Microbiol* **76**:922–926 (2010).
136. The Human Microbiome Jumpstart Reference Strains Consortium. A catalog of reference genomes from the human microbiome. *Science* **328**:994–999 (2010).
137. Tujula NA, Holmstrôm C, Muflmann M, Amann R, Kjelleberg S, Crocetti GR. A CARD-FISH protocol for the identification and enumeration of epiphytic bacteria on marine algae. *J Microbiol Meth* **65**:604–607 (2006).
138. Turnbaugh PJ, Ley RE, Hamady M, Fraser-Liggett CM, Knight R, Gordon JI. The Human Microbiome Project. *Nature* **449**:804–810 (2007).
139. Turnbaugh PJ, Quince C, Faith JJ, McHardy AC, Yatsunenko T, Niazi F, Affourtit J, Egholm M, Henrissat B, Knight R, et al. Organismal, genetic, and transcriptional variation in the deeply sequenced gut microbiomes of identical twins. *Proc Natl Acad Sci USA* **107**:7503–7508 (2010).
140. Turnbaugh PJ, Ridaura VK, Faith JJ, Rey FE, Knight R, Gordon JI. The effect of diet on the human gut microbiome: a metagenomic analysis in humanized gnotobiotic mice. *Sci Transl Med* **1**:6ra14 (2009).
141. Tyagi S, Kramer FR. Molecular beacons: probes that fluoresce upon hybridization. *Nat Biotechnol* **14**:303–308 (1996).
142. Tyson GW, Chapman J, Hugenholtz P, Allen EE, Ram RJ, Richardson PM, Solovyev VV, Rubin EM, Rokhsar DS, Banfield JF. Community structure and metabolism through reconstruction of microbial genomes from the environment. *Nature* **428**:37–43 (2004).
143. Uchiyama T, Miyazaki K. Functional metagenomics for enzyme discovery: Challenges to efficient screening. *Curr Opin Biotechnol* **20**:616–622 (2009).
144. van Duynhoven J, Vaughan EE, Jacobs DM, Kemperman RrA, van Velzen EJJ, Gross G, Roger LC, Possemiers S, Smilde AK, Doré J, et al. Metabolic fate of polyphenols in the human superorganism. *Proc Natl Acad Sci USA* **108**:4531–4538 (2011).
145. VerBerkmoes NC, Denef VJ, Hettich RL, Banfield JF. Systems biology: Functional analysis of natural microbial consortia using community proteomics. *Nat Rev Microbiol* **7**:196–205 (2009).
146. Wagner M, Nielsen PH, Loy A, Nielsen JL, Daims H. Linking microbial community structure with function: fluorescence in situ hybridization-microautoradiography and isotope arrays. *Curr Opin Biotechnol* **17**:83–91 (2006).
147. Wang D, Coscoy L, Zylberberg M, Avila PC, Boushey HA, Ganem D, DeRisi JL. Microarray-based detection and genotyping of viral pathogens. *Proc Natl Acad Sci USA* **99**:15687–15692 (2002).
148. Wang D, Urisman A, Liu Y-T, Springer M, Ksiazek TG, Erdman DD, Mardis ER, et al. Viral discovery and sequence recovery using DNA microarrays. *PLoS Biol* **1**:e2 (2003).
149. Wang Q, Garrity GM, Tiedje JM, Cole JR. Naive Bayesian classifier for rapid assignment of rRNA sequences into the new bacterial taxonomy. *Appl Environ Microbiol* **73**:5261–5267 (2007).
150. Woese CR. Bacterial evolution. *Microbiol Mol Biol Rev* **51**:221–271 (1987).
151. Wooley JC, Godzik A, Friedberg I. A primer on metagenomics. *PLoS Comput Biol* **6**:e1000667 (2010).
152. Woyke T, Teeling H, Ivanova NN, Huntemann M, Richter M, Gloeckner FO, Boffelli D, Anderson IJ, Barry KW, Shapiro HJ, et al. Symbiosis insights through metagenomic analysis of a microbial consortium. *Nature* **443**:950–955 (2006).

153. Wu D, Hugenholtz P, Mavromatis K, Pukall R, Dalin E, Ivanova NN, Kunin V, Goodwin L, Wu M, Tindall BJ, et al. A phylogeny-driven genomic encyclopaedia of bacteria and archaea. *Nature* **462**:1056–1060 (2009).
154. Youssef N, Sheik CS, Krumholz LR, Najar FZ, Roe BA, Elshahed MS. Comparison of species richness estimates obtained using nearly complete fragments and simulated pyrosequencing-generated fragments in 16S rRNA gene-based environmental surveys. *Appl Environ Microbiol* **75**:5227–5236 (2009).
155. Yue JC, Clayton MK. A similarity measure based on species proportions. *Commun Stat-Theory Meth* **34**:2123–2131 (2005).
156. Zaura E, Keijser B, Huse S, Crielaard W. Defining the healthy "core microbiome" of oral microbial communities. *BMC Microbiol* **9**:259 (2009).
157. Zengler K, Toledo G, Rappe M, Elkins J, Mathur EJ, Short JM, Keller M. Cultivating the uncultured. *Proc Natl Acad Sci USA* **99**:15681–15686 (2002).

3

PHYLOARRAYS

EOIN L. BRODIE

Ecology Department, Earth Sciences Division, Lawrence Berkeley National Laboratory, Berkeley, California

SUSAN V. LYNCH

Department of Medicine, University of California, San Francisco, California

Microbes are central to maintenance of life on Earth. In the postgenomic era, it has become abundantly clear that environmental ecosystems and human health are heavily reliant on the functional properties of resident microbiota. However, study of these complex communities has been impeded by the fact that much of microbial diversity remains uncultivated, and until relatively recently, largely undescribed. Building on the wealth of information generated through sequence-based approaches, phylogenetic microarrays, based on discriminatory sequences in specific microbial biomarker genes, have emerged as an efficient and standardized approach for monitoring the dynamics of complex microbial populations. These tools have been used to profile microbiota in diverse settings ranging from radionuclide-contaminated sediments to mammalian gastrointestinal tracts and have dramatically enhanced our ability to rapidly and broadly define features of the microbiome that contribute to ecosystem functioning in these and other disparate niches. In this chapter we discuss the range of phylogenetic arrays available for microbial profiling and, as an example, focus much of our discussion on a specific tool developed at Lawrence Berkeley National Lab, the 16*S* rRNA PhyloChip. This high-density microarray containing approximately 500,000 probes, is capable of monitoring the relative abundance of

The Human Microbiota: How Microbial Communities Affect Health and Disease, First Edition. Edited by David N. Fredricks.

thousands of different bacterial and archaeal taxa in a single assay. We discuss its design, sample preparation, and data analysis and highlight applications of this technology in human microbiome research.

3.1. INTRODUCTION

Famously coined "the great plate count anomaly" by Staley and Konopka in 1985 [1], the finding that orders of magnitude more microbial cells were observed microscopically in environmental samples compared to those cultivated using conventional media, set the stage for a molecular revolution in microbial ecology. Facilitated by pioneering work by Woese [2] and Pace [3] and the parallel development of the polymerase chain reaction (PCR) [4], the application of sequence-based analysis to profile microbial communities without the need for culture led to a dramatic increase in the described diversity of microbial life on our planet [5,6]. Critical to this success were molecular approaches employing targeted amplification of microbial biomarker genes, such as the bacterial 16*S* rRNA gene, and subsequent cloning and sequencing of these clones, which identified a diversity of previously undescribed microbial species and their relative distribution in mixed-species communities.

The utility of these culture-independent approaches is apparent in the exponential increase in the number of biomarker gene sequences deposited to public databases since the 1980s. More recent developments in second generation sequencing technologies (454, Illumina platforms) have further accelerated this discovery and provide excellent tools for microbiome profiling due to their enhanced coverage compared to more traditional serial sequencing approaches, while third-generation technologies (e.g., developed by Pacific Biosciences) that provide longer reads will improve the representation of full-length biomarker sequences from lower abundance taxa and underrepresented biomes. High-density microarrays are an alternative and highly standardized approach to assaying the diverse populations of microbes using phylogenetic biomarker amplicons generated during PCR. The rapid expansion of 16*S* rRNA gene sequence databases has permitted the design of phylogenetic microarrays where hybridization of amplicon fragments to a combination of oligonucleotides can indicate the presence and relative abundance of many thousands of taxa in a single, massively parallel assay. Much like expression arrays, phylogenetic microarrays are ideal for comparing samples across treatment groups to define key structural and discrete taxonomic changes associated with each group. Moreover, the data generated by phylogenetic microarrays, because they are semiquantitative in nature, are amenable to robust statistical analyses in which environmental variables, such as pH or expression of specific inflammatory cytokines, can be related to both gross and discrete taxonomic shifts in microbiota composition related to these parameters.

3.2. PHYLOGENETIC MICROARRAY DESIGN

Phylogenetic microarrays take many forms, but are characterized primarily according to format (e.g., spotted probes or *in situ* synthesized probes), probe density (10s–100s–1000s) and target range (species to domains). For excellent reviews of published microarray applications in microbial ecology, see References 7–9. The ability to spot or print oligonucleotide probes onto an array surface dramatically changed our ability to perform a vast number of assays using a single tool. As the technology progressed, particularly due to advances in the semiconductor field, the density of oligonucleotides that could be attached to array surfaces significantly increased.

For the purpose of this chapter, we will focus on the 16*S* rRNA PhyloChip version G2 as an example of high-density phylogenetic array design. Microarray probe design for this tool was based on a standard approach used previously for differentiating *Staphylococcaceae* [10]. Initially, more than 30,000 16*S* rRNA gene sequences (*Escherichia coli* base pair positions 47 to 1473) at least 600 nucleotides in length were retrieved from publically available databases. This region was used since flanking its extremes are universally conserved sequence segments that permit PCR priming for amplification of bacterial or archaeal signatures using two distinct primer pairs. To ensure that only high-quality nonchimeric sequence data were used for array design, the software package Bellerophon [11] was employed to prevent chimeric sequences from being misconstrued as novel organisms. Quality filtered sequences were clustered on the basis of common 17-mers found in the sequence. The resulting 8741 clusters, each containing organisms with 16*S* rRNA sequences that diverged by no more than 3%, were considered operational taxonomic units (OTUs) representing 121 demarcated prokaryotic orders at the time of design. Taxonomic family assignments for each OTU were assigned by defining placement of their member organisms using Bergey's taxonomic outline [12]. For each OTU a complementary set of perfectly matching (PM) 25-mer probes ($n \geq 11$) targeting loci that collectively distinguish that OTU from all others were designed. To accomplish efficient discriminatory probe design, each 16*S* rRNA sequence in the OTU was separated into overlapping 25-mers, representing potential probe targets that were then matched to as many of the other sequences as possible within that OTU. A subset of the prevalent targets for a given OTU was selected and reverse-complemented to produce the probe sequence. To avoid potential mishybridization to an unintended amplicon, those probes that contained a central 17-mer matching sequences in more than one OTU [13,14] were removed from the design. This ultimate collection of probes for each of the 8741 OTUs detected by the array were termed *PM probes*. In addition, for every PM probe, a control 25-mer [mismatching (MM) probe], identical in all positions except the 13th base [13], was designed. MM probes did not contain a central 17-mer complementary to sequences in any OTU, and together with the PM probe constitute a probe pair for analysis purposes.

Oligonucleotides for the 16*S* rRNA PhyloChip were synthesized by a photolithographic method, at Affymetrix, Inc. (Santa Clara, CA), directly onto a 1.28×1.28 cm glass surface at an approximate density of 10,000 probes per μm^2 [15]. The semiconductor technology used in this process permits high-density oligonucleotides arrays to be produced, facilitating interrogation of substantially more organisms in a single assay. Each oligonucleotide probe sequence was present on the array at approximately 3.2×10^6 copies. Ultimately, this array of 506,944 probes, of which 297,851 were oligonucleotide PM or MM probes targeting 16*S* rRNA gene sequences, was arranged as a square grid of 712 rows and columns. The remaining probes were used for standardized image orientation, normalization controls, or other unrelated analyses.

3.3. SAMPLE PREPARATION FOR PHYLOGENETIC MICROARRAY PROFILING

Samples to be analyzed by PhyloChip undergo extraction and amplification protocols to ensure optimal processing for descriptive phylogenetic profiling. Samples are typically preserved at the point of collection either by snap freezing or using a nucleic acid preservative (e.g., RNALater™) to preserve sample integrity and enhance nucleic acid retrieval. To improve lysis of both Gram-negative and Gram-positive species in the sample, cell lysis is typically enhanced with a short bead-beating step (5.5 m/s for 30 s), prior to nucleic acid extraction with sample-specific commercially available kits. Following extraction and purification, total DNA is used as template for a series of 8–12 PCR reactions performed across a gradient of annealing temperatures designed to optimize the diversity of 16*S* rRNA gene sequences retrieved from the sample and minimize the influence of random primer annealing events that occur during the early cycles of PCR. The products of these reactions are pooled and purified (gel or affinity column) prior to fragmentation by DNase and terminal biotin labeling of the fragments. Internal non-rRNA spikes are included during this process to account for variance introduced during these steps and all subsequent downstream procedures. This labeled material is then hybridized to the PhyloChip as described in Brodie et al. [10], prior to washing, staining, and scanning of the array surface. Those probes that have hybridized to their cognate 16*S* rRNA sequences produce a fluorescent signal that is detected by the confocal laser scanner. This array surface image is transformed using Affymetrix software such that each probe feature is represented by a single fluorescence value. After scaling of the intensities using the response of the internal control spikes, the fluorescence data are then used to first determine the probe pair responses [perfect match (PM) relative to mismatch (MM)] across a probe set for each taxon and to calculate a metric of relative abundance. For detection, a probe set must pass certain criteria. Typically, for a probe pair to be deemed positive, the intensity of the PM probes must be ≥30% higher than the MM probes and the difference between their two intensities should be >130 times the squared noise value. Within a probe set, it is typically required ≥90% of probe pairs pass these criteria for the probe set (taxon) to be considered present. A fluorescence value is then obtained for each taxon by averaging the PM minus MM intensity values after removing the minimum and maximum values. These data may then be normalized to total array fluorescence intensity to account for variations in quantity of 16*S* rRNA amplicons hybridized. Transformation (typically log) of fluorescence intensity data is also performed due to nonhomogeneity of variance across different intensities.

The extensive dataset generated by this PhyloChip permits application of robust statistical approaches to analyze the phylogenetic data generated for each sample. These approaches include, at the very simplest level, calculation of indices of gross microbiota composition (e.g. detected richness, evenness, and diversity) for comparative analysis to determine whether significant compositional changes in the microbiota are associated with changes in environment such as bioremediation processes or host health status. Discovery of biologically meaningful patterns, such as microbiota associations with disease status, can be achieved using a variety of multivariate procedures [16] and are represented by clustering or ordination of distance matrices derived from a number of distance metrics (e.g., Bray–Curtis) that compare patterns of relative abundances of taxa across samples. The G2 PhyloChip has the ability to detect almost 9000 taxa (the next generation G3 PhyloChip detects ~60,000 taxa) and the sensitivity to detect low-abundance taxa even in dominated samples [17,18]. The PhyloChip allows for the rapid detection of changes in relative abundance of taxa and provides candidate taxa for hypothesis testing as to their functional role.

A good example of the utility of this type of basic comparative analysis comes from a study of isogenic mice from two different laboratories in which mice were fed distinct chow formula [19]. Mice from one supplier possessed large populations of Th17 cells in the lamina propria. Th17 cells are a novel subset of proinflammatory T-helper (Th) cells associated with idiopathic autoimmune diseases such as inflammatory bowel disease [20,21]. As these mice were isogenic, the overabundance of Th17 cells in one group was hypothesized to be related to differences in their microbiota. Comparative profiling of gastrointestinal microbiota from these two groups of mice using PhyloChip revealed that approximately 100 taxa exhibited significantly altered relative abundance across the two mouse groups. Two taxa were detected in significantly higher abundance in those mice who possessed high Th17 cell numbers, implicating them as drivers of this host immune response. One of these two species, *Candidatus arthromitus*, is a segmented filamentous bacterium related to Clostridia. Subsequent experiments involving co-housing of mice and monocolonization of non-Th17 producers with *Candidatus arthromitus* resulted in robust accumulation of Th17 cells and expression of their associated cytokines (IL22 and IL17) in the lamina propria of the gut. It should be pointed out that previous attempts using other culture-independent approaches to identify the specific microbial species responsible for driving Th17 accumulation in these animals had not identified these organisms. Thus this study illustrates the utility of the PhyloChip for comparative microbiota profiling and to identify discriminatory taxa that represent potential target species responsible for specific and tractable host phenotypes.

3.4. THE UTILITY OF PHYLOGENETIC MICROARRAYS

Microarray hybridization of target sequences such as 16*S* rRNA PCR amplicons to the surface of an array containing probes specific to unique discriminatory loci on the target biomarker permit assay of a large number of molecules in parallel under standardized conditions [17]. One distinct advantage of a microarray approach is detection of lower-abundance organisms despite the presence of dominant species in the community [17] as described above. This provides a significant advantage, particularly for interrogation of clinical samples, which often possess a relatively small number of dominant species, but whose rare biosphere may play a key role in modulating the behavior of these species. In a sense, because of the ability to detect rare microbiota members as efficiently as dominant members of the community, the PhyloChip and other phylogenetic microarrays such as the microbiota array [22] and HITChip (human intestinal tract chip) [23], amongst others, permit a high-resolution, cross-sectional view of the community present that encompasses lower abundance species (Figure 3.1). This is particularly pertinent given emerging data from a human microbiome research conglomerate in Europe, which demonstrated that functional genes detected in high relative abundance were encoded by relatively rare species in these communities [24]. Hence tools that profile the

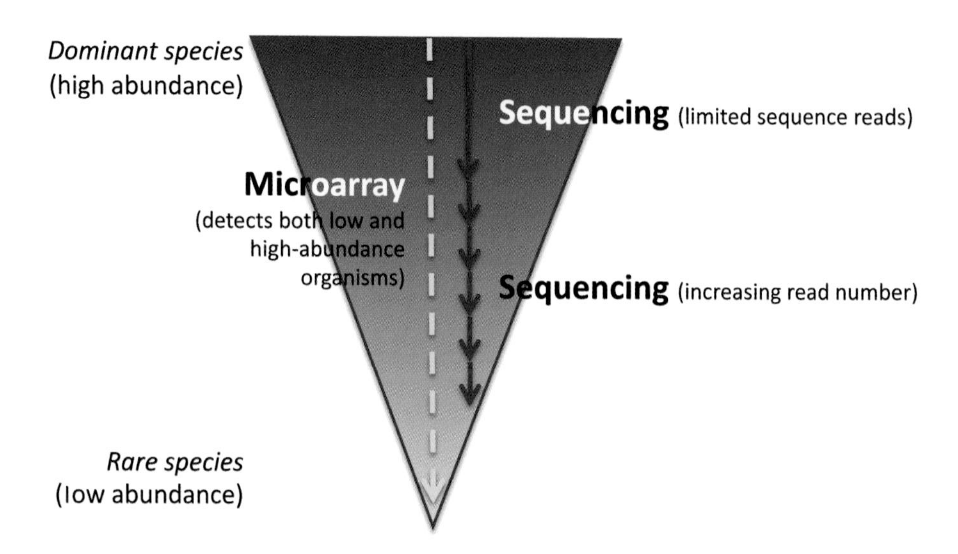

Figure 3.1. Schematic comparison of phylogenetic microarray and sequencing approaches to profile bacterial community composition. Because of its parallel nature, a phylogenetic microarrays can detect lower-abundance species despite the presence of high-abundance dominant organisms, but also may miss community members not represented on the array. In comparison, sequencing—at the depths typically engaged in for microbiome profiling—may detect only dominant members of the community. Detection of lower-abundance membership, requires substantially more extensive sequencing for complex communities, such as colonic microbiota.

presence and relative abundance of such low-abundance species offer a considerable advantage in truly understanding the contribution of these species to consortium assembly and function.

The advantage of this feature of phylogenetic microarray profiling was demonstrated in a study of cystic fibrosis patient airway microbiota using the 16*S* rRNA PhyloChip [25]. Traditionally pulmonary disease in this population is considered to be due primarily to colonization of the airways by a handful of pulmonary pathogens. However, there is increasing evidence that this and other chronic airway diseases are associated with the presence of more complex polymicrobial communities rather than simply by a single microbial species [26–28]. A PhyloChip-based study demonstrated that, compared with pediatric patients, adult CF patients exhibit dramatically restructured airway microbiota [25]. Specifically, a significant correlation existed between airway microbiota diversity and age (a proxy for pulmonary function; $r = 0.33$; $p < 0.004$); older CF patients who exhibited significantly worse pulmonary function possessed less diverse and more phylogenetically related microbial communities compared to younger patients [25]. Moreover, this study identified more than 100 bacterial community members whose abundance was highly correlated with patient age and pulmonary function status, implicating them in specific and distinct pathogenic processes in pediatric and adult CF patients, respectively (Figure 3.2). Because of the predominance of adult CF airway communities by *Pseudomonas aeruginosa* and a small number of other species, previous sequence-based approaches to examining community composition in this niche failed to

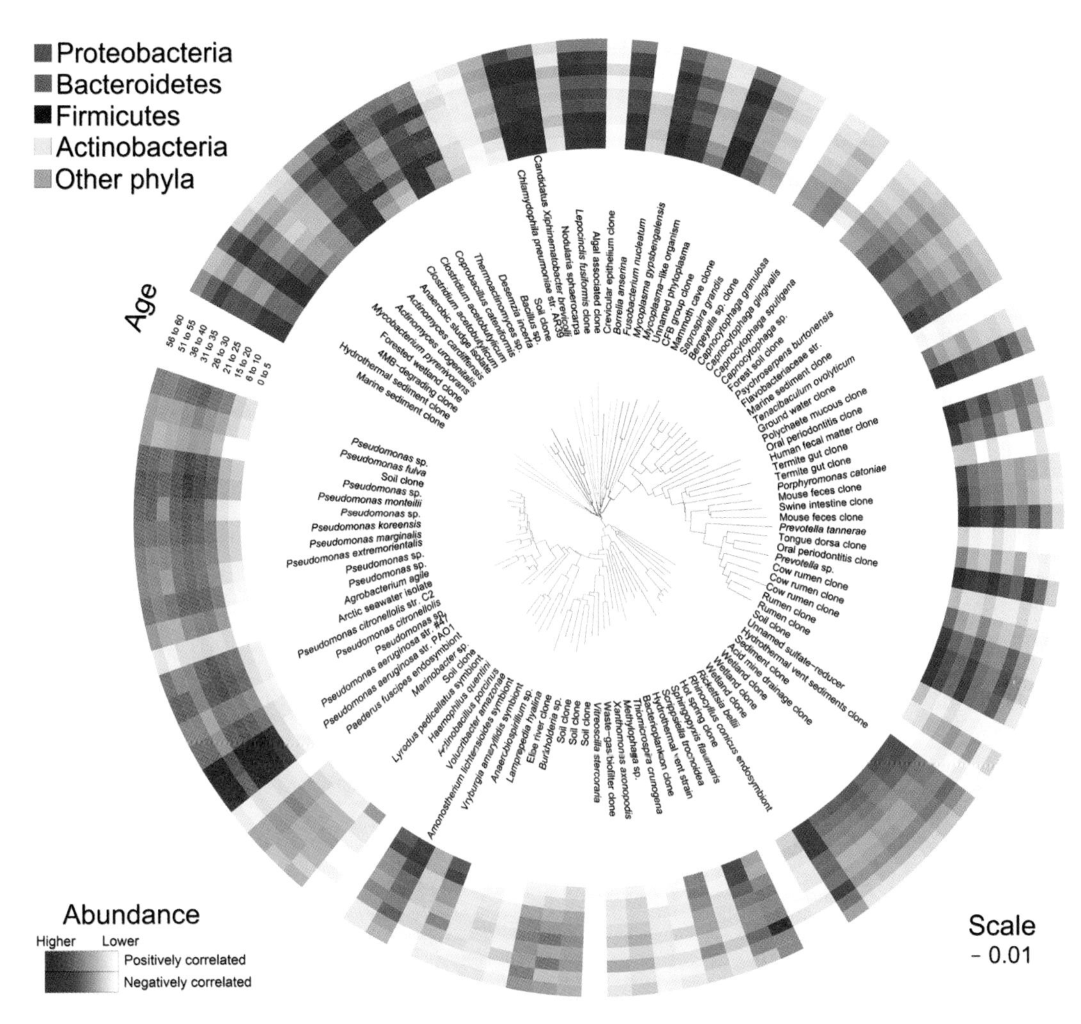

Figure 3.2. **Phylogenetic tree displaying relationship between age and taxon relative abundance.** Taxa exhibiting a significant increase (red) or decrease (blue) in relative abundance with increasing CF patient age are illustrated. These organisms represent community members putatively involved in differential age-related pathogenic processes in CF airways. Scale bar indicates 0.01 nucleotide substitutions per base. (Reproduced from Cox et al. [25].)

identify these age- and pulmonary-function-associated changes in lower-abundance taxa. Evidence from both environmental and human microbial ecology studies indicates that these subtle changes in the relative abundance of members of the rare biosphere may well be important contributors to ecosystem functioning in this niche and merit consideration. Support for this hypothesis comes from several more recent studies, including one demonstrating that treatment of *Clostridium difficile* infected mice with antimicrobials decimated GI bacterial community diversity and promoted development of a "supershedder" mouse phenotype [29]. This study suggests that bacterial community diversity may be instrumental in regulating abundance and activities of members of the microbiota. Hence, simply examining the dominant species in these systems may overlook rare biosphere membership that contributes to overall community diversity and ecosystem functioning.

It is important to acknowledge the strengths and weaknesses of phylogenetic profiling tools and that phylogenetic microarrays, like all other microbiome profiling tools, have their limitations. Its parallel nature and ability to detect low-abundance organisms as efficiently as high-abundance members of the community permits a cross sectional view of the microbial assemblage, providing information on which taxa are present and the relative change in abundance of specific taxa between samples. However, array-based technology is subject to cross-hybridization, potentially producing false positives in the datasets generated. To combat this, for example, the 16*S* rRNA PhyloChip houses a probe set for each taxon detected. This set of probes comprise at least 11 probe pairs; a probe pair comprises a perfect match (PM) probe complementary to the target of interest and a mismatch (MM) probe that differs in sequence at the central (13th) nucleotide of the 25-mer oligonucleotide. At least 11 probe pairs are employed to interrogate a minimum of two unique loci on the 16*S* rRNA gene that discriminate that sequence from all others, together with an algorithm based on the ratio of fluorescence intensity of the PM versus MM probes, to partially account for signal generated due to cross hybridization. In addition, phylogenetic microarrays can only detect taxa for which sequences have been reported for either the specific taxon or a close relative. For this reason, truly novel microorganisms may be overlooked. However, the ability to redesign the array as more sequence data become available helps alleviate this issue, as is evident with the latest third-generation (3G) version of the PhyloChip, which can detect approximately 60,000 bacterial taxa [30]. Ultimately, the power of a standardized assay such as the PhyloChip for routine and consistent comparative microbiomics makes it a highly valuable tool in the discovery of microbie-driven phenotypes in phylogenetically well-characterized ecosystems, particularly when coupled with other technologies to provide a truly comprehensive view of the microbiome present.

ACKNOWLEDGMENTS

Part of this work was carried out at Lawrence Berkeley National Laboratory under contract DE-AC02-05CH11231 between the University of California and the US Department of Energy, Office of Science. SVL is supported by NIH awards, AI075410PO1, HL098964, and AT004732.

REFERENCES

1. Staley JT, Konopka A. Measurement of in situ activities of nonphotosynthetic microorganisms in aquatic and terrestrial habitats. *Annu Rev Microbiol* **39**:321–346 (1985).
2. Woese CR, Fox GE. Phylogenetic structure of the prokaryotic domain: the primary kingdoms. *Proc Natl Acad Sci USA* **74**(11):5088–5090 (1977).
3. Lane DJ, Pace B, Olsen GJ, Stahl DA, Sogin ML, Pace NR. Rapid determination of 16*S* ribosomal RNA sequences for phylogenetic analyses. *Proc Natl Acad Sci USA* **82**(20):6955–6959 (1985).
4. Mullis KB, Faloona FA. Specific synthesis of DNA in vitro via a polymerase-catalyzed chain reaction. *Meths Enzymol* **155**:335–350 (1987).

5. Hugenholtz P, Goebel BM, Pace NR. Impact of culture-independent studies on the emerging phylogenetic view of bacterial diversity. *J Bacteriol* **180**(18):4765–4774 (1998).
6. Hugenholtz P, Pace NR. Identifying microbial diversity in the natural environment: a molecular phylogenetic approach. *Trends Biotechnol* **14**(6):190–197 (1996).
7. Bodrossy L, Sessitsch A. Oligonucleotide microarrays in microbial diagnostics. *Curr Opin Microbiol* **7**(3):245–254 (2004).
8. Gentry TJ, Wickham GS, Schadt CW, He Z, Zhou J. Microarray applications in microbial ecology research. *Microbial Ecol* **52**(2):159–175 (2006).
9. Andersen GL, He Z, DeSantis TZ, Brodie EL, Zhou J. The use of microarrays in microbial ecology: In Liu W-T, Jansson JK, eds., *Environmental Molecular Microbiology*, Horizon Scientific Press, Norwich, UK, (2010).
10. Brodie EL, DeSantis TZ, Joyner DC, Baek SM, Larsen JT, Andersen GL, Hazen TC, Richardson PM, Herman DJ, Tokunaga TK, et al. Application of a high-density oligonucleotide microarray approach to study bacterial population dynamics during uranium reduction and reoxidation. *Appl Environ Microbiol* **72**(9):6288–6298 (2006).
11. Huber T, Faulkner G, Hugenholtz P. Bellerophon: A program to detect chimeric sequences in multiple sequence alignments. *Bioinformatics* **20**(14):2317–2319 (2004).
12. Garrity GM. *Bergey's Manual of Systematic Bacteriology*, Springer-Verlag, New York, 2001.
13. Mei R, Hubbell E, Bekiranov S, Mittmann M, Christians FC, Shen M-M, Lu G, Fang J, Liu W-M, Ryder T, et al. Probe selection for high-density oligonucleotide arrays. *Proc Natl Acad Sci USA* **100**(20):11237–11242 (2003).
14. Urakawa H, Noble PA, El Fantroussi S, Kelly JJ, Stahl DA. Single-base-pair discrimination of terminal mismatches by using oligonucleotide microarrays and neural network analyses. *Appl Environ Microbiol* **68**(1):235–244 (2002).
15. Chee M, Yang R, Hubbell E, Berno A, Huang XC, Stern D, Winkler J, Lockhart DJ, Morris MS, Fodor SP. Accessing genetic information with high-density DNA arrays. *Science* **274**(5287):610–614 (1996).
16. Ramette A. Multivariate analyses in microbial ecology. *FEMS Microbiol Ecol* **62**(2):142–160 (2007).
17. Brodie EL, DeSantis TZ, Parker JP, Zubietta IX, Piceno YM, Andersen GL. Urban aerosols harbor diverse and dynamic bacterial populations. *Proc Natl Acad Sci USA* **104**(1):299–304 (2007).
18. DeSantis TZ, Brodie EL, Moberg JP, Zubieta IX, Piceno YM, Andersen GL. High-density universal 16*S* rRNA microarray analysis reveals broader diversity than typical clone library when sampling the environment. *Microbial Ecol* **53**(3):371–383 (2007).
19. Ivanov II, Atarashi K, Manel N, Brodie EL, Shima T, Karaoz U, Wei D, Goldfarb KC, Santee CA, Lynch SV, et al. Induction of intestinal Th17 cells by segmented filamentous bacteria. *Cell* **139**(3):485–498 (2009).
20. Winer S, Paltser G, Chan Y, Tsui H, Engleman E, Winer D, Dosch HM. Obesity predisposes to Th17 bias. *Eur J Immunol* **39**(9):2629–2635 (2009).
21. Brand S. Crohn's disease: Th1, Th17 or both? The change of a paradigm: New immunological and genetic insights implicate Th17 cells in the pathogenesis of Crohn's disease. *Gut* **58**(8):1152–1167 (2009).
22. Paliy O, Kenche H, Abernathy F, Michail S. High-throughput quantitative analysis of the human intestinal microbiota with a phylogenetic microarray. *Appl Environ Microbiol* **75**(11):3572–3579 (2009).
23. Rajilić-Stojanović M, Heilig HG, Molenaar D, Kajander K, Surakka A, Smidt H, de Vos WM. Development and application of the human intestinal tract chip, a phylogenetic

microarray: Analysis of universally conserved phylotypes in the abundant microbiota of young and elderly adults. *Environ Microbiol* **11**(7):1736–1751 (2009).

24. Arumugam M, Raes J, Pelletier E, Le Paslier D, Yamada T, Mende DR, Fernandes GR, Tap J, Bruls T, Batto JM, et al. Enterotypes of the human gut microbiome. *Nature* **473**(7346):174–180 (2011).
25. Cox MJ, Allgaier M, Taylor B, Baek MS, Huang YJ, Daly RA, Karaoz U, Andersen GL, Brown R, Fujimura KE. Airway microbiota and pathogen abundance in age-stratified cystic fibrosis patients. *Plos ONE* **5**(6):10 (2010).
26. Huang YJ, Kim E, Cox MJ, Brodie EL, Brown R, Wiener-Kronish JP, Lynch SV. A persistent and diverse airway microbiota present during chronic obstructive pulmonary disease exacerbations. *OMICS* **14**(1):9–59 (2010).
27. Hilty M, Burke C, Pedro H, Cardenas P, Bush A, Bossley C, Davies J, Ervine A, Poulter L, Pachter L, et al. Disordered microbial communities in asthmatic airways. *PLoS ONE* **5**(1): p. e8578 (2010).
28. Harris JK, De Groote MA, Sagel SD, Zemanick ET, Kapsner R, Penvari C, Kaess H, Deterding RR, Accurso FJ, Pace NR. Molecular identification of bacteria in bronchoalveolar lavage fluid from children with cystic fibrosis. *Proc Natl Acad Sci USA* **104**(51):20529–20533 (2007).
29. Lawley TD, Bouley DM, Hoy YE, Gerke C, Relman DA, Monack DM. Host transmission of Salmonella enterica serovar typhimurium is controlled by virulence factors and indigenous intestinal microbiota. *Infect Immun* **76**(1):403–416 (2008).
30. Hazen TC, Dubinsky EA, DeSantis TZ, Andersen GL, Piceno YM, Singh N, Jansson JK, Probst A, Borglin SE, Fortney JL, et al. Deep-sea oil plume enriches indigenous oil-degrading bacteria. *Science* **330**(6001):204–208 (2010).

4

MATHEMATICAL APPROACHES FOR DESCRIBING MICROBIAL POPULATIONS: PRACTICE AND THEORY FOR EXTRAPOLATION OF RICH ENVIRONMENTS

MANUEL E. LLADSER

Department of Applied Mathematics, University of Colorado, Boulder, Colorado

ROB KNIGHT

Howard Hughes Medical Institute and Department of Chemistry and Biochemistry, University of Colorado, Boulder, Colorado

4.1. INTRODUCTION: PRACTICE FOR EXTRAPOLATION OF ENVIRONMENTS

Microbial community analysis is undergoing a revolution, as increasingly powerful sequencing methods provide information about millions of sequences in hundreds of communities simultaneously [1–4]. Briefly, the 16*S* rRNA gene is found in all autonomously replicating organisms (which excludes, e.g., viruses), and provides an excellent phylogenetic marker for many different kinds of organisms, allowing relationships to be found among both very distant organisms such as *E. coli* and humans (using slowly evolving parts of the molecule) and very similar organisms such as *E. coli* and *Salmonella* (using rapidly evolving parts of the molecule) [5]. We can thus use a technique called the *polymerase chain reaction* (PCR) to make many copies of this gene from all members of a microbial community, using as handles regions of the DNA that are common to all organisms [6]. Although there are certain biases in DNA extraction and the quality of the match of these handles to the DNA sequence in each individual organism [7], similarities and differences among communities can be effectively revealed using these techniques [8].

The Human Microbiota: How Microbial Communities Affect Health and Disease, First Edition. Edited by David N. Fredricks.

The key input into most mathematical techniques for analyzing communities, whether microbial or macroscopic, is the taxon table: a table in which the columns list different biological samples and the rows, different kinds of organisms. The key difficulty here is deciding which organisms should be grouped together into "kinds." In microbes, definitions of species are problematic because bacteria do not interbreed as part of their lifecycle and DNA can be exchanged even across very distantly related organisms, such as bacteria and archaea [9] and even bacteria and insects [10], so the typical approach is to group organisms into *operational taxonomic units* (OTUs) [11] based on sequence similarity. Even this definition is problematic; although it is typically true that, for example, rRNA sequences within the same well-defined species are ≥97% identical to one another on average across positions, those within the same well-defined genus are ≥95% identical, the converse is not necessarily true. An additional consideration is that the manner in which sequences are grouped together into clusters has a significant effect on both the diversity within each cluster and the number of clusters [12–14]. Sequencing error and chimeras (crossovers between sequences during the PCR step) can also strongly influence the estimated number of kinds of sequences in a given sample [15–18]. Some analyses, such as analyses of so-called α diversity (the number of kinds of organism in a single sample), are very sensitive to these effects, whereas other analyses, such as phylogenetic measures of β diversity (how differences in communities are distributed among environments), are relatively insensitive [19]. Consequently, it is important to consider carefully whether a given analysis will be sensitive or robust to the types of errors introduced by technical limitations, and therefore whether the result will be reproducible as techniques improve.

There are several practical considerations in building the taxonomy table, and several techniques have been successfully employed:

1. *How to Demultiplex the Data.* Typically, many samples are barcoded with a unique DNA tag added to the DNA primer, sequenced together, and then sequences must be assigned to individual samples on the basis of this tag (see Figure 4.1). Formal error-correcting codes are often useful for this step, although other strategies have also been employed with some success [20].
2. *Quality Filtering.* In pyrosequencing, long reads are often error-prone, and errors tend to be clustered such that sequences are more often perfect or more often highly problematic than would be expected from for instance the Poisson distribution. Typically, some combination of filtering based on length of the sequence, errors in the PCR primer sequence (which should match perfectly) or barcode (which should match perfectly to one of the known choices), and quality scores both over the length of the full sequence and in any individual consecutive window within the sequence is employed [21]. The second consideration is whether to relate sequences to an existing taxonomy or whether to pick OTUs de novo. The advantage to taxonomic clustering is that the sequences so clustered have a known identity and can be related to existing knowledge about their relatives. The disadvantage, however, is that most environments are (at the time of writing) undersampled, and so many newly discovered types of microbes will not be assignable to an existing taxon. The considerations of OTU-based clustering are the reverse; additionally, many methods for OTU assignment are stochastic, hindering integration

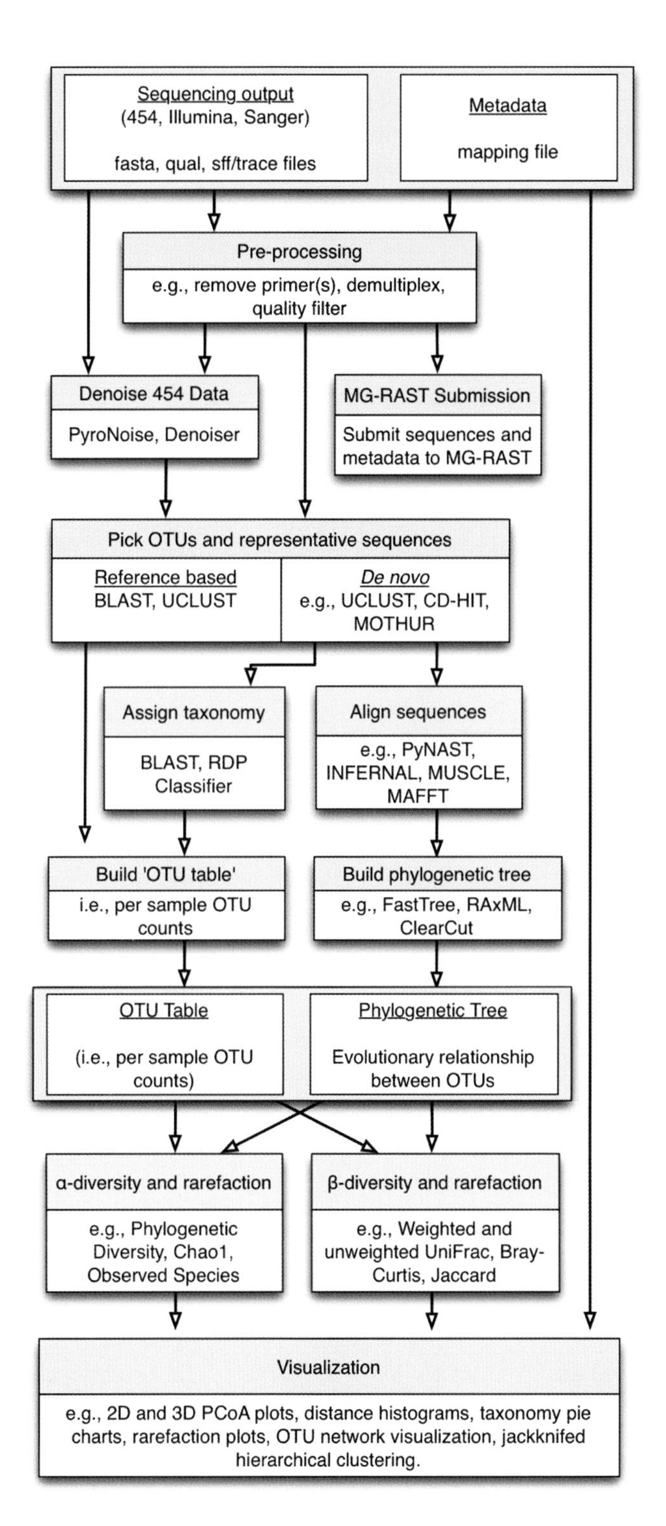

Figure 4.1. Analysis of metagenomic data: pipeline to build an OTU table and associated statistical analysis.

of OTU results across different studies (or even samples within a study). In practice, both kinds of method are able to give useful information about the dominant kinds of microbes in most well-sampled environments, and both have difficulty with environments in which few close relatives of the dominant organisms are represented in existing databases. Popular tools for taxonomy assignment include the RDP classifier; popular tools for OTU clustering of pyrosequencing-scale data include cd-hit [22], uclust [23], slp [13], and mothur [24].

3. *The Level at Which to Cluster the OTUs.* Typically, 97% is considered to be about the species level, 95% about the genus level, and then higher-level designations tend to be more problematic because of heterogeneous rates of sequence evolution in different lineages. These thresholds were trained long ago on databases containing far less sequence information than is available at present, and will likely be reevaluated in the near future. Higher levels of sequence similarity produce more distinct clusters, but also run a higher risk of misclassification due to sequencing error, heterogeneous sequence lengths, or other issues. In practice, it is useful to precluster near-identical sequences (including sequences that are exact prefixes of other sequences) to reduce the size of the dataset. Sequences can either be compared aligned or unaligned: an alignment is ultimately essential if a phylogenetic tree is required, e.g. to calculate phylogenetic diversity [25] or metrics such as UniFrac [26,27]. However, hierarchical alignment methods such as clustal or muscle produce very poor results [28]—methods seeded on an existing alignment, such as NAST [29,30] or infernal [31] are essential for reasonable alpha-diversity results.
4. *How to Assign the Best Taxonomy to the OTUs Once Chosen.* Either the best match to the reference sequence (which might be the longest sequence, the most abundant sequence, the centroid of the cluster, etc.) can be found, or the sequence can be classified according to a statistical procedure such as naive Bayes inference on the frequencies of short words within the sequence, or the consensus of the best n matches to all the sequences in the cluster can be computed and a consensus determined. Taxonomy assignment of some kind is typically crucial for understanding the results, as variation in abundance of an unknown sequence is typically unsatisfying as an explanation of an ecological trend or disease state.

Pipelines such as QIIME [32] and mothur [24] can greatly facilitate this process by providing a single package that combines all these steps, proceeding from files containing the raw sequences and quality scores to a table showing the abundance of each cluster of sequences in each sample.

4.2. THEORY FOR EXTRAPOLATION OF ENVIRONMENTS

4.2.1. Urn Model

One may conceptualize samples from a microbial environment as draws of colored balls from an urn. The urn represents the environment where samples are collected (e.g., the gut of a person), each ball represents a microorganism in that environment, and the different colors represent different OTUs. In this framework, the goal is to use a sample from the urn to *extrapolate* its composition, that is, learn about the colors that compose the urn and the proportions in which they occur in it.

In what follows, draws from the urn are always assumed *with replacement*, where a ball is drawn uniformly at random and, once its color is recorded, it is returned back to the urn. This assumption is reasonable in rich environments, that is, those with a very large number of microorganisms. We use n to denote the *sample size*, that is, the number of draws from the urn.

All the relevant information about the urn contained in the sample is given by the statistics $N(k, n)$, denoting *the number of colors observed k times in a sample of size n*. The logic behind this is that different colors carry no intrinsic meaning besides being different under the urn model. In fact, any particular permutation of the sample is as likely as any other. Thus, order may be conveyed into the sample by applying a random permutation to it. An efficient way to do this is to apply the discrete inverse transform method to draw samples *without replacement* from a finite population [33]. In what follows, $X_1, \ldots, X_n$ denotes the sequence of colors observed when a sample of size n is permuted at random; that is, X_1 is the first color in the permutation, X_2 the second, and so on.

4.2.2. Paradigm Shift

In ecological studies, a traditional approach to characterizing an environment is to estimate the total number of different species present in that environment, namely, its *α diversity* [34]. In microbial communities, however, the abundance of rare species in most actual biological samples suggests that the accurate estimation of α diversity is a very difficult problem.

In terms of our urn model, the α diversity is simply the number of different colors that are found in the urn. Although the estimation of this parameter is an interesting mathematical problem (see Refs. 35, 36, and references cited therein), its utility in microbial studies is less clear [37,38], and perhaps a more useful quantity to target is the *fraction m of balls in the urn with a color represented in the sample*—called the *coverage of the sample* in the literature.

Note that $0 \leq m \leq 1$. The coverage of the sample is the probability that a new observation from the urn is of a color already seen in the sample; in particular, $(1 - m)$ is the probability of discovering a completely new color with one additional draw from the urn. The additional number of samples to observe a new color therefore has a *geometric distribution* [33]. In other words, the probability that exactly k additional samples from the urn are needed to observe a new color is $m^{k-1}(1 - m)$, for all $k \geq 1$. In average, for a same value of m, one would therefore require $1/(1 - m)$ additional draws to observe a color unrepresented in the sample $X_1, \ldots, X_n$.

A sample with low coverage (i.e., in which m is close to zero) has only captured colors that as a whole and hence individually are not dominant in the urn. In contrast, most balls in the urn are colored with a color represented a sample with high coverage (i.e., in which m is close to 1). The coverage of the sample thus indicates its depth; however, it is not informative of the α diversity of the urn because a small fraction of the urn could contain a single or several 100s of different colors.

The coverage of the sample is also important for estimating the proportion of balls in the urn with each of the colors represented in the sample. One might imagine that estimating these proportions is simply a matter of estimating the parameters of a *multinomial distribution* [33], however, this procedure is incorrect in our setting [39]. To clarify ideas, suppose that in a sample of size 10 from an urn, we observe 3 balls colored red and 7 colored blue. If we knew that the urn was composed of just these two colors then our best guess of the number of balls colored red in the urn would be 30%. However, imagine that we knew the urn was actually composed of more than these two colors and in fact, 94% of the urn was colored red or blue. Then 30% would be our best guess of the proportion of red balls relative to the proportion in the urn of red and blue balls. Because only 94% of the urn is composed of these two colors, we should guess that 30% × 94%, that is, that 28.2% of the urn was colored red.

Relative proportions in a sample estimate relative and not absolute proportions in the urn: *if a certain color i occurs in proportion f_i in the sample, then we should estimate its* true *proportion in the urn as $f_i m$, where m is the coverage probability of the sample*. Back in the previous example, $f_{\text{red}} = 0.3$. If the urn only contains red and blue balls, then $m = 1$, and our best guess of the proportion of red balls in the urn would be $f_{\text{red}} \times 1$ (i.e., 30%). However, if red and blue represented only 94% of the urn (i.e., $m = 0.94$), then our best guess of the proportion of red balls in the urn would be instead $f_{\text{red}} \times 0.94$ (i.e., 28.2%).

Of course, to make practical use of these considerations, we need to predict the coverage of the sample! This is the central topic of Sections 4.3 and 4.4. In Section 4.2.3 we revisit the concept of coverage of a sample, and show how to compute its expected value.

To finish this section, we would like to emphasize the difference between *estimation* versus *prediction*. *Estimation* refers to the attempt of guessing a parameter of the urn that does not depend on any sample. For instance, one may try to estimate the α diversity of the urn. On the other hand, *prediction* refers to the attempt of guessing a random quantity related to the urn and possibly a sample from it. For example, one may predict the coverage of a sample. Note that this is in general a random quantity because it depends on the colors observed in the sample, which are random unless the urn is composed by balls of identical colors.

4.2.3. Expected Coverage of a Sample

For each color i, let p_i denote the proportion of balls in the urn of that color. Thus, for example, if no ball in the urn is colored red, then $p_{\text{red}} = 0$; however, if a quarter of the balls in the urn are colored blue, then $p_{\text{blue}} = 0.25$.

The *coverage of a sample* (of size n) corresponds to the quantity.

$$m = \sum_{i \in \{X_1, \ldots, X_n\}} p_i, \tag{4.1}$$

where $X_1, \ldots, X_n$ is the sequence of colors observed when drawing n balls with replacement from the urn. The index i above belongs to the random set $\{X_1, \ldots, X_n\}$; in particular, m is a random quantity. (The only exception to this is when the urn consists only of balls of identical colors, in which case $m = 1$ regardless of the sample size.) To fix ideas, if a sample consists only of the colors red, white, and blue, then $m = p_{\text{red}} + p_{\text{white}} + p_{\text{blue}}$. Since a set is a list of objects where repetitions and order are irrelevant, $m = p_{\text{red}} + p_{\text{white}} + p_{\text{blue}}$ regardless of the number of times each of these colors was observed in the sample and the particular order in which they were discovered.

In what follows, M denotes the *expected value* of m. We refer to this quantity as the *expected coverage of a sample* (of size n). To determine this quantity, one needs to know the distribution of all the colors in the urn. In fact, if I denotes an arbitrary set of at most n different colors, then

$$M = \sum_I p_I \cdot \mathbb{P}[\{X_1, \dots, X_n\} = I], \tag{4.2}$$

where

$$p_I = \sum_{i \in I} p_i, \tag{4.3}$$

and $\mathbb{P}[\{X_1, \dots, X_n\} = I]$ is the probability that all and only the colors belonging to I occur in the sample.

Example 4.1. Let us compute the expected coverage of $n \geq 2$ samples in an urn containing only red and blue balls occurring in proportions p and $(1 - p)$, respectively.

In this case, the index I in Equation (4.2) may take only the values {red}, {blue}, and {red, blue}. Since the probability that all the n draws from the urn are red is p^n, the probability that $\{X_1, \dots, X_n\}$ = {red} is p^n. In symbols, this means that:

$$\mathbb{P}[\{X_1, \dots, X_n\} = \{\text{red}\}] = p^n.$$

Similarly, $\mathbb{P}[\{X_1, \dots, X_n\} = \{\text{blue}\}] = (1-p)^n$.

On the other hand, to obtain $\{X_1, \dots, X_n\}$ = {red, blue}, we need at least one ball of each color in the sample. Since the probability that all balls are of the same color is $p^n + (1 - p)^n$, it follows that $\mathbb{P}[\{X_1, \dots, X_n\} = \{\text{red, blue}\}] = 1 - p^n - (1-p)$. According to Equations (4.2) and (4.3), for this particular urn, the expected coverage of a sample of size n is

$$\begin{aligned} M &= p \cdot p^n + (1-p) \cdot (1-p)^n + 1 \cdot \left(1 - p^n - (1-p)^n\right), \\ &= 1 - p(1-p)^n - (1-p)p^n. \end{aligned}$$

Following an argument by Robbins [40], we may compute the expected coverage of a sample from any urn more directly as follows. For each color i, consider the *indicator function* U_i of the event "color i was not observed in the sample." This means that U_i is a random variable that takes the value 1 when the event under consideration occurs and the value 0 otherwise. More precisely, $U_i = 1$ when $i \notin \{X_1, \dots, X_n\}$, and $U_i = 0$ when $i \in \{X_1, \dots, X_n\}$. Since $\sum_i p_i \cdot U_i$ is the fraction of balls in the urn with colors not represented in the sample, the coverage of the sample is $1 - \sum_i p_i \cdot U_i$. Since the expected value of U_i is $(1 - p_i)^n$, we find that

$$M = 1 - \sum_i p_i (1 - p_i)^n, \text{ equivalently: } (1 - M) = \sum_i p_i (1 - p_i)^n. \tag{4.4}$$

The calculation in Example 4.1 can now be obtained directly from the identity given above. Here is another example.

Example 4.2. Let us compute the expected coverage of a sample of size n in an urn consisting of k different colors occurring in equal proportions. Without loss of generality, we assume that colors are integers between 1 and k; in particular, $p_1 = \cdots = p_k = 1/k$. Using formula (4.4), we find for this urn that

$$M = 1 - \sum_{i=1}^{k} p_i (1-p_i)^n = 1 - \sum_{i=1}^{k} \frac{1}{k}\left(1-\frac{1}{k}\right)^n = 1 - \left(1-\frac{1}{k}\right)^n.$$

4.3. AVERAGE ANALYSIS OF SAMPLE COVERAGE

In this section, we review various methods found in the literature to estimate the expected coverage M of a sample of size n. We aim to estimate M only using the information contained in the sample, without any prior knowledge about the colors that compose the urn or the proportions in which they occur. Note that M is "estimated" because it is a parameter associated with the urn that does not depend on any sample from the latter. In fact, as seen in Equation (4.4), M is determined by the exact proportion of each color in the urn—regardless of the colors we observe in any sample from it.

Robbins [40] proposed that $(1 - M)$ could be estimated by drawing one additional ball from the urn, that is, enlarging the sample size from n to $(n + 1)$. His approach is based on the identity in (4.4), which we may alternatively rewrite as

$$\mathbb{E}(1-m) = \sum_i (1-p_i)^n p_i, \tag{4.5}$$

where $\mathbb{E}$ denotes the *expectation* of the random variable within.

Robbins' key idea is to determine (observable) random variables with expected value $(1 - p_i)^n p_i$, which, when combined linearly, produce a random variable with the same expected value as $(1 - m)$. The need to enlarge the sample is implied by the fact that the probability $(1 - p_i)^n p_i$ consists of $(n + 1)$ factors. Here are the details of his calculation.

First notice that $(1 - p_i)^n p_i$ is the probability that *color i is observed for the first time in the last of $(n + 1)$ draws from the urn*. However, it is also the probability that *color i is observed only on the first of $(n + 1)$ draws from the urn*. More generally, for each $1 \leq k \leq (n + 1)$, $(1 - p_i)^n p_i$ is the probability that *color i is observed only in the kth of $(n + 1)$ draws from the urn*. The probability that color i is observed once in $(n + 1)$ draws from the urn is therefore $(n + 1) \cdot (1 - p_i)^n p_i$. So, if we now define U_i as the indicator of the event "color i is observed once in $(n + 1)$ draws from the urn", then

$$\mathbb{E}\left(\sum_i U_i\right) = \sum_i \mathbb{E}(U_i) = \sum_i (n+1)(1-p_i)^n p_i = (n+1)\sum_i (1-p_i)^n p_i.$$

Alternatively

$$\mathbb{E}\left(\frac{1}{n+1}\sum_i U_i\right) = \sum_i (1-p_i)^n p_i.$$

Hence, using Equation (4.5), we finally obtain the following:

$$\mathbb{E}\left(\frac{1}{n+1}\sum_i U_i\right) = \mathbb{E}(1-m).$$

We have therefore determined a random variable, namely, $\Sigma_i U_i/(n+1)$, that has the same expected value as (1 – m). But there is an important difference. To compute m and hence (1 – m) also, we would need to know in advance the proportions p_i for each color i in the sample. Hence (1 – m) cannot be in general determined unless the urn's composition is known. In contrast, $\Sigma_i U_i/(n+1)$ can be determined directly from the sample without prior knowledge of the urn composition. In fact, note that $\Sigma_i U_i$ is the *number of colors observed exactly once in* (n + 1) *draws from the urn*, that is, $\Sigma_i U_i = N(1, n+1)$, according to the notation introduced in Section 4.2.1.

Robbins' statistic [40] is the random variable:

$$V_1 = \frac{N(1, n+1)}{n+1}. \tag{4.6}$$

In view of the previous discussion, $\mathbb{E}(V_1) = \mathbb{E}(1-m)$; however, and unlike (1 – m), V_1 can be computed directly from the data without prior knowledge of the urn's composition. Because $\mathbb{E}(V_1) = (1-M)$, V_1 is what is called an *unbiased estimator* for (1 – M).

EXAMPLE 4.3. Suppose that in a sample of size 10 from a certain urn we observe the colors and frequencies displayed in Table 4.1; in particular, the coverage probability of the sample is $m = p_{\text{blue}} + p_{\text{brown}} + p_{\text{green}} + p_{\text{red}} + p_{\text{white}} + p_{\text{yellow}}$.

Robbins' estimator of (1 – M) requires to sample one additional ball from the urn. If that ball turned out to be green, then N (1, 11) = 3 as only the colors blue, red, and yellow would have been seen once in the enlarged sample of size 11. In this case, $V_1 = 3/(10+1) = \frac{3}{11}$. Hence, our best guess for M would be $\frac{8}{11}$.

Instead, if the additional sample had been of color white, then $N(1, 11) = 4$, $V_1 = \frac{4}{11}$ and we would estimate M as $\frac{7}{11}$. Finally, if the extra ball had been colored black, then $N(1, 11) = 5$, $V_1 = \frac{5}{11}$, and our estimate of M would be $\frac{6}{11}$.

Robbins' estimator is closely related to the *Good–Turing* estimator [41]:

$$V_0 = \frac{N(1, n)}{n}. \tag{4.7}$$

Thus, for instance, the Good–Turing estimation of the expected coverage probability of the sample summarized in Table 4.1 would be $1 - \frac{4}{10} = \frac{6}{10}$.

TABLE 4.1. Urn Data: Summary of Colors Observed in a Sample of Size 10 from a Certain Urn, to Estimate Expected Coverage Probability of a Sample of the Same Size Using Robbins' Statistic

Color	Blue	Brown	Green	Red	White	Yellow
Frequency	1	2	1	1	4	1

Good attributes the form of the estimator V_0 of $(1 - M)$ to Turing, who derived it by intuition. Unlike Robbins' estimator, the Good–Turing estimator does not require any additional samples from the urn. However, it is in general biased for $(1 - M)$. This is because V_0 corresponds to Robbins' estimator when the original sample size was $(n - 1)$ instead of n. In fact, from Equation (4.4), we obtain

$$\mathbb{E}(V_0)-(1-M)=\sum_i(1-p_i)^{n-1}p_i-\sum_i(1-p_i)^n p_i=\sum_i(1-p_i)^{n-1}p_i^2.$$

In other words, $\mathbb{E}(V_0)>(1-M)$; that is, V_0 overestimates (on average) the true value of $(1 - M)$. (The only exception to this is when the urn consists of balls of identical colors, a case of little interest that is ruled out as soon as two different colors are observed in a sample.) However, according to the last equation above, this bias will usually be negligible when n is large. In fact, the following formula applies:

$$-\frac{1}{n}\leq(V_0-V_1)\leq\frac{2}{n}.$$

In large sample sizes, the Robbins and Good–Turing estimators are therefore indistinguishable on a linear scale.

We have seen this far that $(1 - M)$ can be estimated unbiasedly when the sample can be enlarged with one additional observation from the urn. It is conceivable, however, that we may improve the estimation of this parameter with additional samples from the urn. This approach was considered by Starr [42], who showed that the statistic:

$$V_r=\sum_{k=1}^{r}\frac{\binom{r-1}{k-1}}{\binom{n+r}{k}}\cdot N(k,n+r), \tag{4.8}$$

is the only linear combination (with positive coefficients) of the observable random variables $N(1, n + r), \ldots, N(r, n + r)$ that is unbiased form $(1 - M)$. [Recall that $N(k, n)$ is the number of colors observed k times in a sample of size n.] Note that Starr's estimator is identical to Robbins' when $r = 1$. Furthermore, Starr's methodology estimates the coverage probability of the sample $X_1, \ldots, X_n$ using information from $X_1, \ldots, X_n, X_{n+1}, \ldots, X_{n+r}$, where $X_{n+1}, \ldots, X_{n+r}$ is the sequence of colors observed in r additional draws from the urn.

Example 4.4. Suppose that we want to estimate, using Starr's estimator with $r = 5$, the expected coverage of a sample of size 10 from the urn in Example 4.3. To do so, we would need to sample five additional balls with replacement from it. To fix our ideas, imagine that we observe one black, one brown, one green, and two red balls in the next five draws from the urn. The enlarged sample would then be summarized by the frequency statistics in Table 4.2.

According to this table, $N(1, 15) = 3$, $N(2, 15) = 1$, $N(3, 15) = 2$, $N(4, 15) = 1$, and $N(5, 15) = 0$. As a result, returning to Equation (4.8), we find that

TABLE 4.2. Enlarged Urn Data: Summary of Colors Observed in an Enlarged Sample of Size 15 from the Urn in Example 4.3, to Estimate Expected Coverage Probability of a Sample of Size 10 Using Starr's Statistic

Color	Black	Blue	Brown	Green	Red	White	Yellow
Frequency	1	1	3	2	3	4	1

$$V_5 = \frac{\binom{4}{0}}{\binom{15}{1}} \cdot N(1,15) + \frac{\binom{4}{1}}{\binom{15}{2}} \cdot N(2,15) + \frac{\binom{4}{2}}{\binom{15}{3}} \cdot N(3,15)$$
$$+ \frac{\binom{4}{3}}{\binom{15}{4}} \cdot N(4,15) + \frac{\binom{4}{4}}{\binom{15}{5}} \cdot N(5,15),$$
$$= \frac{1}{15} \cdot 3 + \frac{4}{105} \cdot 1 + \frac{6}{455} \cdot 2 +, \frac{4}{1365} \cdot 1 + \frac{1}{3003} \cdot 0 = \frac{73}{273}.$$

Note that this is an estimate of $(1 - M)$. In particular, Starr's estimator of the expected coverage probability of a sample of size 10, based on 5 additional samples from the urn, is $(1 - 73/273)$ (i.e., ~73.3%).

We note that the Starr and Robbins estimators are indistinguishable on a linear scale when the parameter r is selected much smaller than n. In fact, provided that the sample of size n is enlarged by $r \geq 1$ additional draws from the urn, the following equation applies [43]:

$$|V_r - V_1| \leq \frac{3(r-1)}{n+1}. \tag{4.9}$$

The fact that Starr's estimator is unbiased for $(1 - M)$ follows more directly from the analysis by Clayton and Frees [44], who showed that V_r is in fact the *minimum variance unbiased estimator* of $(1 - M)$ based on r additional observations from the urn.

The end of this section is devoted to showing that V_r is unbiased for $(1 - M)$, specifically, $\mathbb{E}(V_r) = (1 - M)$. For this notice that a sample of size $(n + r)$ contains a total of $\binom{n+r}{n+1}$ possible subsamples of size $(n + 1)$. If in each of these we estimated $(1 - M)$ using Robbins' statistics, then, to better guess $(1 - M)$, we should average the estimates over all the subsamples. To express this more mathematically, let s denote an arbitrary subsample of size $(n + 1)$ from $X_1, \ldots, X_{n+r}$. If $V_1(s)$ denotes Robbins' estimator of $(1 - M)$ based on s, then the average over all the possible subsamples is

$$\frac{1}{\binom{n+r}{n+1}} \sum_s V_1(s).$$

Extending the notation further, let $N(1, s)$ be the number of colors observed once in the subsample s. In particular, if $U_i(s)$ is the indicator of the event "color i is seen once in s" then $N(1,s) = \sum_i U_i(s)$. Furthermore, if $U_{i,k}(s)$ is the indicator of the event "color i is seen once in s and a total of k times in the enlarged sample of size $(n + r)$," then we also have that $U_i(s) = \sum_{k=1}^{n+r} U_{i,k}(s)$. But recall that s represents a generic sample of size $(n + 1)$. In particular, if color i was seen once in s, then it cannot be seen more than r times in the enlarged sample; that is, $U_{i,k}(s) = 0$ when $k > r$.

Following the considerations, mentioned above, Equation (4.6) yields.

$$\begin{aligned}
\frac{1}{\binom{n+r}{n+1}}\sum_s V_1(s) &= \frac{1}{(n+1)\cdot\binom{n+r}{n+1}}\sum_s N(1, s),\\
&= \frac{1}{(n+1)\cdot\binom{n+r}{n+1}}\sum_i\sum_s U_i(s),\\
&= \frac{1}{(n+1)\cdot\binom{n+r}{n+1}}\sum_{k=1}^{r}\sum_i\sum_s U_{i,k}(s).
\end{aligned}$$

The punchline of the calculation is to note that the summation $\sum_s U_{i,k}(s)$ depends only on the number of times that color i was observed in the enlarged sample. In fact, if this number was l, then $\sum_s U_{i,k}(s) = 0$ when $k \neq l$. On the other hand, if $k = l$, then $\sum_s U_{i,k}(s)$ is the number of subsamples of size $(n + 1)$ that contain color i exactly once. Since color i occurs k-times in the enlarged subsample, this number is found to be $k\cdot\binom{n+r-k}{n}$. Finally, since there are $N(k, n + r)$ colors observed k times in the enlarged sample, we conclude that

$$\begin{aligned}
\frac{1}{\binom{n+r}{n+1}}\sum_s V_1(s) &= \frac{1}{(n+1)\cdot\binom{n+r}{n+1}}\sum_{k=1}^{r} k\cdot\binom{n+r-k}{n}\cdot N(k, n+r),\\
&= \sum_{k=1}^{r}\frac{k\cdot\binom{n+r-k}{n}}{(n+1)\cdot\binom{n+r}{n+1}}\cdot N(k, n+r),\\
&= \sum_{k=1}^{r}\frac{\binom{r-1}{k-1}}{\binom{n+r}{k}}\cdot N(k, n+r) = V_r.
\end{aligned}$$

This calculation shows that

$$V_r = \frac{1}{\binom{n+r}{n+1}}\sum_s V_1(s). \tag{4.10}$$

Since $\mathbb{E}(V_1(s)) = (1-M)$, for each subsample s in the summation, and there are a total of $\binom{n+r}{n+1}$ such samples, it follows from the identity calculations above that $\mathbb{E}(V_r) = (1-M)$ as claimed.

4.4. CONDITIONAL ANALYSIS OF SAMPLE COVERAGE

The estimators V_r, with $r \geq 1$, discussed in the previous section are unbiased for the expected coverage of a sample of size n from an urn: $\mathbb{E}(V_r) = \mathbb{E}(1-m)$. In other words, the random variable V_r has the same expected value as $(1 - m)$. Although, in general, this does not enable us to conclude that V_r is a good approximation of $(1 - m)$, Robbins showed that [40]

$$\mathbb{E}\left[(V_1-(1-m))^2\right] \leq \frac{1}{n+1}; \tag{4.11}$$

in other words, the *average quadratic variation distance* between V_1 and $(1 - m)$ is small when n is large. This may be interpreted as follows.

Suppose that we repeat a *very large number* N of times the experiment of drawing n balls with replacement from the urn. Let m_j be the coverage probability of the sample associated with the jth experiment and $V_{1,j}$ be Robbins' statistic associated with that sample. According to the *law of large numbers*, the preceding inequality translates into saying that

$$\frac{1}{N}\sum_{j=1}^{N}(V_{1,j}-(1-m_j))^2 \lesssim \frac{1}{n+1},$$

where $\lesssim$ denotes *approximately less than or equal to*. Clearly this is possible only when the majority of the random variables $V_{1,j}$ and $(1 - m_j)$ are similar to one another. In fact, in the context of Equation (4.9), the same considerations should apply when one replaces $V_{1,j}$ by $V_{r,j}$, Starr's estimator based on r additional draws from the urn in each of the N experiments.

Although the above findings are encouraging for predicting the coverage probability of a sample using Robbins or Starr statistics, Starr showed in [42] that V_1 may be negatively correlated with $(1 - m)$. In particular, because of Equation (4.11), the same holds between V_r and $(1 - m)$ when r is much smaller than the same size [43]. Fortunately, there is a way to construct predictors of $(1 - m)$ that are positively correlated with this quantity based on the recent analysis in [43]. To explain this procedure we review briefly the concept of homogenous Poisson point process [45].

4.4.1. Thinning Property of Poisson Point Processes

A *homogeneous Poisson (point) process* (HPP) with intensity λ on the interval $[0, +\infty)$ is any sequence of random variables of the form E_1, $(E_1 + E_2)$, $(E_1 + E_2 + E_3)$, and so on, where $E_1, E_2, E_3, \ldots$ are independent *exponential random variables* with mean $1/\lambda$ [33]. We usually visualize such random process as a sequence of points in

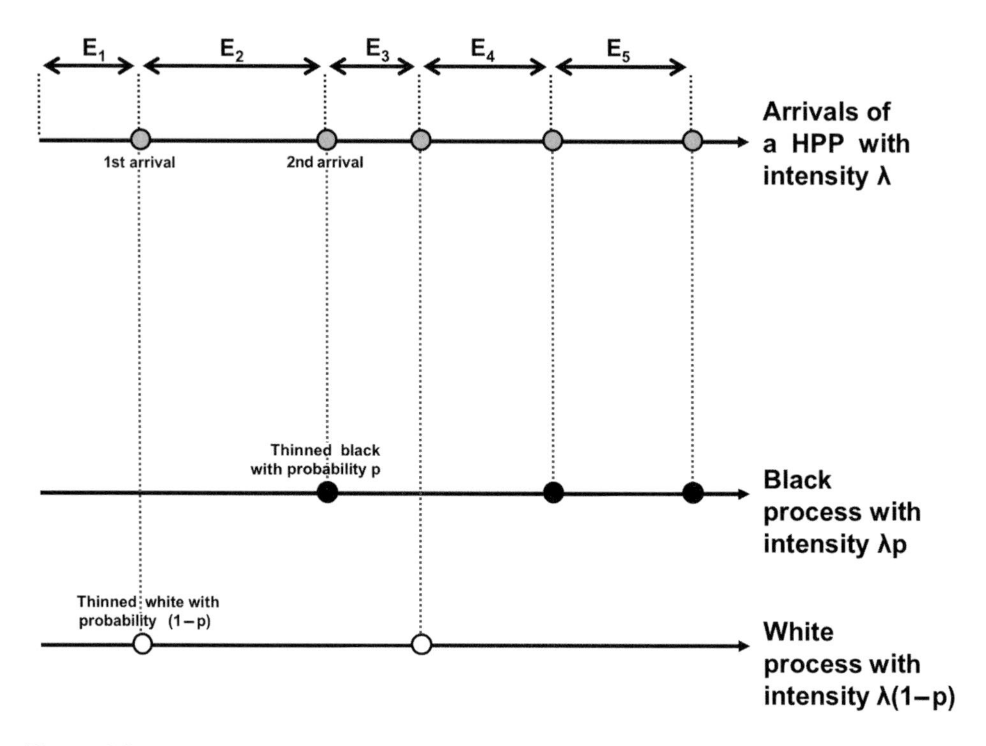

Figure 4.2. Thinning of a homogeneous Poisson process with intensity λ into two independent HPPs of intensities λp and $\lambda(1-p)$, respectively.

the semiinfinite interval $[0, +\infty)$ (see Figure 4.2). We refer to the point closest to the origin as the *first arrival*, the second closest as the *second arrival*, and so on. Note that the average separation between consecutive points is $1/\lambda$; in particular, the average number of points per unit of length is λ. This is the rationale for referring to this parameter as the *intensity* of the process, and the term *homogenous* is used to emphasize that the intensity is constant.

Now consider a number $0 < p < 1$, and imagine that each arrival of a HPP is colored *black* with probability p and *white* with probability $(1 - p)$, independently of all other arrivals. *What's the average separation between black arrivals?* Since each arrival after a black one is black or white with probability p and $(1 - p)$, respectively, the number of draws from the urn between consecutive black arrivals has a geometric distribution. Furthermore, because the expected value of this distribution is $1/p$, and the average separation between consecutive arrivals (of either color) is $1/\lambda$, the average separation between black arrivals is $1/(\lambda p)$. As a result, there are in average λp black arrivals per unit of length i.e. the intensity of the black arrivals is λp.

It turns out that the black-arrivals also conform a HPP just with intensity λp [45]. This process of keeping certain arrivals of a HPP (the black ones in this case) and disregarding others (the white ones) is called *thinning*. For the same reasons, it can be shown that the white-arrivals conform also a HPP but with intensity $\lambda \cdot (1 - p)$. Furthermore, the black and white processes are statistically independent from one another [45].

4.4.2. Embedding Algorithm

Suppose that we would like to predict the proportion of balls in the urn with a color unrepresented in the sample $X_1, \ldots, X_n$. To do so, imagine simulating a HPP with intensity one and coloring the arrivals of the process according to the following scheme: color the ith arrival as *black* if the ith additional draw from the urn is in the set $\{X_1, \ldots, X_n\}$, and otherwise color it *white*. Since the probability of a black arrival corresponds to the coverage probability of the sample $X_1, \ldots, X_n$, the thinning property of HPPs let us conclude that the white process is a HPP with intensity $(1 - m)$. In particular, the average separation between white arrivals is $1/(1 - m)$, which gives a geometric interpretation to the probability of interest.

To turn the calculations and assumptions given above into a method for predicting $(1 - m)$, we select a parameter $r \geq 1$ and proceed to draw balls from the urn until r balls with colors outside the set $\{X_1, \ldots, X_n\}$ are observed. If we used the colors of these balls to thin a HPP with intensity 1 into black and white arrivals as we just described, it would follow that the separation between consecutive white arrivals are independent exponentials with mean $1/(1 - m)$ and, as a result, the distance T_r between the origin and the rth white arrival would have a *γ distribution with shape parameter r and scale parameter* $1/(1 - m)$ [33]. In particular, $(1 - m) \cdot T_r$ has a γ distribution with shape parameter r and scale parameter 1. Since this distribution does not depend on the unknown coverage probability of the sample, we may construct point predictors and prediction intervals for $(1 - m)$ based on the observable random variable T_r. Specifically, the following conditions apply [43]:

1. For each $r \geq 3$, $(r - 1)/T_r$ is conditionally unbiased for $(1 - m)$, with standard deviation $(1-m)/\sqrt{r-2}$.
2. Furthermore, $(r - 1)/T_r$ and $(1 - m)$ are positively correlated.
3. If $0 < \alpha < 1$ and $a < b$ are nonnegative constants such that $a^{r-1}e^{-a} = b^{r-1}e^{-b}$ and $\int_a^b [x^{r-1}/(r-1)!]e^{-x}dx = (1-\alpha)$, then the interval $[a/T_r,\ b/T_r]$ contains $(1 - m)$ with probability $(1 - \alpha)$.

It is immediately evident from condition 1 that the prediction of $(1 - m)$ based on T_r is more robust, the larger the parameter r is. Notice, however, that the calculation of T_r requires a random rather than deterministic number of additional draws from the urn until observing r colors outside the sample $\{X_1, \ldots, X_n\}$. In particular, the closer m is to 1, the more balls will need to be drawn from the urn to compute the random variable T_r. Since in practice all samples are finite, this methodology yields predictions only on the coverage probability of a subsample of the original sample.

Figure 4.3 shows various outputs of the embedding procedure over a sample of size 12,903 from a human-gut [46]. On the basis of the embedding procedure, with parameter $r = 50$, it was estimated that 97 of the 123 species present in this sample represent ~99.4% of that gut environment; in particular, the remaining ~0.6% consists of at least 26 species. To test these predictions, Lladser et al. [43] simulated the rare biosphere as follows. For each $i = 0.06, 0.006, 0.0006$ and $j = 10, 100, 1000$, the plot in row i and column j in the figure is associated with a sample of size 50,000 from an urn constructed as a mixture of two urns: one with the same distribution as the microbes found in the gut dataset, and weighted by the

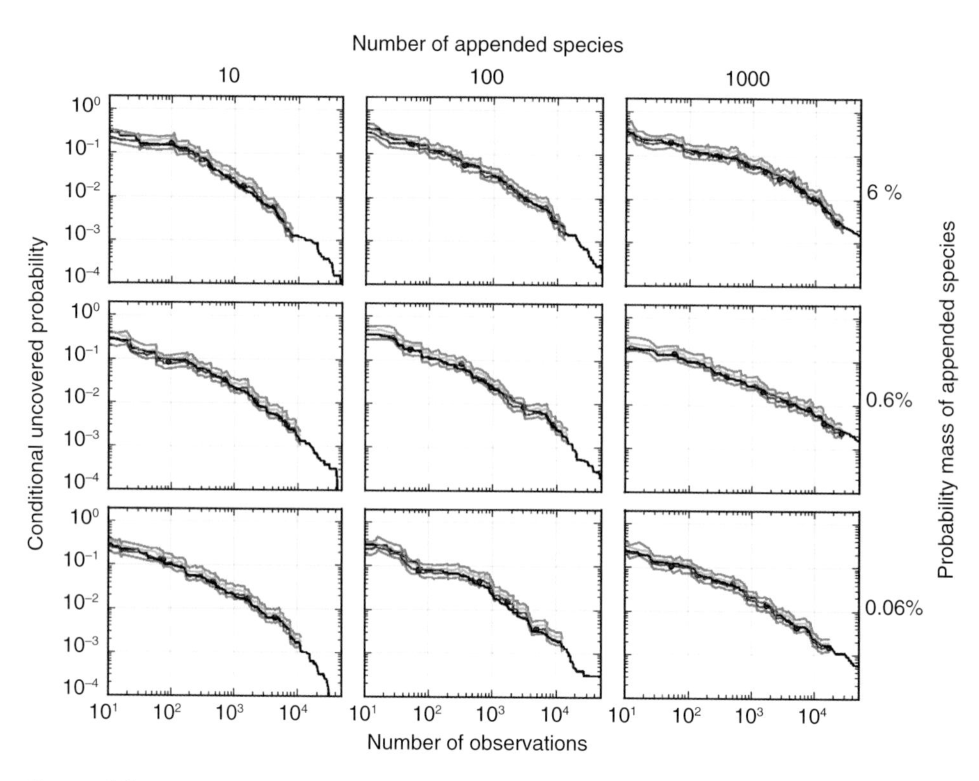

Figure 4.3. Predictions in the rare biosphere using the embedding algorithm of the discovery probability (black curve) as a function of the number of observations, in urns constructed from a human gut sample.

factor $(1 - i)$; and another one consisting of j new colors, with an exponentially decaying rank curve and weighted by the factor i. In the figure, the black curve corresponds to the sequential determination of $(1 - m)$ as a function of the number of draws from the urn. The blue points represent consecutive outputs of the embedding procedure. Instead, the red curve is associated with outputs of the embedding procedure each time a new species was discovered. The orange curve is a 95% upper prediction bound for the black curve, whereas the green curve denotes a 95% bandwidth for the black curve. Observe how in each plot all the predictions end before exhausting the sample; this is because the predictions cannot be continued after less than r (possibly repeated) new colors remain on the tail of the sample. Despite the fact that this methodology yields predictions of $(1 - m)$ only for the first part of the sample, its accuracy—when a prediction is produced—is remarkable for each of the nine scenarios considered. See [43] for comparisons with other prediction methods of the sample coverage of a sample.

4.5. CONCLUSIONS

Thus we have seen that improved mathematical methods can substantially improve our ability to measure the diversity of a complex biological sample, although these diversity estimates also depend sensitively on many steps in collecting and processing the data. In particular, sample coverage should be reported using point predictors such as those due to Lladser et al. [43], with associated upper and lower prediction bounds, especially when the statistical significance of differences in coverage between two samples is of interest, or when the rare biosphere is being examined. It is expected that further improvements in both sequencing technology and analysis techniques will allow us to relate the diversity of a wide range of biological samples to key ecological parameters and/or to human health and disease.

REFERENCES

1. Sogin ML, Morrison HG, Huber JA, Welch MD, Huse SM, Neal PR, Arrieta JM, Herndl GJ. Microbial diversity in the deep sea and the underexplored "rare biosphere." *Proc Natl Acad Sci USA* **103**(32):12115–12120 (2006).
2. Turnbaugh PJ, Hamady M, Yatsunenko T, Cantarel BL, Duncan A, Ley RE, Sogin ML, Jones WJ, Roe BA, Affourtit JP, et al. A core gut microbiome in obese and lean twins. *Nature* **457**(7228):480–484 (2009).
3. Costello EK, Lauber CL, Hamady M, Fierer N, Gordon JI, Knight R. Bacterial community variation in human body habitats across space and time. *Science* **326**(5960):1694–1697 (2009).
4. Caporaso JG, Lauber CL, Costello EK, Berg-Lyons D, Gonzalez A, Stombaugh J, Knights D, Gajer P, Ravel J, Fierer N, et al. Moving pictures of the human microbiome. *Genome Biol* **12**(5):R50 (2011).
5. Pace NR. A molecular view of microbial diversity and the biosphere. *Science* **276**(5313): 734–740 (1997).
6. Schmidt TM, DeLong EF, Pace NR. Analysis of a marine picoplankton community by 16*S* rRNA gene cloning and sequencing. *J Bacteriol* **173**(14):4371–4378 (1991).
7. Reysenbach AL, Giver LJ, Wickham GS, Pace NR. Differential amplification of rRNA genes by polymerase chain reaction. *Appl Environ Microbiol* **58**(10):3417–3418 (1992).
8. Kuczynski J, Costello EK, Nemergut DR, Zaneveld J, Lauber CL, Knights D, Koren O, Fierer N, Kelley ST, Ley RE, et al. Direct sequencing of the human microbiome readily reveals community differences. *Genome Biol* **11**(5):210 (2010).
9. Nelson KE, Clayton RA, Gill SR, Gwinn ML, Dodson RJ, Haft DH, Hickey EK, Peterson JD, Nelson WC, Ketchum KA, et al. Evidence for lateral gene transfer between archaea and bacteria from genome sequence of thermotoga maritima. *Nature* **399**(6734):323–329 (1999).
10. Dunning Hotopp JC, Clark ME, Oliveira DC, Foster JM, Fischer P, Muñoz Torres MC, Giebel JD, Kumar N, Ishmael N, Wang S, et al. Widespread lateral gene transfer from intracellular bacteria to multicellular eukaryotes. *Science* **317**(5845):1753–1756 (2007).
11. Sneath PH, Sokal RR. Numerical taxonomy. *Nature* **193**:855–860 (1962).
12. Schloss PD, Handelsman J. Introducing DOTUR, a computer program for defining operational taxonomic units and estimating species richness. *Appl Environ Microbiol* **71**(3): 1501–1506 (2005).
13. Huse SM, Welch DM, Morrison HG, Sogin ML. Ironing out the wrinkles in the rare biosphere through improved otu clustering. *Environ Microbiol* **12**(7):1889–1898 (2010).
14. Schloss PD. The effects of alignment quality, distance calculation method, sequence filtering, and region on the analysis of 16*S* rRNA gene-based studies. *PLoS Comput Biol* **6**(7):e1000844 (2010).

15. Quince C, Lanzén A, Curtis TP, Davenport RJ, Hall N, Head IM, Read LF, Sloan WT. Accurate determination of microbial diversity from 454 pyrosequencing data. *Nat Meth* **6**(9):639–641 (2009).
16. Kunin V, Engelbrektson A, Ochman H, Hugenholtz P. Wrinkles in the rare biosphere: Pyrose-quencing errors can lead to artificial inflation of diversity estimates. *Environ Microbiol* **12**(1):118–123 (2010).
17. Reeder J, Knight R. Rapidly denoising pyrosequencing amplicon reads by exploiting rank-abundance distributions. *Nat Meth* **7**(9):668–669 (2010).
18. Quince C, Lanzen A, Davenport RJ, Turnbaugh PJ. Removing noise from pyrosequenced amplicons. *BMC Bioinforms* **12**:38 (2011).
19. Hamady M, Knight R. Microbial community profiling for human microbiome projects: Tools, techniques, and challenges. *Genome Res* **19**(7):1141–1152 (2009).
20. Hamady M, Walker JJ, Harris JK, Gold NJ, Knight R. Error-correcting barcoded primers for pyrosequencing hundreds of samples in multiplex. *Nat Meth* **5**(3):235–237 (2008).
21. Huse SM, Huber JA, Morrison HG, Sogin ML, Welch DM. Accuracy and quality of massively parallel dna pyrosequencing. *Genome Biol* **8**(7):R143 (2007).
22. Li W, Godzik A. Cd-hit: A fast program for clustering and comparing large sets of protein or nucleotide sequences. *Bioinformatics* **22**(13):1658–1659 (2006).
23. Edgar RC. Search and clustering orders of magnitude faster than BLAST. *Bioinformatics* **26**(19):2460–2461 (2010).
24. Schloss PD, Westcott SL, Ryabin T, Hall JR, Hartmann M, Hollister EB, Lesniewski RA, Oakley BB, Parks DH, Robinson CJ, et al. Introducing mothur: Open-source, platform-independent, community-supported software for describing and comparing microbial communities. *Appl Environ Microbiol* **75**(23):7537–7541 (2009).
25. Faith DP. Conservation evaluation and phylogenetic diversity. *Biol Conserv* **61**:1–10 (1992).
26. Lozupone C, Knight R. UniFrac: a new phylogenetic method for comparing microbial communities. *Appl Environ Microbiol* **71**(12):8228–8235 (2005).
27. Lozupone C, Lladser ME, Knights D, Stombaugh J, Knight R. UniFrac: An effective distance metric for microbial community comparison. *ISME J* **5**(2):169–172 (2011).
28. Sun Y, Cai Y, Liu L, Yu F, Farrell ML, McKendree W, Farmerie W. ESPRIT: Estimating species richness using large collections of 16*S* rRNA pyrosequences. *Nucleic Acids Res* **37**(10):e76 (2009).
29. DeSantis TZ Jr, Hugenholtz P, Keller K, Brodie EL, Larsen N, Piceno YM, Phan R, Andersen GL. NAST: A multiple sequence alignment server for comparative analysis of 16*S* rRNA genes. *Nucleic Acids Res*, **34**(web server issue):W394–W399 (2006).
30. Caporaso JG, Bittinger K, Bushman FD, DeSantis TZ, Andersen GL, Knight R. PyNAST: A flexible tool for aligning sequences to a template alignment. *Bioinformatics* **26**(2):266–267 (2010).
31. Nawrocki EP, Kolbe DL, Eddy SR. Infernal 1.0: Inference of RNA alignments. *Bioinformatics* **25**(10):1335–1337 (2009).
32. Caporaso JG, Kuczynski J, Stombaugh J, Bittinger K, Bushman FD, Costello EK, Fierer N, Peña AG, Goodrich JK, Gordon JI, et al. QIIME allows analysis of high-throughput community sequencing data. *Nat Methods* **7**(5):335–336 (2010).
33. Ross SM. Simulation. In *Statistical Modeling and Decision Science*, 3rd ed., Academic Press, 2002.
34. Magurran AE. *Measuring Biological Diversity*, Oxford-Blackwell, 2004.
35. Chao A. Nonparametric estimation of the number of classes in a population. *Scand J Stat* **11**:265–270 (1984).

36. Chao A. Estimating the population size for capture-recapture data with unequal catchability. *Biometrics* **43**(4):783–791 (1897).
37. Curtis TP, Head IM, Lunn M, Woodcock S, Schloss PD, Sloan WT. What is the extent of prokaryotic diversity? *Phil Trans R Soc Lond* **361**:2023–2037 (2006).
38. Roesch LF, Fulthorpe RR, Riva A, Casella G, Hadwin AK, Kent AD, Daroub SH, Camargo FA, Farmerie WG, Triplett EW. Pyrosequencing enumerates and contrasts soil microbial diversity. *ISME J* **1**:283–290 (2007).
39. Lladser ME. Prediction of unseen proportions in urn models with restricted sampling. *Proc 5th Workshop Analytic Algorithmics and Combinatorics (ANALCO)*, 2009, pp. 85–91.
40. Robbins HE. On estimating the total probability of the unobserved outcomes of an experiment. *Ann Math Stat* **39**(1):256–257 (1968).
41. Good IJ. The population frequencies of species and the estimation of population parameters. *Biometrika* **40**(3–4):237–264 (1953).
42. Starr N. Linear estimation of the probability of discovering a new species. *Ann Stat* **7**(3):644–652 (1979).
43. Lladser ME, Gouet R, Reeder J. Extrapolation of urn models via poissonization: Accurate measurements of the microbial unknown. *PLoS ONE* **6**(6):e21105 (2011).
44. Clayton MK, Frees EW. Nonparametric estimation of the probability of discovering a new species. *J Am Stat Assoc* **82**(397):305–311 (1987).
45. Durrett R. *Essentials of Stochastic Processes*, Springer Texts in Statistics. Springer, 2010.
46. Turnbaugh PJ, Ridaura VK, Faith JJ, Rey FE, Knight R, Gordon JI. The effect of diet on the human gut microbiome: A metagenomic analysis in humanized gnotobiotic mice. *Sci Transl Med* **1**:6ra14 (2009).

5

TENSION AT THE BORDER: HOW HOST GENETICS AND THE ENTERIC MICROBIOTA CONSPIRE TO PROMOTE CROHN'S DISEASE

DANIEL N. FRANK

School of Medicine, University of Colorado, Mucosal and Vaccine Research Program Colorado (MAVRC), and UC-Denver Microbiome Research Consortium (MiRC), Denver, Colorado

ELLEN LI

Department of Medicine, Stony Brook University, Stony Brook, New York

5.1. INTRODUCTION AND OVERVIEW

Crohn's disease (CD) is a chronic human inflammatory bowel disease (IBD) that arises through complex interactions between environmental, microbiological, and host risk factors. Although commensal microbial communities of the gastrointestinal tract play a role in disease etiology, their exact contributions are unknown. Disturbances in the structure of enteric microbial communities (*dysbiosis*) are evident in IBD, but whether these are causes or consequences of pathogenesis has yet to be determined. Investigating the human enteric microbiota as it relates to CD is a daunting task for several reasons. The steady-state microbial communities of the intestinal tract constitute a dynamic system that is affected by multiple factors, which include host genetics, age, disease state, disease duration, lifestyle (diet, smoking), obesity, medications (antibiotics, immunosuppressive drugs), and previous surgery. To determine whether enteric dysbiosis modifies the incidence and progression of IBD, we have begun to integrate and correlate large-scale surveys of microbial communities with rigorously characterized human genotypes and molecular phenotypes (microarray data, proteomic data, and histologic imaging data) in both CD and

The Human Microbiota: How Microbial Communities Affect Health and Disease, First Edition. Edited by David N. Fredricks.

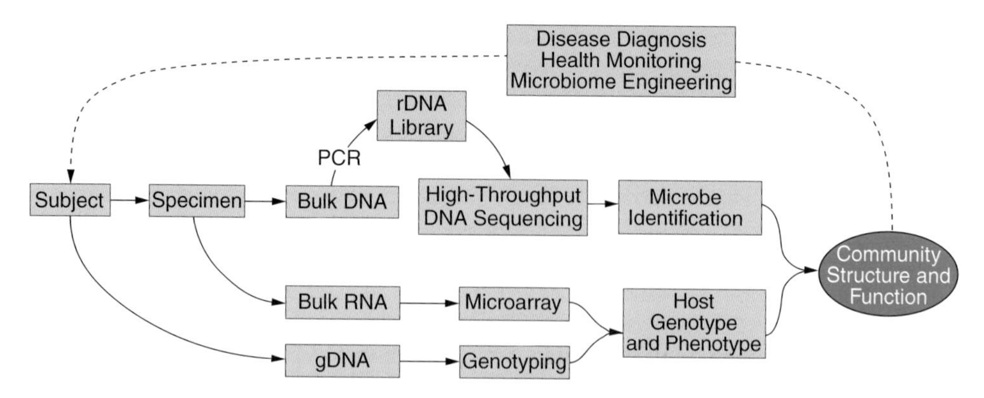

Figure 5.1. Metagenomic tools for studying host–microbiota interactions in Crohn's disease. Bacteria are identified through culture-independent analysis of 16S rRNA genes amplified from surgical specimens by broad-range PCR. Host genotypes and phenotypes are assessed through analysis of genomic DNA and mRNA of surgical specimens. Together, these data provide insight into how host factors and enteric microbial communities interact during development and/or progression of CD. The ultimate goal of these experiments is to use this knowledge to design, deploy, and monitor novel treatment regimes.

non-CD cohorts (Figure 5.1). These data are being integrated with results from genetic mice models in order to dissect the genetic and environmental effects on the intestinal microbiome as it relates to CD. Our hypothesis is that collection of a very large number of well-phenotyped clinical specimens is necessary to accurately link potentially subtle patterns of altered enteric microbiotas with disease occurrence. In particular, we wish to answer the following:

1. Does harboring a CD risk allele correlate with an abnormal microbiota?
2. How do intraindividual variations in the microbiota relate to the interindividual variations in the microbiota observed in earlier studies?
3. Does mucosal inflammation correlate with the presence of an abnormal microbiota?
4. Does inflammation arise in response to altered GI microbiota or vice versa?

5.2. HUMAN INFLAMMATORY BOWEL DISEASES

Human IBD represents a heterogeneous spectrum of chronic inflammatory disorders of the digestive tract [1,2] that can be broadly classified into ulcerative colitis (UC) and Crohn's disease (CD). Ulcerative colitis is limited to the colon and does not involve the small bowel. The pattern of intestinal inflammation in UC extends proximally in a continuous fashion from the rectum and is limited to the mucosal lining. In contrast, CD can occur anywhere along the gastrointestinal tract, but most commonly involves the terminal ileum. In addition, the inflammatory lesions of CD are characterized by discontinuities ("skip" lesions) and transmural involvement of all the layers of the bowel.

Approximately 80% of all patients with CD have ileal involvement and approximately 20% have pure colonic involvement. In certain cases, it may be difficult to distinguish pure Crohn's colitis (L2 [2]) from UC and the patients are classified as having indeterminate colitis. Ileal CD presents a relatively unambiguous phenotype. The majority (80%) of patients with ileal CD eventually undergo resection of diseased bowel, due to failure of medical management, to obstruction caused by stricturing disease, to fistulas, abscesses, or free peritoneal perforation (penetrating disease). The usual surgical procedure is an ileocolic resection (ICR), which entails the removal of disease affected terminal ileum with a grossly unaffected proximal margin of 2–4 cm. The distal margin of the resection is usually in the proximal ascending colon with the cecum and the ileocolic valve included in the resection specimen. Intestinal continuity is restored by creating an ileocolic anastomosis. Unfortunately, surgery is not curative. Evidence of disease in the formerly unaffected neo-terminal disease-affected ileum is often observed within months after resection. Approximately half of the patients report clinical relapse 5 years after surgery. Many patients require repeated resection of diseased bowel, which may lead to the development of short bowel syndrome.

5.3. THE ENTERIC MICROBIOTA AND IBD

The enteric microbiota clearly influences the development and progression of IBD pathogenesis [3–7], including that of CD. In general, intestinal inflammation in animals raised under germ-free conditions is greatly attenuated compared with conventionally housed animals [3,4]. Moreover, antibiotic treatment and diversion of fecal flow ameliorate inflammation in humans in clinical settings [8–11]. In CD, proximal diversion of the fecal stream from the ileocolic anastomosis and postoperative treatment with antibiotics have been shown to prevent early recurrence in the neoterminal ileum [12]. However, whether these phenotypes result from particular microbial pathogens, loss of beneficial taxa, or simply an aberrant response to the presence of any microbe, remains an open question. We review evidence for and against these possibilities in the following sections.

Specific Microbial Pathogens

Thus far, no specific microbial pathogen has been unambiguously associated with CD. Although *Mycobacterium avium* subspecies *paratuberculosis* (MAP) has been investigated extensively as a possible etiological agent, there have been inconsistent reports on whether this organism is associated with CD [13–18] and CD patients fail to respond to antimycobacterial treatment [19,20]. Quantitative PCR assays using MAP-specific primers were negative in our analyses [5], although other groups report positive results [14]. Potentially pathogenic *Escherichia coli* species also have been implicated in CD [21–23]. The adherent invasive *E. coli* (AIEC) are a potentially pathogenic group of strains that has been recovered from the ileum in 22–65% of ileal Crohn's disease biopsies compared to 6% of control ileal biopsies. In a separate study, the number of *E. coli* associated with the ileal mucosa correlated with the degree of inflammation in the mucosa. The prototype AIEC strain, LF82, is characterized by the ability to invade and persist for extended periods in the macrophages without killing the host cell. Other potentially pathogenic commensal organisms include enterotoxigenic *Bacteroides* species and *Enterococcus faecalis*.

Pathogenic Microbial Communities

Dramatic shifts in the composition of enteric microbial communities, as determined by culture-independent surveys, have long been associated with IBD [5,24–37]. For instance, we analyzed 190 GI tissues obtained by surgical resection from roughly equal numbers of CD and UC patients, along with and non-IBD controls [5]. Small subunit rRNA genes were cloned and Sanger-sequenced from each tissue sample after broad range PCR with pan-bacterial primers (Figure 5.1). Regardless of disease state or tissue location, the majority of sequences were associated with four phyla: Firmicutes, Bacteroidetes, Proteobacteria, and Actinobacteria. Sequences representative of the Bacillus subgroup of Firmicutes as well as the phylum Actinobacteria were increased in non-IBD small intestinal samples compared to colonic samples. Bacteroidetes and clostridial sequences were less frequent in small intestinal samples compared to colonic sequences. Bacteroidetes and clostridial sequences were significantly less abundant in UC and CD small intestinal samples compared with non-IBD samples, with concomitantly increased proteobacterial sequences.

Exploratory statistical analysis, independent of disease state or tissue sample location, separated the samples into two primary groups: (1) a control subset including virtually all (except one) of the non-IBD samples, ~47% of the CD samples and ~75% of the UC samples and (2) an "IBD" subset composed of the remaining CD and UC samples. Marked reductions in commensal bacteria, especially the clostridial family Lachnospiraceae (300-fold) and Bacteroidetes (50-fold), were observed in the IBD subset compared to the control subset. In this study, there was no substantial increase in the proteobacteria as measured by quantitative PCR. The disparities in the microbiotas correlated positively with a younger age of surgery and the presence of abscesses in CD patients, but did not correlate with antibiotic or immunosuppressive therapy.

How could dysbiosis contribute to CD pathogenesis? Commensal microbial communities, particularly those of the GI, provide myriad beneficial services to the host, for instance, by transforming otherwise indigestible plant polysaccharides to short-chain fatty acids (SCFA), which are digestible [38]. In addition, enteric microbes stimulate many aspects of intestinal immune development and function and thereby contribute to maintenance of gut homeostasis [39,40]. Loss of particular species, gain of microbes that incite inflammation, or large-scale shifts in community composition (i.e., dysbiosis [41]) could undermine health through disruption of any of these mutualistic interactions.

5.4. HUMAN GENETIC LOCI ASSOCIATED WITH IBD

Since 2001, >40 IBD susceptibility loci have been identified through linkage analysis, association mapping, and candidate gene association studies [1,42–50]. The identification of single-nucleotide polymorphisms (SNPs) that are associated with an increased risk of developing CD has greatly advanced our understanding of CD pathogenesis. The *NOD2/CARD15* and *ATG16L1* loci are of particular interest because some alleles (*risk alleles*) are associated with ileal CD. Moreover, these risk alleles also have been linked to abnormalities in Paneth cell function [42–50]. Paneth cells are specialized cells of the small intestinal lining that are concentrated in the ileum, where they secrete antimicrobial peptides and proteins, such as defensins, into the intestinal lumen. Because of their potential role in containing lumenal bacteria, disruption of normal Paneth cell function could significantly affect the composition and abundance of the intestinal microbiota. Although the *NOD2/CARD15* risk alleles have not been linked to ulcerative colitis, there may be an association with the risk of pouchitis in the ileal pouch in UC patients who have undergone restorative proctocolectomy with ileal pouch anal anastomosis (IPAA).

The *NOD2/CARD15* gene encodes an intracellular receptor that activates NFκB signaling in response to muramyldipeptide (MDP), a component of bacterial cell walls, and is thus linked to the innate immune response. Three SNPs (L1007fs, R702W, and G908R) account for ~80% of the variants observed in CD patients, but are relatively uncommon in the non-IBD population (<5%). These risk alleles are observed in Caucasians with CD but rarely in individuals of Asian or African descent with CD. In the ileum, the NOD2/CARD15 protein is expressed predominantly in Paneth cells and ileal mucosal expression of α-defensins is reduced in Nod2- deficient mice and in patients with ileal CD, particularly those harboring *NOD2/CARD15* risk alleles [47,48].

The ATG16 autophagy related 16-like 1 (*ATG16L1*) gene encodes a protein with an *N*-terminal domain that is homologous to the essential yeast autophagy gene *ATG16*. The mammalian ATG16L1 protein differs from yeast Atg16 by virtue of a *C*-terminal extension that includes a series of WD repeats of unknown function. The T300A polymorphism, which is associated with ileal CD, lies within the WD repeat domain of the ATG16L1 protein. Both the risk allele and the protective allele are well represented in the population (57% CD vs. 51% control). The ATG16L1 protein is important for localization of the autophagy protein LC3 to intracellular *Salmonella typhimurium*, and for generation of intracellular LC3-positive bodies in response to starvation or rapamycin treatment. This indicates that the *Atg16L1* protein is important in autophagy, particularly as it relates to phagocytosis of bacteria. Surprisingly, more recent work indicates that *Nod2* also is a required component of the autophagy pathway through interactions with *Atg16l* [51,52]. Thus the risk alleles identified in NOD2 may manifest pleiotropic effects through impairments in both production of antimicrobial peptides and autophagy-mediated bacterial clearance.

Caldwell et al. [53] reported abnormalities in the secretory granules of Paneth cells in mice deficient in the mouse homolog of *Atg16l*. Similar abnormalities were observed in 100% of unaffected ileal mucosa of ileal CD patients who did not harbor any of the *NOD2/CARD15* risk alleles ($NOD2^{P}$) and were homozygous for the T300A *ATG16L1* risk allele (GG) and none that were homozygous for the protective allele (AA) [53]. These observations suggest that the *ATG16L1* risk allele also is associated with impaired Paneth cell function. Because of their detrimental affects on Paneth cell function, our current working hypothesis is that the *NOD2/CARD15* and *ATG16L1* risk alleles compromise innate immune barrier function in the ileum, which, in turn, leads to defective host containment of commensal bacteria.

5.5. HOW HOST GENETICS SHAPES ENTERIC MICROBIOTA

Our observation that a subset of IBD patients harbored abnormal enteric microbiotas, while the microbiotas of other IBD patients apparently were "normal" raised several questions:

1. Do these differences persist within an individual, or do all IBD patients experience periodic flares of dysbiosis?
2. Did dysbiosis arise from prolonged treatment with antibiotics or other agents that could alter the microbial habitat of the intestinal tract?
3. Does host genetics, particularly the presence of IBD risk alleles affect the incidence or prevalence of dysbiosis in individuals, regardless of disease state?

The correlation of abnormal microbiota with a younger age of surgery [5] was intriguing because the *NOD2/CARD15* and *ATG16L1* risk alleles also are associated with younger age of surgery. Correlation of the abnormal microbiota with abscesses in CD patients [5] also was of interest because the *IBD5* risk alleles have been associated with increased disease severity.

In order to determine whether human genetic factors are associated with the dysbiosis that we previously reported [5], we determined the genotypes of the individuals from which 16*S*-based microbiota data already were collected (Figure 5.1). To date, we have assayed three prevalent *NOD2* (Leu1007fs, R702W, G908R) alleles and the *ATG16L1*T300A allele. The genotype of each sample was determined for the NOD2 and ATG16L1 SNPs listed in Table 5.1 by matrix-assisted laser desorption ionization – time of flight (MALDI-TOF) mass spectrometry in the Washington University Sequenom Technology/Genotyping Core (`http://hg.wustl.edu/info/Sequenom_description.html`) and by TaqMan allelic discrimination assays.

Significant differences in the distributions of enteric microbes were observed as a function of NOD2 and ATG16L1 genotypes (Figure 5.2). For more detailed statistical analysis, NOD2 genotypes were categorized as $NOD2^{R}$ (subjects with at least one of the three major NOD2 risk alleles) and $NOD2^{NR}$ (subjects with none of the three major risk alleles). A multivariate regression analysis (MANCOVA) of clinical phenotype (i.e., CD vs. UC vs. non-IBD), NOD2 and ATG16L1 genotypes, and age of surgery (a confounding variable because non-IBD subjects were significantly older than IBD subjects) was performed against frequency of intestinal bacterial phyla (and subphyla of Firmicutes) to delineate the relative influences of these variables on composition of the intestinal microbiota. Our analysis indicated that

TABLE 5.1. Single-Nucleotide Polymorphisms (SNPs) Associated with an Increased Risk of Crohn's Disease

Gene	dbSNP	Risk Variant	Amino Acid Substitution
NOD2/CARD15	rs2066847	InsC	L1007fs
NOD2/CARD15	rs2066844	G to A	R702W
NOD2/CARD15	rs2066845	G to C	G908W
ATG16L1	rs2241880	A to G	T300A

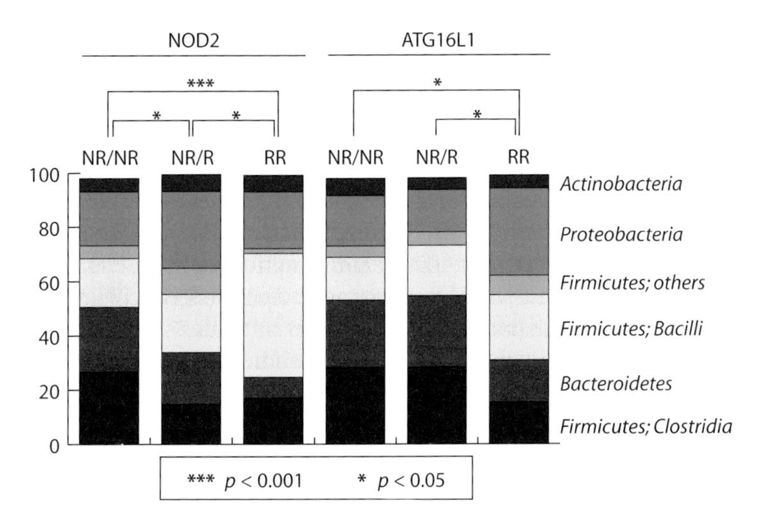

Figure 5.2. Effect of Crohn's disease susceptibility alleles on bacterial biodiversity in surgical specimens; bars heights represent relative percent abundances of bacteria classified by 16S rRNA phylogenetics to the phylum or subphylum level (R—risk allele; NR—nonrisk allele for NOD2 and ATG16L1 genes). (Data from Frank et al. [36].)

disease phenotype ($p = 0.002$), NOD2 composite genotype ($p = 0.024$), and ATG16L1 genotype ($p = 0.011$) are all significantly associated with changes in the composition of enteric communities as a whole [36]. Furthermore, the frequencies of several bacterial genera of Clostridia were significantly associated with NOD2 and/or ATG16L1 genotypes. One notable example, the genus *Faecalibacterium*, which has previously been implicated in protection against CD re-currence [34,35,54], was significantly associated with both disease phenotype and ATG16L1 genotype.

Thus, NOD2 and ATG16L1 alleles have determining effects on the microbiota that are independent of disease itself. An important implication of this result is that the dysbiosis that we, and others, have reported in connection with CD is not solely the result of environmental effects such as treatment history or diet. Interestingly, the abundance of an enteric genus was rarely significantly correlated with both NOD2 and ATG16L1 genotypes, suggesting complex interactions between host genotype and microbial community composition. Moreover, several genera were significantly associated with disease phenotype, but not with NOD2 or ATG16L1 genotype, suggesting either that disease itself influenced the relative frequencies of these genera or that other genetic determinants are involved.

5.6. MODELING DYSBIOSIS AND CROHN'S DISEASE

Human inflammatory bowel diseases such as CD arise through complex interactions between host, environmental, and microbiological factors. Application of high-throughput, molecular–biological technologies to IBD, including genomewide association studies and microbial metagenomics, have begun to illuminate the mechanistic bases for development of these chronic, debilitating diseases. Functional characterization of the CD susceptibility loci *NOD2* and *ATG16L1* indicates that both are critical for proper Paneth cell function in the terminal ileum. Because Paneth cells secrete a variety of antimicrobial compounds into the intestinal lumen in response to bacteria (e.g., lysozyme, α-defensins), it is likely that loss of Nod2 or Atg16L1 activities increase disease risk through abrogation of GI mucosal host defense mechanisms.

Our more recent work has demonstrated that in addition to these host-associated phenotypes, particular alleles of *NOD2* and *ATG16L1* are associated with substantial phylum-level alterations in the enteric microbiota, irrespective of disease phenotype or overt inflammation. Although genetic predisposition suggests that dysbiosis arises as a consequence of the pathogenic process, the question of whether dysbiosis contributes to CD pathogenesis or is an innocuous byproduct remains to be settled. Furthermore, how mucosal barrier dysfunction per se could lead to dysbiosis also is not clear. In the following model, we propose that loss of the normally beneficial services provided by the enteric microbiota is a pivotal event in undermining immune homeostasis in the gut, and is thus a critical determinant of the onset, progression, and severity of CD.

In the healthy GI tract, the immune system maintains a state of watchful waiting with respect to bacterial invaders and their products. Unless enteric microbes breach the epithelial wall and reach regional lymph nodes, the trillions of bacteria in the GI are largely tolerated by the mucosal immune system and ignored by the systemic immune systems [55]. Enteric microbial communities play a critical role in this process by stimulating the development of GI lymphoid tissues [56,57], fortifying the physical epithelial barrier [58], and regulating the quality and magnitude of the mucosal immune response [39,40] to commensal bacteria and potential pathogens (Figure 5.3a).

During an initial, preclinical phase of the disease (Figure 5.3b), we propose that a relatively normally functioning enteric microbiota develops and strengthens mucosal barrier function, through provision of metabolites, such as short-chain fatty acids, and antiinflammatory compounds [59]. However, in opposition to these positive interactions, an intrinsically weaker physicoimmunologic barrier increases the influx of enteric antigens (e.g., microbes, microbial products, food) into the lamina propria and regional lymph nodes, relative to the healthy gut. Indeed, CD patients as well as mouse models of colitis develop high-titer antibodies against commensal flagellin proteins [60], whereas healthy individuals do not mount strong systemic responses to their enteric microbiotas [55,61].

If immune homeostasis cannot be reestablished, perhaps due to overwhelming antigen loads, then acute mucosal inflammation, which arises in response to increased exposure to enteric antigens, will not resolve (Figure 5.3c). We hypothesize that commensal microbial communities are negatively impacted at this stage of disease progression. Remodeling of the intestinal epithelium could, for instance, alter either the physical or innate immunologic environment of the bowel through shifts in pH, nutrient levels (e.g., iron availability), or expression of specific antimicrobial proteins. Any of these effects likely would relieve selective pressure on some microbial groups and thus shift the balance of the GI microbiota. Alternatively, secreted antibodies (sIgA, sIgG) that target particularly immunogenic commensal microbes could inhibit the growth of these organisms within the intestinal lumen. Interestingly, a potential link between adaptive immunity and dysbiosis is evident in members of the *Clostridium* cluster XIVa (including the family Lachnospiraceae), which not only are significantly reduced in abundance in CD [5,36] but also are the primary source of

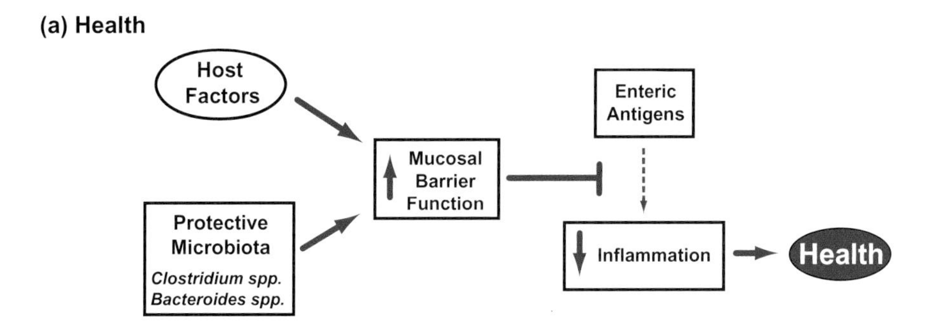

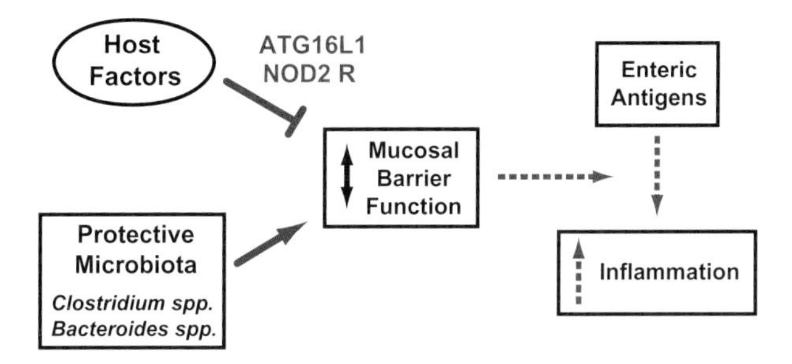

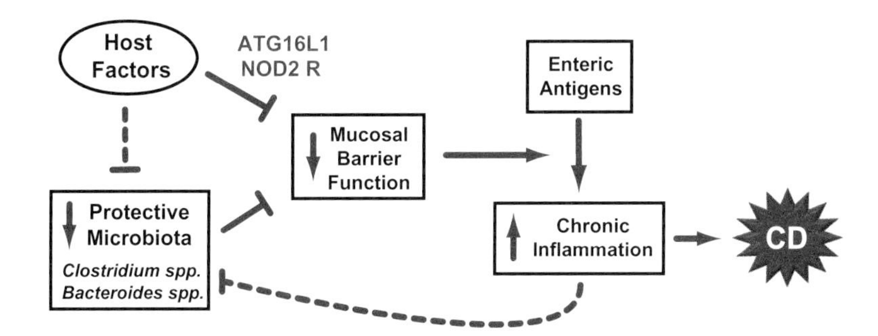

Figure 5.3. Impact of host factors and enteric microbes on development and progression of Crohn's disease: (a) health; (b) preclinical CD; (c) clinical CD. Blue lines indicate health-promoting activities, red lines indicate processes that promote CD. (a) In health, host genetics and members of the enteric microbiota stimulate formation of a strong mucosal barrier that limits the body's exposure to microbes and microbial products. (b) Genetic lesions compromise mucosal defenses such as secretion of antimicrobial compounds and removal of microbes and antigens from the lamina propria. Increased exposure to these antigens increases mucosal inflammation. (c) Loss of key members of the enteric microbiota exacerbates mucosal barrier dysfunction through loss of mutualistic interactions and leads to chronic inflammation and disease. Dysbiosis is hypothesized to arise through reduced levels of antimicrobial peptides in the mucosal layer, inflammatory remodeling of GI epithelium, and an adaptive immune response to commensal microbes within the intestinal lumen.

immunodominant flagellin antigens recognized by serum IgG in CD patients, but not controls [60]. Rather than an autoimmune disease, CD ultimately could be a supraautoimmune disease in the sense that commensal communities (which with the host comprise a supraorganism) are the targets of immune dysregulation.

Regardless of the mechanism by which the enteric microbiota is disrupted during development of CD, loss of mutualistic interactions between host and microbe are likely to amplify disease phenotype in at least two ways:

1. The microorganisms that are diminished in abundance in CD normally provide critical metabolic services to the host, such as conversion of undigestible complex carbohydrates to soluble short-chain fatty acids [58]. In our studies, many of the taxa that were reduced in CD are closely related to cultured microorganisms that produce the SCFA butyrate, a key nutrient for enterocytes [62,63] and a potential stimulatory signal for antimicrobial peptide synthesis in the colon [64]. Loss of any of these functions would be expected to compromise function of the mucosal barrier.
2. Several enteric commensal species of the phyla Firmicutes and Bacteroidetes suppress enteric inflammation in experimental systems. *Bacteroides fragilis*, for example, stimulates development of regulatory T lymphocytes in the gut, which, in turn, release antiinflammatory cytokines [59,65,66]. The molecular mediator of these effects, *B. fragilis* polysaccharide A, can itself suppress experimentally induced colitis in a murine model [66]. Similarly, *Faecalibacterium prausnitzii*, which is of reduced abundance in CD [35,36], suppresses inflammatory signaling *in vitro* and in murine colitis [34]. It is likely that many other uncharacterized enteric commensal species have developed mechanisms to downregulate mucosal immunity during coevolution with their hosts and that these functions are a critical prerequisite of immune homeostasis. Consequently, replacement of normally immunosuppressive microbes by other, more immunogenic species would be expected to promote inflammation.

Overall, then, we propose that dysbiosis arises in CD during the nascent inflammatory state (i.e., before overt disease) that results from genetically mediated dysfunctions in the physico immunologic milieu of the gut mucosa. Dysbiosis is predicted to amplify mucosal inflammation by further degrading mucosal barrier function and thus cause a vicious circle in which inflammation promotes dysbiosis, and dysbiosis promotes inflammation. A consequence of this multifactorial model is that treating either inflammation or dysbiosis alone would not necessarily be effective in the long run because the untreated factor would remain in force. Rather, a complex disease such as CD may require a complex treatment regimen that restores normal function in both the host and its microbiota.

ACKNOWLEDGMENTS

We thank Claire Gustafson for helpful discussions. This work was supported by NIH grants UH2DK083994 (to EL) and R21HG005964 (to DNF), a grant from the Crohn's and Colitis Foundation of America (to EL), and a grant from the Simons

Foundation (to EL). We acknowledge use of the Washington University Digestive Diseases Research Core Center Tissue Procurement Facility (P30 DK52574).

REFERENCES

1. Abraham C, Cho JH. Inflammatory bowel disease. *N Engl J Med* **361**(21):2066–2078 (2009).
2. Satsangi J, Silverberg MS, Vermeire S, Colombel JF. The Montreal classification of inflammatory bowel disease: Controversies, consensus, and implications. *Gut* **55**(6):749–753 (2006).
3. Sartor RB. Microbial influences in inflammatory bowel diseases. *Gastroenterology* **134**(2):577–594 (2008).
4. Elson CO, Cong Y, McCracken VJ, Dimmitt RA, Lorenz RG, Weaver CT. Experimental models of inflammatory bowel disease reveal innate, adaptive, and regulatory mechanisms of host dialogue with the microbiota. *Immunol Rev* **206**:260–276 (2005).
5. Frank DN, St Amand AL, Feldman RA, Boedeker EC, Harpaz N, Pace NR. Molecular-phylogenetic characterization of microbial community imbalances in human inflammatory bowel diseases. *Proc Natl Acad Sci USA* **104**(34):13780–13785 (2007).
6. Peterson DA, Frank DN, Pace NR, Gordon JI. Metagenomic approaches for defining the pathogenesis of inflammatory bowel diseases. *Cell Host Microbe* **3**(6):417–427 (2008).
7. Sokol H, Lay C, Seksik P, Tannock GW. Analysis of bacterial bowel communities of IBD patients: What has it revealed? *Inflamm Bowel Dis* **14**(6):858–867 (2008).
8. de Silva HJ, Millard PR, Soper N, Kettlewell M, Mortensen N, Jewell DP. Effects of the faecal stream and stasis on the ileal pouch mucosa. *Gut* **32**(10):1166–1169 (1991).
9. Harper PH, Lee EC, Kettlewell MG, Bennett MK, Jewell DP. Role of the faecal stream in the maintenance of Crohn's colitis. *Gut* **26**(3):279–284 (1985).
10. Rutgeerts P, Goboes K, Peeters M, Hiele M, Penninckx F, Aerts R, Kerremans R, Vantrappen G. Effect of faecal stream diversion on recurrence of Crohn's disease in the neoterminal ileum. *Lancet* **338**(8770):771–774 (1991).
11. Winslet MC, Allan A, Poxon V, Youngs D, Keighley MR. Faecal diversion for Crohn's colitis: A model to study the role of the faecal stream in the inflammatory process. *Gut* **35**(2):236–242 (1994).
12. Unkart JT, Anderson L, Li E, Miller C, Yan Y, Gu CC, Chen J, Stone CD, Hunt S, Dietz DW. Risk factors for surgical recurrence after ileocolic resection of Crohn's disease. *Dis Colon Rectum* **51**(8):1211–1216 (2008).
13. Chiodini RJ. Crohn's disease and the mycobacterioses: A review and comparison of two disease entities. *Clin Microbiol Rev* **2**(1):90–117 (1989).
14. Feller M, Huwiler K, Stephan R, Altpeter E, Shang A, Furrer H, Pfyffer G, Jemmi T, Baumgartner A, Egger M. Mycobacterium avium subspecies paratuberculosis and Crohn's disease: A systematic review and meta-analysis. *Lancet Infect Dis* **7**:607–613 (2007).
15. Hermon-Taylor J. Mycobacterium avium subspecies paratuberculosis in the causation of Crohn's disease. *World J Gastroenterol* **6**(5):630–632 (2000).
16. Van Kruiningen HJ. Lack of support for a common etiology in Johne's disease of animals and Crohn's disease in humans. *Inflamm Bowel Dis* **5**(3):183–191 (1999).
17. Sartor RB. Does Mycobacterium avium subspecies paratuberculosis cause Crohn's disease? *Gut* **54**(7):896–898 (2005).

18. Parrish NM, Radcliff RP, Brey BJ, Anderson JL, Clark DL Jr, Koziczkowski JJ, Ko CG. Goldberg ND, Brinker DA, Carlson RA, et al. Absence of mycobacterium avium subsp. paratuberculosis in Crohn's patients. *Inflamm Bowel Dis* **15**(4):558–565 (2009).
19. Shafran I, Kugler L, El-Zaatari FA, Naser SA, Sandoval J. Open clinical trial of rifabutin and clarithromycin therapy in Crohn's disease. *Digest Liver Dis* **34**(1):22–28 (2002).
20. Prantera C, Scribano ML. Antibiotics and probiotics in inflammatory bowel disease: Why, when, and how. *Curr Opin Gastroenterol* **25**(4):329–333 (2009).
21. Darfeuille-Michaud A. Adherent-invasive Escherichia coli: A putative new E. coli pathotype associated with Crohn's disease. *Int J Med Microbiol* **292**(3–4):185–193 (2002).
22. Darfeuille-Michaud A, Boudeau J, Bulois P, Neut C, Glasser AL, Barnich N, Bringer MA, Swidsinski A, Beaugerie L, Colombel JF. High prevalence of adherent-invasive Escherichia coli associated with ileal mucosa in Crohn's disease. *Gastroenterology* **127**(2):412–421 (2004).
23. Rolhion N, Darfeuille-Michaud A. Adherent-invasive Escherichia coli in inflammatory bowel disease. *Inflamm Bowel Dis* (2007).
24. Krook A, Lindstrom B, Kjellander J, Jarnerot G, Bodin L. Relation between concentrations of metronidazole and *Bacteroides* spp in faeces of patients with Crohn's disease and healthy individuals. *J Clin Pathol* **34**:645–650 (1981).
25. Giaffer MH, Holdsworth CD, Duerden BI. The assessment of faecal flora in patients with inflammatory bowel disease by a simplified bacteriological technique. *J Med Microbiol* **35**:238–241 (1991).
26. Merwe JPvd, Schroder AM, Wensick F, Hazenberg MP. The obligate anaerobic faecal flora of patients with Crohn's disease and their first-degree relatives. *Scand J Gastroenterol* **23**:1125–1131 (1988).
27. Seksik P, Rigottier-Gois L, Gramet G, Sutren M, Pochart P, Marteau P, Jian R, Dore J. Alterations of the dominant faecal bacterial groups in patients with Crohn's disease of the colon. *Gut* **52**(2):237–242 (2003).
28. Mangin I, Bonnet R, Seksik P, Rigottier-Gois L, Sutren M, Bouhnik Y, Neut C, Collins MD, Colombel J-F, Marteau P, et al. Molecular inventory of faecal microflora in patients with Crohn's disease. *FEMS Microbiol Ecol* **50**:25–36 (2004).
29. Prindiville T, Cantrell M, Wilson KH. Ribosomal DNA sequence analysis of mucosa-associated bacteria in Crohn's disease. *Inflamm Bowel Dis* **10**(6):824–833 (2004).
30. Lepage P, Seksik P, Sutren M, de la Cochetiere MF, Jian R, Marteau P, Dore J. Biodiversity of the mucosa-associated microbiota is stable along the distal digestive tract in healthy individuals and patients with IBD. *Inflamm Bowel Dis* **11**(5):473–480 (2005).
31. Manichanh C, Rigottier-Gois L, Bonnaud E, Gloux K, Pelletier E, Frangeul L, Nalin R, Jarrin C, Chardon P, Marteau P, et al. Reduced diversity of faecal microbiota in Crohn's disease revealed by a metagenomic approach. *Gut* **55**:205–211 (2006).
32. Gophna U, Sommerfeld K, Gophna S, Doolittle WF, Veldhuyzen van Zanten SJ. Differences between tissue-associated intestinal microfloras of patients with Crohn's disease and ulcerative colitis. *J Clin Microbiol* **44**(11):4136–4141 (2006).
33. Scanlan PD, Shanahan F, O'Mahony C, Marchesi JR. Culture-independent analyses of temporal variation of the dominant fecal microbiota and targeted bacterial subgroups in Crohn's disease. *J Clin Microbiol* **44**(11):3980–3988 (2006).
34. Sokol H, Pigneur B, Watterlot L, Lakhdari O, Bermudez-Humaran LG, Gratadoux JJ, Blugeon S, Bridonneau C, Furet JP, Corthier G, et al. Faecalibacterium prausnitzii is an anti-inflammatory commensal bacterium identified by gut microbiota analysis of Crohn disease patients. *Proc Natl Acad Sci USA* **105**(43):16731–16736 (2008).

35. Sokol H, Seksik P, Furet JP, Firmesse O, Nion-Larmurier I, Beaugerie L, Cosnes J, Corthier G, Marteau P, Dore J. Low counts of Faecalibacterium prausnitzii in colitis microbiota. *Inflamm Bowel Dis* **15**(8):1183–1189 (2009).
36. Frank DN, Robertson CE, Hamm CM, Kpadeh Z, Zhang T, Chen H, Zhu W, Sartor RB, Boedeker EC, Harpaz N, et al. Disease phenotype and genotype are associated with shifts in intestinal-associated microbiota in inflammatory bowel diseases. *Inflamm Bowel Dis* **17**(1):179–184 (2011).
37. Qin J, Li R, Raes J, Arumugam M, Burgdorf KS, Manichanh C, Nielsen T, Pons N, Levenez F, Yamada T, et al. A human gut microbial gene catalogue established by metagenomic sequencing. *Nature* **464**(7285):59–65 (2010).
38. Flint HJ, Bayer EA, Rincon MT, Lamed R, White BA. Polysaccharide utilization by gut bacteria: Potential for new insights from genomic analysis. *Nat Rev Microbiol* **6**(2):121–131 (2008).
39. Hooper LV, Macpherson AJ. Immune adaptations that maintain homeostasis with the intestinal microbiota. *Nat Rev Immunol* **10**(3):159–169 (2010).
40. Round JL, Mazmanian SK. The gut microbiota shapes intestinal immune responses during health and disease. *Nat Rev Immunol* **9**(5):313–323 (2009).
41. Tamboli CP, Neut C, Desreumaux P, Colombel JF. Dysbiosis in inflammatory bowel disease. *Gut* **53**(1):1–4 (2004).
42. Barrett JC, Hansoul S, Nicolae DL, Cho JH, Duerr RH, Rioux JD, Brant SR, Silverberg MS, Taylor KD, Barmada MM, et al. Genome-wide association defines more than 30 distinct susceptibility loci for Crohn's disease. *Nat Genet* **40**(8):955–962 (2008).
43. Goyette P, Labbe C, Trinh TT, Xavier RJ, Rioux JD. Molecular pathogenesis of inflammatory bowel disease: Genotypes, phenotypes and personalized medicine. *Ann Med* **39**(3):177–199 (2007).
44. Imielinski M, Baldassano RN, Griffiths A, Russell RK, Annese V, Dubinsky M, Kugathasan S, Bradfield JP, Walters TD, Sleiman P, et al. Common variants at five new loci associated with early-onset inflammatory bowel disease. *Nat Genet* **41**(12):1335–1340 (2009).
45. Hugot JP, Chamaillard M, Zouali H, Lesage S, Cezard JP, Belaiche J, Almer S, Tysk C, O'Morain CA, Gassull M, et al. Association of NOD2 leucine-rich repeat variants with susceptibility to Crohn's disease. *Nature* **411**(6837):599–603 (2001).
46. Ogura Y, Bonen DK, Inohara N, Nicolae DL, Chen FF, Ramos R, Britton H, Moran T, Karaliuskas R, Duerr RH, et al. A frameshift mutation in NOD2 associated with susceptibility to Crohn's disease. *Nature* **411**(6837):603–606 (2001).
47. Ogura Y, Lala S, Xin W, Smith E, Dowds TA, Chen FF, Zimmermann E, Tretiakova M, Cho JH, Hart J, et al. Expression of NOD2 in Paneth cells: A possible link to Crohn's ileitis. *Gut* **52**(11):1591–1597 (2003).
48. Wehkamp J, Harder J, Weichenthal M, Schwab M, Schaffeler E, Schlee M, Herrlinger KR, Stallmach A, Noack F, Fritz P, et al. NOD2 (CARD15) mutations in Crohn's disease are associated with diminished mucosal alpha-defensin expression. *Gut* **53**(11):1658–1664 (2004).
49. Rioux JD, Xavier RJ, Taylor KD, Silverberg MS, Goyette P, Huett A, Green T, Kuballa P, Barmada MM, Datta LW, et al. Genome-wide association study identifies new susceptibility loci for Crohn disease and implicates autophagy in disease pathogenesis. *Nat Genet* **39**(5):596–604 (2007).
50. Prescott NJ, Fisher SA, Franke A, Hampe J, Onnie CM, Soars D, Bagnall R, Mirza MM, Sanderson J, Forbes A, et al. A nonsynonymous SNP in ATG16L1 predisposes to ileal Crohn's disease and is independent of CARD15 and IBD5. *Gastroenterology* **132**(5):1665–1671 (2007).

51. Cooney R, Baker J, Brain O, Danis B, Pichulik T, Allan P, Ferguson DJ, Campbell BJ, Jewell D, Simmons A. NOD2 stimulation induces autophagy in dendritic cells influencing bacterial handling and antigen presentation. *Nat Med* **16**(1):90–97 (2010).
52. Travassos LH, Carneiro LA, Ramjeet M, Hussey S, Kim YG, Magalhaes JG, Yuan L, Soares F, Chea E, Le Bourhis L, et al. Nod1 and Nod2 direct autophagy by recruiting ATG16L1 to the plasma membrane at the site of bacterial entry. *Nat Immunol* **11**(1):55–62 (2010).
53. Cadwell K, Liu JY, Brown SL, Miyoshi H, Loh J, Lennerz JK, Kishi C, Kc W, Carrero JA, Hunt S, et al. A key role for autophagy and the autophagy gene Atg16l1 in mouse and human intestinal Paneth cells. *Nature* **456**(7219):259–263 (2008).
54. Solnick JV. Clinical significance of Helicobacter species other than Helicobacter pylori. *Clin Infect Dis* **36**(3):349–354 (2003).
55. Slack E, Hapfelmeier S, Stecher B, Velykoredko Y, Stoel M, Lawson MA, Geuking MB, Beutler B, Tedder TF, Hardt WD, et al. Innate and adaptive immunity cooperate flexibly to maintain host-microbiota mutualism. *Science* **325**(5940):617–620 (2009).
56. Shroff KE, Meslin K, Cebra JJ. Commensal enteric bacteria engender a self-limiting humoral mucosal immune response while permanently colonizing the gut. *Infect Immun* **63**(10):3904–3913 (1995).
57. Cebra JJ. Influences of microbiota on intestinal immune system development. *Am J Clin Nutr* **69**(5):1046S–1051S (1999).
58. Hooper LV, Midtvedt T, Gordon JI. How host-microbial interactions shape the nutrient environment of the mammalian intestine. *Annu Rev Nutr* **22**:283–307 (2002).
59. Mazmanian SK, Round JL, Kasper DL. A microbial symbiosis factor prevents intestinal inflammatory disease. *Nature* **453**(7195):620–625 (2008).
60. Duck LW, Walter MR, Novak J, Kelly D, Tomasi M, Cong Y, Elson CO. Isolation of flagellated bacteria implicated in Crohn's disease. *Inflamm Bowel Dis* **13**(10):1191–1201 (2007).
61. Konrad A, Cong Y, Duck W, Borlaza R, Elson CO. Tight mucosal compartmentation of the murine immune response to antigens of the enteric microbiota. *Gastroenterology* **130**(7):2050–2059 (2006).
62. Pryde SE, Duncan SH, Hold GL, Stewart CS, Flint HJ. The microbiology of butyrate formation in the human colon. *FEMS Microbiol Lett* **217**(2):133–139 (2002).
63. Scheppach W, Weiler F. The butyrate story: Old wine in new bottles? *Curr Opin Clin Nutr Metab Care* **7**(5):563–567 (2004).
64. Raqib R, Sarker P, Bergman P, Ara G, Lindh M, Sack DA, Nasirul Islam KM, Gudmundsson GH, Andersson J, Agerberth B. Improved outcome in shigellosis associated with butyrate induction of an endogenous peptide antibiotic. *Proc Natl Acad Sci USA* **103**(24):9178–9183 (2006).
65. Mazmanian SK, Liu CH, Tzianabos AO, Kasper DL. An immunomodulatory molecule of symbiotic bacteria directs maturation of the host immune system. *Cell* **122**(1):107–118 (2005).
66. Round JL, Mazmanian SK. Inducible Foxp3+ regulatory T-cell development by a commensal bacterium of the intestinal microbiota. *Proc Natl Acad Sci USA* **107**(27):12204–12209 (2010).

6

THE HUMAN AIRWAY MICROBIOME

EDITH T. ZEMANICK and J. KIRK HARRIS

Department of Pediatrics, University of Colorado Anschutz Medical Campus, Aurora, Colorado

6.1. INTRODUCTION

Human lungs are generally considered sterile, but particulate matter, including microorganisms, are continuously inhaled and challenge the airway. Not only are organisms inhaled from the air [58], but micro-aspiration of mucosal secretions from the oral cavity and upper airway also contribute organisms to the lower airways [26, 47]. In a healthy human, innate defenses in the lung generally prevent infection. Mechanisms of innate immunity include mucociliary clearance, ingestion of foreign material by alveolar macrophages, and opsonization or inhibition of bacteria by soluble airway surface liquid components [49]. In disease, these innate defense mechanisms may be disrupted. For example, in the genetic lung disease cystic fibrosis, dehydration of the airway surface liquid layer and disrupted mucociliary clearance lead to mucus retention and chronic bacterial infection [16]. Chronic and acute inflammation may also alter the function of innate defense systems, which could influence microbial communities by modification of the normal niche.

Traditionally, detection and identification of bacteria have relied on culture techniques using organism-specific cultivation conditions. Thus, only a narrow understanding of the microbial communities present in humans during health and disease was possible as many bacteria resist cultivation by standard methods [2, 56]. Newer, specialized cultivation-based approaches may allow detection of the

The Human Microbiota: How Microbial Communities Affect Health and Disease, First Edition. Edited by David N. Fredricks.

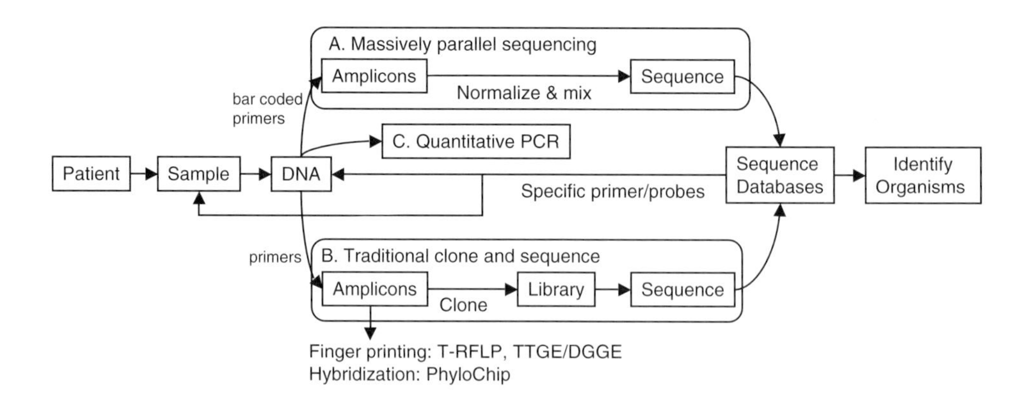

Figure 6.1. Schematic representation of sample processing and molecular biology techniques used to investigate the microbiota.

majority of organisms associated with the human body; however, the cost of this endeavor in terms of both time and money is daunting [35].

The development of molecular approaches capable of identifying many species in parallel (Figure 6.1) through interrogation of the genetic content, or microbiome, of the microorganisms present provides a more comprehensive view of microbial communities [59]. Several indirect approaches, including fingerprinting [denaturing gradient gel electrophoresis (DGGE), terminal restriction fragment length polymorphisms (T-RFLP), temporal temperature gradient gel electrophoresis (TTGE)] and hybridization (colony hybridization, microarray) have been used to examine the microbiota present in clinical samples. These approaches were established as a result of the difficulty and expense of sequence-based approaches [18]. Sequence data are often collected in parallel from a subset of samples to validate these approaches. However, comprehensive validation is not possible without loss of the cost savings of the fingerprinting technique. Sequence-based approaches provide unambiguous data for molecular identification of microbes. Sequence data are discrete entities that can be stored in databases and compared to other sequences in a quantitative manner to determine identity between organisms across studies. Quantitative comparisons between studies are limited with other techniques, particularly fingerprinting approaches, where the identity of the organism each fragment represents is not routinely confirmed. More recent advancements in DNA sequencing capabilities now allow acquisition of sequence data from hundreds of samples in parallel with the adoption of next-generation sequencing platforms [29, 57]. This greatly reduces the cost per sample for sequence-based bacterial identification and enables investigators to obtain sequence data from replicate samples (e.g., longitudinal designs). Thus, the cost savings advantage of fingerprinting and hybridization approaches is now largely absent.

There are several ways that better understanding of the microbiome may contribute to improved clinical care of patients. First, there is evidence that acute disease, especially in early life, can result in increased propensity for chronic disease later in life [33, 37]. Further, chronic disease generally involves alteration to normal host physiology, which creates a different niche for colonization by the resident microbiota. Alterations in host colonization may impact clinical outcomes to the extent that the microbiota influence disease progression. Differences in host microbiota have been observed associated with disease [24, 78], and it is likely that outcomes are influenced by individual differences in the microbiota. However, the large variability between individuals' microbiota make generalization difficult, and adequate power to delineate the role of the microbiota in disease will probably necessitate more extensive studies.

The remainder of the chapter examines the existing lung microbiome data, and the research needed to establish the role of the microbiome during respiratory health and disease. The existing data are heavily skewed toward cystic fibrosis (CF), largely due to the prominent role of chronic airway infection in this disease. Application of microbiome methods is now occurring in additional lung diseases, and will allow us to more broadly determine the importance of the microbiome in health and disease.

6.2. LUNG-CHARACTERISTICS AND CONDITIONS

6.2.1. Normal Lung

The view that normal, healthy airways are sterile came from culture-based analyses of bronchoalveolar lavage fluid and induced sputum samples from healthy subjects [13, 38, 42, 43, 60, 87]. In these studies, the large majority of airway specimens was sterile by culture or contained only small quantities of organisms typical of the upper airway and oral cavity, which were considered oral contamination. Hilty and colleagues challenged the sterility of normal airways by identification of bacteria present in protected brush bronchoscopy samples from the lower airways of eight healthy adults [31]. Protected brush samples were used to minimize the potential for oral contamination. The microbiome was analyzed using Sanger sequencing of cloned ribosomal RNA genes. Bacteroidetes (primarily *Prevotella* spp.) and Firmicutes (*Veillonella* spp.) were the most common bacterial genera detected, with smaller amounts of Proteobacteria and Fusobacteria present. Asthmatic and chronic obstructive pulmonary disease (COPD) subject's bronchoalveolar lavage samples contained higher proportions of Proteobacteria and *Staphylococcus* spp. relative to seven disease control pediatric subjects. Bacterial communities identified in normal lungs were distinct from those of asthmatics. Significantly, the study estimated that 2000 genome equivalents per cm^2 of lung were present, similar to the microbial load in the small intestines. Further studies are needed to understand the airway microbiome in normal hosts; however, the invasive nature of sample collection increases the difficulty of obtaining samples, particularly in pediatric populations.

6.2.2. Cystic Fibrosis

Cystic fibrosis (CF), the most common life-shortening genetic disease in Caucasians, is caused by abnormalities in the gene that codes for the CF transmembrane regulator (CFTR) protein, a chloride channel present in epithelial cells [25]. In the lungs, CFTR dysfunction leads to dehydration of the airway surface liquid layer, impaired mucociliary clearance, and chronic infection [16]. Progressive lung disease, marked by a cycle of infection and inflammation, is responsible for the majority of morbidity and mortality in CF patients [25].

Bacterial pathogens, including *Pseudomonas aeruginosa*, *Staphylococcus aureus*, and members of the *Burkholderia cepacia* complex, are frequently detected from CF airway samples (sputum, oropharyngeal swab, or bronchoalveolar lavage) and contribute to lung disease progression [11, 72]. However, airway cultures may be negative, even in patients diagnosed with an acute pulmonary exacerbation [85] and inflammation in the absence of identified pathogens has been observed in multiple studies [5, 52, 66], suggesting the presence of unsuspected bacteria [44, 69, 70].

More recent studies using stringent anoxic culture techniques demonstrate that anaerobes are frequently present in CF airway samples in quantities comparable to classic CF pathogens [20, 77, 82]. The contributions of these bacteria to lung disease are unclear; however, Worlitzsch and colleagues found high quantities of anaerobic bacteria in sputum samples from CF patients with acute pulmonary exacerbation suggesting anaerobes may be pathogenic [82]. Additional evidence for the presence of anaerobes in CF includes earlier culture studies [9, 36, 76], and *in vivo* measurement of oxygen in mucus plugs showing a steep gradient with anoxic conditions within a few millimeters of the surface of mucus plugs [83]. All these studies suggest that anaerobes may contribute to CF lung disease progression; however, the difficulty of performing anaerobic culture limits this avenue of research to a few dedicated research laboratories. Molecular diagnostic tools offer an efficient way to identify many types of bacteria present in samples in parallel. In the remainder of this section we review what is known about the CF microbiome from the application of molecular diagnostic tools.

The microbiome in CF has been examined by a number of groups and approaches (Table 6.1). The earliest analyses utilized PCR amplified small subunit ribosomal RNA (SSUrRNA) genes cloned into *E. coli*, and colony hybridization with organism-specific probes for typical CF-associated pathogens [79]. The colony hybridization approach was used to interrogate a larger number of clones than was feasible by sequencing at that time. However, the limited sequencing applied to the same samples demonstrated that hybridization resulted in many false negatives. Important results from this study included the presence of many probe-negative colonies suggesting additional diversity beyond standard CF pathogens were present, and the identification of classic CF pathogen DNA present in the absence of culture detection of specific CF pathogens.

Other early efforts to examine the microbiome of CF patients utilized fingerprinting methods (TTGE, LH-PCR, and T-RFLP) to identify bacteria associated with CF airway secretions [40, 61, 64]. These studies identified complex fingerprints, including DNA fragments consistent with classic CF pathogens and others indicative of organisms not classically associated with CF lung disease. Similar to the hybridization approach, the primary limitation of these studies was the sparse sequence data to corroborate the identifications from the fingerprint analysis and provide specific information about the bacteria present. Particular bands identified in a fingerprint often indicate the potential presence of multiple organisms, either close relatives or very disparate types of bacteria [19, 64]. Despite this limitation, these studies were an early indication of the complexity present in the CF airways.

In particular, T-RFLP has been used extensively to study samples from the CF patient population. Sibley and colleagues used a novel approach, where isolates

TABLE 6.1. Overview of Lung Microbiome Studies

Condition	Technique	Number of Subjects	Reference
Normal controls	Sanger	8	31
Asthma	Sanger	24	31
Bronchopulmonary dysplasia	T-RFLP, Sanger	8	73
Cystic fibrosis	Hybridization, Sanger	6	79
	TTGE, Sanger	13	40
	T-RFLP, Sanger	14	64
	T-RFLP, Sanger	34	61
	T-RFLP, Sanger	17	63
	T-RFLP, Sanger	19	62
	Sanger	28	30
	Sanger	25	7
	Pyrosequence, Sanger	1	3
	Sanger	4	27
	PhyloChip	45	—
	DGGE	5	54
	T-RFLP	10	65
Chronic obstructive pulmonary disease	Sanger	5	31
	PhyloChip	8	34
Ventilator-associated pneumonia	PhyloChip	7	21
	Sanger	16	4

from matched samples, were used to identify the organisms responsible for particular T-RFLP fragments. This approach identified an important group of bacteria, the *Streptococcus milleri* group, associated with pulmonary exacerbations in sputum from adults [71]. T-RFLP has also been used to show that the communities present in the oral cavity differ from those in sputum [62], which is an important control for the veracity of organisms identified from lung samples that travel through the mouth. The active fraction of bacteria based on presence of ribosomal RNA versus DNA amplification [63] has also been addressed by T-RFLP. RNA based identification was better suited to identifying the active or viable members of the community than DNA, which may persist longer after cell death. However, there are technical limitations to this approach, including potential differences in efficiency of reverse transcription in different sequences [12, 86], documented instances where RNA content is not proportional to activity of the cell [22], and the potential of processing events to fragment ribosomal RNAs that would limit the ability to recover processed RNAs by amplification. These limitations are poorly quantified across the diversity of bacteria present in the CF airway, and thus introduce assumptions that limit the ability to evaluate these studies.

Microbial community comparisons from expectorated and serial induced sputum samples (five per patient) have also been performed using T-RFLP and found no significant differences between samples, similar to previous studies of inflammation in serial samples [1, 51]. However, more diversity was found over multiple samples, with only 58% of total diversity identified in a given sample. This is likely attributed to regional heterogeneity, which has been observed in BAL by culture [28]. These results suggest that utilization of a single sample for clinical evaluation of bacteriology may not be adequate for routine identification of bacteria in sputum.

We, and others, have utilized Sanger sequencing as the primary approach to identify the bacterial species present in CF airway samples [7, 30]. In our study of bronchoalveolar lavage specimens, we identified 65 species of bacteria from CF patients. Importantly, in 25% of the patients the communities were dominated by anaerobic species, and recognized pathogens were not identified in the community. Additionally, clinical microbiology results for 25% of patients were negative for CF pathogens, and the two groups only partially overlapped with molecular identification of typical CF pathogens in culture-negative samples. We have subsequently shown that culture-negative results during pulmonary exacerbation are not uncommon [85]. Bittar and colleagues conducted a similar study analyzing sputum from 25 adult CF subjects with Sanger sequencing [7]. In this study 53 bacterial species were identified, including a large number of anaerobes. They also used 16*S* sequencing to identify culture isolates from the samples and found that standard phenotypic approaches failed to correctly identify 25% of the isolates. As in our study, sequence based identification found standard CF pathogens (*P. aeruginosa* and *S. aureus*) that were not detected by culture.

More recent studies using the PhyloChip, a proprietary (to Affymetrix) ribosomal RNA gene microarray capable of identification of thousands of bacteria in parallel, to study the microbiome in oropharyngeal samples from 45 children with CF revealed high levels of diversity with >1000 taxa identified in some samples [39]. Lower diversity was associated with increasing age, treatment with oral or inhaled antibiotics, and colonization with *P. aeruginosa*. Bacterial community composition differed by genotype; subjects with at least one copy of the CFTR mutation F508del, deletion of phenylalanine at position 508, were more similar to each other than were those with no copies of F508del. Analysis of community composition demonstrated that individuals colonized by *P. aeruginosa* contained dissimilar communities while patients without *P. aeruginosa* were more similar. The details of this finding may be important; for example, if a particular community is needed for colonization by *P. aeruginosa*, it may offer a new approach for delaying the acquisition of *P. aeruginosa*. In a second study, Cox and colleagues analyzed airway samples (deep throat swab and expectorated sputum samples) from 51 pediatric and adult CF subjects [15]. As in the Klepac–Ceraj study, decreased diversity was associated with advanced age and additionally associated with decreased lung function. Bacterial community structure also varied by age, lung function, and CF genotype. Dominant bacteria within the community shifted with age, with more members of Pseudomonadaceae, Burkholderiaceae, Thermoactinomycetaceae, and Xanthomonadaceae emerging with increased age and decreased lung function.

Next-generation sequencing approaches applied to CF airway specimens have been piloted on small numbers of samples [3, 27]. Armougon and colleagues examined a single sputum sample with culture, Sanger sequencing, and 454 pyrosequencing (GS20, ~100 nucleotide sequences). The increased depth of sequencing identified more diversity than did culture or Sanger, but the short reads used in this study made organism identification difficult. This issue has been largely resolved by the increases in read length with subsequent versions of 454 pyrosequencing [45]. Guss and colleagues applied the second generation of 454 pyrosequencing (FLX) to three samples with extensive data collection (>30,000 sequences per sample). They identified >60 genera of bacteria, including substantial numbers of anaerobes. This study obtained anonymous samples, so no information is available regarding the clinical characteristics of the patients studied.

To date, microbiome studies in CF have generally been small (average 16 patients), single-center studies. Further, clinical information included in most microbiome studies in CF is limited. Going forward larger, multicenter clinical studies, such as those performed with airway cultures [11, 55, 72] should be a focus. It is likely that there will be geographic differences similar to what has been seen by culture for particular groups of CF pathogens [55]. One study using T-RFLP revealed differences in bacterial communities between CF patients in the United States and United Kingdom, supporting this theory [74].

Several additional areas of CF microbiome investigations are currently developing. Studies in CF that expand beyond the lung microbiome are needed. More recent evidence of decreased pulmonary exacerbations after ingestion of probiotic formulations suggests that the gastrointestinal tract may systematically influence inflammation [10, 81]. Another critical need in the CF community is better markers for outcomes in clinical trials [84]. Evaluation of the microbiome, particularly for inhaled antibiotic trials, as a source of new outcome measures should be a priority. We are currently evaluating the microbiome for potential biomarkers of disease progression, with the goal of building prognostic models of CF lung disease to improve clinical care of CF patients.

6.2.3. Ventilator-Associated Pneumonia

Mechanical ventilation is extremely important for support of critically ill patients, but also results in morbidity due to pneumonia caused by nosocomial infections, termed *ventilator-associated pneumonia* (VAP). Ventilator-associated pneumonia is a significant problem, with an incidence of 8–28% and mortality rates of 24–76% depending on the population studied [14]. Many cases of VAP are caused by aspiration of bacterial pathogens that colonize the oropharynx. Entry into the lung is facilitated by the insertion of the endotracheal tube necessary for ventilation. In addition, airway secretions are moved to the endotracheal tube by mucociliary clearance, but cannot exit the airways as intended, disrupting normal host defense mechanisms.

The most common bacterial pathogens associated with VAP are Gram-negative bacteria, including *P. aeruginosa*, Enterobacteriaceae, *Acinetobacter* sp., and Gram-positive bacteria, most commonly *S. aureus* [14]. Anaerobes are infrequently detected in VAP, although one study identified anaerobes in 23% of VAP cases ($n = 130$) [17]. However, in a second study ($n = 143$ patients) only a single case was associated with anaerobic isolates [48]. Differences in sample handling and culture techniques may account for some of the variation between studies.

Sanger sequencing and microarray approaches have been employed for microbiome analysis of VAP patients in two more recent studies. The first focused specifically on patients colonized by *P. aeruginosa* [21], while the second did not require any specific detection of bacterial pathogens as inclusion criteria [4]. In both studies diverse communities were identified that included known pathogens. Bahrani-Mougeot and colleagues analyzed oral and bronchoalveolar lavage samples from 39 adult trauma patients intubated and mechanically ventilated. Using 16*S* rRNA gene amplification and Sanger sequencing, known VAP pathogens were identified from both the oral cavity and lung samples from 14 subjects. In addition, 40 bacterial species, including anaerobes, that had not previously been associated with VAP were detected from lung samples. In nine subjects, known VAP-associated pathogens were detected by molecular analysis but not by culture. This is particularly important given that VAP mortality decreases when appropriate antibiotics are initiated [41, 46].

Flanagan and colleagues employed both Sanger sequencing and hybridization approaches to study VAP patients. There was agreement between the two approaches, but greater diversity was observed by microarray [21]. Antibiotic treatment decreased the diversity present in endotracheal aspirate samples (airway secretions removed from the endotracheal tube by suction), and led to dominance of the community by a single organism. Interestingly, *P. aeruginosa* was enriched even though the *in vitro* susceptibility of isolates indicated that antibiotic treatment was appropriate. Microarray analysis with the PhyloChip identified coordinated change between particular groups of bacteria [8]. Dominance of the community by *P. aeruginosa* came at the expense of loss of organisms from commensal groups (e.g., *Streptococcus*), suggesting that strategies that preserve commensal organisms may improve outcomes by exclusion of opportunistic pathogens from the lung.

6.2.4. Bronchopulmonary Dysplasia

Bronchopulmonary dysplasia (BPD) is a chronic lung disease that primarily affects premature infants, typically following mechanical ventilation for respiratory distress syndrome due to surfactant deficiency. BPD is characterized by decreased lung alveolarization and changes to the pulmonary vasculature [75]. Use of exogenous surfactant protein as a treatment and application of different ventilation modalities have improved the clinical course of BPD patients, but significant morbidity persists. *Ureaplasma* species have been associated with increased risk of BPD in some studies, but the link has not been identified in all studies [80]. Limited work has been performed to document the microbiome in this patient population. Stressmann and colleagues utilized T-RFLP to identify the bacteria in endotracheal aspirates from intubated premature infants with BPD [73]. Multiple terminal restriction fragment bands were observed, and sequence data identified the dominant organism present in each sample along with some minor constituents. These data suggest that the mechanically ventilated preterm infant lung is colonized rapidly after birth, and may explain the observed inflammatory response linked to BPD. However, the etiology is considered multifactorial, with inputs from mechanical ventilation, hyperoxic conditions, dysregulated inflammation, and infection. Thus, large numbers of patients are likely needed to identify the role of the microbiome in this disease.

6.2.5. Asthma

Asthma is a common chronic inflammatory condition characterized by intermittent attacks of shortness of breath, cough, and wheeze secondary to reversible airway bronchoconstriction and mucus hypersecretion [53]. Asthma affects approximately

300 million people worldwide. Infections, (e.g., bacterial, viral, fungal) likely impact the pathogenesis of asthma; however, the mechanisms by which infection influences asthma development and exacerbations are not well understood [68].

There are data suggesting that early colonization influences the onset of asthma, based on cultivation studies in the Copenhagen Prospective Study of Asthma birth cohort [6]. In this study colonization at one month by any of three pulmonary pathogens (*Streptococcus pneumoniae*, *Haemophilus influenzae*, or *Moraxella catarrhalis*) was associated with increased risk of developing asthma. Interestingly, colonization by *Staphylococcus aureus* was not associated with asthma and may be protective, presumably by excluding other organisms from the oropharynx.

Hilty and colleagues examined bronchial brushings from 11 asthmatic adults and bronchoalveolar lavage (BAL) from 13 children with difficult-to-control asthma. Control samples for the adults were obtained from normal volunteers and patients with COPD (discussed below). Disease controls were included for the pediatric samples, and BAL samples were obtained in the absence of antibiotics and clinical evidence of infection. Comparisons were also made between the airways and other adjacent sites (oropharynx and nose) in the adult group. Airway microbial communities differed between asthma, disease controls, and healthy controls. This suggests that clinically relevant information may be present in the microbiome of asthmatics. However, this is based on a small number of patients, and larger studies are indicated. There may be only a subset of individuals where the specific microbiota present affect disease, as seen in inflammatory bowel disease (IBD) [23], which will further increase the number of subjects required to identify the relationship to disease.

6.2.6. Chronic Obstructive Pulmonary Disease (COPD)

COPD is a chronic lung condition primarily caused by cigarette smoke that describes chronic bronchitis and emphysema [32]. Exposure to other sources of smoke, such as open fires for cooking, is also associated with development of this type of lung disease. COPD causes significant morbidity and mortality, and is the only leading cause of death that is increasing in prevalence [50].

The role of infection is controversial in COPD, but bacteria are associated with approximately half of the episodes of pulmonary exacerbation in this group of patients. Typical organisms associated with COPD include *H. influenzae*, *S. pneumoniae*, *M. catarrhalis*, and in advanced disease *P. aeruginosa* [67]. Limited information is available about the microbiome in this disease. A single study of five patients, included as disease controls, based on sequence data is available [31]. The microbiome was studied only during stable pulmonary function, and the most prominent genera were similar to those of the normal controls. Thus, the clinical significance of the microbiome is unclear.

Huang and colleagues examined eight COPD patients hospitalized for pulmonary exacerbation that required mechanical ventilation using the PhyloChip [34]. Hundreds of bacterial taxa were identified, and duration of intubation at time of sampling was inversely associated with the richness of the communities observed. A core set of 75 taxa was present in all patients, and contained a large number of potential human pathogens. Members of Pseudomonodaceae and Enterobacteriaceae dominated patients with lower-diversity communities that had been intubated longer at time of sampling.

Additional studies are needed in this patient population to determine the stability of lung communities, and how they relate to clinical phenotypes. Longitudinal study designs that examine the change with clinical status should have priority. Identification of particular organisms associated with pulmonary exacerbation may lead to new areas of investigation that result in better clinical management for patients with COPD.

6.3. OUTLOOK FOR RESPIRATORY TRACT MICROBIOME STUDIES

Future studies are needed to determine the relationship between airway microbial communities and clinical outcome in a variety of conditions. If, as suggested by early microbiome studies, the normal lung is not sterile, then understanding how microbial communities shift during acute and chronic disease states may lead to better diagnostic and treatment options. Continuous challenge of the airways by environmental and human-associated bacteria does not normally result in disease; however, in the case of chronic lung disease there is increased opportunity for persistent colonization due to change in normal lung physiology. Chronic colonization may result in modification of the clinical course of disease by alteration of the *in vivo* lung environment (e.g., change in inflammatory state). It is also important to recognize that the role of the microbiota may change over time within individuals. Thus, longitudinal analysis of patients should be utilized to determine the role of the microbiota in lung disease.

The selection of methods for study of the airway microbiome is diverse. High cost of sequencing has led to the development of many fingerprinting and hybridization methods, but with per base cost of sequencing decreasing rapidly, sequence-based approaches are now cost competitive (<$100 per sample), and can accommodate 100s of samples in parallel [29]. Sequences are discrete, quantifiable entities that are maintained in public databases and can be incorporated into subsequent analyses across studies. Other types of data can be compared between studies, but not in the quantitative methods possible with sequence data. Further, as sequencing expands beyond the ribosomal RNA gene with metagenomic approaches, fingerprinting methods will not be feasible, because of the large number of genes under investigation.

Investigation of the microbiota is complicated by high variability between communities associated with individuals. Thus, inclusion of adequate patients to address the clinical relevance of the microbiota is critical. Few of the initial studies of the lung microbiota have contained adequate numbers of patients to allow generalization of the results. The amount of clinical information in lung microbiota studies should also be increased. In many studies, clinical relevance was not a primary factor in the design, but translation of microbiome information to improve clinical practice should be a high priority.

Understanding the relevance of bacteria identified will require expanded approaches to demonstrate activity of particular populations within the lung. Efforts to identify active groups by modification of DNA prior to extraction such that only DNA from viable cells can serve as templates for PCR are based on assumptions that need specific investigation. There are little data to determine how distinct lineages of bacteria react to the reagents utilized. Further, it is expected that treatment of the sample to allow processing may alter the viability of certain organisms (e.g., anaerobes), potentially increasing false negatives. An alternative approach, metatranscriptomics, shotgun sequencing cDNA from bulk RNA extracted directly from clinical samples, involves fewer assumptions about how sample manipulation affects the study of active populations. Metatranscriptomics, however, may require stringent collection protocols due to rapid alterations of bacterial RNA populations that would no longer reflect *in vivo* activity. Separation of human RNA will also require protocol development until sequence cost decreases sufficiently to allow parallel characterization of the host–microbiome function. The results of this approach will provide a metabolic snapshot of activity in the lung, and potentially provide insight into clinical management and novel approaches to control the microbiome and disease.

In chronic disease, the use of approaches successfully used to treat acute infections may not prove as effective. In chronic disease, underlying alteration of the host may prevent the complete resolution of detrimental host–microbiome interactions. The ability to maintain stability of the microbial communities and host response may provide better clinical outcomes by maintaining homeostasis. To achieve this, detailed knowledge of the myriad interactions that occur between the microbial communities and host must be determined.

A variety of approaches are needed to describe the respiratory tract microbiome present in health and disease, with the goal of improving human health. The ability to analyze the complex datasets generated by these approaches is another area where much development is needed. Statistical approaches that accurately calculate significance by accounting for novelty in data structure and repeated measurements in longitudinal designs are under development, and are of critical importance for the success of future microbiome studies.

In summary, the human lung contains complex microbial communities. These communities dramatically influence clinical outcome in some diseases (e.g., cystic fibrosis), and likely influence outcomes in other diseases. Because of the difficulty involved in traditional bacterial detection and identification techniques, the influence of these communities in complex chronic disease is not well understood. New methods applied to the microbiome now allow more comprehensive identification of many distinct organisms in parallel. These approaches provide the ability to study and determine even subtle contributions that these communities have on clinical outcomes, and promise novel approaches to control of chronic disease.

REFERENCES

1. Aitken ML, etc. Greene KE, Tonelli MR, Burns JL, Emerson CJ, Goss CH, Gibson RL. Analysis of sequential aliquots of hypertonic saline solution-induced sputum from clinically stable patients with cystic fibrosis. *Chest* **123**(3):792–799 (2003).
2. Amann RI, Ludwig W, Schleifer KH. Phylogenetic identification and in situ detection of individual microbial cells without cultivation. *Microbiol Rev* **59**(1):143–169 (1995).
3. Armougom F, Bittar F, Stremler N, Rolain JM, Robert C, Dubus JC, Sarles J, Raoult D, La Scola B. Microbial diversity in the sputum of a cystic fibrosis patient studied with 16S rDNA pyrosequencing. *Eur J Clin Microbiol Infect Dis* **28**(9):1151–1154 (2009).
4. Bahrani-Mougeot FK, Paster BJ, Coleman S, Barbuto S, Brennan MT, Noll J, Kennedy T, Fox PC, Lockhart PB. Molecular analysis of oral and respiratory bacterial species associated with ventilator-associated pneumonia. *J Clin Microbiol* **45**(5):1588–1593 (2007).

5. Balough K, McCubbin M, Weinberger M, Smits W, Ahrens R, Fick R. The relationship between infection and inflammation in the early stages of lung disease from cystic fibrosis. *Pediatr Pulmonol* **20**(2):63–70 (1995).
6. Bisgaard H, Hermansen MN, Buchvald F, Loland L, Halkjaer LB, Bonnelykke K, Brasholt M, Heltberg A, Vissing NH, Thorsen SV, et al. Childhood asthma after bacterial colonization of the airway in neonates. *N Engl J Med* **357**(15):1487–1495 (2007).
7. Bittar F, Richet H, Dubus JC, Reynaud-Gaubert M, Stremler N, Sarles J, Raoult D, Rolain JM. Molecular detection of multiple emerging pathogens in sputa from cystic fibrosis patients. *PLoS ONE* **3**(8):e2908 (2008).
8. Brodie EL, Desantis TZ, Joyner DC, Baek SM, Larsen JT, Andersen GL, Hazen TC, Richardson PM, Herman DJ, Tokunaga TK, et al. Application of a high-density oligonucleotide microarray approach to study bacterial population dynamics during uranium reduction and reoxidation. *Appl Environ Microbiol* **72**:6288–6298 (2006).
9. Brook I, Fink R. Transtracheal aspiration in pulmonary infection in children with cystic fibrosis. *Eur J Resp Dis* **64**(1):51–57 (1983).
10. Bruzzese E, Raia V, Spagnuolo MI, Volpicelli M, De Marco G, Maiuri L, Guarino A. Effect of Lactobacillus GG supplementation on pulmonary exacerbations in patients with cystic fibrosis: A pilot study. *Clin Nutr* **26**:322–328 (2007).
11. Burns JL, Emerson J, Stapp JR, Yim DL, Krzewinski J, Louden L, Ramsey BW, Clausen CR. Microbiology of sputum from patients at cystic fibrosis centers in the United States. *Clin Infect Dis* **27**(1):156–163 (1998).
12. Bustin SA, Nolan T. Pitfalls of quantitative real-time reverse-transcription polymerase chain reaction. *J Biomol Techn* **15**(3):155–166 (2004).
13. Cabello H, Torres A, Celis R, El-Ebiary M, Puig de la Bellacasa J, Xaubet A, Gonzalez J, Agusti C, Soler N Bacterial colonization of distal airways in healthy subjects and chronic lung disease: A bronchoscopic study. *Eur Resp J* **10**(5):1137–1144 (1997).
14. Chastre J, Fagon JY. Ventilator-associated pneumonia. *Am J Resp Crit Care Med* **165**(7): 867–903 (2002).
15. Cox MJ, Allgaier M, Taylor B, Baek MS, Huang YJ, Daly RA, Karaoz U, Andersen GL, Brown R, Fugimura KE, et al. Airway microbiota and pathogen abundance in age-stratified cystic fibrosis patients. *PLoS ONE* **5**(6):e11044 (2010).
16. Donaldson SH, Boucher, RC. Update on pathogenesis of cystic fibrosis lung disease. *Curr Opin Pulmon Med* **9**(6):486–491 (2003).
17. Dore P, Robert R, Grollier G, Rouffineau J, Lanquetot H, Charriere JM, Fauchere JL. Incidence of anaerobes in ventilator-associated pneumonia with use of a protected specimen brush. *Am J Resp Crit Care Med* **153**:1292–1298 (1996).
18. Doud M, Zeng E, Schneper L, Narasimhan G, Mathee K. Approaches to analyse dynamic microbial communities such as those seen in cystic fibrosis lung. *Human Genom* **3**(3): 246–256 (2009).
19. Dunbar J, Ticknor LO, Kuske CR. Phylogenetic specificity and reproducibility and new method for analysis of terminal restriction fragment profiles of 16s rRNA genes from bacterial communities. *Appl Environ Microbiol* **67**(1):190–197 (2001).
20. Field TR, Sibley CD, Parkins MD, Rabin HR, Surette MG. The genus Prevotella in cystic fibrosis airways. *Anaerobe* **16**(4):337–344 (2010).
21. Flanagan JL, Brodie EL, Weng L, Lynch SV, Garcia O, Brown R, Hugenholtz P, Desantis TZ, Andersen GL, Wiener-Kronish JP, et al. Loss of bacterial diversity during antibiotic treatment of intubated patients colonized with Pseudomonas aeruginosa. *J Clin Microbiol* **45**(6):1954–1962 (2007).

22. Flardh K, Cohen PS, Kjelleberg S. Ribosomes exist in large excess over the apparent demand for protein synthesis during carbon starvation in marine Vibrio sp. Strain CCUG 15956. *J Bacteriol* **174**(21):6780–6788 (1992).

23. Frank DN, St Amand AL, Feldman RA, Boedeker EC, Harpaz N, Pace NR. Molecular-phylogenetic characterization of microbial community imbalances in human inflammatory bowel disease. *Proc Natl Acad Sci USA* **104**(34):13780–13785 (2007).

24. Gao Z, Tseng CH, Strober BE, Pei Z, Blaser MJ. Substantial alterations of the cutaneous bacterial biota in psoriatic lesions. *PLoS ONE* **3**(7):e2719 (2008).

25. Gibson RL, Emerson J, McNamara S, Burns JL, Rosenfeld M, Yunker A, Hamblett N, Accurso J, Dovey M, Hiatt P, et al. Significant microbiological effect of inhaled tobramycin in young children with cystic fibrosis. *Am J Resp Crit Care Med* **167**(6):841–849 (2003).

26. Gleeson K, Eggli DF, Maxwell SL. Quantitative aspiration during sleep in normal subjects. *Chest* **111**(5):1266–1272 (1997).

27. Guss AM, Roeselers G, Newton IL, Young CR, Klepac-Ceraj V, Lory S, Cavanaugh CM. Phylogenetic and metabolic diversity of bacteria associated with cystic fibrosis. *ISME J* **5**:20–29 (2010).

28. Gutierrez JP, Grimwood K, Armstrong DS, Carlin JB, Carzino R, Olinsky A, Robertson CF, Phelan PD. Interlobar differences in bronchoalveolar lavage fluid from children with cystic fibrosis. *Eur Resp J* **17**(2):281–286 (2001).

29. Hamady M, Walker JJ, Harris JK, Gold NJ, Knight R. Error-correcting barcoded primers for pyrosequencing hundreds of samples in multiplex. *Nat Meth* **5**(3):235–237 (2008).

30. Harris JK, De Groote MA, Sagel SD, Zemanick ET, Kapsner R, Penvari C, Kaess H, Deterding RR, Accurso FJ, et al. Molecular identification of bacteria in bronchoalveolar lavage fluid from children with cystic fibrosis. *Proc Natl Acad Sci USA* **104**(51):20529–20533 (2007).

31. Hilty M, Burke C, Pedro H, Cardenas P, Bush A, Bossley C, Davies J, Ervine A, Poulter L, Pachter L, et al. Disordered microbial communities in asthmatic airways. *PLoS ONE* **5**(1):e8578 (2010).

32. Hogg JC, Chu F, Utokaparch S, Woods R, Elliott WM, Buzatu L, Cherniack RM, Rogers RM, Sciurba FC, Coxson HO, et al. The nature of small-airway obstruction in chronic obstructive pulmonary disease. *N Engl J Med* **350**(26):2645–2653 (2004).

33. Holtzman MJ, Tyner JW, Kim EY, Lo MS, Patel AC, Shornick LP, Agapov E, Zhang Y. Acute and chronic airway responses to viral infection: Implications for asthma and chronic obstructive pulmonary disease. *Proc Am Thor Soc* **2**(2):132–140 (2005).

34. Huang YJ, Kim E, Cox MJ, Brodie EL, Brown R, Wiener-Kronish JP, Lynch SV. A persistent and diverse airway microbiota present during chronic obstructive pulmonary disease exacerbations. *OMICS* **14**(1):9–59 (2010).

35. Hugenholtz P. Exploring prokaryotic diversity in the genomic era. *Genome Biol* **3**(2): (2002) (review).

36. Jewes LA, Spencer RC. The incidence of anaerobes in the sputum of patients with cystic fibrosis. *J Med Microbiol* **31**(4):271–274 (1990).

37. Kim JS, Okamoto K, Rubin BK. Pulmonary function is negatively correlated with sputum inflammatory markers and cough clearability in subjects with cystic fibrosis but not those with chronic bronchitis. *Chest* **129**(5):1148–1154 (2006).

38. Kirkpatrick M.B, Bass JB Jr. Quantitative bacterial cultures of bronchoalveolar lavage fluids and protected brush catheter specimens from normal subjects. *Am Rev Resp Dis* **139**(2):546–548 (1989).

39. Klepac-Ceraj V, Lemon KP, Martin TR, Allgaier M, Kembel SW, Knapp AA, Lory S, Brodie EL, Lynch SV, Bohannan BJ, et al. Relationship between cystic fibrosis respiratory

tract bacterial communities and age, genotype, antibiotics and Pseudomonas aeruginosa. *Environ Microbiol* **12**(5):1293–1303 (2010).

40. Kolak M, Karpati F, Monstein HJ, Jonasson J Molecular typing of the bacterial flora in sputum of cystic fibrosis patients. *Int J Med Microbiol* **293**(4):309–317 (2003).
41. Kollef MH, Bock KR, Richards RD, Hearns ML. The safety and diagnostic accuracy of minibronchoalveolar lavage in patients with suspected ventilator-associated pneumonia. *Ann Inter Med* **122**(10):743–748 (1995).
42. Konstan MW, Hilliard KA, Norvell TM, Berger M. Bronchoalveolar lavage findings in cystic fibrosis patients with stable, clinically mild lung disease suggest ongoing infection and inflammation. *Am J Resp Crit Care Med* **150**(2):448–454 (1994).
43. Laurenzi GA, Potter RT, Kass EH. Bacteriologic flora of the lower respiratory tract. *N Engl J Med* **265**:1273–1278 (1961).
44. Lipuma JJ. The changing microbioal epidemiology in cystic fibrosis. *Clin Microbiol Rev* **23**(2):299–323 (2010).
45. Liu Z, Desantis TZ, Andersen GL, Knight R. Accurate taxonomy assignments for 16S rRNA sequences produced by highly parallel pyrosequencers. *Nucleic Acids Res* **36**(18): e120 (2008).
46. Luna CM, Vujacich P, Niederman MS, Vay C, Gherardi C, Matera J, Jolly EC. Impact of BAL data on the therapy and outcome of ventilator-associated pneumonia. *Chest* **111**(3): 676–685 (1997).
47. Marik PE. Aspiration pneumonitis and aspiration pneumonia. *N Engl J Med* **344**(9): 665–671 (2001).
48. Marik PE, Careau P. The role of anaerobes in patients with ventilator-associated pneumonia: A prospective study. *Chest* **115**(1):178–183 (1999).
49. Martin TR, Frevert CW. Innate immunity in the lungs. *Proc Am Thor Soc* **2**(5):403–411 (2005).
50. Mathers CD, Loncar D. Projections of global mortality and burden of disease from 2002 to 2030. *PLoS Med* **3**(11):e442 (2006).
51. Meyer KC, Sharma A. Regional variability of lung inflammation in cystic fibrosis. *Am J Resp Crit Care Med* **156**(5):1536–1540 (1997).
52. Muhlebach MS, Stewart PW, Leigh MW, Noah TL. Quantitation of inflammatory responses to bacteria in young cystic fibrosis and control patients. *Am J Resp Crit Care Med* **160**(1):186–191 (1999).
53. Murphy DM, O'Byrne PM. Recent advances in the pathophysiology of asthma. *Chest* **137**(6):1417–1426 (2010).
54. Nelson A, De Soyza A, Bourke SJ, Perry JD, Cummings SP. Assessment of sample handling practices on microbial activity in sputum samples from patients with cystic fibrosis. *Lett Appl Microbiol* **51**:272–277 (2010).
55. Olivier KN, Weber DJ, Wallace RJ Jr, Faiz AR, Lee JH, Zhang Y, Brown-Elliot BA, Handler A, Wilson RW, Schechter MS, et al. Nontuberculous mycobacteria. I: Multicenter prevalence study in cystic fibrosis. *Am J Resp Crit Care Med* **167**(6):828–834 (2003).
56. Pace NR. A molecular view of microbial diversity and the biosphere. *Science* **276**(5313): 734–740 (1997).
57. Parameswaran P, Jalili R, Tao L, Shokralla S, Gharizadeh B, Ronaghi M, Fire AZ. A pyrosequencing-tailored nucleotide barcode design unveils opportunities for large-scale sample multiplexing. *Nucleic Acids Res* **35**(19):e130 (2007).
58. Perkins SD, Mayfield J, Fraser V, Angenent LT. Potentially pathogenic bacteria in shower water and air of a stem cell transplant unit. *Appl Environ Microbiol* **75**(16):5363–5372 (2009).

59. Relman DA. New technologies, human-microbe interactions, and the search for previously unrecognized pathogens. *J Infect Dis* **186**(Suppl 2):S254–S258 (2002).

60. Riise GC, Larsson S, Larsson P, Jeansson S, Andersson BA. The intrabronchial microbial flora in chronic bronchitis patients: A target for N-acetylcysteine therapy? *Eur Resp J* **7**(1):94–101 (1994).

61. Rogers GB, Carroll MP, Serisier DJ, Hockey PM, Jones G, Bruce KD. Characterization of bacterial community diversity in cystic fibrosis lung infections by use of 16S ribosomal DNA terminal restriction fragment length polymorphism profiling. *J Clin Microbiol* **42**(11):5176–5183 (2004).

62. Rogers GB, Carroll MP, Serisier DJ, Hockey PM, Jones G, Kehagia V, Connett GJ, Bruce KD. Use of 16S rRNA profiling by terminal restriction fragment length polymorphism analysis to compare bacterial communities in sputum and mouthwash samples from patients with cystic fibrosis. *J Clin Microbiol* **44**(7):2601–2604 (2006).

63. Rogers GB, Carroll MP, Serisier DJ, Hockey PM, Kehagia V, Jones GR, Bruce KD. Bacterial activity in cystic fibrosis lung infections. *Resp Res* **6**(6):49 (2005).

64. Rogers GB, Hart CA, Mason JR, Hughes M, Walshaw MJ, Bruce KD. Bacterial diversity in cases of lung infection in cystic fibrosis patients: 16S ribosomal DNA (rDNA) length heterogeneity PCR and 16S rDNA terminal restriction fragment length polymorphism profiling. *J Clin Microbiol* **41**(8):3548–3558 (2003).

65. Rogers GB, Skelton S, Serisier J, van der Gast CJ, Bruce KD. Determining cystic fibrosis-affected lung microbiology: Comparison of spontaneous and serially induced sputum samples by use of terminal restriction fragment length polymorphism profiling. *J Clin Microbiol* **48**(1):78–86 (2010).

66. Rosenfeld M, Gibson RL, McNamara S, Emerson J, Burns JL, Castile R, Hiatt P, McCoy K, Wilson CB, Inglis A, et al. Early pulmonary infection, inflammation, and clinical outcomes in infants with cystic fibrosis. *Pediatr Pulmonol* **32**(5):356–366 (2001).

67. Sethi S, Murphy TF. Infection in the pathogenesis and course of chronic obstructive pulmonary disease. *New England Journal of Medicine* **359**(22):2355–2365 (2008).

68. Sevin CM, Peebles RS Jr. Infections and asthma: New insights into old ideas. *Clin Exp Allergy* **40**(8):1142–1154 (2010).

69. Sibley CD, Parkins MD, Rabin HR, Surette MG. The relevance of the polymicrobial nature of airway infection in the acute and chronic management of patients with cystic fibrosis. *Curr Opin Investig Drugs* **10**(8):787–794 (2009).

70. Sibley CD, Rabin HR, Surette MG. Cystic fibrosis: A polymicrobial infectious disease. *Future Microbiol* **1**(1):53–61 (2006).

71. Sibley CD, Sibley KA, Leong TA, Grinwis ME, Parkins MD, Rabin HR, Surette MG. The Streptococcus milleri population of a cystic fibrosis clinic reveals patient specificity and intraspecies diversity. *J Clin Microbiol* **48**(7):2592–2594 (2010).

72. Spicuzza L, Sciuto C, Vitaliti G, Di Dio G, Leonardi S, La Rosa M. Emerging pathogens in cystic fibrosis: Ten years of follow-up in a cohort of patients. *Eur J Clin Microbiol Infect Dis* **28**(2):191–195 (2009).

73. Stressmann FA, Connett GJ, Goss K, Kollamparambil TG, Patel N, Payne MS, Puddy V, Legg J, Bruce KD, Rogers GB. The use of culture-independent tools to characterize bacteria in endo-tracheal aspirates from pre-term infants at risk for bronchopulmonary dysplasia. *J Perinat Med* **38**(3):333–337 (2010).

74. Stressmann FA, Rogers GB, Klem ER, Lilley AK, Donaldson SH, Daniels TW, Carroll MP, Patel N, Forbes B, Boucher RC, et al. Analysis of the bacterial communities present in lungs of patients with cystic fibrosis from American and british centers. *J Clin Microbiol* **49**(1):281–291 (2011).

75. Thebaud B, Abman SH. Bronchopulmonary dysplasia: Where have all the vessels gone? Roles of angiogenic growth factors in chronic lung disease. *Am J Resp Crit Care Med* **175**(103):978–985 (2007).
76. Thomassen MJ, Klinger JD, Badger SJ, van Heeckeren DW, Stern, RC. Culture of the thoracotomy specimens confirm usefulness of sputum cultures in cystic fibrosis. *J Pediatr* **104**:352–356 (1984).
77. Tunney MM, Field TR, Moriarty TF, Patrick S, Doering G, Muhlebach MS, Wolfgang MC, Boucher R, Gilpin DF, McDowell A, et al. Detection of anaerobic bacteria in high numbers in sputum from patients with cystic fibrosis. *Am J Resp Crit Care Med* **177**(9): 995–1001 (2008).
78. Turnbaugh PJ, Ley RE, Mahowald MA, Magrini V, Mardis ER, Gordon JI. An obesity-associated gut microbiome with increased capacity for energy harvest. *Nature* **444**(7122): 1027–1031 (2006).
79. van Belkum A, Renders NH, Smith S, Overbeek SE, Vergrugh HA. Comparison of conventional and molecular methods for the detection of bacterial pathogens in sputum samples from cystic fibrosis patients. *FEMS Immunol Med Microbiol* **27**(1):51–57 (2000).
80. Viscardi RM, Hasday JD. Role of Ureaplasma species in neonatal chronic lung disease: Epidemiologic and experimental evidence. *Pediatr Res* **65**(5 Pt 2):84R–90R (2009).
81. Weiss B, Bujanover Y, Yahav Y, Vilozni D, Fireman E, Efrati O. Probiotic supplementation affects pulmonary exacerbation in patients with cystic fibrosis: A pilot study. *Pediatr Pulmonol* **45**(6):536–540 (2010).
82. Worlitzsch D, Rintelen C, Bohm K, Wollschlager B, Merkel N, Borneff-Lipp M, Doring G. Antibiotic-resistant obligate anaerobes during exacerbation of cystic fibrosis patients. *Clin Microbiol Infect* **15**(5):454–460 (2009).
83. Worlitzsch D, Tarran R, Ulrich M, Schwab U, Cekici A, Meyer KC, Birrer P, Bellon G, Berger J, Weiss T, et al. Effects of reduced mucus oxygen concentration in airway Pseudomonas infection of cystic fibrosis patients. *J Clin Investig* **109**(3):317–325 (2002).
84. Zemanick ET, Harris JK, Conway S, Konstan MW, Marshall B, Quittner AL, Retsch-Bogart G, Saiman L, Accurso FJ. Measuring and improving respiratory outcomes in cystic fibrosis lung disease: Opportunities and challenges to therapy. *J Cystic Fibrosis* **9**(1):1–16 (2009).
85. Zemanick ET, Wagner BD, Harris JK, Wagener J, Accurso F, Sagel SD. Pulmonary exacerbations in cystic fibrosis with negative bacterial cultures. *Pediatr Pulmonol* **45**(6):569–577 (2010).
86. Zhang J, Byrne CD. Differential priming of RNA templates during cDNA synthesis markedly affects both accuracy and reproducibility of quantitative competition reverse-transcriptase PCR. *Biochem J* **337**(Pt 2):231–241 (1999).
87. Zhang M, Li Q, Zhang XY, Ding X, Zhu D, Zhou X. Relevance of lower airway bacterial colonization, airway inflammation, and pulmonary function in the stable stage of chronic obstructive disease. *Eur J Clin Microbiol Infect Dis* **29**(12):1487–1493 (2010).

7

MICROBIOTA OF THE MOUTH: A BLESSING OR A CURSE?

ANGELA H. NOBBS, DAVID DYMOCK, and HOWARD F. JENKINSON

School of Oral and Dental Sciences, University of Bristol, United Kingdom

7.1. INTRODUCTION

The human mouth, or oral cavity, provides a unique environment for microbial colonization, distinct from other body sites. The ecosystem consists of a variety of microbial niches that have evolved together with the host. The human oral cavity and nasopharynx are home to 100s of different microorganisms, including viruses, bacteria, fungi and protozoa. More than 600 species of bacteria alone have been identified by cultivation or molecular means as potential colonizers, and many additional phylotypes of bacteria have been discovered [26]. In a healthy subject, these oral microbial communities live in relative harmony with the host. They provide many benefits including protection against incoming pathogens by biological exclusion, and continuous low-level stimulation of the immune system so that it is poised to respond rapidly when challenged, for example, by an invasive microorganism. The communities also harbor potential pathogens that can proliferate to cause hard or soft tissue destruction when the conditions become conducive for these activities. The oral microbiome represents one of the first detailed analyses of microbial populations within a human body niche. However, this voyage of discovery is tempered by the problem that a microbial genome sequence alone, without the cultivated organism, provides limited functional information. Therefore, studies of oral microbial interactions leading to the development of communities, and of site-specific

The Human Microbiota: How Microbial Communities Affect Health and Disease, First Edition. Edited by David N. Fredricks.

communities, remain of critical significance in better understanding the factors that influence health or disease [63].

The oral cavity contains a range of mucosal surfaces, including the gingivae or gums, tongue, throat, and buccal mucosa or cheeks. Each of these has their own special characteristics and, with the exception of young infants and older edentulous subjects, the mouth also contains hard, nonshedding enamel surfaces of the teeth. These provide a unique surface for microorganisms to become anchored to, as constituents of dental plaque. Microbial growth is generally controlled by mechanical flushing of surfaces with saliva that contains a spectrum of antimicrobial compounds. However, saliva also promotes microbial colonization through deposition of proteinaceous films or pellicles onto surfaces to which bacteria can then adhere. This chapter considers the ecology and diversity of microorganisms present within the oral cavity, and means by which microbes can colonize the host. The human microbiota is vital for maintenance of health, but under certain conditions can contribute to oral or systemic diseases. Some of the oral conditions are very common, causing pain and discomfort to the subject, while systemic conditions are rarer but may be life-threatening.

7.1.1. Anatomy

The oral cavity is sterile at birth but is immediately exposed to bacteria from the mother and others with whom the infant has close contact. Bacteria become established on one or more of the mucosal surfaces present, since teeth do not erupt until ≥ 6 months. A stratifying squamous epithelium covers all the soft tissues of the oral cavity. The gingivae and hard palate, which are subject to masticatory forces, have a keratinized epithelium. The floor of the mouth and buccal regions, which require functional flexibility for chewing or speech, have a nonkeratinized epithelium. A specialized epithelium, consisting of a mosaic of keratinized and nonkeratinized epithelia, covers the dorsum of the tongue. Interactions of microorganisms with the oral mucosa are influenced by the nature of the epithelium at different sites and by the mechanical forces to which these tissues are exposed. The oral mucosa has a turnover rate of 14–24 days [137], while the clearance rate of surface cells is likely to be within the range of hours rather than days [75]. In contrast, the teeth provide a stable platform for microbial growth and proliferation. Tooth surfaces have a much higher microbial load than do mucosal surfaces, although the papillary structure of the tongue allows for a higher level of microbial colonization.

The unique anatomic features of the mouth thus provide a range of different environments in which microbial communities are able to develop (Figure 7.1a). The communities that colonize the teeth, coated with a salivary pellicle, are quite different from the communities found on the tongue, and these in turn are different from those found at other sites [13]. Streptococci are the major components of early dental plaque [100], while *Neisseria*, *Actinomyces*, and *Gemella* are primary colonizers of mucosal surfaces [86]. *Streptococcus salivarius* is a major colonizer of the tongue dorsum. Anaerobic bacteria such as *Fusobacterium nucleatum* and *Porphyromonas gingivalis* are often found in higher numbers at gum disease sites (see Figure 7.1). To a major extent it is the anatomy that influences the composition of these communities, through provision of different receptors for microbial adhesion and modulation of salivary flow rate. For example, pits and fissures that are present on the

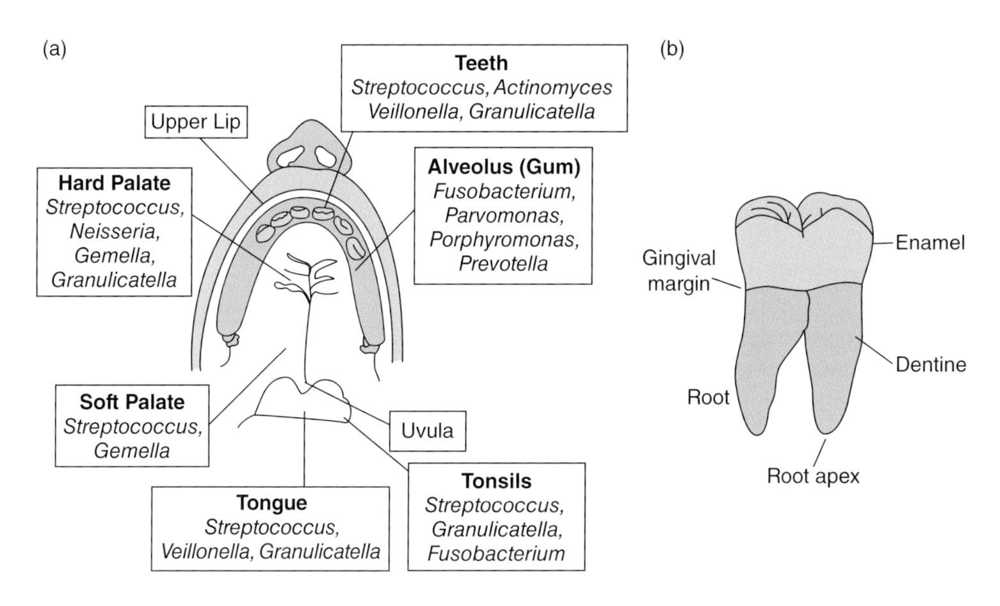

Figure 7.1. Diagrammatic representation of the human oral cavity (a) depicting some of the various surfaces and structures that become colonized by different communities of bacteria in a healthy subject. Each of the structures or sites shown associated with a human tooth (b) may be colonized by bacteria and result in infection.

crowns of the teeth (Figure 7.1b), and the regions between the teeth (interproximal) provide excellent nesting areas, protected from shear forces and salivary flow, for bacterial communities to develop. Moreover, the salivary glycoproteins that make up the salivary pellicle provide receptors for bacterial adhesion that are very different from the receptors expressed by epithelial cells, thus directing specificity of microbial adhesion. This is well exemplified by molecular studies of the predominant microbial populations present on these different surfaces in healthy mouths [2]. *Streptococcus mitis* showed a wide distribution across all oral surfaces, while *Streptococcus australis* was found mainly on the tongue along with *S.salivarius*. On tooth surfaces, *Streptococcus sanguinis* and *Streptococcus gordonii* were present among multiple subjects, while *Streptococcus intermedius* was found only subgingivally.

7.1.2. Saliva

The major host defence system that microorganisms have to contend with in colonizing the oral cavity is saliva. Saliva is a mixture of secretions from three pairs of salivary glands: the submandibular and sublingual glands, located in the floor of the mouth, and parotid glands, toward the upper rear portion of the mouth (the mandibular ramus). The glandular secretions vary in composition and differentially contribute to salivary composition depending on whether glandular secretion is stimulated by mastication. Oral microorganisms in their natural lifestyle on mucosal or enamel surfaces gain nutritional sustenance from the glycoproteins, proteins, and polysaccharides in saliva. More recent research has utilized highly sensitive proteomics techniques to identify more than 2000 proteins that may be present in salivary secretions [83]. Taken in conjunction with evidence for over 1000 microbial taxa being present, the oral cavity ecosystem is universally complex.

Many of the salivary glycoproteins are serum-derived, while the major innate defence molecules present include mucins, lysozyme, lactoferrin, lactoperoxidase, and antimicrobial peptides including human β-defensins. Saliva contains two types of mucin; a high-molecular-mass highly glycosylated molecule designated MG1 (MUC5B), and a lower-molecular-weight mucin MG2 (MUC7). The viscosity of saliva is provided mainly by MG2 [57] and has a major effect on susceptibility of individuals to tooth decay (dental caries). Paradoxically, saliva also provides the microorganisms with a major source of nutrients and a pellicle on teeth for them to adhere to. The microbial communities have evolved to utilize salivary glycoproteins very efficiently as nutrients, breaking down the polysaccharide chains through secretion of glycan hydrolases, such as neuraminidase, β-glucosidases, and the polypeptide backbone with proteases. Many of the bacteria, especially streptococci, carry a large repertoire of genes encoding nutrient transport systems and enzymatic pathways for metabolism of multiple sugars. This underlines the metabolic versatility of oral microorganisms and the fact that most of them grow and survive with saliva as the sole source of nutrients, until of course the host provides them with a larger meal.

7.2. FROM COLONIZATION TO COMMUNITIES

Colonization of the oral cavity is a defined process involving a diverse range of microorganisms that follows a general spatiotemporal model (Figure 7.2). It begins with initial attachment of pioneer bacterial colonizers, which recognize specific host receptors displayed on epithelial cells and within the salivary pellicle of the tooth surface. Interbacterial interactions with this layer of early colonizers then facilitate incorporation of further species, promoting both expansion and microbial diversity of the developing community. As the attached population grows, mutualistic and antagonistic relationships act to further define the microbiota, together with the exchange of signaling molecules that can exert changes in gene expression. Secretion of polymeric substances such as complex carbohydrates then leads to formation of a protective matrix around the microbial community and thus, the multispecies biofilm constituting dental plaque is established. Understanding the critical components at each stage of this process should facilitate the development of novel preventative and therapeutic regimens targeted against oral disease and related systemic complications.

7.2.1. Initial Attachment

"Stick or be swallowed" is the motto that underpins successful microbial colonization and persistence within the oral cavity. Microorganisms entering the mouth must attach to a surface to evade clearance in the salivary flow from the mouth to the digestive tract. Most oral cavity microorganisms have therefore developed the capacity to bind constituents of the salivary film (acquired pellicle) that continuously bathes both soft and hard tissues within the mouth. The precise composition of the acquired pellicle is host-dependent and varies according to the surface on which it forms. The pellicle found on the enamel surface of teeth comprises mucins,

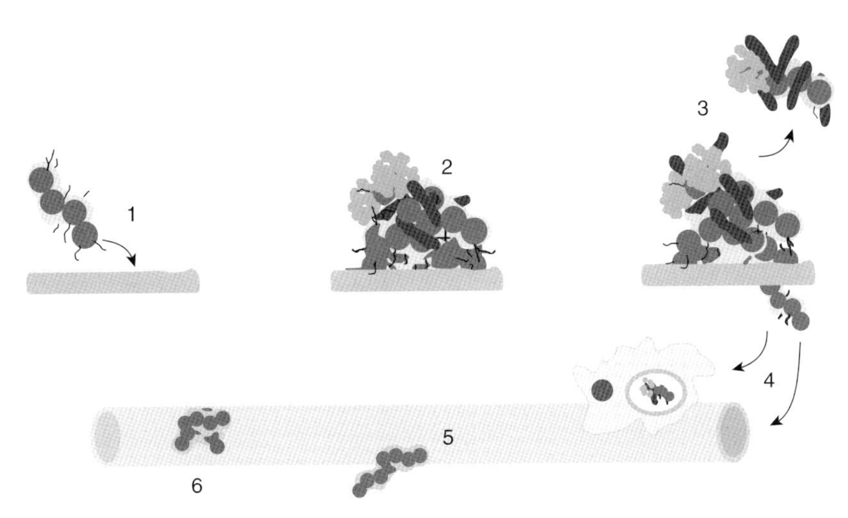

Figure 7.2. Microbial colonization of the oral cavity involves establishment of microbial communities that may then provide a source for systemic spread of microorganisms in the host. Initial deposition (1) of microorganisms onto saliva-coated or epithelial cell surfaces involves longer-range interactions mediated by pili or fimbriae, and shorter-range molecular interactions between bacterial cell surface proteins and substratum receptors. Nonencapsulated bacteria have often been shown to initially adhere better than capsulated organisms (1). Microbial community development then ensues (2) as a result of microbe–microbe interactions (coaggregation). Host macromolecules, such as mucous glycoproteins and collagen, assist in cementing the communities, and highly efficient metabolic co-operations develop. When environmental conditions become less conducive to growth and survival, microorganisms break away from the biofilm communities either singly or in clusters (3), becoming dispersed to colonize other sites. Internalization of microbial cells (4) by host tissues may occur, either through epithelial cell activation or via breaches in tissue integrity. Microbial cells may enter the bloodstream or be taken up by macrophages and become disseminated (4). Recolonization of endothelium can lead to invasion and infection of tissues, for example, the placenta, leading to the possibility of preterm birth or low birth weight (5). Localization of bacteria, especially viridans streptococci, to damaged endothelium may lead to platelet aggregation and development of thrombotic vegetations (6) that restrict bloodflow and promote infective endocarditis.

agglutinins, proline-rich proteins (PRPs), phosphate-rich proteins such as statherin, and enzymes such as α-amylase. Salivary pellicle, together with any associated microbiota, may be removed from the mouth as a result of continual shedding of the epithelial surface or mechanical abrasion (e.g., toothbrushing). Nonetheless, acquired pellicle on enamel re-forms just 30 seconds after professional tooth cleaning, and extensive microbial attachment is visible within a few hours [107].

Bacterial and host surfaces are negatively charged. To initiate attachment, bacteria must first overcome these repulsive forces and target thermodynamically favorable sites. To facilitate this process, many bacteria express surface fibrillar appendages that facilitate long-range ($\geq$1 μm) adhesion. Such elongated structures also promote the likelihood of bacteria making contact with a desirable substratum as they are carried in the salivary flow. These interactions are first driven by noncovalent (e.g., van der Waals), electrostatic, and hydrophobic forces and then, as the distance decreases, are stabilized by hydrogen bonding and divalent cation bridges. Long-range interactions are not, however, sufficiently strong to allow bacterial persistence at the site. For this, higher affinity adhesion is required. This is mediated by surface-expressed bacterial adhesins (usually proteins), which recognize complementary or cognate receptors within the pellicle (often sugars/oligosaccharides). Such adhesin–receptor mechanisms impart specificity on this process, resulting in a temporal succession of microbial colonization. Oral streptococci constitute up to 80% of the bacteria that colonize teeth within the first few hours after professional cleaning, together with *Actinomyces* and *Veillonella* species [27,29]. The particular success of streptococci as so-called early or pioneer colonizers is related, at least in part, to their expression of a vast range of surface adhesins that target multiple salivary pellicle constituents. These include adhesins that recognize salivary agglutinin glycoprotein gp340 (AgI/II family polypeptides), α-amylase (amylase-binding proteins), sialic acid carbohydrate chains (serine-rich repeat proteins), and glucose polymers (glucan-binding proteins) [97].

7.2.2. Physical Interbacterial Interactions

Pioneer colonizers compete for pellicle binding sites, those with the highest-affinity adhesin-receptor interactions under the prevailing environmental conditions succeeding in outperforming other bacteria for available receptors. This results in an initial layer of early colonizers formed on the salivary pellicle. These bacteria are not, however, restricted to targeting salivary receptors, and may also persist by binding receptors presented on the surface of neighboring microbes through a series of coadhesion or coaggregation reactions. Such events enable the attachment and incorporation of additional microbes, and so the multispecies community that constitutes dental plaque develops. These attachment events occur only between compatible partner microorganisms, imposing a distinct, temporal order on plaque formation. Interbacterial interactions begin among early colonizers. For example, type 2 fimbriae expressed on the surface of *Actinomyces* recognize receptor polysaccharide of several streptococci [20]. As the plaque community develops, secondary/late colonizers appear. *F. nucleatum* is considered a particularly important "bridging" organism, as it has the most promiscuous coaggregation capabilities identified to date, interacting with both early and late colonizers. *F. nucleatum* has been shown to bind pioneer colonizer *S. sanguinis* via arginine-specific protein RadD [66], while coaggregation with a number of Gram-negative bacteria is mediated by specific lectin-carbohydrate interactions [72]. These latter microorganisms include *P. gingivalis*, which is also able to bind pioneer streptococci and actinomycetes. In the case of *S. gordonii*, two distinct mechanisms have been shown to underpin this coadhesion. The major (FimA) fimbriae of *P. gingivalis* recognizes streptococcal surface-exposed glyceraldehyde-3-phosphate dehydrogenase [85], while *P. gingivalis* minor (Mfa) fimbriae target *S. gordonii* AgI/II family polypeptide SspB [109].

7.2.3. Metabolic Interbacterial Interactions

In addition to providing a surface for attachment, neighboring microorganisms can exhibit mutualistic relationships that promote growth and community development based on compatible metabolic requirements. For example, as dental plaque accumulates, facultative anaerobes such as oral streptococci deplete oxygen from the immediate surroundings. This can lead to the development of anoxic pockets in which the growth of obligate anaerobes such as *Treponema denticola* and *P. gingivalis* is promoted. Alternatively, intermicrobial interactions may have a nutritional basis, whereby one microorganism excretes a metabolite or breaks down a substrate to release products that can be utilized by a second microorganism. This is elegantly demonstrated when considering a three-species community comprising early colonizers *Streptococcus* and *Veillonella* and late colonizer *P. gingivalis*. Utilizing flow cell culture with saliva as the sole nutrient source, none of these bacteria grow individually as a biofilm. However, significant growth and biofilm formation is achieved when all three species are cultured simultaneously [111]. *Veillonella* lack a fully functional glycolytic pathway and so utilize hydroxyl acids such as lactate as a major energy source. Lactate is excreted as a metabolic waste product by streptococci, and so *Veillonella* is able to grow by consuming the lactate. Concomitantly, depletion of lactate from the immediate environment by *Veillonella* increases the flux of glucose to lactate, which promotes streptococcal growth [45]. *P. gingivalis* is auxotrophic for vitamin K. This compound is not synthesized in humans, but an analogue of vitamin K, menaquinone (vitamin K_2), has been reported in *Veillonella* supernatants [52]. It is proposed, therefore, that *Veillonella* can promote growth of *P. gingivalis* through secretion of this essential compound [53].

A nutrition-based relationship has also been shown for *S. gordonii* and *Actinomyces oris* [60]. In this instance, the presence of *A. oris* enables *S. gordonii* to grow under low-arginine conditions. *A. oris* achieves this by regulating expression of several *S. gordonii* genes, including those associated with stabilization of arginine biosynthesis. This effect is dependent on coaggregation between the two bacterial species.

7.2.4. Antagonistic Interbacterial Interactions

Adhesion mechanisms and physiological dependences are critical to shaping the precise composition of multispecies communities. Nonetheless, antagonistic interactions also play an important role in this process. As mentioned previously, microorganisms within the same environmental niche will likely compete for available binding sites and nutrients. However, some oral bacteria also produce specific inhibitory compounds to enhance their competitive advantage. One example of this is seen with the production of bacteriocins. These are bactericidal peptides produced by bacteria to inhibit the growth of closely related species. *Aggregatibacter actinomycetemcomitans* expresses actinobacillin, which is toxic to streptococci and other actinomycetes [46]. Likewise, *Streptococcus mutans* produces a range of bacteriocins, designated *mutacins*, which target a number of oral streptococci and other Gram-positive bacteria [34]. Several oral streptococci are also able to impair growth of neighboring microorganisms through the production of hydrogen peroxide (H_2O_2) [61,73]. This compound is able to cross bacterial membranes and, once in the cytoplasm, generates free radicals and oxidative stress by oxidation of intracellular macromolecules. As an oxidizing agent, H_2O_2 is also able to impair growth of anaerobic bacteria.

Excretion of acids resulting from glycolysis by oral streptococci can antagonize the growth of certain bacteria [119]. Although waste products are beneficial to

Veillonella, their release can reduce the pH of the local environment, thus impairing the growth of less aciduric microorganisms. It should be noted, however, that not all competition mechanisms require release of a specific compound. *Streptococcus cristatus* is able to inhibit biofilm formation by periodontal pathogen *P. gingivalis*. This is a contact-dependent mechanism, and requires an interaction between *P. gingivalis* and surface-expressed arginine deiminase (ArcA) of *S. cristatus*. This triggers a signaling cascade within *P. gingivalis*, the result of which is the downregulation of major (FimA) fimbriae expression [150].

7.2.5. Interbacterial Signaling

An additional factor that has been shown to influence the composition and architecture of a developing dental plaque biofilm is the stimulation of changes in gene expression following recognition of secreted signaling molecules. One example of this has been reported for the lactate-based mutualistic relationship between *Streptococcus* and *Veillonella* described previously. Similarly, *S. gordonii* upregulates expression of *amyB*, encoding α-amylase, in response to a short-range diffusible signal by *Veillonella atypica* [32]. This enzyme mobilizes intracellular carbohydrate reserves within *S. gordonii*, providing additional fermentation substrates (lactic acid) for *V. atypica*.

The modification of gene expression is often related to the density of a bacterial cell population; changes in gene expression are induced only once a critical threshold concentration of signaling molecule has been reached. This process is known as *quorum sensing*. Two quorum-sensing mechanisms that have been particularly well studied with respect to dental plaque formation are those based on signaling molecules autoinducer-2 (AI-2) and competence-stimulating peptide (CSP). AI-2 is the collective term for a number of molecules that are formed when 4,5-dihydroxy-2,3-pentanedione (DPD) is dissolved in water. The synthesis of AI-2 is catalyzed by the enzyme LuxS, encoded by the *luxS* gene, which is highly conserved among both Gram-positive and Gram-negative bacteria [53]. AI-2 can be recognized by a number of bacteria, independent of its source, making it an interspecies signaling molecule. A *P. gingivalis* Δ*luxS* mutant is impaired in production of Arg- and Lys-gingipains and exhibits altered hemin and iron uptake compared to the parent strain [53]. Furthermore, AI-2 plays a role in biofilm formation. *P. gingivalis* is typically able to form substantial biofilm on salivary pellicle with *S. gordonii*, but this process is impaired on deletion of *luxS* from these two species [88]. Likewise, AI-2 is essential for mutualistic biofilm growth of *S. gordonii* and *A. oris* [120]. There is also evidence that AI-2 plays a role in intergeneric interactions. A *S. gordonii luxS* deletion mutant is impaired in its ability to induce hypha formation in yeast *Candida albicans*, a process that has important implications for candidal pathogenesis [6]. While AI-2 biosynthesis is understood, little is known about AI-2 detection. The major exception is *A. actinomycetemcomitans*, for which two receptors, LsrB and RbsB, have been identified [128].

In contrast to AI-2, CSP is a species- or strain-specific quorum sensing system. It was first described for *Streptococcus pneumoniae*, but is present in a number of oral streptococcal species. In *S. pneumoniae*, CSP is a 21-amino acid residue peptide encoded by *comC*. This gene forms part of an operon with *comD* and *comE*, which together encode a two-component histidine-aspartate phosphorelay system that detects extracellular CSP. ComE controls *comA* and *comB*, which are responsible for export of CSP, and also *comX*, which encodes a regulator of a number of specific target genes [132]. Similar genetic arrangements are found in many mitis and anginosus group streptococci [132]. Important to the accretion of dental plaque, the CSP system has been shown to play a major role in biofilm formation for a number of oral streptococci [97]. For *S. mutans*, CSP-mediated communication is also associated with induction of competence, bacteriocin production, and autolysis. It is speculated that such effects might both promote variability within a biofilm through genetic exchange, and provide structural stability to the biofilm through the release of extracellular DNA (eDNA) [97].

7.2.6. Biofilm Matrix

All the interactions described above work in concert to define the microbial composition of dental plaque. Contributing to the overall structure of this community is the biofilm matrix. This matrix comprises a mix of extracellular polymeric substances (EPS) whose components are either directly secreted by members of the microbiota or are released following cell lysis. The matrix functions to provide structural integrity, and may also act as an ion exchange resin that influences the flow of molecules throughout the biofilm. Extensive research has demonstrated that, for dental plaque, a principal component of the biofilm matrix is polysaccharide. *S. mutans* and several other oral streptococci are able to synthesize extracellular glucose or fructose polymers from sucrose provided in the host diet [7]. This is performed by glucosyltransferases (GTFs) and fructosyltransferases (FTFs), which are secreted by the streptococci and hydrolyze sucrose to yield glucose and fructose. The GTFs/FTFs then polymerize their respective monomers to produce glucans/fructans. Depending on their structure, glucans might be either soluble, with predominantly α-1,6 *O*-glycosidic bonds, or insoluble, with α-1,6 and α-1,3 linkages between sugar residues.

Insoluble glucans are important to plaque biofilm development as they are sticky, providing attachment sites for the oral microbiota and "cementing" them together. Among the oral streptococci, this adhesion is mediated by surface expression of specific glucan-binding proteins (GBPs) [7]. This process is crucial for the incorporation of *S. mutans* into dental plaque and is positively associated with the risk of dental caries [93]. Extracellular polysaccharide production in dental plaque is not, however, restricted to oral streptococci. A major component of the biofilm matrix formed by *A. actinomycetemcomitans* is poly-β-1,6-*N*-acetyl-D-glucosamine (PGA). This matrix was also found to contain eDNA. This macromolecule has been shown to be an integral component of biofilm matrices produced by a number of bacteria, including *Pseudomonas aeruginosa* [147] and *Enterococcus faecalis* [43]. It likely plays a similar role in dental biofilms, although this has yet to be proved definitively.

7.3. COMMUNITIES IN HEALTH AND DISEASE

7.3.1. Health

For many years research has focused on microorganisms in the oral cavity that have been associated with the major disease processes of dental caries and periodontitis (Figure 7.3). There is strong evidence that *S. mutans* and *P. gingivalis* are active in

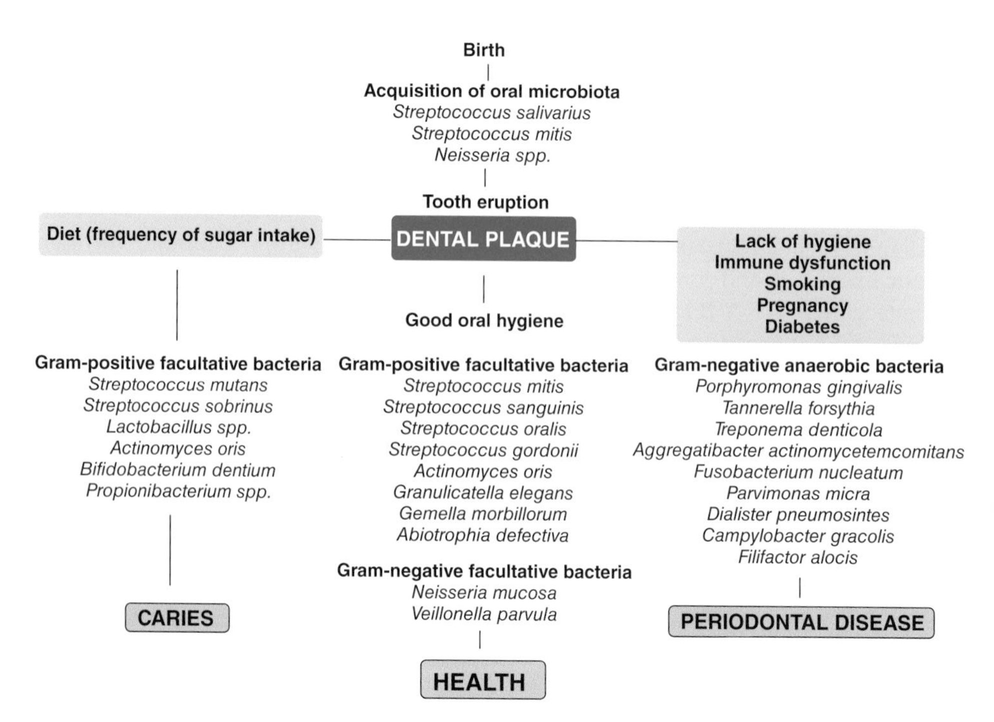

Figure 7.3. Oral bacteria in health and disease. Shortly after birth an initial oral microbiota is acquired on the mucosa comprising mainly streptococci. A new microbiota develops after eruption of first teeth chiefly in the form of dental plaque. The diagram depicts various changes in the composition of the oral microbiota that occur in the induction of dental caries, on the one hand, and in the development of periodontal disease on the other. (Data compiled from several sources [1,22,65,144].)

mediating dental caries and periodontitis, respectively. However, there is a reasonably valid argument that these microorganisms might not be generally as important in these diseases as we have come to believe. The argument is based on the notion that we are yet to understand the full complement of microorganisms that make up a disease-related community, and that it is inevitably the polymicrobial community that initiates or advances the disease assisted by the activities of specific microbial species within that community. This does not mean that the idea of an ecological enrichment for a specific group of organisms, for example, aciduric streptococci in dental caries, is not tenable. Indeed, it is well proven that sugars and acidic drinks in the diet help enrich for the mutans group streptococci and lactobacilli, which produce acids that dissolve enamel. However, it seems more prudent now to think of oral microbial communities as health- or disease-associated, as there is no clear dividing line at this stage, and no specific proportions of individual taxa allocated.

The healthy microbial communities on the tooth surface, tongue, buccal mucosa, and hard palate consist principally of streptococci (e.g., *S. mitis*, *Streptococcus oralis*, *S. australis*). These organisms produce a repertoire of adhesins and consequently are versatile in colonization of different sites. When disease-associated sites are sampled, it is sometimes very clear that a shift in composition of the microbial population has occurred. For example, the development of oral malodor (halitosis) in one study was shown to be associated with a shift in streptococcal cell populations on the tongue, with the major component *S. salivarius* eliminated and replaced by *Atopobium parvulum*, an unknown *Streptococcus* and fusobacteria [68]. There is no better example of population shift being associated with disease as the transition from healthy to diseased gingival tissues. The transition from a mixed microbial community comprising Gram-positive and Gram-negative bacteria, to a community comprising principally Gram-negative bacteria, is the main indicator of periodontal disease. The desire of the microbiologist to attribute a disease to a specific microorganism, or a small group of microorganisms, has clouded our interpretations of the disease processes. By focusing on just a few cultivable bacteria such as *P. gingivalis*, *Tannerella forsythia*, and *T. denticola*, which were regularly co-identified in sub-gingival plaque, we have been gradually convinced that these are the main pathogens. Indeed, these bacteria are pathogenic in the disease models that are available, but it seems prudent to take a step back. It should be acknowledged that these bacteria are among those regularly identified at diseased sites, but that their presence does not seem to be essential for development of periodontal disease [42]. We are left, therefore, with a very strong message: that microbial communities show considerable variation in their components and only by further studies of their composition in conjunction with their functional activities will we further understand of the advantages and disadvantages of the oral microbiota.

7.3.2. Disease

One contentious issue in oral microbiological research surrounds the etiology of infections of the tooth roots (dentine and the root canal system). The tooth pulp–root canal system is usually sterile, but infections may occur through breaching of the enamel and dentine by bacteria, often as a result of dental or root caries, or cracked-tooth restorations (Figure 7.4). The infections result in inflammation and necrosis at the root apex (Figure 7.1), often with bone destruction, and may be asymptomatic or symptomatic with acute pain. The microbiota of infected root canals is very complex, and the most prevalent taxa cultivated from such infections include *Fusobacterium*, *Porphyromonas*, *Pseudoramibacter*, *Parvimonas* and *Streptococcus*. However, it has been estimated that 50% of the bacteria associated with infected root canals will not grow on artificial media. Thus, when molecular methods are employed to define the microbial communities, some of the other more prevalent organisms detected are *Tannerella*, *Treponema*, *Filofactor*, *Dialister*, *Olsenella*, and *Atopobium*. While cultivation methods seriously underestimate the diversity of the microbiota, molecular methods probably overestimate the complexity of the active microbiota because they detect only the presence of DNA. The microbial community within an infected root canal develops successively over a long period of time, sometimes for months or even years, and so DNA footprints may be left all along the way. The primary infections of root canals have been shown to contain between 10 and 30 microbial species, while persistent infections seem to contain fewer species. Despite an immense body of work, no single species is recognized as a pathogen and there is general disagreement regarding the composition of microbial communities associated with clinical symptoms (symptomatic vs. asymptomatic). It appears that different combinations of bacteria can result in similar disease symptoms, making these infections truly polymicrobial diseases.

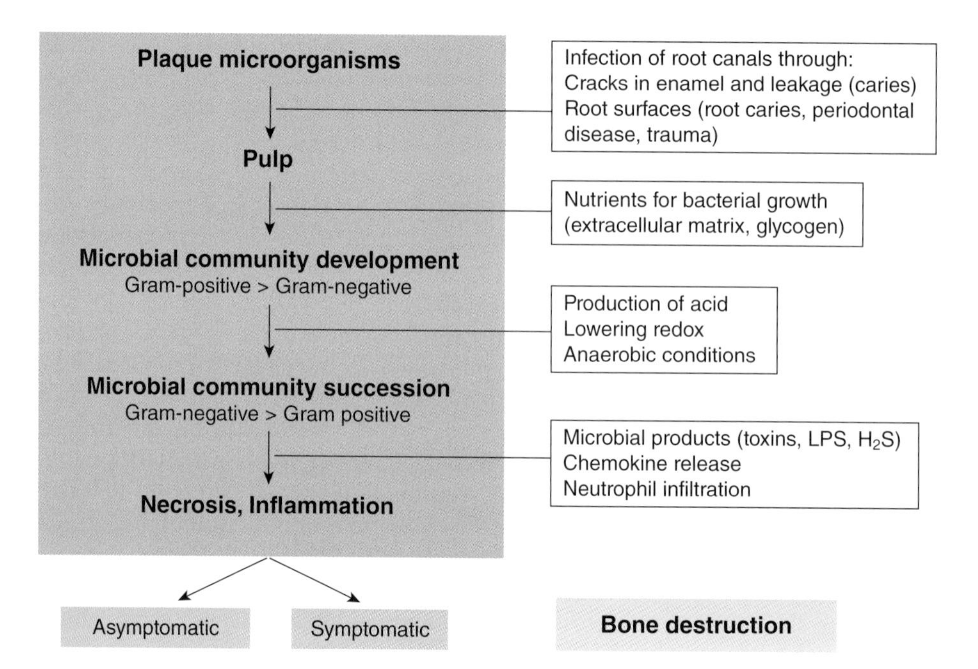

Figure 7.4. The course of endodontic infection. Microorganisms enter the tooth pulp, where they grow and survive in the rich nutritional environment, developing into an anaerobic microbial community that stimulates inflammatory responses. Long-term (chronic) infection leads to apical bone resorption.

7.4. ORAL COMMUNITIES AND SYSTEMIC DISEASE

As our understanding of the complexity of the oral microbiota advances, so has the realization that diseases associated with these microorganisms are not restricted to those of the oral cavity. Oral microbes have been linked with a wide range of systemic and end organ diseases, including infections of the cardiac, respiratory, and central nervous systems. Furthermore, the causative agents are not restricted to periodontal pathogens, as even commensals have the propensity to cause infection on translocation from the oral cavity to an alternative niche (Figure 7.5).

7.4.1. Bacteremia

For systemic diseases of oral origin, the first stage in the infection process is often entry of bacteria into the bloodstream. Within the oral cavity, this opportunity is provided by the highly vascularized periodontium and oral mucosa. Continuous shedding of the uppermost layer of cells of the oral epithelium impedes direct microbial growth on this surface. However, the development of biofilm communities on nonshedding surfaces such as teeth, particularly at the dentogingival junction, or on prosthetic devices, can bring the oral microbiota in close proximity to soft tissues. Some bacteria, such as *P. gingivalis*, express virulence factors that function to disrupt

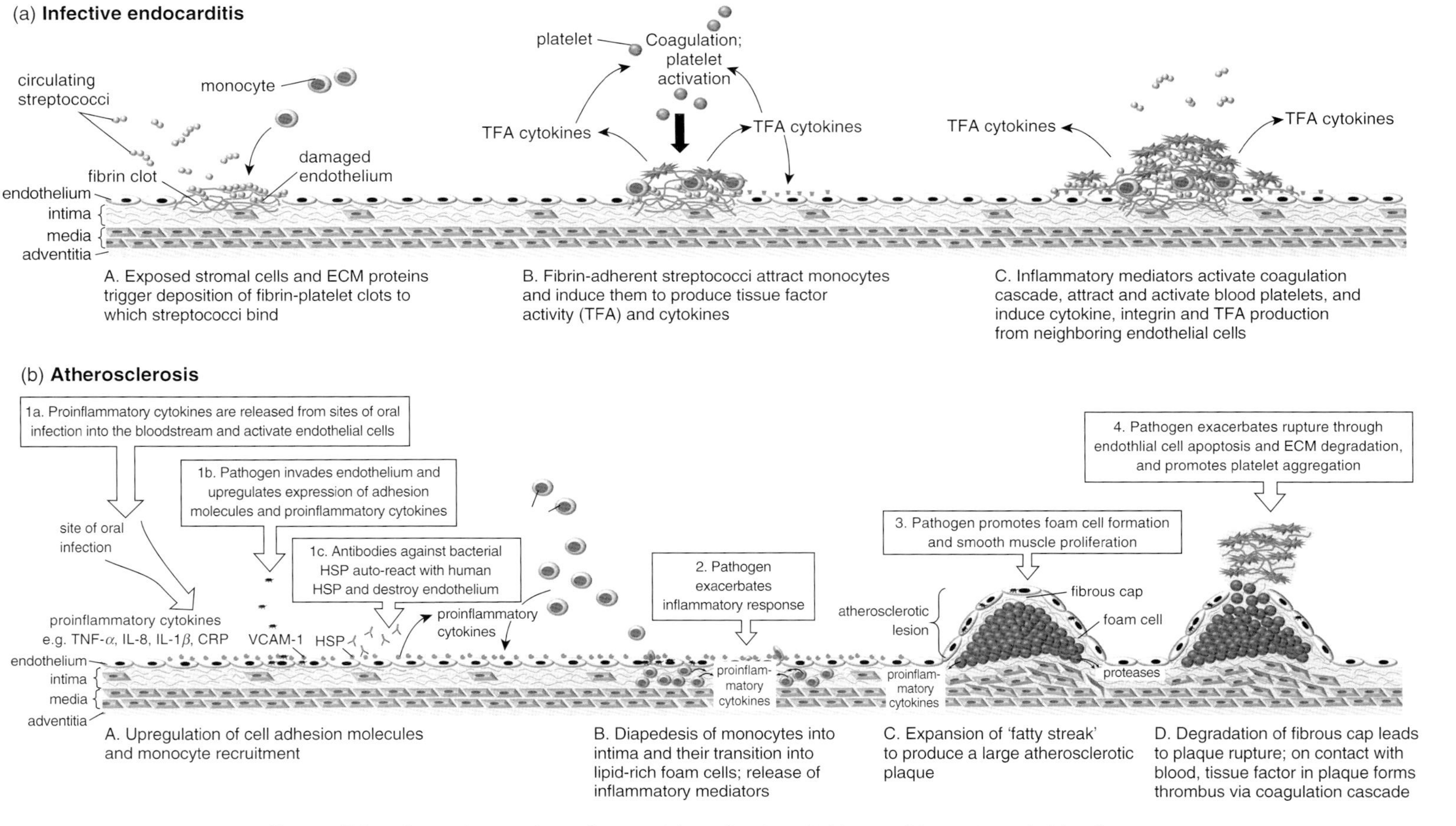

Figure 7.5. Schematic overview of potential mechanisms linking oral bacteria with (a) infective endocarditis and (b) atherosclerosis. (Reproduced with permission from Moreillon and Que [92].)

epithelial integrity. These microorganisms may therefore enter capillaries following active invasion of gingival tissue. Alternatively, any procedure that physically disrupts the delicate oral mucosa potentially exposes the underlying circulatory system to entry by the oral biofilm community. Bacteria within plaque may also translocate down the root canal of a damaged tooth or via a periapical lesion to access the alveolar blood vessels.

Given the link with systemic disease, much research effort has focused on determining the risk of bacteremia derived from oral practices [106,108], ranging from the relatively innocuous (e.g., toothbrushing) to the more traumatic (e.g., tooth extraction). There is some evidence that chewing and personal oral hygiene measures can result in a transient bacteremia [12,21,44,122]. The likelihood of this has been correlated with gingival health status, with increased risk reported for individuals with periodontitis compared to healthy subjects or those suffering gingivitis [37]. Nonetheless, relatively small subject numbers and individual variation makes such studies difficult to evaluate, and there are several examples of conflicting data [37,48,94]. Further research in this area is therefore required to draw a definitive conclusion.

Several dental procedures have been associated with an increased risk of developing bacteremia. In the case of periodontal probing [23,24,70] and root scaling [37,49,70], this likely reflects the physical trauma to which the periodontium is exposed during the procedure. Of the orthodontic treatments, only the use of tooth separators has been associated with significant bacteremia [84]. Damage to the oral epithelium during tooth extractions is considered another risk factor for bacteremia [49,123], and this risk correlates with extent and duration of surgery and with blood loss [104]. As one might predict, the incidence of bacteremia arising from each of these dental procedures is positively associated with bacterial load within the patient mouth [14,37,139], which, in turn, reflects patient oral health status. Furthermore, tissue inflammation that typifies conditions such as gingivitis and periodontitis results in dilation of the periodontal vasculature, which, in turn, increases the surface area potentially accessible to the oral microbiota [157]. For the most part, bacteremia arising from any oral procedure is transient, as a significant proportion of the detectable microbiota is eliminated by host innate and adaptive immune responses after the first few minutes. Nonetheless, a small population of bacteria have been shown to persist in the circulation for an average of 30 mins [108], providing ample opportunity for bacteria to target other parts of the body.

7.4.2. Infective Endocarditis

Infective endocarditis (IE) is a life-threatening disease that typically results from a combination of bacteremia and a predisposing cardiac condition. IE is defined as an infection of the lining of the heart chambers (endocardium) and usually involves the heart valves. The role of oral microorganisms as causative agents of this disease was first realized by Horder in 1909, who stated that "oral sepsis" was an important factor in IE pathogenesis [54]. At that time, the most common predisposing factors for IE were congenital heart defects and chronic rheumatic heart disease [98]. However, with improved healthcare in the industrialized world, the occurrence of IE in these at-risk individuals has declined, to be replaced by other vulnerable groups [92]. To reflect this, IE cases are nowadays classified according to four categories: native valve IE, prosthetic valve IE, IV drug users IE, and nosocomial IE, each with a distinct profile of microbial pathogens. Across the general population there are currently 1.7–6.2 IE cases per 100,000 patient years, with men approximately twice as susceptible as women, and incidence progressively increasing with age [11].

Significant advances have been made over the past few decades in the diagnosis and treatment of IE and, in 1954, antibiotic prophylaxis prior to surgical or dental procedures for individuals with preexisting cardiac conditions was introduced. There is increasing debate, however, over the utility of such an approach. Several more recent studies imply that, in many cases, the likelihood of prophylaxis successfully preventing IE may be outweighed by the potential risks of an adverse patient drug reaction or the promotion of microbial antibiotic resistance [31]. Consequently, the American Heart Association issued revised guidelines in 2007 stating that antibiotic prophylaxis is recommended prior to dental procedures that disrupt the oral mucosa only for individuals at the highest risk of adverse outcome from IE [149]. Nonetheless, IE is lethal if left untreated, and in-hospital mortality rates remain at ~20% [31,92]. There is therefore much research effort focused on the development of novel preventive and therapeutic strategies against IE, much of which is based on determining the molecular mechanisms of IE pathogenesis.

There are two main mechanisms for the development of IE [92]. The first involves microbial attachment to damaged endocardium. Such damage can arise in a number of ways, including sequelae of chronic rheumatic heart disease or previous IE, aberrant bloodflow across a congenital defect, or insertion of a prosthetic valve. This damage exposes subendothelial components such as extracellular matrix (ECM) proteins and stromal cells, which, in turn, trigger the deposition of sterile fibrin–platelet clots. Bacteria introduced into the bloodstream through a transient bacteremia may then bind clot constituents and attract and activate monocytes, leading to initiation of the coagulation cascade. This induces further cytokine release and the incorporation and activation of platelets into the growing vegetation (coagulum).

Alternatively, some bacteria (e.g., *Staphylococcus aureus*) are able to bind endocardium at sites of local inflammation. These might occur as a result of regular injections with impure materials by IV drug users, or due to microulceration in cases of degenerative valve lesions. Inflamed endothelial cells upregulate expression of fibronectin-binding integrin receptors and thus the level of surface-bound fibronectin (Fn). This can then act as a bridging molecule between Fn-binding adhesins on the surface of bacteria and the endocardium. Once attached, bacteria such as *S. aureus* can trigger their internalization into endothelial cells, which in response secrete various cytokines and induce the coagulation cascade, leading to coagulum formation, as mentioned above. For both mechanisms, the subsequent step is bacterial persistence within and maturation of the coagulum. Through continued recruitment and activation of platelets, bacteria can become enclosed within the coagulum and thus be protected from host immune defenses in the external environment. As bacteria replicate, many release toxins that damage host tissues further. This leads to local myocardial abscess formation and inhibition of valvular function, resulting in congestive heart failure. Furthermore, pieces of infected coagulum (emboli) may break off and be carried in the bloodstream to infect distant organs such as the brain, kidney, and spleen.

Bacteria are the most common causative microorganisms for IE and together, *S. aureus*, *E. faecalis*, and *Streptococcus* species account for >80% of all IE cases [92]. Of the streptococci, it is members of the oral viridans group that are predominantly associated with IE, which include *S. sanguinis*, *S. mitis*, *S. oralis*, *S. mutans*, and *S. gordonii*. Interestingly, most of these streptococci exist as commensals within the mouth, whereas those bacteria considered periodontopathogens, such as *P. gingivalis*, are rarely or have never been isolated from patients with IE [58].

The success of bacteria in causing IE is derived from their capacity to interact effectively with cardiac tissues and host immune cells. Viridans streptococci are predominantly associated with native valve IE in which the endocardium has been previously damaged. An ability to bind ECM components is therefore likely to promote bacterial colonization of these sites. A number of streptococcal surface proteins have been implicated in attaching to immobilized Fn, including pili of *S. sanguinis* [105], SmFnB of *S. mutans* [90], and Hsa, CshA/B, and SspA/B of *S. gordonii* [59]. Furthermore, through binding to soluble Fn via surface adhesin AtlA, *S. mutans* is able to evade phagocytosis, a mechanism that increases its virulence in a rat model of IE [64].

Once attached to the site of endocardial damage, the next critical step for streptococci in IE is to promote growth of the coagulum. This requires recruitment and activation of host polymorphonuclear leukocytes (PMNs) and platelets. For *S. mutans*, this is achieved, at least in part, by the actions of its glucosyltransferases (GTFs). These induce IL-6 production and ICAM-1 expression by endothelial cells, which, in turn, lead to the recruitment of monocytes [129,155]. Alternatively, *S. gordonii* adhesin Hsa has been shown to bind directly to PMNs via receptors CD11b, CD43, and CD50 [143]. Interactions of bacteria with platelets in IE vegetations occur via a two-step process [36]. The first requires binding of bacteria to platelets, which triggers a signaling cascade that results in platelet activation. In the second step, activated platelets upregulate surface expression of integrin GPIIb/IIIa, which binds soluble fibrinogen and ultimately leads to platelet aggregation.

A number of streptococci mediate attachment to platelets through recognition of sialylated oligosaccharide chains of specific glycoproteins. For example, bacteriophage-encoded PblA and PblB of *S. mitis* have been shown to bind α-2,8-linked sialic acid residues on ganglioside GD3 [91], while sialic acid-binding adhesin SrpA of *S. sanguinis* targets platelet receptor GPIbα [114]. Collagen-binding adhesin CbpA of *S. sanguinis* has been implicated in platelet aggregation [50], but no receptor has yet been identified. Likewise, the mechanism(s) by which *S. mutans* GTFs [133] and AgI/II family polypeptide PAc [87] facilitate platelet aggregation is unknown. It is becoming increasingly apparent, however, that bacteria–platelet interactions may often be quite complex, involving multiple receptors. For *S. gordonii*, a mechanism involving three distinct sets of adhesins has been postulated [112]. When platelets initially contact *S. gordonii* within a growing vegetation, sialic acid–binding adhesin Hsa binds platelets via receptor GPIbα, causing them to slow under shear. This slowing enables a second interaction between *S. gordonii* PadA and platelet integrin GPIIb/IIIa, serving to hold the platelet in place and leading to platelet activation. AgI/II family polypeptides SspA and SspB then act in concert with Hsa to induce platelet aggregation, although the receptor(s) targeted by SspA/B has (have) yet to be identified.

By identifying the bacterial and host receptors involved in IE pathogenesis, inhibitors might be generated for effective prevention and control of IE. Some progress is already being made in this area. For example, a vaccine against *S. parasanguinis* lipoprotein FimA significantly reduced the incidence of IE in rats that were challenged with *S. mitis*, *S. mutans*, and *S. salivarius* [71]. However, the ability of bacteria to utilize multiple cell surface components may limit such strategies. Indeed, using a series of knockout mutants, it was shown that no single cell wall protein of *S. sanguinis* was essential for development of IE in a rabbit model [142]. Aside from targeting bacteria directly, other areas of development include the use of drugs to disrupt vegetation formation, and the development of novel biomaterials for prosthetic valves that are less readily bound by bacteria [92].

7.4.3. Atherosclerosis

Atherosclerosis is characterized by the accumulation of lipids and fibrous elements within medium-size and large arteries, and is the predominant precursor to cardiovascular disease (CVD), the leading cause of death and morbidity in the developed world. There has long been an association between atherosclerosis and elevated levels of circulating lipids such as cholesterol [124], but work since 2001 has highlighted the crucial role that inflammation also plays in the progression of this condition [80].

The first step in the formation of an atheroma is activation of the endothelial cells that line the arterial lumen. In a "normal" state, these cells resist mediation of firm attachment to leukocytes from the bloodstream. On activation, however, these cells upregulate expression of several surface proteins, including vascular cell adhesion molecule-1 (VCAM-1), which bind and recruit passing monocytes and T lymphocytes to the arterial wall. These inflammatory cells subsequently undergo diapedesis over a chemoattractant gradient to pass between endothelial cells into the underlying tunica intima layer. On entry to the intimal layer, macrophage-colony stimulating factor (M-CSF) induces monocytes to transform into lipid-rich foam cells. These characterize the initial atherosclerotic lesion, known as the "fatty streak." Should inflammatory conditions persist, this fatty streak can expand and develop into a more complex lesion, with the associated clinical complications of atherosclerosis. Plaque (atheroma) disruption ultimately leads to thrombosis and results most commonly from rupture of the plaque's fibrous cap. This can lead to a heart attack (myocardial infarction), or cause ischemia and symptoms such as *angina pectoris* following restricted bloodflow through the vessel.

It is widely acknowledged that injury to the arterial endothelial monolayer triggers the inflammatory activation of these cells that initiates atherogenesis. What triggers this injury is, however, less clear. Predisposing factors for atherosclerosis include high levels of LDL cholesterol and smoking, but ≤50% of patients with this condition lack any of the recognized risk criteria [145]. This implies that other causes exist, and one major area of current interest is the potential role of microbial infection. Evidence for the ability of oral bacteria to initiate or promote progression of atherosclerosis began with epidemiological studies that found a correlation between the incidence of periodontitis and CVD [9,126]. Further support comes from more recent work in which strong correlations have been found between the presence of serum antibodies (Abs) to periodontal pathogens and the occurrence of CVD

[10,118]. Several reports have also demonstrated the presence of DNA from periodontal pathogens within human atherosclerotic plaques [16,38,117,156], although viable bacteria have yet to be isolated from such lesions. Oral bacteria associated most frequently with atheromas include *A. actinomycetemcomitans*, *T. denticola*, *Prevotella intermedia*, and *P. gingivalis*. As periodontal pathogens, these microorganisms are known to trigger strong innate immune and inflammatory reactions within the oral cavity, and similar capabilities are thought to play a role in the promotion of atherogenesis. Using *P. gingivalis* as the model bacterium, Gibson et al. have postulated four putative mechanisms for pathogen-accelerated atherosclerosis [40]:

1. The first model requires direct microbial invasion of endothelial cells, as might occur during a transient bacteremia. *P. intermedia* [30], *A. actinomycetemcomitans* [127] and *P. gingivalis* [116] have all been shown to invade vascular endothelial cells and, for *P. gingivalis*, this process is dependent on surface expression of major fimbriae. Following uptake, *P. gingivalis* has been shown to stimulate endothelial cells to upregulate expression of cell adhesion molecules, cytokines, and chemokines [69,95]. These include VCAM-1, IL-1β, and MCP-1, all of which play a prominent role in early atherosclerotic plaque formation. *P. gingivalis* is also able to induce proliferation of vascular smooth muscle cells, a process that might promote fibrous cap thickening and subsequent artery constriction [56].
2. Rather than direct invasion, the second proposed mechanism involves delivery of *P. gingivalis* to the site of atheroma development within host immune cells. Extensive tissue damage during periodontitis stimulates a strong inflammatory response, bringing phagocytic cells to the site of infection, where they may ingest the pathogens present. *P. gingivalis* fimbriae enable this bacterium to persist within macrophages [146]. It is possible, therefore, that should pathogen-laden macrophages return from the periodontal pocket to the circulation, they might carry bacteria to the site of a developing atherosclerotic lesion. If released, these bacteria could then mediate their proinflammatory effects, as described above.
3. The third model does not require the presence of periodontal pathogens at the site of atherogenesis. Rather, the proinflammatory effects are triggered by mediators released into the bloodstream from sites of local oral infection. It is known that gingival crevicular fluid within active periodontal disease sites contains high levels of cytokines and chemokines, including *C*-reactive protein (CRP), IL-1β, IL-6, TNF-α, and MCP-1 [130]. These inflammatory mediators may access the circulation and, if continuously released over the course of a chronic oral infection, lead to systemic changes in the host immune response, resulting in endothelial cell activation and progression to atherosclerosis.
4. The fourth mechanism involves induction of an autoimmune response via molecular mimicry. Some bacteria express proteins with a high level of homology to those found in their human host. Should an immune response be mounted against the bacterial protein, this can inadvertently trigger an autoimmune attack on those host tissues bearing cross-reactive epitopes. One such candidate found in *P. gingivalis* is heat shock protein (HSP), a molecule

expressed by both humans and several bacteria. It is possible, therefore, that *P. gingivalis* infection could elicit an autoimmune response against host HSP, resulting in localized endothelial tissue damage and the initiation of atherogenesis. Supporting this notion is the observation that patients suffering from periodontitis and atherosclerosis apparently possess elevated serum Ab levels to both human and *P. gingivalis* HSPs [19,154].

7.4.4. Diabetes

Diabetes mellitus is a collective term for a group of disorders affecting metabolism of carbohydrates, lipids, and proteins, and is characterized by elevated blood glucose levels. As a chronic condition, diabetes is also associated with a number of clinical complications, including long-term damage to the eyes, kidneys, nerves, and vascular system. Diabetes currently affects an estimated 24 million people within the United States alone, with associated medical costs in excess of $174 billion [125], and its incidence is rapidly increasing in association with rising obesity rates. It is estimated that by 2030, 4.4% of the world's population will be diabetic [148]. Type 1 diabetes is typically found in children and adolescents, and results from autoimmune destruction of pancreatic islet cells, leading to total loss of insulin secretion. Type 2 diabetes is the predominant form and is commonly seen in adults, although its prevalence in younger individuals is increasing in conjunction with childhood obesity issues. This form arises from impaired utilization of endogenously produced insulin (insulin resistance), together with the inability of pancreatic β cells to secrete sufficient insulin to compensate.

There is strong evidence that individuals with diabetes are more likely to suffer severe periodontal disease than those without [18,89]. This relationship is not, however, all one way, and there is increasing evidence that periodontal disease can adversely affect the progression of diabetes, particularly type 2. Given the huge financial burden of diabetes on society, the mechanisms underlying this phenomenon, and the potential for periodontal therapy to positively impact the incidence of diabetes, are areas of significant research. Several studies have found an association between periodontitis and the risk of diabetes-related complications [89], while diabetics suffering periodontal disease were shown to be at a six-fold higher risk of poor glycemic control than those without [135]. It has been proposed that these effects are related to the ability of periodontal pathogens to perpetuate a systemic inflammatory state [96].

As noted previously, the highly vascularized periodontium enables cytokines and other inflammatory mediators induced during periodontitis to access the general circulation. Insulin resistance in type 2 diabetes is associated with high serum levels of CRP, IL-6, and fibrinogen. Elevated serum levels of these molecules are also seen in individuals suffering periodontitis, particularly those colonized by *P. gingivalis*, *Tannerella forsythia*, and *P. intermedia* [89]. Furthermore, severe periodontal disease is associated with high levels of circulating TNF-α [33], a strong stimulant of acute-phase reactant production in the liver and subsequent insulin resistance [113]. Cytokines TNF-α, IL-6, and IL-1β are also known insulin antagonists, all of which play major roles in the pathogenesis of periodontitis [138].

Taken collectively, these data imply that through the chronic release of pro-inflammatory mediators into the bloodstream, periodontal disease is able to significantly promote the development of insulin resistance and thus type 2 diabetes. To further support this hypothesis, several intervention studies have investigated the effects of periodontal treatment on glycemic control. More recent meta-analyses of such trials have shown a modest but significant improvement in glycemic control on therapy, as determined by glycohemoglobin (HbA1c) levels [25,131,136]. However, heterogeneity across studies means that further evidence is required before definitive conclusions can be drawn.

Epidemiological studies also indicate a possible association with periodontitis and gestational diabetes. This condition affects approximately 4% of pregnant women in the United States each year and is associated with significantly increased risks of maternal and infant morbidity. Although the number of specific clinical trials is relatively small, current evidence implies that pregnant women with periodontal disease are at higher risk of suffering gestational diabetes than those without [99,152,153]. Similar to type 2 diabetes, it has been proposed that this risk is associated with the persistent release of proinflammatory cytokines (e.g., TNF-α, IL-1β, IL-6, CRP) from inflamed periodontal tissues into the circulation. This stimulates a maternal systemic inflammatory reaction that, coupled with the insulin resistance commonly associated with pregnancy itself [74], triggers the onset of gestational diabetes.

7.4.5. Obesity

Countries across the world are currently experiencing an obesity epidemic. Since 1980, obesity in adults has increased by >75%, affecting 65% of the adult population in the United States. Worryingly, this trend is also being replicated among children, for whom obesity rates have tripled, with approximately 16% of children and adolescents in this country classified as overweight [103,121]. Obesity is a cause of significant morbidity and mortality, with serious health consequences, including CVD, type 2 diabetes, liver disease, hypertension, and cancer. The socio-economic implications of this epidemic are therefore overwhelming.

Several studies have identified obesity as a risk factor for periodontitis [17,121]. Similar to diabetes, the mechanisms underlying this effect are thought to be associated with an altered systemic inflammatory response. However, a more recent concept of growing interest in the health community is that the human microbiota may influence the prevalence of obesity. Studies to date have focused principally on those microorganisms found within the gut [28,77]. Obese animals have been shown to consistently display a greater representation of bacterial phylum *Firmicutes* and fewer *Bacteroidetes* compared to their lean littermates [78]. Furthermore, this pattern is replicated in human subjects [79,140]. Such changes have been shown to affect the overall metabolic activity of the microbiota, which, in turn, impacts an individual's abilities to extract energy (calories) from the diet and to store them in adipose (fat) tissue [5,141].

A second mechanism via which the gut microbiota are proposed to promote obesity is through chronic low-grade stimulation of systemic inflammation, leading to a subsequent increase in host insulin resistance and weight gain. This can be achieved by LPS stimulation of a CD14/TLR4-dependent innate immune response, resulting in the expression of inflammatory cytokines such as TNF-α, IL-1, and IL-6 [15]. Furthermore, this effect is exacerbated by a high-fat diet, which promotes LPS adsorption, leading to elevated LPS serum levels [3,15,39].

Gut bacteria pass through the oral cavity en route to the digestive tract. It has been proposed that members of this community may transiently occupy or be seeded from the mouth, leading to a potential association between the oral microbiota, gut microbiota, and obesity. To investigate this hypothesis, salivary bacterial populations of 313 overweight women were compared with existing data on equivalent populations of 232 healthy individuals [41]. Of the 12 *Firmicutes* members monitored, 10 had a significantly greater median percentage in the overweight individuals, mimicking somewhat the obesity-related microbiota shifts reported in the gut. Furthermore, the presence of a single Gram-negative bacterial species, *Selenomonas noxia*, was capable of distinguishing 98.4% of the overweight women from the healthy group. Originally associated with the initiation of periodontal disease [134], it is likely that *S. noxia* is capable of stimulating a host inflammatory response. It is possible, therefore, that *S. noxia* may promote obesity via mechanisms similar to those described for members of the gut microbiota, although further evidence is required to justify this claim. Nonetheless, as for the gut, it is apparent that the composition of the oral microbiota alters in parallel with an individual's transition from lean to obese. Whether this is in response to the changing host environment, or represents a causal relationship, remains to be defined.

7.4.6. Adverse Pregnancy Outcomes

A series of epidemiological and interventional studies have implicated that poor periodontal status might result in adverse pregnancy outcomes, including preterm delivery, low birth weight, miscarriage, and preeclampsia [151]. A case–control study of 124 postpartum or pregnant women found that those individuals with periodontal disease were significantly more likely to give birth prematurely or to deliver low-weight babies [102]. Likewise, a cohort study of 1300 women demonstrated an association between moderate or severe periodontitis and the risk of preterm birth [62]. However, many of these reports are limited, and there are also studies for which no significant associations have been found [125]. More robust investigations are therefore required to definitively demonstrate that periodontal disease increases the risk of adverse pregnancy outcome.

Two main mechanisms have been proposed to explain how periodontal disease might promote adverse pregnancy outcomes [115]. The first mechanism results from the systemic dissemination of inflammatory mediators and prostaglandins from the site of periodontitis, followed by their hematogenous passage through the placenta and initiation of intrauterine inflammation. In support of this model, Offenbacher et al. [101] found that women with periodontitis who gave birth to premature or low-weight babies had significantly higher gingival crevicular fluid levels of prostaglandin E_2 (PGE_2) than those whose births were without complications.

The second model is associated with the transient bacteremias often seen with periodontal disease. It is postulated that oral bacteria and/or their products (e.g., LPS) are then able to pass from the bloodstream into the fetoplacental tissue, where, utilizing mechanisms seen in the pathogenesis of periodontal disease, they cause tissue damage and stimulate intrauterine inflammatory responses. Evidence for this mechanism derives, in part, from the fact that periodontal pathogens, including *P. gingivalis* and *F. nucleatum*, have been isolated from the placenta and/or amniotic fluid of women suffering pregnancy complications such as preterm birth and preeclampsia [8,47,51,67,76].

Animal models have also been developed to study the effects of these intrauterine infections in greater detail, and have corroborated the hypothesis that bacterial manipulation of maternal immune and inflammatory responses promotes pregnancy complications. For *P. gingivalis*, systemic infection of pregnant mice was associated with an elevation in proinflammatory TNF-α serum levels, coupled with suppression of antiinflammatory IL-10 levels. Such effects stimulated the maternal inflammatory response and were correlated strongly with fetal growth restriction [81]. Periodontal pathogens were also shown to manipulate inflammatory responses by targeting TLRs. These receptors are critical components of the innate immune response and play important roles in pregnancy maintenance, placental immune protection, and delivery initiation [110]. For *F. nucleatum*, infection was found to stimulate a TLR4-mediated intrauterine inflammatory response that resulted in fetal death [82]. Likewise, oral infection of pregnant mice with both *Campylobacter rectus* and *P. gingivalis* significantly reduced fecundity and was associated with elevated placental expression levels of TLR4 [4].

It is clear, therefore, that oral bacteria are at least capable of inducing adverse pregnancy outcomes. Furthermore, the association of pregnancy complications and the oral microbiota might not be restricted to periodontal pathogens. By injecting pregnant mice with pooled saliva or subgingival plaque samples, Fardini et al. observed 11 different bacterial genera belonging to 4 phyla undergo hematogenous transmission to the placenta, the majority being commensals [35]. Ongoing research therefore aims to further define the mechanisms utilized by the oral microbiota to trigger pregnancy complications so that intervention strategies might be devised. To this end, surface fibrillar protein FadA has been identified as playing an important role in mediating placental colonization by *F. nucleatum*, and may represent a therapeutic target to prevent intrauterine infection [55].

7.5. SUMMARY

This chapter commenced with the implication that the oral microbiota was of some benefit to the human host. The potential benefits included low-level priming of the immune system and providing additional innate defence. Clearly, within the oral environment, the microbiota is normally of very little concern to those subjects who maintain a good level of oral hygiene and cleanliness. The risk factors for the development of dental caries and of periodontal disease are very well defined, as depicted in Figure 7.3, and many of these risk factors may be easily avoided. However, components of the oral microbiota can become a major problem when they move to a foreign environment, such as the tooth pulp, and especially when they enter the bloodstream. In these situations the very same molecules that are utilized by the microorganisms to benignly colonize the surfaces present in the mouth appear to come into new action by binding platelets and extracellular matrix molecules, and stimulating the production of proinflammatory cytokines. These features of oral microorganisms indicate an even greater versatility of the oral microbiota to remain in close contact with the human host, even when out of the natural habitat of the human mouth. But it's generally all quiet at home.

REFERENCES

1. Aas JA, Griffen AL, Dardis SR, Lee AM, Olsen I, Dewhirst FE, Leys EJ, Paster BJ. Bacteria of dental caries in primary and permanent teeth in children and young adults. *J Clin Microbiol* **46**:1407–1417 (2008).
2. Aas JA, Paster BJ, Stokes LN, Olsen I, Dewhirst FE. Defining the normal bacterial flora of the oral cavity. *J Clin Microbiol* **43**:5721–5732 (2005).
3. Amar J, Burcelin R, Ruidavets JB, Cani PD, Fauvel J, Alessi MC, Chamontin, B, Ferrieres J. Energy intake is associated with endotoxemia in apparently healthy men. *Am J Clin Nutr* **87**:1219–1223 (2008).
4. Arce RM, Barros SP, Wacker B, Peters B, Moss K, Offenbacher S. Increased TLR4 expression in murine placentas after oral infection with periodontal pathogens. *Placenta* **30**:156–162 (2009).
5. Backhed F, Ding H, Wang T, Hooper LV, Koh GY, Nagy A, Semenkovich CF, Gordon JI. The gut microbiota as an environmental factor that regulates fat storage. *Proc Natl Acad Sci U S A* **101**:15718–15723 (2004).
6. Bamford CV, d'Mello A, Nobbs AH, Dutton LC, Vickerman MM, Jenkinson HF. *Streptococcus gordonii* modulates *Candida albicans* biofilm formation through intergeneric communication. *Infect Immun* **77**:3696–3704 (2009).
7. Banas JA, Vickerman MM. Glucan-binding proteins of the oral streptococci. *Crit Rev Oral Biol Med* **14**:89–99 (2003).
8. Barak S, Oettinger-Barak O, Machtei EE, Sprecher H, Ohel G. Evidence of periopathogenic microorganisms in placentas of women with preeclampsia. *J Periodontol* **78**:670–676 (2007).
9. Beck J, Garcia R, Heiss G, Vokonas PS, Offenbacher S. Periodontal disease and cardiovascular disease. *J Periodontol* **67**:1123–1137 (1996).
10. Beck JD, Eke P, Heiss G, Madianos P, Couper D, Lin D, Moss K, Elter J, Offenbacher S. Periodontal disease and coronary heart disease: a reappraisal of the exposure. *Circulation* **112**:19–24 (2005).
11. Beynon RP, Bahl VK, Prendergast BD. Infective endocarditis. *Br Med J* **333**:334–339 (2006).
12. Bhanji S, Williams B, Sheller B, Elwood T, Mancl L. Transient bacteremia induced by toothbrushing a comparison of the Sonicare toothbrush with a conventional toothbrush. *Pediatr Dent* **24**:295–299 (2002).
13. Bik EM, Long CD, Armitage GC, Loomer P, Emerson J, Mongodin EF, Nelson KE, Gill SR, Fraser-Liggett CM, Relman DA. Bacterial diversity in the oral cavity of 10 healthy individuals. *ISME J* **4**:962–974 (2010).
14. Brennan MT, Kent ML, Fox PC, Norton HJ, Lockhart PB. The impact of oral disease and nonsurgical treatment on bacteremia in children. *J Am Dent Assoc* **138**:80–85 (2007).
15. Cani PD, Amar J, Iglesias MA, Poggi M, Knauf C, Bastelica D, Neyrinck AM, Fava F, Tuohy KM, Chabo C, et al. Metabolic endotoxemia initiates obesity and insulin resistance. *Diabetes* **56**:1761–1772 (2007).
16. Cavrini F, Sambri V, Moter A, Servidio D, Marangoni A, Montebugnoli L, Foschi F, Prati C, Di BR, Cevenini R. Molecular detection of *Treponema denticola* and *Porphyromonas gingivalis* in carotid and aortic atheromatous plaques by FISH: Report of two cases. *J Med Microbiol* **54**:93–96 (2005).
17. Chaffee BW, Weston SJ. The association between chronic periodontal disease and obesity: a systematic review with meta-analysis. *J Periodontol* **81**:1708–1724 (2010).

18. Chavarry NG, Vettore MV, Sansone C, Sheiham A. The relationship between diabetes mellitus and destructive periodontal disease: A meta-analysis. *Oral Health Prevent Dent* **7**:107–127 (2009).
19. Chung SW, Kang HS, Park HR, Kim SJ, Kim SJ, Choi JI. Immune responses to heat shock protein in *Porphyromonas gingivalis*-infected periodontitis and atherosclerosis patients. *J Periodont Res* **38**:388–393 (2003).
20. Cisar JO, Sandberg AL, Abeygunawardana C, Reddy GP, Bush CA. Lectin recognition of host-like saccharide motifs in streptococcal cell wall polysaccharides. *Glycobiology* **5**:655–662 (1995).
21. Cobe HM. Transitory bacteremia. *Oral Surg Oral Med Oral Pathol* **7**:609–615 (1954).
22. Colombo AP, Boches SK, Cotton SL, Goodson JM, Kent R, Haffajee AD, Socransky SS, Hasturk H, Van Dyke TE, Dewhirst F, et al. Comparisons of subgingival microbial profiles of refractory periodontitis, severe periodontitis, and periodontal health using the human oral microbe identification microarray. *J Periodontol* **80**:1421–1432 (2009).
23. Daly C, Mitchell D, Grossberg D, Highfield J, Stewart D. Bacteraemia caused by periodontal probing. *Austral Dent J* **42**:77–80 (1997).
24. Daly CG, Mitchell DH, Highfield JE, Grossberg DE, Stewart D. Bacteremia due to periodontal probing: A clinical and microbiological investigation. *J Periodontol* **72**:210–214 (2001).
25. Darre L, Vergnes JN, Gourdy P, Sixou M. Efficacy of periodontal treatment on glycaemic control in diabetic patients: A meta-analysis of interventional studies. *Diabetes Metab* **34**:497–506 (2008).
26. Dewhirst FE, Chen T, Izard J, Paster BJ, Tanner AC, Yu WH, Lakshmanan A, Wade WG. The human oral microbiome. *J Bacteriol* **192**:5002–5017 (2010).
27. Diaz PI, Chalmers NI, Rickard AH, Kong C, Milburn CL, Palmer RJ Jr, Kolenbrander PE. Molecular characterization of subject-specific oral microflora during initial colonization of enamel. *Appl Environ Microbiol* **72**:2837–2848 (2006).
28. DiBaise JK, Zhang H, Crowell MD, Krajmalnik-Brown R, Decker GA, Rittmann BE. Gut microbiota and its possible relationship with obesity. *Mayo Clin Proc* **83**:460–469 (2008).
29. Dige I, Nyengaard JR, Kilian M, Nyvad B. Application of stereological principles for quantification of bacteria in intact dental biofilms. *Oral Microbiol Immunol* **24**:69–75 (2009).
30. Dorn BR, Dunn WA, Jr, Progulske-Fox A. Invasion of human coronary artery cells by periodontal pathogens. *Infect Immun* **67**:5792–5798 (1999).
31. Duval X, Leport C. Prophylaxis of infective endocarditis: current tendencies, continuing controversies. *Lancet Infect Dis* **8**:225–232 (2008).
32. Egland PG, Palmer RJ, Jr, Kolenbrander PE. Interspecies communication in *Streptococcus gordonii-Veillonella atypica* biofilms: Signaling in flow conditions requires juxtaposition. *Proc Natl Acad Sci USA* **101**:16917–16922 (2004).
33. Engebretson S, Chertog R, Nichols A, Hey-Hadavi J, Celenti R, Grbic J. Plasma levels of tumour necrosis factor-alpha in patients with chronic periodontitis and type 2 diabetes. *J Clin Periodontol* **34**:18–24 (2007).
34. Fabio U, Bondi M, Manicardi G, Messi P, Neglia R. Production of bacteriocin-like substances by human oral streptococci. *Microbiologica* **10**:363–370 (1987).
35. Fardini Y, Chung P, Dumm R, Joshi N, Han YW. Transmission of diverse oral bacteria to murine placenta: Evidence for the oral microbiome as a potential source of intrauterine infection. *Infect Immun* **78**:1789–1796 (2010).
36. Fitzgerald JR, Foster TJ, Cox D. The interaction of bacterial pathogens with platelets. *Nat Rev Microbiol* **4**:445–457 (2006).

37. Forner L, Larsen T, Kilian M, Holmstrup P. Incidence of bacteremia after chewing, tooth brushing and scaling in individuals with periodontal inflammation. *J Clin Periodontol* **33**:401–407 (2006).
38. Gaetti-Jardim E Jr, Marcelino SL, Feitosa AC, Romito GA, Vila-Campos MJ. Quantitative detection of periodontopathic bacteria in atherosclerotic plaques from coronary arteries. *J Med Microbiol* **58**:1568–1575 (2009).
39. Ghoshal S, Witta J, Zhong J, de Villiers W, Eckhardt E. Chylomicrons promote intestinal absorption of lipopolysaccharides. *J Lipid Res* **50**:90–97 (2009).
40. Gibson FC, III, Yumoto H, Takahashi Y, Chou HH, Genco CA. Innate immune signaling and *Porphyromonas gingivalis*-accelerated atherosclerosis. *J Dent Res* **85**:106–121 (2006).
41. Goodson JM, Groppo D, Halem S, Carpino E. Is obesity an oral bacterial disease? *J Dent Res* **88**:519–523 (2009).
42. Griffen AL, Lyons SR, Becker MR, Moeschberger ML, Leys EJ. *Porphyromonas gingivalis* strain variability and periodontitis. *J Clin Microbiol* **37**:4028–4033 (1999).
43. Guiton PS, Hung CS, Kline KA, Roth R, Kau AL, Hayes E, Heuser J, Dodson KW, Caparon MG, Hultgren SJ. Contribution of autolysin and Sortase A during *Enterococcus faecalis* DNA-dependent biofilm development. *Infect Immun* **77**:3626–3638 (2009).
44. Guntheroth WG. How important are dental procedures as a cause of infective endocarditis? *Am J Cardiol* **54**:797–801 (1984).
45. Hamilton IR, Ng SKC. Stimulation of glycolysis through lactate consumption in a resting cell mixture of *Streptococcus salivarius* and *Veillonella parvula*. *FEMS Microbiol Lett* **20**:61–65 (1983).
46. Hammond BF, Lillard SE, Stevens RH. A bacteriocin of *Actinobacillus actinomycetemcomitans*. *Infect Immun* **55**:686–691 (1987).
47. Han YW, Fardini Y, Chen C, Iacampo KG, Peraino VA, Shamonki JM, Redline RW. Term stillbirth caused by oral *Fusobacterium nucleatum*. *Obstet Gynecol* **115**:442–445 (2010).
48. Hartzell JD, Torres D, Kim P, Wortmann G. Incidence of bacteremia after routine tooth brushing. *Am J Med Sci* **329**:178–180 (2005).
49. Heimdahl A, Hall G, Hedberg M, Sandberg H, Soder PO, Tuner K, Nord CE. Detection and quantitation by lysis-filtration of bacteremia after different oral surgical procedures. *J Clin Microbiol* **28**:2205–2209 (1990).
50. Herzberg MC, Nobbs A, Tao L, Kilic A, Beckman E, Khammanivong A, Zhang Y. Oral streptococci and cardiovascular disease: Searching for the platelet aggregation-associated protein gene and mechanisms of *Streptococcus sanguis*-induced thrombosis. *J Periodontol* **76**:2101–2105 (2005).
51. Hill GB. Preterm birth: Associations with genital and possibly oral microflora. *Ann Periodontol* **3**:222–232 (1998).
52. Hojo K, Nagaoka S, Murata S, Taketomo N, Ohshima T, Maeda N. Reduction of vitamin K concentration by salivary *Bifidobacterium* strains and their possible nutritional competition with *Porphyromonas gingivalis*. *J Appl Microbiol* **103**:1969–1974 (2007).
53. Hojo K, Nagaoka S, Ohshima T, Maeda N. Bacterial interactions in dental biofilm development. *J Dent Res* **88**:982–990 (2009).
54. Horder TJ. Infective endocarditis with analysis of 150 cases and with special reference to the chronic form of the disease. *Q J Med* **2**:289–324 (1909).
55. Ikegami A, Chung P, Han YW. Complementation of the *fadA* mutation in *Fusobacterium nucleatum* demonstrates that the surface-exposed adhesin promotes cellular invasion and placental colonization. *Infect Immun* **77**:3075–3079 (2009).
56. Inaba H, Hokamura K, Nakano K, Nomura R, Katayama K, Nakajima A, Yoshioka H, Taniguchi K, Kamisaki Y, Ooshima T, et al. Upregulation of S100 calcium-binding

protein A9 is required for induction of smooth muscle cell proliferation by a periodontal pathogen. *FEBS Lett* **583**:128–134 (2009).

57. Inoue H, Ono K, Masuda W, Inagaki T, Yokota M, Inenaga K. Rheological properties of human saliva and salivary mucins. *J Oral Biosci* **50**:134–141 (2008).
58. Ito HO. Infective endocarditis and dental procedures: evidence, pathogenesis, and prevention. *J Med Investig* **53**:189–198 (2006).
59. Jakubovics NS, Brittan JL, Dutton LC, Jenkinson HF. Multiple adhesin proteins on the cell surface of *Streptococcus gordonii* are involved in adhesion to human fibronectin. *Microbiology* **155**:3572–3580 (2009).
60. Jakubovics NS, Gill SR, Iobst SE, Vickerman MM, Kolenbrander PE. Regulation of gene expression in a mixed-genus community: Stabilized arginine biosynthesis in *Streptococcus gordonii* by coaggregation with *Actinomyces naeslundii*. *J Bacteriol* **190**:3646–3657 (2008a).
61. Jakubovics NS, Gill SR, Vickerman MM, Kolenbrander PE. Role of hydrogen peroxide in competition and cooperation between *Streptococcus gordonii* and *Actinomyces naeslundii*. *FEMS Microbiol Ecol* **66**:637–644 (2008b).
62. Jeffcoat MK, Geurs NC, Reddy MS, Cliver SP, Goldenberg RL, Hauth JC. Periodontal infection and preterm birth: Results of a prospective study. *J Am Dent Assoc* **132**:875–880 (2001).
63. Jenkinson HF, Lamont RJ. Oral microbial communities in sickness and in health. *Trends Microbiol* **13**:589–595 (2005).
64. Jung CJ, Zheng QH, Shieh YH, Lin CS, Chia JS. *Streptococcus mutans* autolysin AtlA is a fibronectin-binding protein and contributes to bacterial survival in the bloodstream and virulence for infective endocarditis. *Mol Microbiol* **74**:888–902 (2009).
65. Kanasi E, Dewhirst FE, Chalmers NI, Kent R Jr, Moore A, Hughes CV, Pradhan N, Loo CY, Tanner AC. Clonal analysis of the microbiota of severe early childhood caries. *Caries Res* **44**:485–497 (2010).
66. Kaplan CW, Lux R, Haake SK, Shi W. The *Fusobacterium nucleatum* outer membrane protein RadD is an arginine-inhibitable adhesin required for inter-species adherence and the structured architecture of multispecies biofilm. *Mol Microbiol* **71**:35–47 (2009).
67. Katz J, Chegini N, Shiverick KT, Lamont RJ. Localization of *P. gingivalis* in preterm delivery placenta. *J Dent Res* **88**:575–578 (2009).
68. Kazor CE, Mitchell PM, Lee AM, Stokes LN, Loesche WJ, Dewhirst FE, Paster BJ. Diversity of bacterial populations on the tongue dorsa of patients with halitosis and healthy patients. *J Clin Microbiol* **41**:558–563 (2003).
69. Khlgatian M, Nassar H, Chou HH, Gibson FC III, Genco CA. Fimbria-dependent activation of cell adhesion molecule expression in *Porphyromonas gingivalis*-infected endothelial cells. *Infect Immun* **70**:257–267 (2002).
70. Kinane DF, Riggio MP, Walker KF, MacKenzie D, Shearer B. Bacteraemia following periodontal procedures. *J Clin Periodontol* **32**:708–713 (2005).
71. Kitten T, Munro CL, Wang A, Macrina FL. Vaccination with FimA from *Streptococcus parasanguis* protects rats from endocarditis caused by other viridans streptococci. *Infect Immun* **70**:422–425 (2002).
72. Kolenbrander PE, London J. Adhere today, here tomorrow: Oral bacterial adherence. *J Bacteriol* **175**:3247–3252 (1993).
73. Kreth J, Zhang Y, Herzberg MC. Streptococcal antagonism in oral biofilms: *Streptococcus sanguinis* and *Streptococcus gordonii* interference with *Streptococcus mutans*. *J Bacteriol* **190**:4632–4640 (2008).

74. Kuhl C. Insulin secretion and insulin resistance in pregnancy and GDM. Implications for diagnosis and management. *Diabetes* **40**:18–24 (1991).
75. Kvidera A, Mackenzie IC. Rates of clearance of the epithelial surfaces of mouse oral mucosa and skin. *Epithel Cell Biol* **3**:175–180 (1994).
76. Leon R, Silva N, Ovalle A, Chaparro A, Ahumada A, Gajardo M, Martinez M, Gamonal J. Detection of *Porphyromonas gingivalis* in the amniotic fluid in pregnant women with a diagnosis of threatened premature labor. *J Periodontol* **78**:1249–1255 (2007).
77. Ley RE. Obesity and the human microbiome. *Curr Opin Gastroenterol* **26**:5–11 (2010).
78. Ley RE, Backhed F, Turnbaugh P, Lozupone CA, Knight RD, Gordon JI. Obesity alters gut microbial ecology. *Proc Natl Acad Sci USA* **102**:11070–11075 (2005).
79. Ley RE, Turnbaugh PJ, Klein S, Gordon JI. Microbial ecology: Human gut microbes associated with obesity. *Nature* **444**:1022–1023 (2006).
80. Libby P. Inflammation in atherosclerosis. *Nature* **420**:868–874 (2002).
81. Lin D, Smith MA, Champagne C, Elter J, Beck J, Offenbacher S. *Porphyromonas gingivalis* infection during pregnancy increases maternal tumor necrosis factor alpha, suppresses maternal interleukin-10, and enhances fetal growth restriction and resorption in mice. *Infect Immun* **71**:5156–5162 (2003).
82. Liu H, Redline RW, Han YW. *Fusobacterium nucleatum* induces fetal death in mice via stimulation of TLR4-mediated placental inflammatory response. *J Immunol* **179**:2501–2508 (2007).
83. Loo JA, Yan W, Ramachandran P, Wong DT. Comparative human salivary and plasma proteomes. *J Dent Res* **89**:1016–1023 (2010).
84. Lucas VS, Omar J, Vieira A, Roberts GJ. The relationship between odontogenic bacteraemia and orthodontic treatment procedures. *Eur J Orthod* **24**:293–301 (2002).
85. Maeda K, Nagata H, Nonaka A, Kataoka K, Tanaka M, Shizukuishi S. Oral streptococcal glyceraldehyde-3-phosphate dehydrogenase mediates interaction with *Porphyromonas gingivalis* fimbriae. *Microbes Infect* **6**:1163–1170 (2004).
86. Mager DL, Ximenez-Fyvie LA, Haffajee AD, Socransky SS. Distribution of selected bacterial species on intraoral surfaces. *J Clin Periodontol* **30**:644–654 (2003).
87. Matsumoto-Nakano M, Tsuji M, Inagaki S, Fujita K, Nagayama K, Nomura R, Ooshima T. Contribution of cell surface protein antigen c of *Streptococcus mutans* to platelet aggregation. *Oral Microbiol Immunol* **24**:427–430 (2009).
88. McNab R, Ford SK, El-Sabaeny A, Barbieri B, Cook GS, Lamont RJ. LuxS-based signaling in *Streptococcus gordonii*: Autoinducer 2 controls carbohydrate metabolism and biofilm formation with *Porphyromonas gingivalis*. *J Bacteriol* **185**:274–284 (2003).
89. Mealey BL, Oates TW. Diabetes mellitus and periodontal diseases. *J Periodontol* **77**:1289–1303 (2006).
90. Miller-Torbert TA, Sharma S, Holt RG. Inactivation of a gene for a fibronectin-binding protein of the oral bacterium *Streptococcus mutans* partially impairs its adherence to fibronectin. *Microbial Pathogen* **45**:53–59 (2008).
91. Mitchell J, Sullam PM. *Streptococcus mitis* phage-encoded adhesins mediate attachment to α2–8-linked sialic acid residues on platelet membrane gangliosides. *Infect Immun* **77**:3485–3490 (2009).
92. Moreillon P, Que YA. Infective endocarditis. *Lancet* **363**:139–149 (2004).
93. Munro CL, Michalek SM, Macrina FL. Sucrose-derived exopolymers have site-dependent roles in *Streptococcus mutans*-promoted dental decay. *FEMS Microbiol Lett* **128**:327–332 (1995).
94. Murphy AM, Daly CG, Mitchell DH, Stewart D, Curtis BH. Chewing fails to induce oral bacteraemia in patients with periodontal disease. *J Clin Periodontol* **33**:730–736 (2006).

95. Nassar H, Chou HH, Khlgatian M, Gibson FC, III, Van Dyke TE, Genco CA. Role for fimbriae and lysine-specific cysteine proteinase gingipain K in expression of interleukin-8 and monocyte chemoattractant protein in *Porphyromonas gingivalis*-infected endothelial cells. *Infect Immun* **70**:268–276 (2002).
96. Nishimura F, Iwamoto Y, Soga Y. The periodontal host response with diabetes. *Periodontology 2000* **43**:245–253 (2007).
97. Nobbs AH, Lamont RJ, Jenkinson HF. *Streptococcus* adherence and colonization. *Microbiol Mol Biol Rev* **73**:407–450 (2009).
98. Normand J, Bozio A, Etienne J, Sassolas F, Le BH. Changing patterns and prognosis of infective endocarditis in childhood. *Eur Heart J* **16**:28–31 (1995).
99. Novak KF, Taylor GW, Dawson DR, Ferguson JE, Novak MJ. Periodontitis and gestational diabetes mellitus: Exploring the link in NHANES III. *J Public Health Dent* **66**:163–168 (2006).
100. Nyvad B, Kilian M. Comparison of the initial streptococcal microflora on dental enamel in caries-active and in caries-inactive individuals. *Caries Res* **24**:267–272 (1990).
101. Offenbacher S, Jared HL, O'Reilly PG, Wells SR, Salvi GE, Lawrence HP, Socransky SS, Beck JD. Potential pathogenic mechanisms of periodontitis associated pregnancy complications. *Ann Periodontol* **3**:233–250 (1998).
102. Offenbacher S, Katz V, Fertik G, Collins J, Boyd D, Maynor G, McKaig R, Beck J. Periodontal infection as a possible risk factor for preterm low birth weight. *J Periodontol* **67**:1103–1113 (1996).
103. Ogden CL, Yanovski SZ, Carroll MD, Flegal KM. The epidemiology of obesity. *Gastroenterology* **132**:2087–2102 (2007).
104. Okabe K, Nakagawa K, Yamamoto E. Factors affecting the occurrence of bacteremia associated with tooth extraction. *Int J Oral Maxillofac Surg* **24**:239–242 (1995).
105. Okahashi N, Nakata M, Sakurai A, Terao Y, Hoshino T, Yamaguchi M, Isoda R, Sumitomo T, Nakano K, Kawabata S, et al. Pili of oral *Streptococcus sanguinis* bind to fibronectin and contribute to cell adhesion. *Biochem Biophys Res Commun* **391**:1192–1196 (2010).
106. Olsen I. Update on bacteraemia related to dental procedures. *Transfus Apher Sci* **39**:173–178 (2008).
107. Palmer RJ, Jr, Diaz PI, Kolenbrander PE. Rapid succession within the *Veillonella* population of a developing human oral biofilm in situ. *J Bacteriol* **188**:4117–4124 (2006).
108. Parahitiyawa NB, Jin LJ, Leung WK, Yam WC, Samaranayake LP. Microbiology of odontogenic bacteremia: beyond endocarditis. *Clin Microbiol Rev* **22**:46–64 (2009).
109. Park Y, Simionato MR, Sekiya K, Murakami Y, James D, Chen W, Hackett M, Yoshimura F, Demuth DR, Lamont RJ. Short fimbriae of *Porphyromonas gingivalis* and their role in coadhesion with *Streptococcus gordonii*. *Infect Immun* **73**:3983–3989 (2005).
110. Patni S, Flynn P, Wynen LP, Seager AL, Morgan G, White JO, Thornton CA. An introduction to Toll-like receptors and their possible role in the initiation of labour. *Br J Obstet Gynecol* **114**:1326–1334 (2007).
111. Periasamy S, Kolenbrander PE. Central role of the early colonizer *Veillonella* sp. in establishing multispecies biofilm communities with initial, middle, and late colonizers of enamel. *J Bacteriol* **192**:2965–2972 (2010).
112. Petersen HJ, Keane C, Jenkinson HF, Vickerman MM, Jesionowski A, Waterhouse JC, Cox D, Kerrigan SW. Human platelets recognize a novel surface protein, PadA, on *Streptococcus gordonii* through a unique interaction involving fibrinogen receptor GPIIbIIIa. *Infect Immun* **78**:413–422 (2010).

113. Pickup JC, Crook MA. Is type II diabetes mellitus a disease of the innate immune system? *Diabetologia* **41**:1241–1248 (1998).

114. Plummer C, Wu H, Kerrigan SW, Meade G, Cox D, Ian Douglas CW. A serine-rich glycoprotein of *Streptococcus sanguis* mediates adhesion to platelets via GPIb. *Br J Haematol* **129**:101–109 (2005).

115. Pretorius C, Jagatt A, Lamont RF. The relationship between periodontal disease, bacterial vaginosis, and preterm birth. *J Perinat Med* **35**:93–99 (2007).

116. Progulske-Fox A, Kozarov E, Dorn B, Dunn W, Jr, Burks J, Wu Y. *Porphyromonas gingivalis* virulence factors and invasion of cells of the cardiovascular system. *J Periodontal Res* **34**:393–399 (1999).

117. Pucar A, Milasin J, Lekovic V, Vukadinovic M, Ristic M, Putnik S, Kenney EB. Correlation between atherosclerosis and periodontal putative pathogenic bacterial infections in coronary and internal mammary arteries. *J Periodontol* **78**:677–682 (2007).

118. Pussinen PJ, Jousilahti P, Alfthan G, Palosuo T, Asikainen S, Salomaa V. Antibodies to periodontal pathogens are associated with coronary heart disease. *Arterioscler Thromb Vasc Biol* **23**:1250–1254 (2003).

119. Quivey RG, Jr, Kuhnert WL, Hahn K. Adaptation of oral streptococci to low pH. *Adv Microbiol Physiol* **42**:239–274 (2000).

120. Rickard AH, Palmer RJ, Jr, Blehert DS, Campagna SR, Semmelhack MF, Egland PG, Bassler BL, Kolenbrander PE. Autoinducer 2: A concentration-dependent signal for mutualistic bacterial biofilm growth. *Mol Microbiol* **60**:1446–1456 (2006).

121. Ritchie CS. Obesity and periodontal disease. *Periodontolog 2000* **44**:154–163 (2007).

122. Roberts GJ, Holzel HS, Sury MR, Simmons NA, Gardner P, Longhurst P. Dental bacteremia in children. *Pediatr Cardiol* **18**:24–27 (1997).

123. Roberts GJ, Jaffray EC, Spratt DA, Petrie A, Greville C, Wilson M, Lucas VS. Duration, prevalence and intensity of bacteraemia after dental extractions in children. *Heart* **92**:1274–1277 (2006).

124. Ross R, Harker L. Hyperlipidemia and atherosclerosis. *Science* **193**:1094–1100 (1976).

125. Scannapieco FA, Dasanayake AP, Chhun N. Does periodontal therapy reduce the risk for systemic diseases? *Dent Clin North Am* **54**:163–181 (2010).

126. Scannapieco FA, Genco RJ. Association of periodontal infections with atherosclerotic and pulmonary diseases. *J Periodont Res* **34**:340–345 (1999).

127. Schenkein HA, Barbour SE, Berry CR, Kipps B, Tew JG. Invasion of human vascular endothelial cells by *Actinobacillus actinomycetemcomitans* via the receptor for platelet-activating factor. *Infect Immun* **68**:5416–5419 (2000).

128. Shao H, Lamont RJ, Demuth DR. Autoinducer 2 is required for biofilm growth of *Aggregatibacter* (*Actinobacillus*) *actinomycetemcomitans*. *Infect Immun* **75**:4211–4218 (2007).

129. Shun CT, Lu SY, Yeh CY, Chiang CP, Chia JS, Chen JY. Glucosyltransferases of viridans streptococci are modulins of interleukin-6 induction in infective endocarditis. *Infect Immun* **73**:3261–3270 (2005).

130. Silva TA, Garlet GP, Fukada SY, Silva JS, Cunha FQ. Chemokines in oral inflammatory diseases: Apical periodontitis and periodontal disease. *J Dent Res* **86**:306–319 (2007).

131. Simpson TC, Needleman I, Wild SH, Moles DR, Mills EJ. Treatment of periodontal disease for glycaemic control in people with diabetes. *Cochrane Database Syst Rev* **5**: CD004714 (2010).

132. Suntharalingam P, Cvitkovitch DG. Quorum sensing in streptococcal biofilm formation. *Trends Microbiol* **13**:3–6 (2005).

133. Taniguchi N, Nakano K, Nomura R, Naka S, Kojima A, Matsumoto M, Ooshima T. Defect of glucosyltransferases reduces platelet aggregation activity of *Streptococcus mutans*: Analysis of clinical strains isolated from oral cavities. *Arch Oral Biol* **55**:410–416 (2010).

134. Tanner A, Maiden MF, Macuch PJ, Murray LL, Kent RL Jr. Microbiota of health, gingivitis, and initial periodontitis. *J Clin Periodontol* **25**:85–98 (1998).

135. Taylor GW, Burt BA, Becker MP, Genco RJ, Shlossman M, Knowler WC, Pettitt DJ. Severe periodontitis and risk for poor glycemic control in patients with non-insulin-dependent diabetes mellitus. *J Periodontol* **67**:1085–1093 (1996).

136. Teeuw WJ, Gerdes VE, Loos BG. Effect of periodontal treatment on glycemic control of diabetic patients: A systematic review and meta-analysis. *Diabetes Care* **33**:421–427 (2010).

137. Thomson PJ, Potten CS, Appleton DR. Mapping dynamic epithelial cell proliferative activity within the oral cavity of man: A new insight into carcinogenesis? *Br J Oral Maxillofac Surg* **37**:377–383 (1999).

138. Tilg H, Moschen AR. Inflammatory mechanisms in the regulation of insulin resistance. *Mol Med* **14**:222–231 (2008).

139. Tomas I, Alvarez M, Limeres J, Potel C, Medina J, Diz P. Prevalence, duration and aetiology of bacteraemia following dental extractions. *Oral Dis* **13**:56–62 (2007).

140. Turnbaugh PJ, Hamady M, Yatsunenko T, Cantarel BL, Duncan A, Ley RE, Sogin ML, Jones WJ, Roe BA, Affourtit JP, et al. A core gut microbiome in obese and lean twins. *Nature* **457**:480–484 (2009).

141. Turnbaugh PJ, Ley RE, Mahowald MA, Magrini V, Mardis ER, Gordon JI. An obesity-associated gut microbiome with increased capacity for energy harvest. *Nature* **444**:1027–1031 (2006).

142. Turner LS, Kanamoto T, Unoki T, Munro CL, Wu H, Kitten T. Comprehensive evaluation of *Streptococcus sanguinis* cell wall-anchored proteins in early infective endocarditis. *Infect Immun* **77**:4966–4975 (2009).

143. Urano-Tashiro Y, Yajima A, Takashima E, Takahashi Y, Konishi K. Binding of the *Streptococcus gordonii* DL1 surface protein Hsa to the host cell membrane glycoproteins CD11b, CD43, and CD50. *Infect Immun* **76**:4686–4691 (2008).

144. Ventura M, Turroni F, Zomer A, Foroni E, Giubellini V, Bottacini F, Canchaya C, Claesson MJ, He F, Mantzourani M, et al. The *Bifidobacterium dentium* Bd1 genome sequence reflects its genetic adaptation to the human oral cavity. *PLoS Genet* **5**: e1000785 (2009).

145. Vita JA, Loscalzo J. Shouldering the risk factor burden: Infection, atherosclerosis, and the vascular endothelium. *Circulation* **106**:164–166 (2002).

146. Wang M, Shakhatreh MA, James D, Liang S, Nishiyama S, Yoshimura F, Demuth DR, Hajishengallis G. Fimbrial proteins of *Porphyromonas gingivalis* mediate in vivo virulence and exploit TLR2 and complement receptor 3 to persist in macrophages. *J Immunol* **179**:2349–2358 (2007).

147. Whitchurch CB, Tolker-Nielsen T, Ragas PC, Mattick JS. Extracellular DNA required for bacterial biofilm formation. *Science* **295**:1487 (2002).

148. Wild S, Roglic G, Green A, Sicree R, King H. Global prevalence of diabetes: Estimates for the year 2000 and projections for 2030. *Diabetes Care* **27**:1047–1053 (2004).

149. Wilson W, Taubert KA, Gewitz M, Lockhart PB, Baddour LM, Levison M, Bolger A, Cabell CH, Takahashi M, Baltimore RS, et al. Prevention of infective endocarditis: Guidelines from the American Heart Association: A guideline from the American Heart Association Rheumatic Fever, Endocarditis, and Kawasaki Disease Committee, Council on Cardiovascular Disease in the Young, and the Council on Clinical Cardiology, Council

on Cardiovascular Surgery and Anesthesia, and the Quality of Care and Outcomes Research Interdisciplinary Working Group. *Circulation* **116**:1736–1754 (2007).

150. Xie H, Lin X, Wang BY, Wu J, Lamont RJ. Identification of a signalling molecule involved in bacterial intergeneric communication. *Microbiology* **153**:3228–3234 (2007).
151. Xiong X, Buekens P, Fraser WD, Beck J, Offenbacher S. Periodontal disease and adverse pregnancy outcomes: A systematic review. *Br J Obstet Gynecol* **113**:135–143 (2006).
152. Xiong X, Buekens P, Vastardis S, Pridjian G. Periodontal disease and gestational diabetes mellitus. *Am J Obstet Gynecol* **195**:1086–1089 (2006).
153. Xiong X, Elkind-Hirsch KE, Vastardis S, Delarosa RL, Pridjian G, Buekens P. Periodontal disease is associated with gestational diabetes mellitus: A case-control study. *J Periodontol* **80**:1742–1749 (2009).
154. Yamazaki K, Ohsawa Y, Itoh H, Ueki K, Tabeta K, Oda T, Nakajima T, Yoshie H, Saito S, Oguma F, et al. T-cell clonality to *Porphyromonas gingivalis* and human heat shock protein 60s in patients with atherosclerosis and periodontitis. *Oral Microbiol Immunol* **19**:160–167 (2004).
155. Yeh CY, Chen JY, Chia JS. Glucosyltransferases of viridans group streptococci modulate interleukin-6 and adhesion molecule expression in endothelial cells and augment monocytic cell adherence. *Infect Immun* **74**:1273–1283 (2006).
156. Zaremba M, Gorska R, Suwalski P, Kowalski J. Evaluation of the incidence of periodontitis-associated bacteria in the atherosclerotic plaque of coronary blood vessels. *J Periodontol* **78**:322–327 (2007).
157. Zoellner H, Chapple CC, Hunter N. Microvasculature in gingivitis and chronic periodontitis: Disruption of vascular networks with protracted inflammation. *Microsc Res Tech* **56**:15–31 (2002).

8

MICROBIOTA OF THE GENITOURINARY TRACT

LAURA K. SYCURO and DAVID N. FREDRICKS

Vaccine and Infectious Disease Division, Fred Hutchinson Cancer Research Center, Seattle, Washington

8.1. INTRODUCTION

Most healthy individuals live in blissful ignorance of the microbial communities residing in their bodies, and this is particularly true for the genitourinary tract. This lack of awareness is both physical and intellectual, stemming from the usual absence of symptoms when colonized by resident genitourinary microbes and a degree of cultural discomfort with the very notion of microbial inhabitants in the nether regions. Even within the human microbiome research community, genitourinary sites are given less attention, as they do not harbor the largest, most diverse, or most easily accessed human ecosystems. Nevertheless, more recent molecular studies have revealed that the microbes inhabiting genitourinary niches are highly dynamic and critical for reproductive health, as we shall document in this chapter.

8.2. MICROBIAL ECOSYSTEMS OF THE FEMALE REPRODUCTIVE TRACT: A BIOGEOGRAPHIC PERSPECTIVE OF THEIR IMPACT ON REPRODUCTION

A blossoming awareness of the intricate coevolutionary relationships between humans and microbes is compelling scientists to reconceptualize many aspects of

The Human Microbiota: How Microbial Communities Affect Health and Disease, First Edition. Edited by David N. Fredricks.

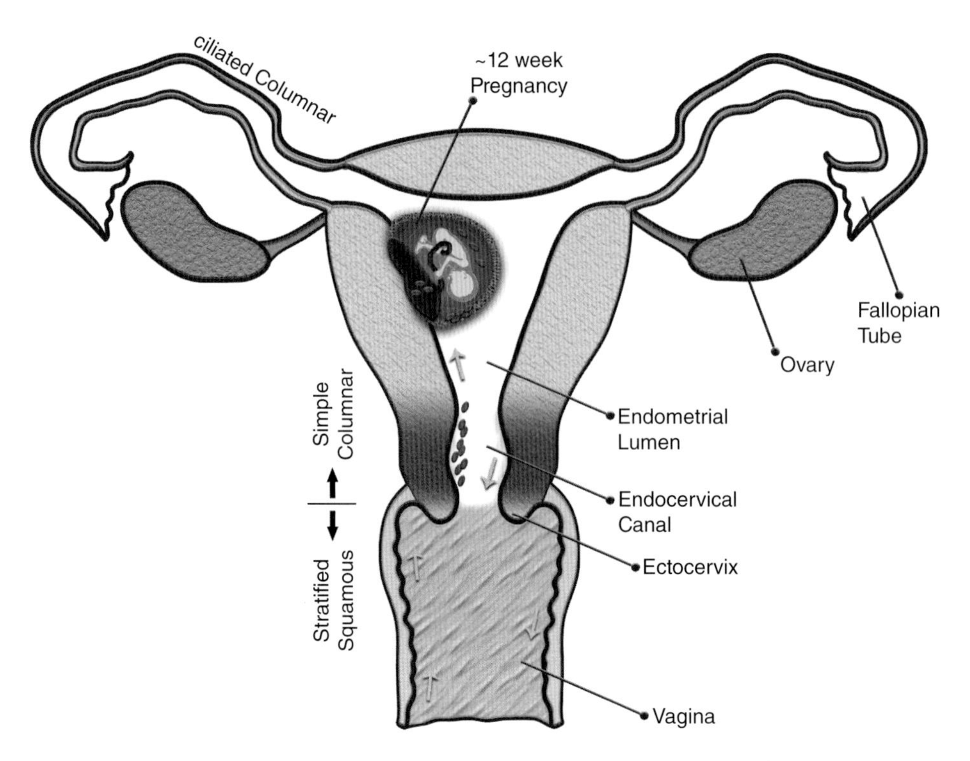

Figure 8.1. Anatomical features of the female reproductive tract with relevance to genitourinary microbes. Regions colonized by resident microbes are the vagina, ectocervix (lower tips of uterus, shaded lightly), and endocervical canal (section of lower uterus, shaded darker gray). The endometrial lumen, any associated fetal tissues, and the fallopian tubes are sometimes colonized by bacteria displaced from the vagina or cervix (round ovals), resulting in pathological infection. Ascension of microbes into the upper genital tract may be facilitated by smooth muscle contractions that together with gravity direct a bidirectional flow of cervicovaginal mucus (arrows). Boundaries of the three types of epithelium lining the female reproductive tract—stratified squamous, simple columnar, and ciliated columnar—are also indicated.

human biology, including reproduction. Although this area of research is still nascent, it is not unthinkable that we may someday view human-associated microbes as important to conception, pregnancy, and birth as they are to digestion. Here we will introduce the various niches of the female reproductive tract and consider the potential of their resident microbes to impact maternal and fetal health.

Extending from the vaginal opening to the fallopian fimbriae, the female reproductive organs form a continuous mucosal epithelium with four key regions: the vagina, endocervical canal, endometrium (uterus), and fallopian tubes (oviduct), each uniquely structured and surfaced to serve specialized functions in reproduction (Figure 8.1). The vagina is the most heavily colonized genitourinary organ, containing an estimated 10^8–10^9 bacterial cells/mL vaginal fluid [14], a level comparable to that of the small intestine [142]. As discussed in detail later in this chapter, the microbial inhabitants of the vagina vary among individuals and can be highly dynamic within individuals as well [133,162,175]. The most common configuration of the bacterial population is one dominated by lactobacilli and consisting of 4–12 species overall [49,133]. However, on the other extreme the microbial population may contain as many as 60–70 species, many of which are obligate or facultative anaerobes [50]. This highly diversified state of the vaginal microbiota is characteristic of a clinical condition known as *bacterial vaginosis* (BV). Found in approximately 30% of women, BV is the most common cause of vaginal symptoms (namely, abnormal discharge) prompting women to seek healthcare [89]. In the United States, more women consult their physicians for vaginal complaints than for any other illness, resulting in more than 10 million physician visits annually and highlighting the importance of understanding the microbial ecology of this body site [134].

One of the primary reproductive functions of the vagina is to receive sperm during vaginal intercourse. Vaginal microbes have the potential to impact the likelihood of conception by influencing the local production of proinflammatory cytokines [1,138,154] and affecting sperm survival and motility in cervical mucus [95,171]. Several studies have detected a comparatively high prevalence of BV among women seeking care for infertility, particularly among those with malfunctioning fallopian tubes [99,190] or no known cause of reproductive failure [1,101]. Further research is needed to determine whether members of the vaginal microbiota play a causal role in infertility.

Vaginal microbes also have a bearing on pregnancy. Lactobacilli may promote successful pregnancy by populating the cervicovaginal niche and providing colonization resistance to more pathogenic microbes capable of ascending infection and induction of preterm labor [54,153]. Vaginal *Lactobacillus* species are thought to modulate the vaginal ecosystem in a way that promotes their own survival, such as through the production of lactic acid [58]. Moreover, lactobacilli are nutritionally supported by the hormonal shifts that occur during pregnancy, feasting on glycogen that is deposited in vaginal epithelial cells with rising estrogen levels [139]. Microbes associated with BV may threaten a pregnancy and have been linked to second trimester miscarriage [35,56,60,127] and preterm labor [36,55,66]. Finally, vaginal bacteria serve as the first reservoir of microbes to inoculate the neonate as it emerges through the birth canal. Microbial populations resembling those of the mother's vagina have been found on the neonate's skin [34,100], as well as in gastric aspirates [19] and meconium [34], although their long-term survival and significance are still in question.

Like the vagina, the cervix is thought to be colonized by a resident microbial community along both the outer ectocervix, a stratified squamous epithelium continuous with that of the vagina [83,98] and the simple columnar epithelium that forms the endocervical canal [141,171]. Using cultivation techniques, Tan and colleagues estimated the number of organisms in homogenized cervical mucus aspirates to be approximately 10^6 cells/mL [171]. More recent culture-independent methods have compared the amount of bacterial DNA (number of copies of bacterial 16*S* rRNA genes) in ectocervical and endocervical swabs to that in vaginal swabs by quantitative PCR. These studies confirmed that the bacterial loads of both the ectocervix and endocervix are several orders of magnitude lower than that in the vagina, likely in the range of 10^5–10^7 cells/mL [98]. It remains unknown how closely a woman's cervical microbial population resembles her vaginal population. One study of only eight women detected differences between anatomic sites, but found the degree of difference varied by the individual and sampling method, making it difficult to draw a definitive conclusion [84]. It seems probable that superficial microbes are mixed and turned together by the slow bidirectional flow of cervicovaginal mucus, therein forming one continuous ecosystem. At the cell surface, however, the vaginal and cervical niches may diverge as a result of their differing epithelial cell types and presentation of receptors for bacterial adhesion molecules.

The frequency and degree to which the endometrium and fallopian tubes are colonized by bacteria remain uncertain. The endocervical canal's constrictive morphology and copious production of highly viscous mucus have long been thought to impede bacterial trafficking to the endometrium and oviduct. Yet sonography studies have shown that radiolabeled albumin microspheres (comparable in size to the head of human spermatozoa or small aggregates of bacteria) are rapidly and efficiently taken up from the vagina into the uterus of nonpregnant women during both the follicular and luteal phases of the menstrual cycle [193]. Likely generated by peristaltic smooth muscle contractions of the uterus and fallopian tubes to facilitate the ascension of sperm, this upward transcervical flow could provide resident cervicovaginal microbes regular access to the endometrium and oviduct.

Attempts to culture bacteria from the endometrium have yielded variable results. Some endometrial cultivation studies took the minimally invasive approach of accessing the uterus through the vagina and cervix using catheters threaded with retractable brushes that were only unsheathed within the endometrial lumen. These authors reported bacterial growth in 60–90% of cultures from asymptomatic women [16,63] and 20% of cultures from women with unexplained irregular bleeding [87]. As any form of transcervical sampling invites the possibility of contaminating the uterus with cervicovaginal bacteria, other researchers chose to sample the uterus during intraoperative hysterectomies and tubal ligations. Some of these studies yielded moderate colonization frequencies of 25–31% [28,116], while others reported much lower rates, in the range of 0–7% [5,158,173]. Positive cultures typically yielded one to seven species, 70–75% of which were also commonly detected in the vagina and cervix [16,63,116]. One study that quantified endometrial bacterial load in women without overt signs of infection generally detected fewer than 10^5 cells/mL [87]. Taken together, these data suggest that the endometrium is likely not a sterile environment, although it may be colonized primarily with transient populations that are cleared by local mucosal immune defenses. Further support for this view comes from cross-sectional studies of women whose uteri were cultured at various timepoints following an event that may have facilitated bacterial entry, such as the insertion of an intrauterine contraceptive device (IUD) or childbirth [16,114]. Both studies observed higher rates of bacterial recovery within 1–2 months of the event.

The fallopian tubes are lined by a ciliated columnar epithelium and are open at the end distal to the peritoneal cavity, a fluid-lined compartment that houses the uterus, ovaries, and other abdominal organs. Researchers have cultured peritoneal fluid and fallopian tube biopsies from asymptomatic women undergoing laparoscopic surgery, usually for elective tubal ligation, finding both to be sterile [34,64]. These findings suggest immunologic tolerance of microbial residence in the female genital tract diminishes beyond the cervix, likely to the point of exclusion in the oviduct and where conception occurs. However, further research is needed to substantiate this conclusion, due to the highly variable sampling methods and patient populations in the abovementioned studies. Additionally, cultivation-independent approaches have not yet been applied to these difficult to access, but important body sites.

Although the question of whether some microbes permanently reside in the upper genital tract is an important one for human microbial ecologists, clinical assessment of upper genital tract colonization has traditionally focused on detecting acute pathogenic infections. Sexually transmitted pathogens such as *Chlamydia trachomatis* and *Neisseria gonorrhoeae* are adept at accessing the upper genital tract, thwarting the immune system, and overwhelming the niche with bacterial densities as high as 10^{10} cells/mL. Inflammatory disease caused by colonization with these microbes may occur at the cervix (cervicitis), endometrium (endometritis), fallopian tubes (salpingitis), and the peritoneal space (peritonitis). The anatomic continuum of these infections, known as *pelvic inflammatory disease* (PID), is a major cause of tubal obstruction and ectopic pregnancy [151]. Although PID is frequently associated with abdominal pain, fever, and abnormal vaginal discharge or bleeding, it can also persist in an asymptomatic state that is termed "silent" or "subclinical" due to mild symptoms and failure to detect the condition upon standard gynecological examination [151].

Bacterial vaginosis–associated organisms found in the upper genital tract have also been etiologically associated with both acute and silent PID [64,152,186]. If these ascending infections occur in a woman who is pregnant, bacteria may infect the fetal membranes (chorioamnionitis), the umbilical cord (funisitis), and the amniotic fluid (amnionitis) [54]. More recent studies indicate that the presence of bacteria in the pregnant uterus may not be uncommon. Cultivation of the amniotic membranes from women who delivered by Cesarean section before membrane rupture revealed nearly 20% were colonized by bacteria [54]. Another group of researchers utilized fluorescence *in situ* hybridization (FISH) to examine the fetal membranes of women who had elective Cesarean births at term and found bacteria present in 70% [164]. In contrast, full penetration of the amniotic membranes resulting in bacterial growth within the amniotic fluid appears to be rare in term pregnancies, but may play a causal role in preterm labor. Both cultivation-dependent and -independent studies have shown an inverse relationship between gestational age at delivery and evidence of bacteria in the amniotic fluid, with amniotic bacteria detected in 45% of pregnancies ending before 25–26 weeks [32]. Inflammatory responses to amniotic infection are thought to ignite a cascade of labor-inducing signals and vary in strength according to which bacteria are present, with several BV-associated species among the most proinflammatory [43]. Interestingly, many other proinflammatory pathogens capable of ascending the female genital tract, namely, *C. trachomatis*, *N. gonorrhoeae*, and *E. coli*, are seldom detected in fetal membranes or amniotic fluid [32,69,183]. Also puzzling is the fact that aside from vaginal BV-associated bacteria and mycoplasmas, the other major group of bacteria seen in amniotic fluid are members of the oral microbiota [69,183]. These organisms likely gain access to the uterus via the blood, but may also be transferred during orogenital contact. How and why particular BV-associated species ascend the genital tract in some individuals is unknown, as is the time in pregnancy at which this usually occurs. However, as some studies have associated the detection of BV in early pregnancy with increased risk of first- and second-trimester fetal loss as well as preterm birth, the vaginal microbiota that are present early on, perhaps even at conception and implantation, could have an impact on the outcome of the pregnancy [36,86].

8.3. SCIENTIFIC CONTROVERSY REGARDING BACTERIAL VAGINOSIS

The vagina is unusual, if not unique, among human microbial habitats in its capacity to house a dichotomy of radically different microbial communities: those dominated by only a few Gram-positive *Lactobacillus* species and those exhibiting a high level of species diversity with numerous Gram-negative anaerobes presiding. In 1892, Professor Albert Döderlein was the first to describe the former bacterial configuration as the most common in women of reproductive age. Döderlein isolated the large nonmotile Gram-positive rods he observed in vaginal smears and demonstrated that these organisms were capable of inhibiting the growth of pathogens [33]. Döderlein found that these microbes produced lactic acid and attributed their competitive advantage to their ability to produce and thrive in an acidic environment. In honor of their discoverer, these organisms were termed *Döderlein's bacilli*, but many years later were classified to the *Lactobacillus* genus.

As observational techniques improved, it was recognized that Döderlein's bacilli were not the only organisms in the vagina and were, in fact, not even predominant in all women. Efforts to describe this variation led to the proposal of three classes of vaginal microbiota: type I, consisting of acidophilic Döderlein's bacilli; type III, typified by the absence of Döderlein's bacilli and the presence of morphologically diverse organisms such as cocci, diphtheroids, and vibrios within more neutral secretions; and type II, representing intermediate flora [30]. By the early 1930s it was generally accepted that Döderlein's bacilli (type I flora) were indicative of vaginal health, largely because their prevalence was found to peak during the late stages of pregnancy, when their protective function was considered most critical [92]. More debated was whether type III flora were linked to symptomatic changes in vaginal discharge [30,184]. More than 75 years later our understanding of the composition of the vaginal microbiota has changed dramatically as a result of two technological revolutions: anaerobic cultivation in the late 1970s and cultivation-independent techniques in the 2000s [48,51,80]. The condition wherein diverse populations of vaginal bacteria occur with the absence, or at least low relative abundance, of vaginal lactobacilli (type III flora) eventually became known as *bacterial vaginosis*, yet remarkably, the question of whether these communities are pathogenic or simply a normal variant of the healthy vaginal ecosystem is now more hotly contested than ever. In this section we will discuss the ongoing controversy of what does and does not constitute a normal vaginal microbiota, highlighting several important underpinnings of the dispute: the complex polymicrobial etiology of BV, its high rate of treatment failure, variation in how it is diagnosed, and the observation that asymptomatic women can have a BV-like microbiota.

It may seem quite illogical that the diverse bacterial communities associated with BV are considered "healthy" by some when BV increases the likelihood of adverse outcomes in pregnancy as we described in Section 8.2. At the heart of this apparent contradiction is our incomplete understanding of BV etiology and the biological mechanisms that lead to its more serious sequelae, as well as the lack of longlasting cures for BV. Although the precise cause of BV is not known, it is now widely accepted that BV etiology is polymicrobial, meaning that the condition generally involves the gain and/or loss of multiple species as opposed to the more traditional concept of a single pathogenic species causing an infection. Precisely which species are present with BV varies from woman to woman and with time, compelling us to unremittingly use the term *BV-associated bacteria* (BVAB) to describe species that are frequently, but not always found in women with BV. Treatment with antibiotics (usually metronidazole or clindamycin) that aim to eradicate anaerobic BVABs is standard treatment for women who are experiencing symptoms of increased and/or malodorous vaginal discharge. Although most women initially respond to treatment with a reduction in BVABs and resolution of symptoms [162], there are high rates of relapse for reasons that are not well understood [18,88]. It has been proposed that treatment does not effectively eradicate all BVABs due to the formation of bacterial biofilms [169], or that the organisms are reinoculated into the vagina from extravaginal reservoirs (i.e., the gut) [37,74] or genital reservoirs in their sex partners [105,166]. However, some investigators have argued that treatment for BV fails in some women because their microbial communities represent a stable and healthy biological state that is naturally restored [195], although this ignores the observation that for many women, a return of BVABs is accompanied by a return of symptoms. Treatment for BV has also largely failed to show efficacy in the prevention of adverse health outcomes. For example, most studies have not shown a reduction in preterm birth with treatment of BV in pregnancy [109,149]; one notable exception was a study of women with a history of preterm birth who were treated for BV early in pregnancy, resulting in a lower rate of preterm birth in the treatment arm (18%) compared to women who received placebo (39%) [118]. It is unclear whether the conflicting data are related to differences in the timing and/or effectiveness of treatment with different antibiotics, host factors, or other variables. Thus, in spite of epidemiological data linking BV and preterm birth, there is ongoing debate regarding whether BV is, in fact, a significant cause of preterm birth [103].

In the absence of a defined pathogen or cause, there is considerable variability in how BV is diagnosed in both the research community and the clinic. The current gold standard method of BV diagnosis does not attempt to identify particular species, but instead assesses bacterial diversity through visualization of bacterial morphologies and cell wall properties on Gram staining. This method is reminiscent of the early morphology-based classification system described above, but differs in its utilization of a weighted scoring matrix developed by Nugent et al. (Figure 8.2) [124]. The Nugent score incorporates the relative abundance of lactobacilli and two other microbial morphotypes that correlate with known BV-associated taxa: small Gram-negative/variable rods or coccobacilli (*Gardnerella vaginalis* and/or *Bacteroidetes* spp.) and curved Gram-negative/variable rods (*Mobiluncus* spp. and/or BVAB1). Low scores between 0 and 3 indicate normal microbiota, 4–6 intermediate microbiota, and scores ranging from 7 to 10 denote BV. Nugent's method of systematically interpreting Gram stains has proved to be the most reliable BV diagnostic in terms of minimizing intercenter variability and enabling interstudy comparisons to be made [46,124]. However, Gram stain interpretation of vaginal smears requires training and experience in this particular area as well as the use of a high-powered compound microscope (1000× magnification), which is not always available in

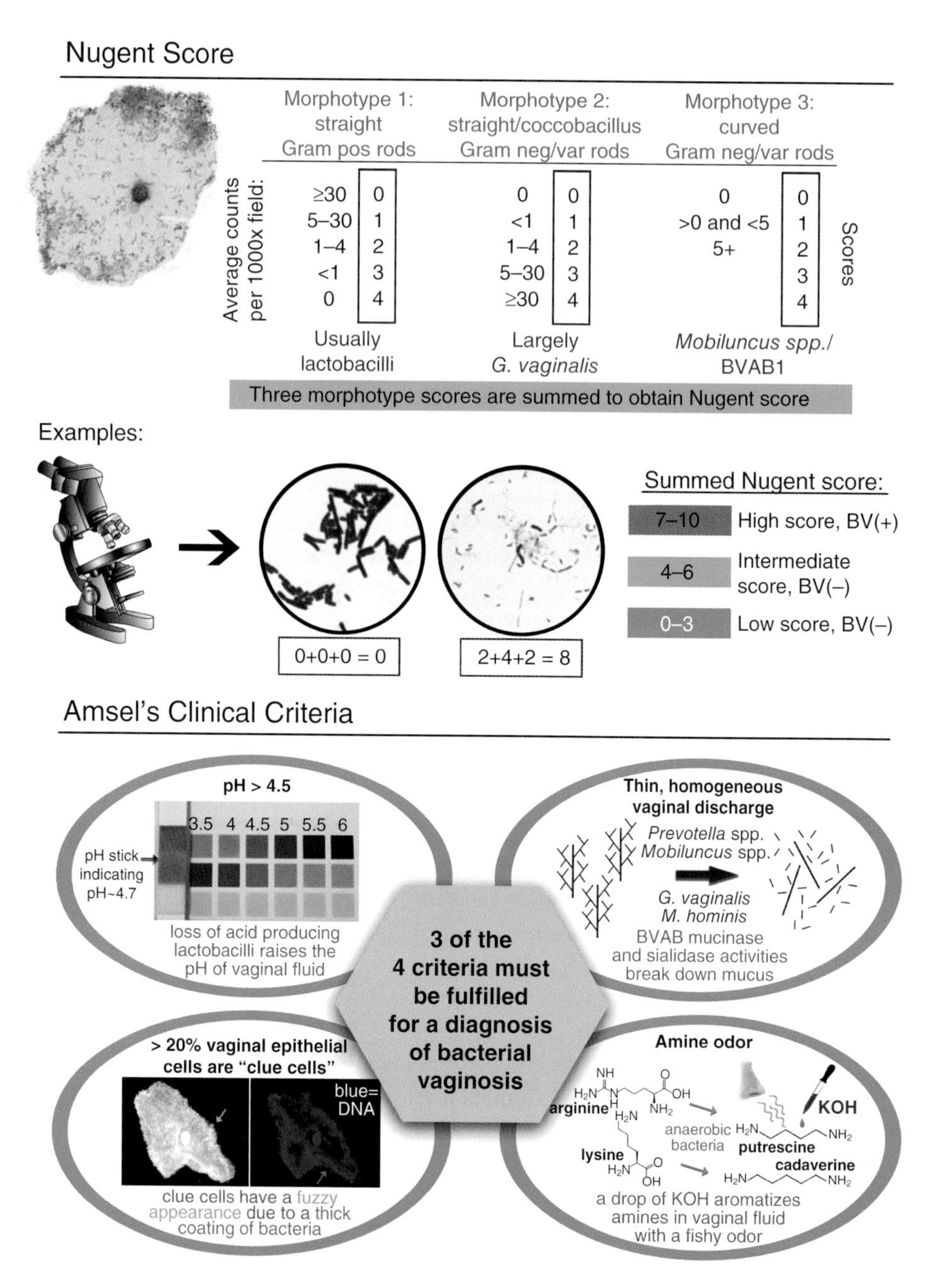

Figure 8.2. Current gold standard diagnostics for bacterial vaginosis. Graphical description of how a Nugent score, the standard classifier of BV in research studies, is calculated. Briefly, three bacterial morphotypes are enumerated in a Gram-stained vaginal smear at 1000× magnification. Each morphotype is assigned a score, and the three morphotype scores are summed to determine the Nugent score. Amsel's criteria are outlined in the lower part of the diagram, specifically, the four characteristics of vaginal discharge that signify the presence of BV to clinicians. Note that access to a compound microscope is essential for evaluation of clue cells.

clinical and resource-limited settings. Knowing how to perform and read a basic Gram stain is not sufficient to generate reliable Nugent scores.

The need for a BV diagnostic that can be performed quickly and routinely in primary care clinics is most often fulfilled by Amsel's criteria [4,41]. This series of tests detects changes in vaginal discharge that frequently accompany the complex bacterial populations present with BV (Figure 8.2). Three of the following criteria must be positive for a woman to be diagnosed as having BV: (1) appearance of vaginal discharge is thin and homogenous; (2) vaginal pH is >4.5, (3) >20% of shed vaginal epithelial cells are "clue cells," that is profusely covered in bacteria such that the edges are nondistinct; and (4) a fishy amine odor is released on the addition of 10% potassium hydroxide to vaginal fluid (usually on a slide). The biological processes responsible for these changes in the vaginal discharge are not well-defined, and a number of bacterial species may be contributing to them. Consequently, a diagnosis by Amsel's criteria does not provide any specific information about the bacterial taxa present in the vaginal community. The diagnostic is also prone to inconsistency among different clinicians due to the acumen and approximation involved in identifying and tabulating clue cells, as well as its reliance on practitioners' highly variable sense of smell [27,90,115]. Additionally, employment of Amsel's criteria still requires a low-power compound microscope (200–400×) to identify clue cells, the lack of which is often an impediment to differentiating BV from other forms of vaginitis. Surprisingly, this is not only a problem in resource limited settings, but also in developed countries, as one US study found that microscopic evaluation of vaginal discharge is not used in over a third of primary care workups for vaginitis [188], potentially leading to misdiagnosis and delayed treatment.

Even when the methods of Nugent and/or Amsel are employed correctly, other factors can still blur the diagnostic outcome. For instance, while the Nugent score and Amsel's criteria usually exhibit moderate sensitivity (rate of false-negative results) and specificity (rate of false-positive results) when compared to each other, they are not always in agreement. In 5–20% of women they are discordant owing to the complex nature of BV and the aforementioned challenges in the accurate collection of Amsel's data [115,145]. Furthermore, in some populations such as women with HIV infection, the sensitivity of Amsel's criteria using Nugent as gold standard is only 36% [147]. No guidelines for resolving such conflicts are currently in existence.

Even more ambiguity arises from the inconsistent handling of intermediate Nugent scores. Intermediate scores are generally thought to signify the presence of a transitional state, but the standard method is to classify these women BV-negative [20,21,68]. However, it is also not uncommon for women with intermediate scores to be classified BV-positive or excluded from the analysis altogether, practices that can skew study findings as well as diminish their applicability to real-world settings. Finally, there is a recent blossoming of other point-of-care BV diagnostics on the market, including tests for vaginal pH, bacterial sialidase activity, and amine compounds that are evolved through anaerobic metabolism [180]. BVBlue, a colorimetric test for sialidase, is one of the few that have been shown to perform nearly as well as Nugent score and Amsel's criteria [18,119,165]. Other tests utilize conventional or quantitative PCR to identify *G. vaginalis* alone or in combination with three to five other BVABs, many of which are also common in healthy women, albeit at lower concentrations. Virtually no data exist on the performance of these newer commercially available PCR tests in the typical clinical setting, nor is it known just how extensively they are being used in place of microscopy-based evaluations. As we will discuss in the following sections, molecular surveys of the vaginal microbiota have illuminated several novel species that are highly specific for BV and are being actively pursued as targets for improved PCR-based diagnostics. However, in order for any of these next-generation BV diagnostics to significantly improve the precision of research studies and/or patient outcomes, care must be taken to require demonstrated efficacy under standardized conditions with relevant and generalizable study populations.

Another unresolved question is what to do with women who have asymptomatic BV. In most developed countries, asymptomatic women are not screened for BV in primary care clinics, while a report of symptoms may motivate a diagnostic workup for BV. However, in research studies where a cross-section of all women in the population are screened for BV, symptomatic vaginal discharge and malodor are not reported by 40% of the women who have BV according to Amsel's criteria or Nugent score [4,49], revealing a substantial fraction with asymptomatic BV. Should women with asymptomatic BV be considered "healthy" as suggested by some investigators [133]? It is not yet clear why some women with BV are symptomatic and others are not. One unassailable factor is that when it comes to vaginal discharge, women's perceptions of what is normal are highly subjective and are often based solely on what they have experienced with their own bodies [81]. Likewise, the degree to which they are aware of any change therein varies widely for numerous cultural and psychosocial reasons. As we have discussed, BV is linked to preterm birth and pelvic inflammatory disease, and has also been associated with a woman's risk of acquiring HIV infection and developing infections after gynecological surgery; in many of these studies, BV cases and healthy controls were defined by Nugent score or Amsel's criteria, not symptoms [72,107,153,170]. Thus, while BV might be considered a common variant of the human vaginal microbiota, it is one with many associated health risks. In this sense, one might compare BV to hypertension. Although systolic blood pressures >120 are common in the US population, these levels should not be considered "normal" in the sense that they are associated with excess mortality from cardiovascular disease and stroke. Debates about BV focused on the concept of normality will not be resolved since the very definition of normal is subject to interpretation. What is not debatable is that BV, as defined by Nugent score or Amsel's criteria, is associated with multiple adverse health outcomes.

8.4. PERSPECTIVES ON THE VAGINAL MICROBIOME IN THE AGE OF HIGH-THROUGHPUT SEQUENCING

Hope of resolving the scientific controversy surrounding BV resides in specifically differentiating the components of vaginal microbial communities that promote healthy vaginal physiology from those that are suboptimal or harmful. Toward this end, the use of cultivation-independent molecular techniques to identify the microbial species present in women with and without BV has advanced our understanding of the diversity and complexity of the vaginal ecosystem. In this section we will look in depth at what the most recent high-throughput sequencing data are revealing about the vaginal microbiome.

Most cultivation-independent analyses of the vaginal microbiome have surveyed 16*S* rRNA gene sequences using PCR, although other targets such as the chaperonin 60 gene (*cpn60*) have also been used [143]. After amplifying the 16*S* rRNA genes of most or all bacteria in a sample of vaginal fluid with broad-range primers (see Chapter 2), various techniques are applied to classify and identify represented taxa. These include electrophoretically sorting the PCR products according to size and G/C content (DGGE) or sizing fragments generated by restriction site polymporphisms (T-RFLP). A 16*S* rRNA gene product that is representative of each type can then be sequenced to obtain taxonomic identification. Another approach is to generate clone libraries containing 16*S* rRNA amplicons and sequencing a limited number of clones, which may be selected using restriction site polymorphisms [known as *amplified ribosomal DNA restriction analysis* (ARDRA)] to maximize diversity. Cost and time constraints generally limit the number of sequenced clones to a few hundred per sample and <2000 per study [49,179,194], although one study sequenced a remarkable 1000 clones for each of 20 subjects [78]. The more recent application of high-throughput sequencing to 16*S* rRNA amplicons has increased the number of sequences for which taxonomic inferences can be made to upward of 10,000 per sample and millions per study, bringing unprecedented depth and clarity to our picture of the vaginal microbiome. For instance, we now know that at least 30% of vaginal species have not been cultivated or properly identified using standard clinical microbiology protocols, and vaginal species diversity is probably 2–4 times greater than earlier estimates based on 16*S* rRNA clone libraries [76,80,128,144]. However, cultivation-based studies are still an effective means of identifying species that readily grow in standard culture media but are too low in abundance to be detected using molecular methods, as was demonstrated by two studies that applied both techniques to the same samples and found 13–34% of species only by culture [114,179]. Additionally, clone libraries typically contain longer 16*S* rRNA sequences than those generated by high-throughput sequencing, enabling more reliable species-level identifications [49,78,128,179,194]. In light of these findings, we will focus our presentation of vaginal bacterial community composition on the most recent high throughput sequencing data, but we will also consider the limitations and biases of these data and supplement them with information gleaned from cultivation studies and clone libraries.

Three studies have used high-throughput sequencing of broad-range 16*S* rRNA gene PCR products to describe the vaginal microbiome in cohorts of ≥100 women residing on three different continents. In one study, Hummelen et al. used Illumina sequencing to analyze variable region 6 (V6) of the 16*S* rRNA gene in 132 HIV-positive Tanzanian women, 67 of whom had BV [76]. Ling et al. used the Roche 454 FLX pyrosequencing platform to sequence variable region 3 (V3) of the 16*S* rRNA gene in 100 Chinese women, half of which were diagnosed with BV according to both Nugent score and Amsel's criteria [97]. Also utilizing the Roche technology, Ravel et al. sequenced variable regions 1 and 2 (V1–V2) of the 16*S* rRNA gene in samples from 396 healthy, asymptomatic American women, 97 of whom nonetheless had Nugent scores in the range of 7–10, suggesting the presence of BV [133]. Despite the methodological differences between these studies, the picture of each population's vaginal microbiome, when painted with the broad taxonomic strokes of phyla and orders, is largely consistent. As shown in Figure 8.3a, the vaginal microbiome is dominated by the phyla Firmicutes, Actinobacteria, and Bacteroidetes, with Fusobacteria, Tenericutes, and Proteobacteria also sometimes detected at ≥0.1% relative abundance. All three studies verified the dominance of Firmicutes, in particular lactobacilli, in women with low Nugent scores. Another clear trend was the decline in lactobacilli from a relative abundance of 53–89% in the different populations of women with low Nugent scores to 18–32% in women with high Nugent scores. Accordingly, Firmicutes were significantly reduced in women with high Nugent scores, but still remained numerically dominant (35–52%) due to the expansion of other orders within the phylum, namely, the Clostridiales and Selenomonadales. The Bacteroidales, Bifidobacteriales, Coriobacteriales, and Fusobacteriales also displayed increased relative abundance in women with high Nugent scores, reaching 7–25% overall relative abundance. All three studies further demonstrated a significant increase in the Shannon index, a measure of bacterial diversity, in women with high Nugent scores compared to those with low scores, confirming the findings of 16*S* rRNA clone library studies [128]. It should be noted, however, that estimates of diversity based on high-throughput sequencing data account for the top 99.9–99.99% of the organisms present in an individual at current read depths, and that more "rare" species with abundances below this detection threshold may still be present at ≤10^5 cells/mL in vaginal fluid. Consequently, it remains unknown whether most women with BV have been recently colonized with new species that flourish and displace lactobacilli, or whether BV is more often manifested by an expansion of species already present, albeit at densities below our level of detection. Data from highly sensitive quantitative PCR studies suggest that at least some BVABs tend to be absent in women without BV.

One of the major differences in the population-level data obtained from these three studies is the absence of the Bifidobacteriales in the Ravel study [133]. The major Bifidobacteriales species in the vagina is *Gardnerella vaginalis*, which is commonly cultivated from women with and without BV and has been shown to reach high relative abundance in 16*S* rRNA clone libraries [49,80]. In line with these findings, Hummelen et al. and Ling et al. detected *G. vaginalis* at 18–21% relative abundance, while in the study by Ravel et al. it represented <1% of amplified taxa [76,97,133]. This is almost certainly an example of amplification bias, wherein particular taxa are underestimated because of mismatches between the target 16*S* rRNA gene sequence and the broad-range primers used in the study. *In silico* analyses suggest the 27F primer used by Ravel et al. to be a poor match for species in the Bifidobacteriales, Chlamydiales, and *Borrelia* genus, and experimental evidence has shown that it inefficiently amplifies *G. vaginalis* [47,161]. All three of these studies, by virtue of their reliance on broad-range PCR to capture bacterial diversity, were subject to this type of bias, although Hummelen et al. were the only authors to address this in their manuscript. *In silico* simulations predicted that their L-V6 (forward) primer sequence was biased against amplification of species belonging to the genera *Sneathia*, *Leptotrichia*, *Ureaplasma*, and *Mycoplasma*. Indeed, while Ling et al. and Ravel et al. determined that *Sneathia* sequences accounted for 10–14% of reads obtained from women with high Nugent scores, Hummelen et al. detected this genus at only 2%. Hummelen et al. was also the only study that failed to detect Mycoplasmatales at ≥0.1% relative abundance in women with high Nugent scores. Ultimately, all PCR primers and methods have some form of bias, and continuation

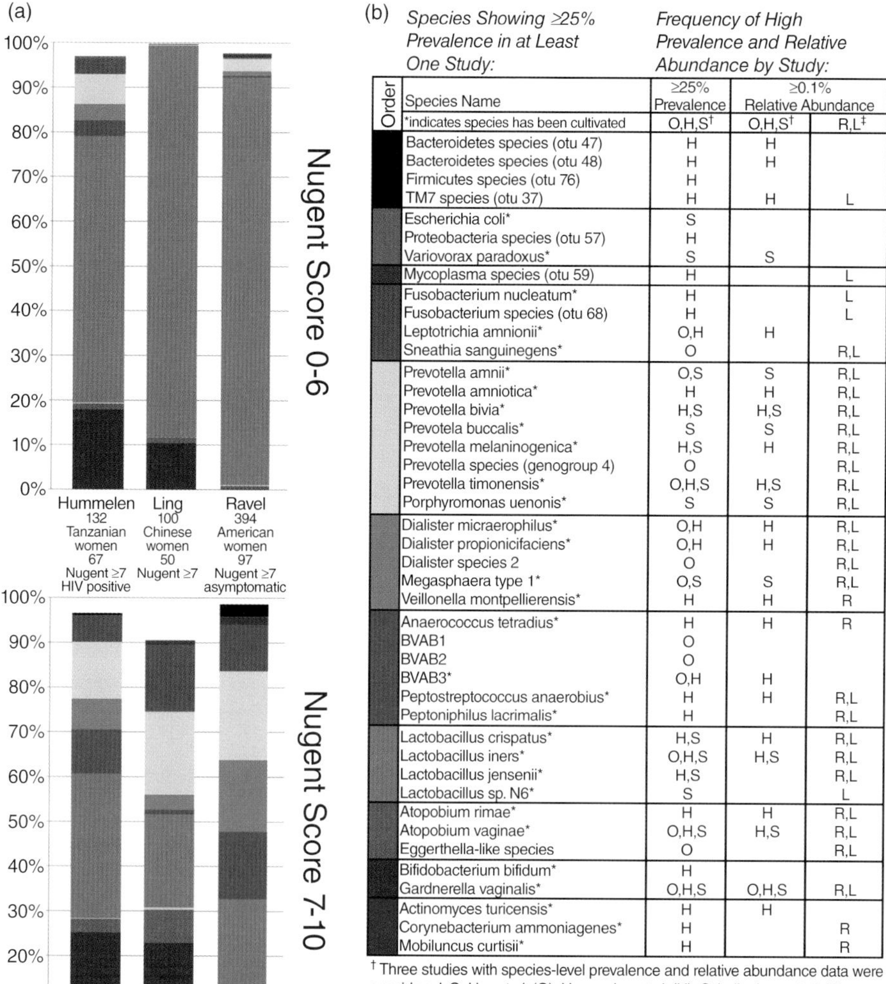

Species Name *indicates species has been cultivated	≥25% Prevalence O,H,S†	≥0.1% Relative Abundance O,H,S†	 R,L‡
Bacteroidetes species (otu 47)	H	H	
Bacteroidetes species (otu 48)	H	H	
Firmicutes species (otu 76)	H		
TM7 species (otu 37)	H	H	L
Escherichia coli*	S		
Proteobacteria species (otu 57)	H		
Variovorax paradoxus*	S	S	
Mycoplasma species (otu 59)	H		L
Fusobacterium nucleatum*	H		L
Fusobacterium species (otu 68)	H		L
Leptotrichia amnionii*	O,H	H	
Sneathia sanguinegens*	O		R,L
Prevotella amnii*	O,S	S	R,L
Prevotella amniotica*	H	H	R,L
Prevotella bivia*	H,S	H,S	R,L
Prevotela buccalis*	S	S	R,L
Prevotella melaninogenica*	H,S	H	R,L
Prevotella species (genogroup 4)	O		R,L
Prevotella timonensis*	O,H,S	H,S	R,L
Porphyromonas uenonis*	S	S	R,L
Dialister micraerophilus*	O,H	H	R,L
Dialister propionicifaciens*	O,H	H	R,L
Dialister species 2	O		R,L
Megasphaera type 1*	O,S	S	R,L
Veillonella montpellierensis*	H	H	R
Anaerococcus tetradius*	H	H	R
BVAB1	O		
BVAB2	O		
BVAB3*	O,H	H	
Peptostreptococcus anaerobius*	H	H	R,L
Peptoniphilus lacrimalis*	H		R,L
Lactobacillus crispatus*	H,S	H	R,L
Lactobacillus iners*	O,H,S	H,S	R,L
Lactobacillus jensenii*	H,S		R,L
Lactobacillus sp. N6*	S		L
Atopobium rimae*	H	H	R,L
Atopobium vaginae*	O,H,S	H,S	R,L
Eggerthella-like species	O		R,L
Bifidobacterium bifidum*	H		
Gardnerella vaginalis*	O,H,S	O,H,S	R,L
Actinomyces turicensis*	H	H	
Corynebacterium ammoniagenes*	H		R
Mobiluncus curtisii*	H		R

† Three studies with species-level prevalence and relative abundance data were considered: Oakley et al. (O), Hummelen et al. (H), Schellenberg et al. (S)
‡ Two additional studies that provided population-level relative abundance data at the genus level were also considered: Ravel et al. (R) and Ling et al. (L)

Figure 8.3. Molecular perspectives of the vaginal microbiota. (a) Vaginal bacteria detected to 0.1% relative abundance by deep sequencing of broad-range 16*S* rRNA PCR products in three different populations of over 100 women. Identified operational taxonomic units (OTUs) are presented in stacked bar charts at the order level by color group (see legend). Higher-order OTUs (class, phylum, or division level) are shown as a single group (black). Microbial populations are segregated according to Nugent scores that indicate the presence (7–10) or absence (0–6) of BV. (b) Prevalent bacterial species in women with BV. Shown are species that were detected by molecular techniques in at least 25% of women diagnosed with BV by Nugent score in three studies: Oakley et al. [128] (16 species with ≥25% prevalence identified from clone libraries generated for 21 women with BV), Hummelen et al. [76] (30 species with ≥25% prevalence reported in pyrosequencing study of 67 women with BV and HIV), and Schellenberg et al. [144] (15 species with ≥25% prevalence detected by pyrosequencing of the *cpn60* gene in 20 women with BV and mixed HIV status). In each study, species were classified by ≥95% 16*S* rRNA sequence identity. Species are taxonomically grouped by order and color-coded as in part (a). Stars denote species with at least one cultured isolate. A breakdown of the studies in which each prevalent species was detected and whether they also exhibited ≥0.1% relative abundance in the population is provided to the right.

of a frank dialog regarding these potential biases will be critical for gaining a complete understanding of the constitution of the vaginal microbiome.

Another prominent difference among the three in-depth sequencing studies is the lower relative abundance of Lactobacillales among HIV-positive women with low Nugent scores [76]. The question of how the vaginal microbiota may differ in women with HIV is of high interest because BV occurs more frequently in HIV-positive women and has been shown to increase the likelihood of both acquiring and transmitting HIV [11,25,79,146,170]. Among women with low Nugent scores, lactobacilli accounted for <60% of the bacterial population in the Hummelen et al. HIV-positive Tanzanian cohort compared to 85–90% of the bacterial population in the Ling et al. and Ravel et al. HIV-negative Chinese and American cohorts, respectively. The diminished numbers of lactobacilli were offset by larger numbers of organisms belonging to orders typically expanded with BV, including Clostridiales, Bacteroidales, and Bifidobacteriales.

The question then arises as to whether the reduction in lactobacilli seen in the Tanzanian cohort is related more to the race of the study population or their HIV status. A fourth high-throughput sequencing study has characterized the vaginal microbiota in 12 HIV-positive and 36 HIV-negative commercial sex workers from Kenya [144]. This study used the Roche 454 platform to sequence a region of *cpn60* designated a universal target due to its high sequence divergence. Approximately half of the women in each HIV serostatus group had low Nugent scores, but like the Tanzanian cohort, the HIV-positive women tended to have fewer lactobacilli, with a maximum relative abundance of *Lactobacillus* operational taxonomic units (OTUs) <80%. In contrast, 25% of the HIV-negative women were dominated by >95% *Lactobacillus* OTUs. This trend appeared to be driven predominantly by a lower prevalence of *Lactobacillus crispatus* in women with HIV (8% compared to 59% in HIV-negative women). Another small 16*S* rRNA pyrosequencing study looking at 36 HIV-positive and 10 HIV negative American women, most of whom were African-Americans, found the median relative abundance of *Lactobacillus* was twice as high in HIV-negative women as HIV-positive women [155]. These authors further observed a direct relationship between low $CD4^+$ T-cell counts (<200 cells/mm^3) and low *L. crispatus* levels, but did not control for BV status in these analyses. Additional studies have shown that the relative abundance of lactobacilli typically exceeds 90% in healthy African-American women with low Nugent scores [185] and that approximately 40% of vaginal communities in this group are dominated by *L. crispatus*, similar to what is seen for Caucasian women (50%) [133]. Thus, although African and African-American women have a high prevalence of BV [31,89,170], black race does not appear to be associated with a significantly lower abundance of lactobacilli among women with low Nugent scores. Together, these findings suggest that HIV infection is specifically associated with a lower relative abundance of lactobacilli, possibly due in part to poor colonization by *L. crispatus*. While further confirmation in larger cohorts is needed, these early data suggest that a dearth of lactobacilli, perhaps *L. crispatus* in particular, in HIV-positive women may contribute to their higher susceptibility to BV and sexually transmitted infections.

Although contemplation of high-throughput sequencing data in aggregate and at high taxonomic levels is well suited for identifying interstudy trends and differences, the development of improved diagnostics and experimental models of BV will require a species-level understanding of the vaginal microbiome. Several research groups have developed custom 16*S* rRNA sequence analysis pipelines that reference a few different classification schemes: NCBI, Greengenes, or the Ribosomal Database Project (RDP). However, because of variable analysis procedures and limited phylogenetic resolution, as well as other methodological differences in sample preparation, target amplification, and sequencing platform, there is extreme heterogeneity in the data reported at fine taxonomic levels. Although a detailed discussion of the number and quality of species-level assignments that have come from individual studies is beyond the scope of this chapter, we will peek through the window of data that have been put forth to get an early, if yet imperfect, glimpse of who's who in the vaginal microbial universe.

One overarching theme is that, similar to what is observed in other human body sites, interpersonal variation in vaginal microbial community composition at the genus–species level is large [97,133,144]. More recent evidence suggests, however, that most of the "uniqueness" is found among low-abundance organisms. In their 2011 pyrosequencing study, Schellenberg et al. described in detail the relationship between species prevalence and population/community relative abundance [144]. Their use of the *cpn60* target, a gene present in single copy within all bacterial genomes, enables direct linkage of sequence abundance with organism abundance; this is not possible with the 16*S* rRNA gene target since it is variably present at 1–15 copies per genome [85]. They categorized groups of sequences with sufficient similarity to represent a single species (OTUs with ≥97% sequence similarity) by population-level "tiers of abundance" as follows: tier 1 (>1%), tier 2 (0.1–0.99%), tier 3 (0.01–0.099%), and tier 4 (<0.01%). Using this classification, they found that 79% of their 248 pyrosequencing OTUs were rare in terms of relative abundance within their population as a whole (tier 3–4). While a significant number of these OTUs were prevalent in over 25% of women, only a handful were found in over 50% of women. Moreover, relatively few (<30) of these tier 3–4 OTUs ever represented >1% of the community in individual women. The challenge with interpretation of such pyrosequencing data is the tendency to inflate the number of OTUs on the basis of sequencing errors in this platform. Nevertheless, these data suggest that most species in the vaginal microbiome are relatively rare and tend to encompass a relatively small proportion of the community within individuals. In contrast, most of the pyrosequencing OTUs that were more abundant in the population (tier 1–2) were prevalent in >25% of women, and all OTUs present in 60% or more women showed tier 1–2 relative abundance. All tier 1–2 OTUs constituted >1% of the community in at least some individuals, and all OTUs that represented ≥10% of the community in any individual were tier 1–2. Taken together, these analyses indicate a relatively limited number of species are abundant (≥1% relative abundance) within individual communities and that these same species also tend to be the most prevalent. Although we do not yet understand the principles that govern community membership within the individual, the observation that some species consistently colonize more frequently and fully than others suggests that the selection is not random. Understanding the significance of rare microbes in the vaginal microbiome, as in the rest of the biosphere, is complicated by the above-noted observation that sequencing errors can generate some false OTUs. Thus, it is important to demonstrate that rare microbes are truly present in multiple samples and studies, although this will likely take some effort given their low prevalence.

A more controversial subject is whether there is a *core microbiome*, or subset of taxa unequivocally present in all women. Some high-throughput sequencing studies have detected *G. vaginalis* and *Lactobacillus iners* in nearly all women, suggesting that they may constitute core species [76,144], but others have argued against the notion of any particular species necessarily playing a central role in vaginal microbial community structure and/or function [133]. At this juncture it may simply be too early to draw a definitive conclusion from high-throughput sequencing data due to limitations in the depth of sequencing as well as the fact that most prevalence data are based on OTUs, which in some cases may represent different strains of the same species. In particular, the use of OTUs may underestimate the true prevalence of species such as *G. vaginalis*, which is known to have a high degree of sequence heterogeneity throughout its genome [192]. On the other hand, it is noteworthy that in one study using highly sensitive targeted PCR, *G. vaginalis* was not identified in every woman with BV, and neither *L. crispatus* nor *L. iners* was present in every woman without BV [50]. These data suggest that there is no core microbiome shared by all women, although there are some bacteria such as *L. iners* that are commonly found in women with and without BV.

Another consistent finding in the vaginal microbiome high-throughput sequencing literature is that the degree of OTU/species diversity within each of the orders shown in Figure 8.3 is highly variable, but conserved. More specifically, most of the orders shown in Figure 8.3 have only 2–10 vaginal species represented, with the exception of the Bacteroidales, Clostridiales, and Lactobacillales, which together have nearly 100 species that have been identified in the vagina by cultivation-independent methods. Overall, considering the number of OTUs identified in the largest pyrosequencing study to date (282) [133], and a study of >1500 16*S* rRNA clone library sequences [128], there are likely to be 200–300 species that inhabit the vagina. With further consideration of the numerous species of *Enterococcus*, *Staphylococcus*, *Streptococcus*, and *Lactobacillus* that have been cultivated from the vagina, but are rarely sufficiently abundant to be detected by cultivation-independent methods, the total number of vaginal species could approach 400.

In women without BV, two species of lactobacilli clearly stand apart in the high-throughput sequencing data as the most prevalent, and at a population level, the most abundant organisms in the vagina: *L. iners* and *L. crispatus* [76,133,144]. Prior to the late 1990s, *L. iners* was not known to exist in the vaginal microbiota because of its inability to grow on agar media typically used to isolate lactobacilli from clinical specimens [42]. More recently, however, numerous cultivation-independent studies have detected *L. iners* in 80–100% of women [50,76,133,144,155,196], suggesting that it is, in fact, the most common vaginal species worldwide [91]. In contrast, *L. crispatus* is less prevalent, detected in 58–66% and 33–37% of women with low and high Nugent scores, respectively [76,133]. *L. iners* is also typically quite abundant irrespective of BV status, representing 42–53% and 16–30% of the total high-throughput sequencing reads from women with low and high Nugent scores, respectively [76,133], whereas *L. crispatus* typically achieves high titer only in women without BV. *Lactobacillus jensenii* and *Lactobacillus gasseri* are the third and fourth most commonly detected *Lactobacillus* species by high-throughput sequencing [76,133,144,155]; together with *L. iners* and *L. crispatus*, these four species usually account for ≥98% of the total *Lactobacillus* high-throughput sequencing reads obtained from BV-negative women [76,133,144]. In any given

BV-negative woman, however, only one or—more rarely—two of these *Lactobacillus* species will dominate the community, with *L. iners* or *L. crispatus* dominating more frequently than *L. jensenii* or *L. gasseri* [6,76,133,155,194]. How and why one species of *Lactobacillus* maintains this dominance within the individual remains one of the great ecological mysteries of the vaginal niche. Cultivation studies have corroborated the findings outlined above, but have also identified more than 20 other species of vaginal lactobacilli that are rare and/or present at very low concentration such that they are not frequently identified by high-throughput sequencing studies. These *Lactobacillus* species include *L. vaginalis*, *L. johnsonii*, *L. gallinarium*, and *L. acidophilus* [6,91,178]. In addition, several novel species of *Lactobacillus* have been identified by high-throughput sequencing [76,133], suggesting that we may yet have much to learn about the biology of these important organisms.

Although lactobacilli are clearly the major players in the healthy vagina, any clinical microbiologist who has cultured a vaginal swab will concede that they also are clearly not alone. *G. vaginalis* is likely the most common non-*Lactobacillus* species in the vagina, with a prevalence as high as 70–85% in BV-negative women [50,144,196]. It is not uncommon for *G. vaginalis* to be detected by quantitative PCR at a moderately high concentration in these women (>10^5 16*S* rRNA copies per 10 ng DNA or per swab) [68,196], a finding corroborated by a review of 39 cultivation studies that found the mean concentration of *G. vaginalis* in BV-negative women to be 10^7 cells/mL vaginal fluid [80]. All of these studies also found *G. vaginalis* loads to be 2–4 logarithms greater in BV-positive women. A similar story is told by high-throughput sequencing, with *G. vaginalis* representing 10–18% of the total reads obtained from BV-negative women and 23–25% of reads from BV-positive women [76,97]. The increased cell density of *G. vaginalis* in BV-positive women generally coincides with an increase in prevalence to 85–100% [50,76,80,128,144,196], rendering detection of *G. vaginalis* by PCR a highly sensitive marker for BV, although it lacks specificity [50,196].

Other species detected in BV-negative women include a group of organisms that seem to colonize the vagina stochastically, rarely comprising a significant proportion of the vaginal community or showing much consistency or strength of association with BV status. Most of the species in this group belong to genera in the order Lactobacillales, specifically *Streptococcus*, *Enterococcus*, and *Aerococcus*, but the Firmicutes genera *Gemella* and *Staphylococcus*, as well as the Actinobacterial genus *Corynebacterium* and the Tenericutes genus *Ureaplasma*, may also be included [57,76,80,128]. The contingencies of Proteobacteria that are frequently detected in the vagina, usually at low relative abundance, may also be characterized as variably or neutrally associated with BV [76,97,133,144]. As Proteobacterial species profiles are widely disparate across different studies (with the exception of *E. coli*), they might represent transient colonizers carried over from the gut or PCR contaminants. One final group of organisms seen in BV-negative women are those that are often present at high concentrations in women with BV, but are present at very low densities (<0.1% relative abundance) in some healthy women. These may include *Prevotella*, *Peptoniphilus*, *Peptostreptococcus*, *Anaerococcus*, *Veillonella*, *Megasphaera*, *Leptotrichia*, *Sneathia*, or *Atopobium* species. The potential of these organisms, when present as a minority, to impact the stability and function of vaginal bacterial communities is poorly understood.

To examine the identities of the most prevalent and abundant organisms in BV, we consulted the two high-throughput sequencing studies that reported species-level information across broad taxonomic groups [76,144], as well as a clone library study that performed extensive phylogenetic analyses on 829 sequences from 21 BV-positive women [128]; each of these studies conferred species assignments to ≥60% of identified OTUs. As shown in Figure 8.3b, all three of these studies identified *L. iners*, *G. vaginalis*, *Atopobium vaginae*, and *Prevotella timonensis* in >25% of BV-positive women. *A. vaginae's* high prevalence is paralleled by moderately high relative abundance; the species *A. vaginae* or genus *Atopobium*, which includes at least two other more rare vaginal species, *Atopobium rimae* and *Atopobium minutum*, were detected in women with high Nugent scores at ≥3% relative abundance in each of the three largest high-throughput sequencing studies to date [76,97,133]. These findings suggest that *A. vaginae* plays a key role in the pathogenesis of BV.

Apart from the trio of *L. iners*, *G. vaginalis*, and *A. vaginae*, the most pervasive organisms in women with BV are rod-shaped Gram-negative anaerobes belonging to the genus *Prevotella*. By high-throughput sequencing, *Prevotella* sequences were found to constitute 10–18% of reads obtained from three independent populations of women with high Nugent scores [76,97,133]. Cultivation, clone libraries, quantitative PCR, and high-throughput sequencing data are all in agreement that *Prevotella* spp. can dominate the microbial community within some individuals, although how frequently this occurs is still subject to debate [80,128,133,196]. The epidemiology of distinct *Prevotella* species has yet to be fully appreciated, due to the vast species diversity that has been observed in the vagina and the fact that many species are novel and uncultivated. Oakley et al. identified 21 OTUs in the *Prevotella* group, 13 of which were novel, including four species designated genogroups 1–4 that each displayed >25% prevalence in their population [133]. Genogroups 1 and 2 are now recognized as the more recently cultivated species *Prevotella amnii* and *Prevotella timonensis* [53,93], while genogroups 3 and 4 remain uncultivated. In addition to *P. amnii* and *P. timonensis*, cultivated *Prevotella* species showing high prevalence and abundance in high-throughput sequencing studies include *P. bivia*, *P. buccalis*, and *P. melaninogenica*. Further understanding of the significance of individual *Prevotella* species within the vaginal ecosystem awaits systematic assessment of their genetic variation and functional redundancy through comparative and functional genomics.

Although *Prevotella* may be the single most diverse genus of BV-associated microbes, the phylogenetic diversity that is characteristic of BV usually involves numerous genera, including many within the phylum Firmicutes. Most members of the Firmicutes have a Gram-positive cell wall, but some have porous or thin walls that cause them to stain Gram-negative. One such group of organisms are those belonging to the order Selenomonadales, which represent the majority of Gram-negative anaerobic cocci found in BV-positive women. Species belonging to three Selenomonadales genera are significant in the vaginal niche: *Megasphaera*, *Veillonella*, and *Dialister*. A novel *Megasphaera* species designated as type 1 (~95% 16*S* rRNA sequence similarity to *Megasphaera elsdenii* [49]) is highly associated with BV and has been shown by PCR and high-throughput sequencing to be present in 85–100% of women with BV [80,144,197]. A second novel *Megasphaera* species, designated as type 2 (~95% 16*S* rRNA sequence similarity to *Megasphaera* *micronuciformis* [49]), is less prevalent, but highly specific for BV, as it is seldom detected in women with low Nugent scores [50,197]. Sequences classified as *Megasphaera* spp. represented >1% of reads from women with high Nugent scores in three of the four large high-throughput sequencing studies [80,133,144]. *Veillonella* species have been detected in two high-throughput sequencing studies at moderately high relative abundance [76,133], and have been shown to attain higher concentrations in BV-positive women than BV-negative women by culture and quantitative PCR [15,80]. However, species-level descriptions of these organisms are still lacking, and with the exception of one study that detected *Veillonella montpellierensis* in every BV-positive woman [76], their prevalence has generally been reported at <16% [14,80,128]. Multiple species of *Dialister*, including *D. micraerophilus* (formerly species α [49]) and *D. propionicifaciens*, are more prevalent than *Veillonella* spp., but similarly show only moderate abundance such that they never dominate the community [76,97,128,133]. The strength and specificity of association between these organisms and BV, as well as their prevalence in a wider array of study populations, remains to be determined.

The principal contributor of Gram-positive anaerobic cocci to the vaginal bacterial communities observed in BV is the order Clostridiales. With the expansion of 16*S* rRNA and whole-genome sequence databases, a significant amount of taxonomic reclassification has occurred within this order since 2001. Numerous vaginal organisms once classified in the genera *Peptococcus* and *Peptostreptococcus* are now housed in a variety of genera, and as a consequence of their shifting nomenclature, our understanding of their association with BV is incomplete [44]. High-throughput sequencing has revealed that both *Peptostreptococcus* and one of the newly described genera, *Peptoniphilus*, commonly expand to >0.1% relative abundance in populations of women with high Nugent scores [76,97,133]. PCR studies that targeted uncultivated *Peptostreptococcus* and *Peptoniphilus* species have also shown them to be positively associated with BV [50,196]. However, as consistent naming schemes are not in place for these organisms, we can only speculate as to their relative importance in BV communities. As for cultivated organisms in these genera, limited sequencing data suggest *Peptostreptococcus anaerobius* and *Peptoniphilus lacrimalis* are most prevalent in the vagina [76]. Some of the other Gram-positive anaerobic cocci that have been associated with BV in high-throughput sequencing studies are *Anaerococcus tetradius*, *Parvimonas micra*, and *Finegoldia magna* [76,97], although additional species such as *Peptoniphilus asaccharolyticus* and *Anaerococcus prevotii* have been detected in the vagina by cultivation [14,80].

Another morphotype of Clostridiales bacteria seen in BV-positive women are Gram-negative rods. Included in this group are three novel species found in a 2005 study to be highly prevalent in clone libraries generated from BV-positive women [49]. Each bore <93% 16*S* rRNA sequence identity to any known organism, precluding even genus-level taxonomic assignment and necessitating their provisional naming as BVAB1 (BV-associated bacterium 1), BVAB2, and BVAB3. Targeted PCR studies subsequently demonstrated each to be strongly associated with BV, and their presence was confirmed in vaginal smears by fluorescence *in situ* hybridization (FISH) [49,50,196]. Of the three, BVAB2 exhibits the highest prevalence, detected in 85–100% of BV-positive women by PCR, while BVAB1 and BVAB3 are less common, found in 26–94% of BV-positive women [50,196]. BVAB1 can be present at high abundance and even dominate individual sample libraries, whereas

BVAB2 and BVAB3 are each, in turn, less abundant and usually minority species [49]. Despite the availability of full-length 16*S* rRNA gene sequences for these organisms, high-throughput sequencing studies have largely failed to identify them; the one exception is the Hummelen et al. report of prevalent (>25%) and moderately abundant BVAB3 reads in their African cohort of BV-positive women [76]. However, all of the large high-throughput sequencing studies that we summarized in Figure 8.3a assigned a large proportion (50–90%) of their Clostridiales sequences to a family or genus without performing species identifications [76,97,133]. It is probable that BVAB1,2,3 are present in these datasets but are unrecognized as such by current analysis pipelines. Indeed, when full-length 16*S* rRNA gene sequences for these bacteria are processed through Greengenes and RDP high-throughput sequencing analysis pipelines, these bacteria are not identified. On the other hand, given the bounty of Clostridiales species that have already been identified, it is equally likely that some of these sequences do, in fact, represent new organisms that await discovery.

While the Clostridiales may be unequaled as a hotbed of novel BVABs, other novel organisms scattered throughout the Bacterial domain are also consistently found in association with BV. Of these, the most prevalent are a pair of species in the order Fusobacteriales: *Leptotrichia amnionii* and *Sneathia sanguinegens*. Because of these species' very similar 16*S* rRNA sequences, they are not always distinguished by molecular methods; however, by targeted PCR their combined prevalence has been found to range from 74% to 100% in BV-positive women [50,148,196]. By means of high-throughput sequencing, *Leptotrichia* and *Sneathia* sequences together accounted for 6–14% of the reads obtained from three separate populations of women with high Nugent scores [76,97,133]. Although both *L. amnionii* and *S. sanguinegens* have been cultivated [26,148], their extremely fastidious nature has slowed progress toward understanding their biology and functional significance within vaginal bacterial communities. Other Fusobacteriales species relevant in BV are *Fusobacterium nucleatum* and possibly a novel species of Fusobacteria (Figure 8.3b), although again, little work has been done to determine the pathogenic potential of these organisms in the vagina. Apart from *Atopobium* species such as *A. vaginae*, the other major Coriobacteriales species that has been reported in the vagina is a novel uncultivated organism distantly related to *Eggerthella hongkongensis* (92% 16*S* rRNA sequence identity, referred to as *Eggerthella-like species* since the true genus is unknown [49]). Like many other BVABs that we have described, it is common in BV-positive women when assayed by sensitive targeted PCR, showing 89–100% prevalence [50,196]. Two high-throughput sequencing studies have reported sequences in the genus *Eggerthella* at ≥0.1% relative abundance [97,133]. One final novel bacterium to note is a species belonging to TM7, a newly proposed division of Bacteria represented by numerous human, animal, and environmental species that have been identified by molecular methods, but have not yet been successfully isolated in culture [75]. The TM7-BVAB is one of the more rare BVABs, found in only 25% of BV-positive women, but it is also one of the few species almost never detected in BV-negative women [50]. Despite its low prevalence, two high-throughput sequencing studies reported TM7 reads at >0.1% relative abundance in BV-positive women [76,97].

The final two genera of BV-associated organisms that we will discuss are not at all related phylogenetically, but are nonetheless united in controversy. Both

Mobiluncus and *Mycoplasma* species have been frequently detected in BV-positive women by Gram stain (for the former) and in culture at concentrations upward of 10^7 bacteria/mL, but are often scarce in molecular surveys [80]. *Mobiluncus* became recognized as a member of the vaginal microbiota in the 1980s during the early Gram stain and culture-based characterizations of the microbial communities associated with BV, or *nonspecific vaginitis* (NSV), as it was termed then [73,131,150,159]. It garnered attention because of its unusual curved-rod morphology, as well as the fact that aside from the occasional Proteobacterial species, it is the only known vaginal organism that is flagellated and motile. These early studies identified motile curved rods in wet mounts in 20–30% of women with NSV [73,150], although not all curved rods were observed to be motile [131]. Later reports using Gram stains found curved Gram-negative or Gram-variable rods in 39–68% of women diagnosed BV-positive by clinical criteria [29,67,137,156]. However, success in culturing the organisms meeting this physical description was variable, with only 20–24% of curved-rod-positive specimens returning cultured curved rods in some studies [156,176], yet 60–77% yielding cultured curved rods in others [73,137]. Eventually, DNA hybridization techniques, the use of antibodies, and 16*S* rRNA gene sequencing helped confirm the identity of curved-rod organisms in vaginal fluid and culture as *Mobiluncus curtisii*, a short (1.7 μm) Gram-variable curved rod with tapered ends, or *Mobiluncus mulieris*, a longer (2.9 μm) Gram-negative curved rod with pointed ends [45,129,137]. Discrepancies between Gram stain and culture were attributed to the fastidiousness of the organism [67,156]. The characteristic cell shape of *Mobiluncus* and specificity of curved rods for BV (which are typically observed in <5% of BV-negative specimens [67,137,156,176]) led to *Mobiluncus* morphotypes becoming an integral component of the Nugent diagnostic scoring matrix for BV [124]. Now fast-forward 30 years and into the molecular era for the twist in this story: clone libraries have surprisingly yielded *Mobiluncus* sequences in <15% of BV-positive women [78,80,128]. Pyrosequencing findings have also been inconsistent; some studies detected *Mobiluncus* species in >25% of women with high Nugent scores [76,133] and/or at ≥0.1% relative abundance [133,144], while others did not [97]. Some have argued that this discrepancy between Gram stain and molecular findings is due to amplification bias, although no definitive evidence of *Mobiluncus* mismatches with broad-range primers has been put forth [80]. However, this does not preclude the possibility of bias stemming from some other aspect of the molecular methodology, such as the efficiency of cell lysis or the G/C content of the amplicon [51]. Indeed, one must bear in mind that all metagenomic sequencing data and the conclusions that they engender are potentially skewed by these kinds of biases, which are difficult to detect and overcome. When *Mobiluncus* species are targeted by species-specific primers, they have been detected in as many as 56–81% of BV-positive women, suggesting that the bacteria are frequently present, but may simply not be so abundant as to be detected in clone libraries or high-throughput sequencing surveys [50,196]. One final piece to the *Mobiluncus* story that has recently come to light is that the morphology of BVAB1 is that of a small curved rod, very similar to that of *M. curtisii*. Thus, given that the prevalence and specificity of BVAB1 in BV are quite high, it is possible that some of the Gram-negative curved rods observed in vaginal smears are BVAB1 [49].

Although *Mycoplasma* species lack the morphological flair of *Mobiluncus*, *Mycoplasma hominis* in particular shares a similarly inconsistent epidemiological history. *M. hominis* has, on average, been detected in 64% of BV-positive women by culture, compared to 13% of BV-negative women [80,82,94,140]. It also exhibits a massive increase in concentration with BV (three logarithms, on average), reaching nearly 10^7 cells/mL vaginal fluid [80,172]. However, *M. hominis* exhibits low prevalence (<10%) in clone library studies [128], and although *Mycoplasma* reads totaled 0.8% relative abundance in one high-throughput sequencing study [97], species-specific recognition of *M. hominis* in high-throughput sequencing data has not yet occurred. In contrast, *M. hominis* is captured in 50–81% of BV-positive women by more sensitive targeted PCR, again suggesting that this species may be present as a minority constituent of these vaginal bacterial communities or that an undefined bias leads to the underestimation of this organism's burden when surveyed by metagenomic methods [110,196].

In summary, high-throughput sequencing technologies are painting an exciting new picture of the vaginal microbiome's constitution at higher taxonomic levels, and increasingly at the level of genus and species. As we have discussed, these are still early days in the application of this technique and there are several caveats in the existing data, including methodological variation and known and unknown biases that can potentially impact all high-throughput sequencing studies. Yet despite these limitations, estimates of many taxa's prevalence and relative abundance are showing interstudy congruence. Moreover, most of the disease associations previously shown using cultivation and earlier cultivation-independent methods are upheld. As these queries continue with ever-deepening sequencing thresholds and improved taxonomic inference tools, we will see further refinement of the microbial ecology and strengthening confidence in newly identified epidemiological associations. Our final discussion of the female side of the genitourinary microbiome will thus look ahead to see what these new discoveries portend for improved BV diagnostics and therapeutics.

8.5. VAGINAL HEALTH IN 2025: HOW UNDERSTANDING THE MICROBIOME WILL SHAPE THE FUTURE OF WOMEN'S HEALTHCARE

Molecular diagnostics that detect nucleic acid sequences or proteins unique to a pathogenic organism are desirable because they are often highly sensitive, meaning that they are capable of detecting very low levels of the pathogen, and generally require little sample material. Additionally, most have exceptional specificity, or a very low chance of yielding false-positive results. A number of target species have emerged as possible indicators of BV, and work is underway to develop a panel of molecular diagnostics that performs robustly despite the interpatient variability that is inherent in this polymicrobial condition. Efforts to better understand the structure and function of whole vaginal microbial communities are also ongoing, and perhaps will one day allow physicians to profile an individual's vaginal microbial community "type" and accordingly customize treatment regimens, as well as identify women at greatest risk for poor outcomes. Finally, new research is shifting the direction of thought as to why lactobacillus-dominated communities are vital for health and leading researchers to reenvision how best to promote them. In this section we will highlight more recent progress in these areas and discuss some of the outstanding questions that will be critical to answer for microbiome-focused healthcare to begin.

As described in the previous section, numerous BVABs identified using cultivation and/or broad-range PCR have now had their prevalence, abundance, and strength of association with BV confirmed with species-specific PCR. Systematic inquiry into which of these BVAB species might be suitable targets for PCR-based diagnostics was undertaken in two prospective studies of symptomatic and asymptomatic women attending research clinics [50,110]. On assessing 17 vaginal bacterial species in 264 samples by conventional PCR, Fredricks et al. observed excellent agreement with established diagnostics when the presence of BV was signified by amplification of *Megasphaera* type 1 or BVAB2 (sensitivity 99/96%, specificity 89/94% compared to Amsel's criteria/Nugent score) [50]. The Menard group's quantitative PCR assays targeted five bacterial species, as well as the genus *Lactobacillus* and the fungus *Candida albicans* in a cohort of 167 pregnant women. Their most promising diagnostic indicator of BV was detection of $\geq 10^9$ *G. vaginalis* 16*S* rRNA gene copies/mL in combination with $\geq 10^8$ *A. vaginae* 16*S* rRNA gene copies/mL in vaginal fluid suspension (sensitivity 95%, specificity 99% compared to Nugent score) [110]. Subsequent confirmatory studies by the same group using a new cohort of >150 pregnant women also showed excellent performance of this diagnostic using both clinician-collected and self-collected swabs [111,112]. It should be noted, however, that women with intermediate microbiota on Gram stain were not grouped with the BV-negative women in the analysis of performance, but were instead considered BV-positive, which counters convention and may have skewed their results. Early findings in other populations, namely, nonpregnant Chinese and American women, suggest that detection of high *G. vaginalis* and *A. vaginae* bacterial loads by PCR may be a universal indicator of BV [98,106]. However, additional prospective studies are needed to verify whether this quantitative assay, or perhaps a less expensive presence/absence test for two or more species that are highly specific for BV, will perform robustly in women of different racial and cultural backgrounds, different ages, and different sexual behavior risk groups. In addition, the vaginal microbiotas in pregnant women may differ from those in cycling women, so performance characteristics in one population cannot be inferred for another.

High-throughput sequencing has also enabled profiling of entire vaginal microbial communities through the use of phylogenetic clustering algorithms that mathematically capture community structures and allow them to be statistically compared. Using this approach, researchers aim to classify distinct bacterial community types, some of which may be useful as predictors of vaginal health or undesirable outcomes such as treatment failure or preterm birth. However, it is important to note that this type of microbial profiling is still in its infancy, and findings should be interpreted with caution until cross-cohort reproducibility is demonstrated. Efforts to identify vaginal microbial community types have thus far focused on women that are asymptomatic or otherwise unselected for BV status. Perhaps not surprisingly, computational segregations of the bacterial community types present in these populations have formed largely, although not explicitly, along the lines of the Nugent score category [133,144]. In the largest of these analyses, Ravel et al. identified five vaginal bacterial community types in their population of 396 asymptomatic women. Types I, II, III, and V were dominated by *L. crispatus*, *L. gasseri*, *L. iners*, and *L. jensenii*, respectively, and of these, only the *L. iners*-dominant group (type III) contained a significant number of women with high Nugent scores (13 women, or ~10% of the total number who had a Nugent score ≥7; one woman with a type II community also had a Nugent score of 7) [133]. The vast majority of women with high Nugent scores fell into the fifth grouping, community type IV, which lacked dominant lactobacilli and exhibited much greater compositional heterogeneity than the other community types. Interestingly, 10% and 13% of women with low and intermediate Nugent scores, respectively, also had type IV communities. This analysis demonstrates that even among asymptomatic women, vaginal microbial architectures are demarcated by the presence or absence of one of several dominant *Lactobacillus* species. One area that has yet to be fully explored is whether the only factor driving separation of type I, II, III, and V communities is their top-ranking lactobacillus, or whether assembly of the rest of the community differs according to which *Lactobacillus* species presides. Two observations in favor of the latter are the broader range of Nugent scores within the *L. iners*-dominant group (type III) and the frequent concurrence of some species within certain lactobacillus-dominated community types, but not others.

From a clinical perspective, the most interesting vaginal microbial community types will be those found in BV-positive women who exhibit differential strengths of association with its sequelae. Thus, an important question to address at this juncture is whether restricting community typing analyses to BV-positive women identifies specific community types within this subpopulation. On examining a very limited number of BV communities from the clone library literature, Kalra et al. identified two factors that appeared to differentiate BV microbial community types: the presence/absence of *L. iners* as a minority constituent of the community and the presence/absence of BVAB1 [80]. We are currently in the process of analyzing high-throughput sequencing data from a cohort of >100 BV-positive women that will confirm or refute these observations.

To begin deciphering which components of vaginal microbial communities may be important contributors to the deleterious health consequences of BV, we and others used PCR-based methodologies to assess epidemiological associations between the presence and/or abundance of particular BVABs and BV symptoms and sequelae. Significant findings from these studies include positive associations between *Leptotrichia/Sneathia* spp., *A. vaginae*, BVAB1, or *G. vaginalis* and the presence of >20% clue cells in vaginal fluid (an indicator of bacterial biofilms) [57]; *A. vaginae* or BVAB1 and amine odor [57,76]; high *G. vaginalis* or *M. hominis* concentrations and vaginal HIV shedding [147]; high *G. vaginalis* concentrations and spontaneous preterm birth [120]; and BVAB1,2,3, *Megasphaera* type 2, or *P. lacrimalis* and recurrent BV [104]. The limitations of these studies are that only a small number of BVABs were tested; small sample sizes often impeded detection of statistically significant associations, particularly if corrections were made for multiple testing; and most of the observed associations have not yet been confirmed in independent cohorts. Nevertheless, these exploratory analyses have yielded a wealth of hypotheses that will not only inform future clinical studies but are well suited for further exploration in cell culture models. As an example, laboratory-based studies of certain BVAB's biofilm-forming behaviors could uncover promising new strategies for promoting long-term cure in women with recurrent BV [130,152,168]. Moreover, sufficient evidence of a given species playing a causal role in BV symptoms or sequelae may illuminate antibiotic therapy options that specifically and more effectively target the offending organism.

The other side of this story is that vaginal lactobacilli are more than simply a benign alternative to the organisms associated with BV; in many of these same PCR-based epidemiological studies, lactobacilli were shown to protect against outcomes such as preterm birth [120], HIV shedding [147], and the acquisition of sexually transmitted infections [187]. As we discussed earlier, the idea that vaginal lactobacilli are capable of thwarting pathogens dates back to the work of Döderlein and colleagues around the turn of the twentieth century. What is new is that molecular studies of the vaginal microbiota are now clarifying which *Lactobacillus* species offer the best protection and sophisticated cultivation studies are mandating that we rethink our simplistic view of the mechanism. Recall the Ravel group's finding that some asymptomatic women with Nugent scores in the range of 7–10 were dominated by *L. iners* or *L. gasseri*, but never *L. crispatus* or *L. jensenii* [133]. Cultivation studies corroborate this observation, rarely diagnosing BV in women who are colonized with *L. crispatus* or *L. jensenii* compared to women who are colonized with *L. iners* or *L. gasseri* [6,189]. Several studies have further proposed that a key factor underlying this association is the ability of some *Lactobacillus* species to produce hydrogen peroxide, a potent oxidizing compound that can inactivate bacterial cells. The mechanism by which hydrogen peroxide acts is thought to involve interactions with other molecules such as ferrous iron and myeloperoxidase (MPO), which produce more potently toxic substances, namely, hydroxyl radicals and hypochlorous acid with respect to the abovementioned examples. Isolates of *L. crispatus* and *L. jensenii* are more likely to produce hydrogen peroxide and also tend to produce it in larger quantities than *L. gasseri* [6,40,59,70,108,189]. Interestingly, *L. crispatus* and *L. iners*, the two most common vaginal lactobacilli, are at opposite extremes of this spectrum; typically 90–100% of *L. crispatus* isolates are robust hydrogen peroxide producers [6,108,189], while *L. iners* appears to be incapable of producing hydrogen peroxide [117]. However, it may not necessarily be a foregone conclusion that hydrogen peroxide is a major factor contributing to these species' differing abilities to protect against BV. More recent studies have shown that vaginal fluid and semen contain potent inhibitors of hydrogen peroxide and that the low levels of hydrogen peroxide that lactobacilli maintain in vaginal fluid are not sufficient to rapidly kill HSV2, *N. gonorrhoeae*, or numerous BVABs when challenged in an anaerobic atmosphere [125].

One alternative explanation for the differing capacities of *L. crispatus* and *L. iners* to protect against colonization by BVABs relates to pH and lactic acid production. Studies have shown that vaginal pH tends to be lower in women colonized heavily by *L. crispatus*, compared to *L. iners* [76,133]. While this could result from *L. iners*-dominated communities having greater buffering capacity, it could also stem from *L. iners* producing less lactic acid than *L. crispatus*. The potential for lactic acid to impact the dynamics of the vaginal microbiota was elegantly demonstrated in a study by O'Hanlon et al. [125]. The investigators found that lowering the pH of nutrient-rich growth media to 4.5 with hydrochloric acid was mildly toxic to a variety of Gram-negative and Gram-positive BVABs in pure culture (i.e., grown individually). In contrast, adding physiologic levels of lactic acid to identical cultures and adjusting the pH to 4.5 resulted in complete loss of viability within 2 h in anaerobic conditions [125]. Surprisingly, adding comparable levels of acetic acid, a similarly small/lipid soluble acid, and adjusting to pH 4.5 had no effect beyond that of pH. Unlike what was seen with hydrogen peroxide, the toxic effects of lactic acid were not abrogated by the addition of fresh sterile-filtered vaginal fluid and did not impact the viability of any of the four major vaginal *Lactobacillus* species. These findings argue that lactic acid is a critical mediator of competitive inhibition in the vagina and that its role extends beyond that of simply maintaining a low vaginal pH. However, several important questions remain.

1. What is the mechanism by which lactic acid kills bacterial cells? One study revealed that lactic acid destabilizes the cell wall of Gram-negative pathogens, but whether this also occurs with Gram-positive bacteria has yet to be tested [3].
2. Do the impacts of hydrogen peroxide and lactic acid on bacterial communities differ from those on organisms growing in pure culture? One of the reasons why this issue will be critical to future studies is that BV-associated microbes may have greater resistance to hydrogen peroxide and lactic acid when growing in a biofilm [130]. Another consideration is that concentrations of hydrogen peroxide or lactic acid that were too low to produce bactericidal effects during the short incubation periods used in the aforementioned studies could have bacteriostatic effects sufficient to shift the balance of organisms growing in direct competition.
3. What is the potential for multiple toxins to act in concert? Since hydrogen peroxide is generated by lactobacilli only when oxygen is introduced into the vaginal environment, typically during menses and sexual intercourse [65,181,182], further study of hydrogen peroxide's bactericidal activity in vaginal fluid should consider potential interactions with other environmental factors that accompany these events, especially oxidative stress, nitrosative stress, and high iron levels. Moreover, one 2010 study found levels of lactic acid that were too low for appreciable bactericidal activity on their own enhanced the bactericidal effects of hydrogen peroxide against *G. vaginalis* [12]. This suggests these molecules may have cooperative modes of action.
4. What biochemical pathways are responsible for removing hydrogen peroxide from genital secretions, and what factors drive this process? If future studies demonstrate the involvement of bacterial peroxidases, they could serve as targets for novel therapies that boost the protective potential of indigenous lactobacilli.
5. Does *L. iners* actually produce less lactic acid than *L. crispatus in vivo*? Or is it hydrogen peroxide, or perhaps another trait linked to the ability to produce hydrogen peroxide that determines these species' differing roles in vaginal health? In spite of its weak bactericidal potential in vaginal fluid [125], the fact that hydrogen peroxide production is dependably correlated with persistent lactobacillus colonization and the absence of BV suggests that there is a critical biological phenomenon related to this molecule that we have yet to fully understand [2,24,40,113,177].

One of the primary goals of research on vaginal lactobacilli is to identify "optimal" strains for development as probiotics. Broadly, probiotic therapy aims to deliver a transient or resident population of bacteria that has been specifically selected to carry out health-promoting functions. In addition to providing general colonization resistance through competitive mechanisms and the production of antimicrobials, probiotic lactobacilli could, in theory, be tailored to antagonize specific classes of vaginal pathogens, such as group B streptococcus and *Staphylococcus aureus*, both of which can cause life-threatening infections in pregnant women and their newborns. Probiotic lactobacilli may also be utilized to modulate cytokine signaling and inflammation in the reproductive tract, which are integral to the success of implantation and an appropriately timed induction of labor. As we will discuss later, lactobacilli could prove efficacious for preventing symptoms of vaginal dryness and atrophy in postmenopausal women, and someday might even serve as vehicles for sustained and localized delivery of other medicines such as hormones, antiretrovirals, or chemotherapy. Toward these ends, traits considered desirable in probiotic lactobacilli include lactic acid and hydrogen peroxide production, bacteriocin production (small peptides with bactericidal activity), the ability to adhere to the epithelium and prevent other microbes from adhering through competitive inhibition, and/or the production of biosurfactants, as well as coaggregative and anti-inflammatory properties [96,160].

Current probiotics research is aimed at repopulating vaginal lactobacilli during or following an episode of symptomatic BV, to either cure the condition or augment standard antibiotic therapy and reduce the rate of recurrence. Several studies have tested vaginal or oral probiotic regimens of intestinal *Lactobacillus* species such as *L. rhamnosus* and *L. reuteri* (which rarely reside in the vagina) with some success in promoting cure and preventing recurrence up to 30 days following treatment in African women [9,10,13]. However, given what new research is revealing about the high degree of *Lactobacillus* species selectivity in the vaginal niche and the varying protective ability of different species, more recent focus has turned to the probiotic potential of the seemingly "optimal" vaginal species, *L. crispatus*. Strain CTV05, a hydrogen peroxide producing *L. crispatus* isolate from the vagina of a healthy woman, has been developed as a vaginal probiotic under the trade name Lactin-V by Osel Inc. (Santa Clara, CA). When delivered intravaginally to healthy women or women coming off antibiotic treatment for BV, Lactin-V detectably colonized 69% and 61% of women at one or more follow-up visits, and was maintained at one month in 59% and 44% of women, respectively [7,62]. Women were significantly less likely to be colonized with Lactin-V if they were already colonized by *L. crispatus*, but the presence of other *Lactobacillus* species did not impact Lactin-V colonization, suggesting that there may be species-specific competition for adherence sites [7,123]. In women who had just been treated with antibiotics for BV, only 1 of 11 women colonized with Lactin-V had a recurrence of BV at one month (9%), compared with 3 of the 4 who were not colonized (75%) [62]. Quantitative PCR studies subsequently demonstrated the women colonized with Lactin-V had lower levels of several BVABs and higher levels of *L. crispatus* at one month, further demonstrating Lactin-V's ability to support the development of a healthy vaginal microbota in women recovering from BV [123]. Most recently, a randomized placebo-controlled trial of 100 women with a history of recurrent urinary tract infection (UTI) showed that a 10-week regimen of Lactin-V reduced UTI recurrence by 50% compared to placebo [163]. It is suspected that vaginal lactobacilli mediate this protection through a reduction in the vaginal carriage of uropathogenic *E. coli*, but it was not reported in this study whether this was indeed the case. In summary, Lactin-V is a promising preventative therapy for BV and UTI, but larger and more definitive efficacy trials are still needed, as are investigations of which genotypic and phenotypic traits are crucial for its protective function. Also awaiting careful examination are the reasons why Lactin-V failed to colonize a significant proportion of women in each trial. The fact that many of these women were instead colonized by indigenous lactobacilli implies that the "optimal" lactobacillus is likely to vary from woman to woman.

Clinical implementation of Lactin-V, or any of the novel diagnostics and therapeutics that we have discussed here, will require continued progress in several key research areas. Of these, perhaps most important is the natural history of the vaginal microbiota. How do these communities develop and change over the lifetime of the individual, and how does this impact the suitability and efficacy of microbiome-focused interventions? Essentially all of the data discussed in this chapter come from studies of women in their peak reproductive years (~18–40), as this is the age range in which most symptomatic BV occurs. During this phase of a woman's life, the presence of estrogen is thought to select for lactobacilli by making nutritive glycogen and its catabolite glucose available in the vaginal epithelium [30,191]. After menopause, women not taking hormone replacement therapy typically retain some of their vaginal lactobacilli, but are less likely to have a lactobacillus-dominated community [22,61,71,77]. Postmenopausal women may have appreciable colonization by coliform bacteria and BVABs [71,77], but these organisms are not necessarily present in the absence of lactobacilli as is the case for reproductive women [23]. Instead, many postmenopausal women have scant vaginal bacteria or morphotypes not specified in the Nugent scoring matrix, causing them to be misclassified by this diagnostic as having "intermediate" flora when they are actually normal [23]. Vaginal pH is also elevated in postmenopausal women, precluding the use of this measure in the clinical diagnosis of BV by Amsel's criteria. However, by fulfillment of the three remaining Amsel criteria, BV was identified as the etiologic cause of symptomatic vaginitis in 10% of postmenopausal women attending a vaginitis clinic [157]. Why lactobacilli seem to thrive in some postmenopausal women and languish in others is unknown, but their absence has been correlated with vaginal dryness, a symptom of the epithelial atrophy that causes significant irritation and soreness in 25–50% of postmenopausal women [77]. Taken together, these findings suggest that postmenopausal women do experience BV and would benefit from a molecular diagnostic specifically designed to differentiate healthy versus suboptimal vaginal microbiotas in the absence of estrogenic cycling. Although additional surveys are needed, the new findings regarding vaginal atrophy serve as a reminder of the potential these metagenomic studies have to significantly increase our understanding of how the vaginal microbiota contribute to a woman's health throughout her lifetime. Moreover, research that is focused on postmenopausal women and another understudied group, adolescent women, may uncover novel ways for vaginal probiotics, including those already in development for BV and UTI, to be of benefit in gynecologic healthcare.

Another important natural history question is that of the vaginal microbiota's more short-term stability on the order of days, weeks, and months. Which the factors predict shifts in community composition or indicate a stable state? How frequently do the microbiota convert to a BV state and spontaneously resolve back to a lactobacillus-dominated state? Knowledge of these dynamics will be critical for determining who should be screened and treated for BV, how often, and when. In two studies that longitudinally monitored Nugent scores using self-collected vaginal swabs, rapid shifts between healthy and BV states (often via the intermediate state) frequently occurred in just a few days to a week's time [20,175]. Of 211 sexually experienced Ugandan women who were followed for nearly 2 years, 202 (96%) experienced at least one episode of BV [175], as did 24 (73%) of 33 American women who were followed for 4 months [20]. In both studies, the length of each episode was highly variable, with many lasting months, but others only 1–2 weeks. Additionally, very few of the women developed symptomatic BV and received antibiotics, yet spontaneous remission was common. In these and other prospective studies, menses and factors related to sex (e.g., report of a new partner, recent sex, lubricant use, semen exposure) have been identified as predictors of incident BV [20,52,68,105,122,174]. In the Ugandan cohort, age at first sex (mean of 15–16) also showed a trending association with total time spent in a BV state, suggesting that early sexual debut might impact the establishment of stable lactobacillus populations [175].

A criticism of many longitudinal studies of the vaginal microbiota has been their failure to address whether changes in the abundance of specific morphotypes (Nugent score) corresponded with changes in community phylogenies and/or functions [106]. However, similar to what has been described in women using Nugent scores, we observed abrupt changes in the concentrations of 11 vaginal taxa (measured with quantitative PCR) occurring in association with shifts into or out of clinical BV (assessed by Nugent score and Amsel's criteria) in the span of just a few days [68]. In contrast, high throughput sequencing of biweekly samples obtained from postmenopausal women found their microbial profiles were largely consistent over a period of 10 weeks [77]. This suggests that greater stability may be seen in certain populations of women or with the use of more global community typing methods that measure relative abundance instead of absolute quantities. Finally, longitudinal studies looking at metagenomes, metabolomic signatures, and host responses are needed to assess whether these factors exhibit greater stability than the microbial community's phylogenetic composition. If this is the case, it would imply that there are significant genetic and/or functional redundancies among different bacterial taxa and indicate that detection of certain genes or metabolites might serve as better diagnostics for BV than detection of bacterial species. In sum, a growing body of evidence in women of reproductive age points to the vaginal microbiota being highly dynamic. A more complete understanding of the factors driving shifts in community composition and their consequences for the host will be required to properly design and interpret clinical trials of novel BV diagnostics and therapeutics. In addition, longitudinal studies with frequent sampling intervals, as well as mechanistic studies in cell culture models, will be needed to learn how variably present BVABs may play causal roles in adverse health outcomes.

One last feature of vaginal microbial dynamics that we will discuss is the impact of extravaginal reservoirs of BVABs, specifically those found in the gut (rectum) and in the urogenital tract of one's sex partner. How often are these microbes harbored at these other body sites, and what is the likelihood that they will reinoculate the vagina and play a causal role in BV relapse? Evidence of gut microbes having the potential to navigate the perianal region that adjoins the rectum and vagina comes from cultivation studies that identified the same species, and even identical strains (typed by rapid amplification of polymorphic DNA, or RAPD), in 36% of pregnant women [37]. Additional support is lent by oral probiotic lactobacilli having a demonstrated ability to transit the gut and colonize the vagina [135,136]. We also know that the predominant vaginal lactobacilli, with the exception of *L. gasseri*, exhibit much stronger tropism for the vagina than the rectum [8,37,109], but having both vaginal and rectal reservoirs of H_2O_2-producing species is more protective against BV than either alone [8]. The inverse is also true: very few gut microbes display a natural tropism for the vagina [167], but the fact that several of the key players in BV, including *G. vaginalis*, *M. hominis*, and *Mobiluncus* spp., are often detected in the rectums of women with BV highlights a potential for recolonization if treatment does not effectively eradicate the organisms from both reservoirs [74]. It has been documented that in women who have sex with women (WSW), report of a new BV-positive sex partner incurs a threefold higher risk of BV acquisition [105], and that 77% of WSW monogamous for at least 3 months share identical lactobacillus strains [109]. Together, these findings suggest resident vaginal bacteria are sexually transmitted in WSW, but whether this also occurs in heterosexual couples is more controversial. In Section 8.6 we will describe more recent advances in our understanding of the male genitourinary microbiota and the potential for male–female crosstalk to be a significant factor in both incident and relapsing BV.

8.6. THE MALE MICROBIOTA: TWO SIDES OF THE SAME COIN OR A DIFFERENT CURRENCY?

Different microbial communities are found in different anatomically defined niches of the male genitourinary tract (Figure 8.4). For example, the microbiota of the penis is affected by the presence of foreskin, with circumcised men having bacterial

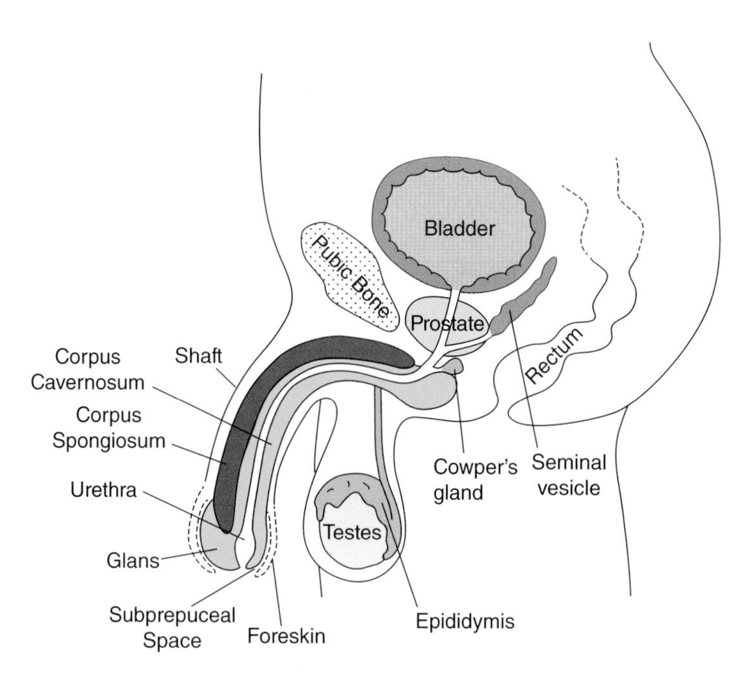

Figure 8.4. Anatomy of the male reproductive tract. Artistry kindly provided by Kyoko Kurosawa.

communities that differ from those of uncircumcised men. Some pathogens are capable of ascending the genitourinary tract to produce inflammation and disease in internal structures such as the urethra, epididymis, prostate, bladder, or kidney. A few studies have used molecular cultivation-independent methods to characterize the male genitourinary tract microbiota, the findings from which are highlighted here.

Price et al. used broad range 16*S* rRNA gene PCR with tagged pyrosequencing to characterize bacterial communities present on the coronal sulcus of the penis [132]. The study population consisted of 12 men participating in a trial of circumcision to prevent HIV transmission, and swabs were obtained from the coronal sulcus before circumcision and one year after. Note that the coronal sulcus is enrobed by the foreskin in uncircumcised men, and the subpreputial area may create an anaerobic environment that favors growth of some bacterial species. The hypothesis that circumcision alters the penile microbiota was confirmed when using PerMANOVA ($p = 0.007$) to assess changes in overall microbiota. There were clear decreases in several putative anaerobic bacterial families after circumcision ($p = 0.014$), such as Prevotellaceae and Clostridium family XI. Several bacteria that are residents of the human vagina were detected more commonly in uncircumcised men, suggesting that the subpreputial space may be a reservoir for bacteria such as *G. vaginalis* and *Sneathia* species. Bacterial families that are associated with normal human skin became more abundant after circumcision, including the Staphylococcaeceae and the Corynebacteriaceae, and the bacterial communities appeared more homogeneous. There was no clear evidence of a change in species richness or diversity after circumcision. The limitations of this study include the relatively a small number of sequence reads per sample (387) and the inability to assign most sequences to a bacterial species. Indeed, only 65.9% of sequences could be assigned to the genus level. In addition, the large number of *Pseudomonas* and *Comamonas* sequences in this dataset could be real, but alternatively these sequences may represent contaminants. As *Pseudomonas* and *Comamonas* DNA may be found in DNA extraction and PCR reagents, these results deserve confirmation with more rigorously defined experimental controls. This study is important because it offers some hypotheses regarding how circumcision may reduce the risk of several sexually transmitted diseases, such as HIV, HSV, and HPV by altering the penile microbiota and local immune milieu.

Nelson et al. used urine samples from 19 men attending an STD clinic to assess the male urethral microbiota by broad range 16*S* rRNA gene PCR with cloning and Sanger sequencing of inserts [121]. The subjects did not have urethral symptoms or evidence of urethritis as defined by having <5 leukocytes per high-powered field on light microscopy. Sequencing results revealed a high degree of intersubject variability in composition of the microbiota. *Lactobacillus iners* was the most common bacterial 16*S* rRNA sequence identified, but represented only 14.3% of sequences. Many bacterial genera found in the human vagina were also detected in this study, including bacteria linked to bacterial vaginosis such as *Sneathia*, *Leptotrichia*, *Atopobium*, and *Prevotella*. Note that the 27F primer used in this study for 16*S* rRNA gene PCR is known to inefficiently amplify *G. vaginalis*, so these sequences are likely underrepresented for technical reasons. A bacterial vaginosis-like microbiota was associated with detection of an STD pathogen in the same urine sample, although the exact STD pathogens detected in each sample are not described. This observation is quite intriguing, but one does not know whether the more diverse anaerobic microbiota seen in men with STDs precedes acquisition of the STD. Another limitation of this study is the fact that urine samples reveal the condition of the entire urinary tract and therefore may not just represent the urethral microbiota. Given major differences in approach and technique, it is not very useful to compare the microbiota of coronal sulcus samples obtained by Price et al. to the urethral microbiota studied by Nelson et al.

Finally, Eren et al. used single-nucleotide differences in the *G. vaginalis* 16*S* rRNA gene to demonstrate that male–female sexual partners tend to share similar sequence types of *G. vaginalis* on genital surfaces [38]. No particular *G. vaginalis* sequence type was associated with BV. These data suggest that the male genital microbiota and female microbiota are in many ways two sides of the same coin, with transfer of similar bacterial communities resulting in persistent colonization.

In summary, very different microbial communities are found on male genital surfaces from different men, and circumcision may alter the bacterial communities found on the penis. Molecular methods demonstrate that there are diverse communities of bacteria in the male genitourinary tract that frequently resemble cutaneous or vaginal bacterial communities. Additional studies are necessary to characterize microbial populations on male genital surfaces from larger populations of men since the studies cited above are small and focused on very discreet populations. In addition, longitudinal studies are needed to determine how male genital microbial communities change with host and environmental factors.

8.7. CONCLUSION

Although science and medicine have celebrated the presence of vaginal lactobacilli as a marker of health for well over a century, we are only now beginning to appreciate the breadth and dynamic nature of the urogenital microbiota. As we have described in this chapter, high-throughput sequencing technology is not only unveiling a wide diversity of urogenital organisms but also informing us that a very small number of *Lactobacillus* species tend to dominate the vaginal bacterial community in women worldwide. Bacterial vaginosis remains a poorly understood condition associated with increased risks to a woman's health. However, the clinical diagnostics that define this condition are becoming outdated in the molecular era and their limitations increasingly apparent. Here we have described the latest epidemiological and ecological findings regarding the vaginal microbiota firmly in the context of the current gold standard diagnostic for BV, the Nugent score, as this remains the best tool we have for peering across studies and drawing the most generalizable conclusions. However, we also noted the fact that Nugent score can be an imperfect descriptor of what is happening with the microbiota. As sequencing costs continue to drop and methods for comparing and classifying microbial communities are improved and standardized, bacterial metagenomics will undoubtedly yield a variety of risk assessment tools for BV and its sequelae. What we do know today is that many of the bacterial species found in women with BV are novel, and although a number remain uncultivated, studies using molecular detection methods are linking them to troublesome symptoms of odorous vaginal discharge and adverse health outcomes such as preterm birth and HIV shedding. Although many challenging questions remain, particularly with regard to the dynamic fluctuations of the vaginal microbiota, the significance of urogenital microbes in niches other than the vagina, and the mechanisms by which these complex communities interact with the host, progress in these areas may have high payoff. Someday, with the ability to profile human genitourinary microbes and medically manipulate them to prevent disease, help produce a healthy full-term baby, or perhaps even improve a woman's sex life after menopause, today's blissful ignorance of the genitourinary microbiota may very well be replaced by informed empowerment.

REFERENCES

1. Aboul Enien WM, El Metwally HA. Association of abnormal vaginal flora with increased cervical tumour necrosis factor—alpha and interferon—gamma levels in idiopathic infertility. *Egypt J Immunol* **12**:53–59 (2005).
2. Al-Mushrif S, Jones BM. A study of the prevalence of hydrogen peroxide generating Lactobacilli in bacterial vaginosis: The determination of H2O2 concentrations generated, in vitro , by isolated strains and the levels found in vaginal secretions of women with and without infection. *J Obstet Gynaecol* **18**:63–67 (1998).
3. Alakomi HL, Skytta E, Saarela M, Mattila-Sandholm T, Latva-Kala K, Helander IM. Lactic acid permeabilizes gram-negative bacteria by disrupting the outer membrane. *Appl Environ Microbiol* **66**:2001–2005 (2000).
4. Amsel R, Totten PA, Spiegel CA, Chen KC, Eschenbach D, Holmes KK. Nonspecific vaginitis. Diagnostic criteria and microbial and epidemiologic associations. *Am J Med* **74**:14–22 (1983).
5. Ansbacher R, Boyson WA, Morris JA. Sterility of the uterine cavity. *Am J Obstet Gynecol* **77**:394–396 (1967).
6. Antonio MA, Hawes SE, Hillier SL. The identification of vaginal Lactobacillus species and the demographic and microbiologic characteristics of women colonized by these species. *J Infect Dis* **180**:1950–1956 (1999).
7. Antonio MA, Meyn LA, Murray PJ, Busse B, Hillier SL. Vaginal colonization by probiotic Lactobacillus crispatus CTV-05 is decreased by sexual activity and endogenous Lactobacilli. *J Infect Dis* **199**:1506–1513 (2009).
8. Antonio MA, Rabe LK, Hillier SL. Colonization of the rectum by Lactobacillus species and decreased risk of bacterial vaginosis. *J Infect Dis* **192**:394–398 (2005).
9. Anukam K, Osazuwa E, Ahonkhai I, Ngwu M, Osemene G, Bruce AW, Reid G. Augmentation of antimicrobial metronidazole therapy of bacterial vaginosis with oral probiotic Lactobacillus rhamnosus GR-1 and Lactobacillus reuteri RC-14: Randomized, double-blind, placebo controlled trial. *Microbes Infect* **8**:1450–1454 (2006).
10. Anukam KC, Osazuwa E, Osemene GI, Ehigiagbe F, Bruce AW, Reid G. Clinical study comparing probiotic Lactobacillus GR-1 and RC-14 with metronidazole vaginal gel to treat symptomatic bacterial vaginosis. *Microbes Infect* **8**:2772–2776 (2006).
11. Atashili J, Poole C, Ndumbe PM, Adimora AA, Smith JS. Bacterial vaginosis and HIV acquisition: A meta-analysis of published studies. *AIDS* **22**:1493–1501 (2008).
12. Atassi F, Servin AL. Individual and co-operative roles of lactic acid and hydrogen peroxide in the killing activity of enteric strain Lactobacillus johnsonii NCC933 and vaginal strain Lactobacillus gasseri KS120.1 against enteric, uropathogenic and vaginosis-associated pathogens. *FEMS Microbiol Lett* **364**:29–38 (2010).
13. Barrons R, Tassone D. Use of Lactobacillus probiotics for bacterial genitourinary infections in women: A review. *Clin Ther* **30**:453–468 (2008).

14. Bartlett JG, Onderdonk AB, Drude E, Goldstein C, Anderka M, Alpert S, McCormack WM. Quantitative bacteriology of the vaginal flora. *J Infect Dis* **136**:271–277 (1977).
15. Biagi E, Vitali B, Pugliese C, Candela M, Donders GG, Brigidi P. Quantitative variations in the vaginal bacterial population associated with asymptomatic infections: A real-time polymerase chain reaction study. *Eur J Clin Microbiol Infect Dis* **28**:281–285 (2009).
16. Bollinger CC. Bacterial flora of the nonpregnant uterus: A new culture technic. *Obstet Gynecol* **23**:251–255 (1964).
17. Bradshaw CS, Morton AN, Garland SM, Horvath LB, Kuzevska I, Fairley CK. Evaluation of a point-of-care test, BVBlue, and clinical and laboratory criteria for diagnosis of bacterial vaginosis. *J Clin Microbiol* **43**:1304–1308 (2005).
18. Bradshaw CS, Morton AN, Hocking J, Garland SM, Morris MB, Moss LM, Horvath LB, Kuzevska I, Fairley CK. High recurrence rates of bacterial vaginosis over the course of 12 months after oral metronidazole therapy and factors associated with recurrence. *J Infect Dis* **193**:1478–1486 (2006).
19. Brook I, Barrett CT, Brinkman CR 3rd, Martin WJ, Finegold SM. Aerobic and anaerobic bacterial flora of the maternal cervix and newborn gastric fluid and conjunctiva: A prospective study. *Pediatrics* **63**:451–455 (1979).
20. Brotman RM, Ravel J, Cone RA, Zenilman JM. Rapid fluctuation of the vaginal microbiota measured by Gram stain analysis. *Sex Transm Infect* **86**:297–302 (2010).
21. Bump RC, Zuspan FP, Buesching WJ 3rd, Ayers LW, Stephens TJ. The prevalence, six-month persistence, and predictive values of laboratory indicators of bacterial vaginosis (nonspecific vaginitis) in asymptomatic women. *Am J Obstet Gynecol* **150**:917–924 (1984).
22. Burton JP, Reid G. Evaluation of the bacterial vaginal flora of 20 postmenopausal women by direct (Nugent score) and molecular (polymerase chain reaction and denaturing gradient gel electrophoresis) techniques. *J Infect Dis* **186**:1770–1780 (2002).
23. Cauci S, Driussi S, De Santo D, Penacchioni P, Iannicelli T, Lanzafame P, De Seta F, Quadrifoglio F, de Aloysio D, Guaschino S. Prevalence of bacterial vaginosis and vaginal flora changes in peri- and postmenopausal women. *J Clin Microbiol* **40**:2147–2152 (2002).
24. Cherpes TL, Hillier SL, Meyn LA, Busch JL, Krohn MA. A delicate balance: Risk factors for acquisition of bacterial vaginosis include sexual activity, absence of hydrogen peroxide-producing lactobacilli, black race, and positive herpes simplex virus type 2 serology. *Sex Transm Dis* **35**:78–83 (2008).
25. Cohen CR, Duerr A, Pruithithada N, Rugpao S, Hillier S, Garcia P, Nelson K. Bacterial vaginosis and HIV seroprevalence among female commercial sex workers in Chiang Mai, Thailand. *AIDS* **9**:1093–1097 (1995).
26. Collins MD, Hoyles L, Tornqvist E, von Essen R, Falsen E. Characterization of some strains from human clinical sources which resemble "Leptotrichia sanguinegens": Description of Sneathia sanguinegens sp. nov., gen. nov. *Syst Appl Microbiol* **24**:358–361 (2001).
27. Coppolillo EF, Perazzi BE, Famiglietti AM, Cora Eliseht MG, Vay CA, Barata AD. Diagnosis of bacterial vaginosis during pregnancy. *J Low Genit Tract Dis* **7**:117–121 (2003).
28. Cowling P, McCoy DR, Marshall RJ, Padfield CJ, Reeves DS. Bacterial colonization of the non-pregnant uterus: A study of pre-menopausal abdominal hysterectomy specimens. *Eur J Clin Microbiol Infect Dis* **11**:204–205 (1992).
29. Cristiano L, Coffetti N, Dalvai G, Lorusso L, Lorenzi M. Bacterial vaginosis: Prevalence in outpatients, association with some micro-organisms and laboratory indices. *Genitourin Med* **65**:382–387 (1989).

30. Cruickshank R, Sharman A. The biology of the vagina in the human subject. *Br J Obstet Gynecol* **41**:208–226, 369–384 (1934).
31. Demba E, Morison L, van der Loeff MS, Awasana AA, Gooding E, Bailey R, Mayaud P, West B. Bacterial vaginosis, vaginal flora patterns and vaginal hygiene practices in patients presenting with vaginal discharge syndrome in The Gambia, West Africa. *BMC Infect Dis* **5**:12 (2005).
32. DiGiulio DB, Romero R, Amogan HP, Kusanovic JP, Bik EM, Gotsch F, Kim CJ, Erez O, Edwin S, Relman DA. Microbial prevalence, diversity and abundance in amniotic fluid during preterm labor: A molecular and culture-based investigation. *PLoS ONE* **3**:e3056 (2008).
33. Döderlein A. *Die Scheidensekret, Leipzig*, 1892.
34. Dominguez-Bello MG, Costello EK, Contreras M, Magris M, Hidalgo G, Fierer N, Knight R. Delivery mode shapes the acquisition and structure of the initial microbiota across multiple body habitats in newborns. *Proc Natl Acad Sci USA* **107**:11971–11975 (2010).
35. Donders GG, Van Bulck B, Caudron J, Londers L, Vereecken A, Spitz B. Relationship of bacterial vaginosis and mycoplasmas to the risk of spontaneous abortion. *Am J Obstet Gynecol* **183**:431–437 (2000).
36. Donders GG, Van Calsteren K, Bellen G, Reybrouck R, Van den Bosch T, Riphagen I, Van Lierde S. Predictive value for preterm birth of abnormal vaginal flora, bacterial vaginosis and aerobic vaginitis during the first trimester of pregnancy. *Br J Obstet Gynecal* **116**:1315–1324 (2009).
37. El Aila NA, Tency I, Claeys G, Verstraelen H, Saerens B, Santiago GL, De Backer E, Cools P, Temmerman M, Verhelst R, et al. Identification and genotyping of bacteria from paired vaginal and rectal samples from pregnant women indicates similarity between vaginal and rectal microflora. *BMC Infect Dis* **9**:167 (2009).
38. Eren AM, Zozaya M, Taylor CM, Dowd SE, Martin DH, Ferris MJ. Exploring the diversity of Gardnerella vaginalis in the genitourinary tract microbiota of monogamous couples through subtle nucleotide variation. *PLoS ONE* **6**:e26732 (2011).
39. Eschenbach DA, Buchanan TM, Pollock HM, Forsyth PS, Alexander ER, Lin JS, Wang SP, Wentworth BB, MacCormack WM, Holmes KK. Polymicrobial etiology of acute pelvic inflammatory disease. *N Engl J Med* **293**:166–171 (1975).
40. Eschenbach DA, Davick PR, Williams BL, Klebanoff SJ, Young-Smith K, Critchlow CM, Holmes KK. Prevalence of hydrogen peroxide-producing Lactobacillus species in normal women and women with bacterial vaginosis. *J Clin Microbiol* **27**:251–256 (1989).
41. Eschenbach DA, Hillier S, Critchlow C, Stevens C, DeRouen T, Holmes KK. Diagnosis and clinical manifestations of bacterial vaginosis. *Am J Obstet Gynecol* **158**:819–828 (1988).
42. Falsen E, Pascual C, Sjoden B, Ohlen M, Collins MD. Phenotypic and phylogenetic characterization of a novel Lactobacillus species from human sources: description of Lactobacillus iners sp. nov. *Int J Syst Bacteriol* **49**(Pt 1):217–221 (1999).
43. Fichorova RN, Onderdonk AB, Yamamoto H, Delaney ML, DuBois AM, Allred E, Leviton A. Maternal microbe-specific modulation of inflammatory response in extremely low-gestational-age newborns. *mBio* **2**:e00280–e00210 (2011).
44. Finegold SM, Song Y, Liu C. Taxonomy—General comments and update on taxonomy of Clostridia and Anaerobic cocci. *Anaerobe* **8**:283–285 (2002).
45. Fohn MJ, Lukehart SA, Hillier SL. Production and characterization of monoclonal antibodies to Mobiluncus species. *J Clin Microbiol* **26**:2598–2603 (1988).

46. Forsum U, Jakobsson T, Larsson PG, Schmidt H, Beverly A, Bjornerem A, Carlsson B, Csango P, Donders G, Hay P, et al. An international study of the interobserver variation between interpretations of vaginal smear criteria of bacterial vaginosis. *APMIS* (*Acta Pathologica, Microbiologica, et Immunologica*) **110**:811–818 (2002).
47. Frank JA, Reich CI, Sharma S, Weisbaum JS, Wilson BA, Olsen GJ. Critical evaluation of two primers commonly used for amplification of bacterial 16S rRNA genes. *Appl Environ Microbiol* **74**:2461–2470 (2008).
48. Fredricks DN. Molecular methods to describe the spectrum and dynamics of the vaginal microbiota. *Anaerobe* **17**:191–195 (2011).
49. Fredricks DN, Fiedler TL, Marrazzo JM. Molecular identification of bacteria associated with bacterial vaginosis. *N Engl J Med* **353**:1899–1911 (2005).
50. Fredricks DN, Fiedler TL, Thomas KK, Oakley BB, Marrazzo JM. Targeted PCR for detection of vaginal bacteria associated with bacterial vaginosis. *J Clin Microbiol* **45**:3270–3276 (2007).
51. Fredricks DN, Marrazzo JM. Molecular methodology in determining vaginal flora in health and disease: Its time has come. *Curr Infect Dis Rep* **7**:463–470 (2005).
52. Gallo MF, Warner L, King CC, Sobel JD, Klein RS, Cu-Uvin S, Rompalo AM, Jamieson DJ. Association between semen exposure and incident bacterial vaginosis. *Infect Dis Obstet Gynecol* **2011**:842652 (2011).
53. Glazunova OO, Launay T, Raoult D, Roux V. Prevotella timonensis sp. nov., isolated from a human breast abscess. *Int J Syst Evol Microbiol* **57**:883–886 (2007).
54. Goldenberg RL, Hauth JC, Andrews WW. Intrauterine infection and preterm delivery. *N Engl J Med* **342**:1500–1507 (2000).
55. Gravett MG, Nelson HP, DeRouen T, Critchlow C, Eschenbach DA, Holmes KK. Independent associations of bacterial vaginosis and Chlamydia trachomatis infection with adverse pregnancy outcome. *JAMA* **256**:1899–1903 (1986).
56. Guerra B, Ghi T, Quarta S, Morselli-Labate AM, Lazzarotto T, Pilu G, Rizzo N. Pregnancy outcome after early detection of bacterial vaginosis. *Eur J Obstet Gynecol Reprod Biol* **128**:40–45 (2006).
57. Haggerty CL, Totten PA, Ferris M, Martin DH, Hoferka S, Astete SG, Ondondo R, Norori J, Ness RB. Clinical characteristics of bacterial vaginosis among women testing positive for fastidious bacteria. *Sex Transm Infect* **85**:242–248 (2009).
58. Hammes WP, Hertel C. The genera *Lactobacillus* and *Carnobacterium*. In Dworkin M, Falkow S, Rosenberg E, Schleifer K-H, Stackebrandt E, eds. *Prokaryotes*, Springer, 2007, pp. 320–403.
59. Hawes SE, Hillier SL, Benedetti J, Stevens CE, Koutsky LA, Wolner-Hanssen P, Holmes KK. Hydrogen peroxide-producing lactobacilli and acquisition of vaginal infections. *J Infect Dis* **174**:1058–1063 (1996).
60. Hay PE, Lamont RF, Taylor-Robinson D, Morgan DJ, Ison C, Pearson J. Abnormal bacterial colonisation of the genital tract and subsequent preterm delivery and late miscarriage. *Br Med J* **308**:295–298 (1994).
61. Heinemann C, Reid G. Vaginal microbial diversity among postmenopausal women with and without hormone replacement therapy. *Can J Microbiol* **51**:777–781 (2005).
62. Hemmerling A, Harrison W, Schroeder A, Park J, Korn A, Shiboski S, Foster-Rosales A, Cohen CR. Phase 2a study assessing colonization efficiency, safety, and acceptability of Lactobacillus crispatus CTV-05 in women with bacterial vaginosis. *Sex Transm Dis* **37**:745–750 (2010).
63. Hemsell DL, Obregon VL, Heard MC, Nobles BJ. Endometrial bacteria in asymptomatic, nonpregnant women. *J Reprod Med* **34**:872–874 (1989).

64. Henry-Suchet J, Catalan F, Loffredo V, Serfaty D, Siboulet A, Perol Y, Sanson MJ, Debache C, Pigeau F, Coppin R, et al. Microbiology of specimens obtained by laparoscopy from controls and from patients with pelvic inflammatory disease or infertility with tubal obstruction: Chlamydia trachomatis and Ureaplasma urealyticum. *Am J Obstet Gynecol* **138**:1022–1025 (1980).
65. Hill DR, Brunner ME, Schmitz DC, Davis CC, Flood JA, Schlievert PM, Wang-Weigand SZ, Osborn TW. In vivo assessment of human vaginal oxygen and carbon dioxide levels during and post menses. *J Appl Physiol* **99**:1582–1591 (2005).
66. Hill GB. Preterm birth: Associations with genital and possibly oral microflora. *Ann Periodontol* **3**:222–232 (1998).
67. Hillier SL, Critchlow CW, Stevens CE, Roberts MC, Wolner-Hanssen P, Eschenbach DA, Holmes KK. Microbiological, epidemiological and clinical correlates of vaginal colonisation by Mobiluncus species. *Genitourin Med* **67**:26–31 (1991).
68. Hillier SL, Kiviat NB, Hawes SE, Hasselquist MB, Hanssen PW, Eschenbach DA, Holmes KK. Role of bacterial vaginosis-associated microorganisms in endometritis. *Am J Obstet Gynecol* **175**:435–441 (1996).
69. Hillier SL, Krohn MA, Cassen E, Easterling TR, Rabe LK, Eschenbach DA. The role of bacterial vaginosis and vaginal bacteria in amniotic fluid infection in women in preterm labor with intact fetal membranes. *Clin Infect Dis* **20**(Suppl 2):S276–S278 (1995).
70. Hillier SL, Krohn MA, Rabe LK, Klebanoff SJ, Eschenbach DA. The normal vaginal flora, H2O2-producing lactobacilli, and bacterial vaginosis in pregnant women. *Clin Infect Dis* **16**(Suppl 4):S273–S281 (1993).
71. Hillier SL, Lau RJ. Vaginal microflora in postmenopausal women who have not received estrogen replacement therapy. *Clin Infect Dis* **25**(Suppl 2):S123–S126 (1997).
72. Hillier SL, Nugent RP, Eschenbach DA, Krohn MA, Gibbs RS, Martin DH, Cotch MF, Edelman R, Pastorek JG 2nd, Rao AV, et al. Association between bacterial vaginosis and preterm delivery of a low-birth-weight infant. The Vaginal Infections and Prematurity Study Group. *N Engl J Med* **333**:1737–1742 (1995).
73. Hjelm E, Hallen A, Forsum U, Wallin J. Anaerobic curved rods in vaginitis. *Lancet* **2**:1353–1354 (1981).
74. Holst E. Reservoir of four organisms associated with bacterial vaginosis suggests lack of sexual transmission. *J Clin Microbiol* **28**:2035–2039 (1990).
75. Hugenholtz P, Tyson GW, Webb RI, Wagner AM, Blackall LL. Investigation of candidate division TM7, a recently recognized major lineage of the domain Bacteria with no known pure-culture representatives. *Appl Environ Microbiol* **67**:411–419 (2001).
76. Hummelen R, Fernandes AD, Macklaim JM, Dickson RJ, Changalucha J, Gloor GB, Reid G. Deep sequencing of the vaginal microbiota of women with HIV. *PLoS ONE* **5**:e12078 (2010).
77. Hummelen R, Macklaim JM, Bisanz JE, Hammond JA, McMillan A, Vongsa R, Koenig D, Gloor GB, Reid G. Vaginal microbiome and epithelial gene array in post-menopausal women with moderate to severe dryness. *PLoS ONE* **6**:e26602 (2011).
78. Hyman RW, Fukushima M, Diamond L, Kumm J, Giudice LC, Davis RW. Microbes on the human vaginal epithelium. *Proc Natl Acad Sci USA* **102**:7952–7957 (2005).
79. Jamieson DJ, Duerr A, Klein RS, Paramsothy P, Brown W, Cu-Uvin S, Rompalo A, Sobel J. Longitudinal analysis of bacterial vaginosis: Findings from the HIV epidemiology research study. *Obstet Gynecol* **98**:656–663 (2001).
80. Kalra A, Palcu CT, Sobel JD, Akins RA. Bacterial vaginosis: Culture- and PCR-based characterizations of a complex polymicrobial disease's pathobiology. *Curr Infect Dis Rep* **9**:485–500 (2007).

81. Karasz A, Anderson M. The vaginitis monologues: Women's experiences of vaginal complaints in a primary care setting. *Soc Sci Med* **56**:1013–1021 (2003).
82. Keane FE, Thomas BJ, Gilroy CB, Renton A, Taylor-Robinson D. The association of Mycoplasma hominis, Ureaplasma urealyticum and Mycoplasma genitalium with bacterial vaginosis: Observations on heterosexual women and their male partners. *Int J Sex Transm Dis AIDS* **11**:356–360 (2000).
83. Keith L, England D, Bartizal F, Brown E, Fields C. Microbial flora of the external os of the premenopausal cervix. *Br J Vener Dis* **48**:51–56 (1972).
84. Kim TK, Thomas SM, Ho M, Sharma S, Reich CI, Frank JA, Yeater KM, Biggs DR, Nakamura N, Stumpf R, et al. Heterogeneity of vaginal microbial communities within individuals. *J Clin Microbiol* **47**:1181–1189 (2009).
85. Klappenbach JA, Saxman PR, Cole JR, Schmidt TM. rrndb: The Ribosomal RNA Operon Copy Number Database. *Nucleic Acids Res* **29**:181–184 (2001).
86. Klebanoff MA, Hillier SL, Nugent RP, MacPherson CA, Hauth JC, Carey JC, Harper M, Wapner RJ, Trout W, Moawad A, et al. Is bacterial vaginosis a stronger risk factor for preterm birth when it is diagnosed earlier in gestation? *Am J Obstet Gynecol* **192**:470–477 (2005).
87. Knuppel RA, Scerbo JC, Dzink J, Mitchell GW Jr, Cetrulo CL, Bartlett J. Quantitative transcervical uterine cultures with a new device. *Obstet Gynecol* **57**:243–248 (1981).
88. Koumans EH, Markowitz LE, Hogan V. Indications for therapy and treatment recommendations for bacterial vaginosis in nonpregnant and pregnant women: A synthesis of data. *Clin Infect Dis* **35**:S152–S172 (2002).
89. Koumans EH, Sternberg M, Bruce C, McQuillan G, Kendrick J, Sutton M, Markowitz LE. The prevalence of bacterial vaginosis in the United States, 2001–2004; associations with symptoms, sexual behaviors, and reproductive health. *Sex Transm Dis* **34**:864–869 (2007).
90. Krohn MA, Hillier SL, Eschenbach DA. Comparison of methods for diagnosing bacterial vaginosis among pregnant women. *J Clin Microbiol* **27**:1266–1271 (1989).
91. Lamont RF, Sobel JD, Akins RA, Hassan SS, Chaiworapongsa T, Kusanovic JP, Romero R. The vaginal microbiome: New information about genital tract flora using molecular based techniques. *Br J Obstet Gynecol* **118**:533–549 (2011).
92. Lash AF, Kaplan B. A study of Döderlein's vaginal bacillus. *J Infect Dis* **38**:333–340 (1926).
93. Lawson PA, Moore E, Falsen E. Prevotella amnii sp. nov., isolated from human amniotic fluid. *Int J Syst Evol Microbiol* **58**:89–92 (2008).
94. Lefevre JC, Averous S, Bauriaud R, Blanc C, Bertrand MA, Lareng MB. Lower genital tract infections in women: Comparison of clinical and epidemiologic findings with microbiology. *Sex Transm Dis* **15**:110–113 (1988).
95. Leppaluoto P. Vaginal flora and sperm survival. *J Reprod Med* **12**:99–107 (1974).
96. Li J, McCormick J, Bocking A, Reid G. Importance of vaginal microbes in reproductive health. *Reprod Sci* **19**:235–242 (2012).
97. Ling Z, Kong J, Liu F, Zhu H, Chen X, Wang Y, Li L, Nelson KE, Xia Y, Xiang C. Molecular analysis of the diversity of vaginal microbiota associated with bacterial vaginosis. *BMC Genom* **11**:488 (2010).
98. Ling Z, Liu X, Chen X, Zhu H, Nelson KE, Xia Y, Li L, Xiang C. Diversity of cervicovaginal microbiota associated with female lower genital tract infections. *Microbial Ecol.* **61**:704–714 (2011).
99. Liversedge NH, Turner A, Horner PJ, Keay SD, Jenkins JM, Hull MG. The influence of bacterial vaginosis on in-vitro fertilization and embryo implantation during assisted reproduction treatment. *Human Reprod* **14**:2411–2415 (1999).

100. Mandar R, Mikelsaar M. Transmission of mother's microflora to the newborn at birth. *Biol Neonate* **69**:30–35 (1996).
101. Mania-Pramanik J, Kerkar SC, Salvi VS. Bacterial vaginosis: A cause of infertility? *Int J Sex Transm Dis AIDS* **20**:778–781 (2009).
102. Marrazzo JM, Antonio M, Agnew K, Hillier SL. Distribution of genital Lactobacillus strains shared by female sex partners. *J Infect Dis* **199**:680–683 (2009).
103. Marrazzo JM, Martin DH, Watts DH, Schulte J, Sobel JD, Hillier SL, Deal C, Fredricks DN. Bacterial vaginosis: Identifying research gaps proceedings of a workshop sponsored by DHHS/NIH/NIAID. *Sex Transm Dis* **37**:732–744 (2010).
104. Marrazzo JM, Thomas KK, Fiedler TL, Ringwood K, Fredricks DN. Relationship of specific vaginal bacteria and bacterial vaginosis treatment failure in women who have sex with women. *Ann Intern Med* **149**:20–28 (2008).
105. Marrazzo JM, Thomas KK, Fiedler TL, Ringwood K, Fredricks DN. Risks for acquisition of bacterial vaginosis among women who report sex with women: A cohort study. *PLoS ONE* **5**:e11139 (2010).
106. Martin DH. The microbiota of the vagina and its influence on women's health and disease. *Am J Med Sci* **343**:2–9 (2012).
107. Martin HL, Richardson BA, Nyange PM, Lavreys L, Hillier SL, Chohan B, Mandaliya K, Ndinya-Achola JO, Bwayo J, Kreiss J. Vaginal lactobacilli, microbial flora, and risk of human immunodeficiency virus type 1 and sexually transmitted disease acquisition. *J Infect Dis* **180**:1863–1868 (1999).
108. Martin R, Suarez JE. Biosynthesis and degradation of H2O2 by vaginal lactobacilli. *Appl Environ Microbiol* **76**:400–405 (2010).
109. McDonald HM, Brocklehurst P, Gordon A. Antibiotics for treating bacterial vaginosis in pregnancy. *Cochrane Database Syst Rev* CD000262 (2007).
110. Menard JP, Fenollar F, Henry M, Bretelle F, Raoult D. Molecular quantification of Gardnerella vaginalis and Atopobium vaginae loads to predict bacterial vaginosis. *Clin Infect Dis* **47**:33–43 (2008).
111. Menard JP, Fenollar F, Raoult D, Boubli L, Bretelle F. Self-collected vaginal swabs for the quantitative real-time polymerase chain reaction assay of Atopobium vaginae and Gardnerella vaginalis and the diagnosis of bacterial vaginosis. *Eur J Clin Microbiol Infect Dis* **29**:1547–1552 (2010).
112. Menard JP, Mazouni C, Fenollar F, Raoult D, Boubli L, Bretelle F. Diagnostic accuracy of quantitative real-time PCR assay versus clinical and Gram stain identification of bacterial vaginosis. *Eur J Clin Microbiol Infect Dis* **29**:1547–1552 (2010).
113. Mijac VD, Dukic SV, Opavski NZ, Dukic MK, Ranin LT. Hydrogen peroxide producing lactobacilli in women with vaginal infections. *Eur J Obstet Gynecol Reprod Biol* **129**:69–76 (2006).
114. Mishell DR Jr, Bell JH, Good RG, Moyer DL. The intrauterine device: A bacteriologic study of the endometrial cavity. *Am J Obstet Gynecol* **96**:119–126 (1966).
115. Modak T, Arora P, Agnes C, Ray R, Goswami S, Ghosh P, Das NK. Diagnosis of bacterial vaginosis in cases of abnormal vaginal discharge: Comparison of clinical and microbiological criteria. *J Infect Dev Countries* **5**:353–360 (2011).
116. Moller BR, Kristiansen FV, Thorsen P, Frost L, Mogensen SC. Sterility of the uterine cavity. *Acta Obstet Gynecol Scand* **74**:216–219 (1995).
117. Moncla BJ, Hillier SL. Why nonoxynol-9 may have failed to prevent acquisition of Neisseria gonorrhoeae in clinical trials. *Sex Transm Dis* **32**:491–494 (2005).
118. Morales WJ, Schorr S, Albritton J. Effect of metronidazole in patients with preterm birth in preceding pregnancy and bacterial vaginosis: A placebo-controlled, double-blind study. *Am J Obstet Gynecol* **171**:345–347; discussion 348–349 (1994).

119. Myziuk L, Romanowski B, Johnson SC. BVBlue test for diagnosis of bacterial vaginosis. *J Clin Microbiol* **41**:1925–1928 (2003).

120. Nelson DB, Hanlon A, Hassan S, Britto J, Geifman-Holtzman O, Haggerty C, Fredricks DN. Preterm labor and bacterial vaginosis-associated bacteria among urban women. *J Perinat Med* **37**:130–134 (2009).

121. Nelson DE, Van Der Pol B, Dong Q, Revanna KV, Fan B, Easwaran S, Sodergren E, Weinstock GM, Diao L, Fortenberry JD. Characteristic male urine microbiomes associate with asymptomatic sexually transmitted infection. *PLoS ONE* **5**:e14116 (2010).

122. Ness RB, Kip KE, Soper DE, Stamm CA, Rice P, Richter HE. Variability of bacterial vaginosis over 6- to 12-month intervals. *Sex Transm Dis* **33**:381–385 (2006).

123. Ngugi BM, Hemmerling A, Bukusi EA, Kikuvi G, Gikunju J, Shiboski S, Fredricks DN, Cohen CR. Effects of bacterial vaginosis-associated bacteria and sexual intercourse on vaginal colonization with the probiotic Lactobacillus crispatus CTV-05. *Sex Transm Dis* **38**:1020–1027 (2011).

124. Nugent RP, Krohn MA, Hillier SL. Reliability of diagnosing bacterial vaginosis is improved by a standardized method of gram stain interpretation. *J Clin Microbiol* **29**:297–301 (1991).

125. O'Hanlon DE, Lanier BR, Moench TR, Cone RA. Cervicovaginal fluid and semen block the microbicidal activity of hydrogen peroxide produced by vaginal lactobacilli. *BMC Infect Dis* **10**:120 (2010).

126. O'Hanlon DE, Moench TR, Cone RA. In vaginal fluid, bacteria associated with bacterial vaginosis can be suppressed with lactic acid but not hydrogen peroxide. *BMC Infect Dis* **11**:200 (2011).

127. Oakeshott P, Kerry S, Hay S, Hay P. Bacterial vaginosis and preterm birth: A prospective community-based cohort study. *Br J Gen Pract* **54**:119–122 (2004).

128. Oakley BB, Fiedler TL, Marrazzo JM, Fredricks DN. Diversity of human vaginal bacterial communities and associations with clinically defined bacterial vaginosis. *Appl Environ Microbiol* **74**:4898–4909 (2008).

129. Pahlson C, Hallen A, Forsum U. Curved rods related to Mobiluncus—phenotypes as defined by monoclonal antibodies. *Acta Pathol Microbiol Immunol Scand B* **94**:117–125 (1986).

130. Patterson JL, Girerd PH, Karjane NW, Jefferson KK. Effect of biofilm phenotype on resistance of Gardnerella vaginalis to hydrogen peroxide and lactic acid. *Am J Obstet Gynecol* **197**:170–177 (2007).

131. Phillips I, Taylor E. Anaerobic curved rods in vaginitis. *Lancet* **1**:221 (1982).

132. Price LB, Liu CM, Johnson KE, Aziz M, Lau MK, Bowers J, Ravel J, Keim PS, Serwadda D, Wawer MJ, et al. The effects of circumcision on the penis microbiome. *PLoS ONE* **5**:e8422 (2010).

133. Ravel J, Gajer P, Abdo Z, Schneider GM, Koenig SS, McCulle SL, Karlebach S, Gorle R, Russell J, Tacket CO, et al. Microbes and Health Sackler Colloquium: Vaginal microbiome of reproductive-age women. *Proc Natl Acad Sci USA* **108**(Suppl 1):4680–4687 (2011).

134. Reef SE, Levine WC, McNeil MM, Fisher-Hoch S, Holmberg SD, Duerr A, Smith D, Sobel JD, Pinner RW. Treatment options for vulvovaginal candidiasis, 1993. *Clin Infect Dis* **20**(Suppl 1):S80–S90 (1995).

135. Reid G, Bruce AW, Fraser N, Heinemann C, Owen J, Henning B. Oral probiotics can resolve urogenital infections. *FEMS Immunol Med Microbiol* **30**:49–52 (2001).

136. Reid G, Charbonneau D, Erb J, Kochanowski B, Beuerman D, Poehner R, Bruce AW. Oral use of Lactobacillus rhamnosus GR-1 and L. fermentum RC-14 significantly alters

vaginal flora: Randomized, placebo-controlled trial in 64 healthy women. *FEMS Immunol Med Microbiol* **35**:131–134 (2003).

137. Roberts MC, Hillier SL, Schoenknecht FD, Holmes KK. Comparison of gram stain, DNA probe, and culture for the identification of species of Mobiluncus in female genital specimens. *J Infect Dis* **152**:74–77 (1985).
138. Robertson SA. Control of the immunological environment of the uterus. *Rev Reprod* **5**:164–174 (2000).
139. Rogosa M, Sharpe ME. Species differentiation of human vaginal lactobacilli. *J Gen Microbiol* **23**:197–201 (1960).
140. Rosenstein IJ, Morgan DJ, Sheehan M, Lamont RF, Taylor-Robinson D. Bacterial vaginosis in pregnancy: Distribution of bacterial species in different gram-stain categories of the vaginal flora. *J Med Microbiol* **45**:120–126 (1996).
141. Ross JM, Needham JR. Genital flora during pregnancy and colonization of the newborn. *J R Soc Med* **73**:105–110 (1980).
142. Savage DC. Microbial ecology of the gastrointestinal tract. *Annu Rev Microbiol* **31**:107–133 (1977).
143. Schellenberg J, Links MG, Hill JE, Dumonceaux TJ, Peters GA, Tyler S, Ball TB, Severini A, Plummer FA. Pyrosequencing of the chaperonin-60 universal target as a tool for determining microbial community composition. *Appl Environ Microbiol* **75**:2889–2898 (2009).
144. Schellenberg JJ, Links MG, Hill JE, Dumonceaux TJ, Kimani J, Jaoko W, Wachihi C, Mungai JN, Peters GA, Tyler S, et al. Molecular definition of vaginal microbiota in East African commercial sex workers. *Appl Environ Microbiol* **77**:4066–4074 (2011).
145. Schwebke JR, Hillier SL, Sobel JD, McGregor JA, Sweet RL. Validity of the vaginal gram stain for the diagnosis of bacterial vaginosis. *Obstet Gynecol* **88**:573–576 (1996).
146. Sewankambo N, Gray RH, Wawer MJ, Paxton L, McNaim D, Wabwire-Mangen F, Serwadda D, Li C, Kiwanuka N, Hillier SL, et al. HIV-1 infection associated with abnormal vaginal flora morphology and bacterial vaginosis. *Lancet* **350**:546–550 (1997).
147. Sha BE, Zariffard MR, Wang QJ, Chen HY, Bremer J, Cohen MH, Spear GT. Female genital-tract HIV load correlates inversely with Lactobacillus species but positively with bacterial vaginosis and Mycoplasma hominis. *J Infect Dis* **191**:25–32 (2005).
148. Shukla SK, Meier PR, Mitchell PD, Frank DN, Reed KD. Leptotrichia amnionii sp. nov., a novel bacterium isolated from the amniotic fluid of a woman after intrauterine fetal demise. *J Clin Microbiol* **40**:3346–3349 (2002).
149. Simcox R, Sin WT, Seed PT, Briley A, Shennan AH. Prophylactic antibiotics for the prevention of preterm birth in women at risk: a meta-analysis. *Austral N Z J Obstet Gynaecol* **47**:368–377 (2007).
150. Skarin A, Mardh PA. Comma-shaped bacteria associated with vaginitis. *Lancet* **1**:342–343 (1982).
151. Soper DE. Pelvic inflammatory disease. *Obstet Gynecol* **116**:419–428 (2010).
152. Soper DE, Brockwell NJ, Dalton HP, Johnson D. Observations concerning the microbial etiology of acute salpingitis. *Am J Obstet Gynecol* **170**:1008–1014; discussion 1014–1007 (1994).
153. Soper DE, Bump RC, Hurt WG. Bacterial vaginosis and trichomoniasis vaginitis are risk factors for cuff cellulitis after abdominal hysterectomy. *Am J Obstet Gynecol* **163**:1016–1021; discussion 1021–1013 (1990).
154. Spandorfer SD, Neuer A, Giraldo PC, Rosenwaks Z, Witkin SS. Relationship of abnormal vaginal flora, proinflammatory cytokines and idiopathic infertility in women undergoing IVF. *J Reprod Med* **46**:806–810 (2001).

155. Spear GT, Gilbert D, Landay AL, Zariffard R, French AL, Patel P, Gillevet PM. Pyrosequencing of the genital microbiotas of HIV-seropositive and -seronegative women reveals Lactobacillus iners as the predominant Lactobacillus species. *Appl Environ Microbiol* **77**:378–381 (2011).
156. Spiegel CA, Eschenbach DA, Amsel R, Holmes KK. Curved anaerobic bacteria in bacterial (nonspecific) vaginosis and their response to antimicrobial therapy. *J Infect Dis* **148**:817–822 (1983).
157. Spinillo A, Bernuzzi AM, Cevini C, Gulminetti R, Luzi S, De Santolo A. The relationship of bacterial vaginosis, Candida and Trichomonas infection to symptomatic vaginitis in postmenopausal women attending a vaginitis clinic. *Maturitas* **27**:253–260 (1997).
158. Spore WW, Moskal PA, Nakamura RM, Mishell DR Jr. Bacteriology of postpartum oviducts and endometrium. *Am J Obstet Gynecol* **107**:572–577 (1970).
159. Sprott MS, Pattman RS, Ingham HR, Short GR, Narang HK, Selkon JB. Anaerobic curved rods in vaginitis. *Lancet* **1**:54 (1982).
160. Spurbeck RR, Arvidson CG. Lactobacilli at the front line of defense against vaginally acquired infections. *Future Microbiol* **6**:567–582 (2011).
161. Srinivasan S, Fredricks DN. The human vaginal bacterial biota and bacterial vaginosis. *Interdiscipl Perspect Infect Dis* **2008**:750479 (2008).
162. Srinivasan S, Liu C, Mitchell CM, Fiedler TL, Thomas KK, Agnew KJ, Marrazzo JM, Fredricks DN. Temporal variability of human vaginal bacteria and relationship with bacterial vaginosis. *PLoS ONE* **5**:e10197 (2010).
163. Stapleton AE, Au-Yeung M, Hooton TM, Fredricks DN, Roberts PL, Czaja CA, Yarova-Yarovaya Y, Fiedler T, Cox M, Stamm WE. Randomized, placebo-controlled phase 2 trial of a Lactobacillus crispatus probiotic given intravaginally for prevention of recurrent urinary tract infection. *Clin Infect Dis* **52**:1212–1217 (2011).
164. Steel JH, Malatos S, Kennea N, Edwards AD, Miles L, Duggan P, Reynolds PR, Feldman RG, Sullivan MH. Bacteria and inflammatory cells in fetal membranes do not always cause preterm labor. *Pediatr Res* **57**:404–411 (2005).
165. Sumeksri P, Koprasert C, Panichkul S. BVBLUE test for diagnosis of bacterial vaginosis in pregnant women attending antenatal care at Phramongkutklao Hospital. *J Med Assoc Thai* **88**(Suppl 3):S7–S13 (2005).
166. Swidsinski A, Doerffel Y, Loening-Baucke V, Swidsinski S, Verstraelen H, Vaneechoutte M, Lemm V, Schilling J, Mendling W. Gardnerella biofilm involves females and males and is transmitted sexually. *Gynecol Obstet Invest* **70**:256–263 (2010).
167. Swidsinski A, Dorffel Y, Loening-Baucke V, Mendling W, Schilling J, Patterson JL, Verstraelen H. Dissimilarity in the occurrence of Bifidobacteriaceae in vaginal and perianal microbiota in women with bacterial vaginosis. *Anaerobe* **16**:478–482 (2010).
168. Swidsinski A, Mendling W, Loening-Baucke V, Ladhoff A, Swidsinski S, Hale LP, Lochs H. Adherent biofilms in bacterial vaginosis. *Obstet Gynecol* **106**:1013–1023 (2005).
169. Swidsinski A, Mendling W, Loening-Baucke V, Swidsinski S, Dorffel Y, Scholze J, Lochs H, Verstraelen H. An adherent Gardnerella vaginalis biofilm persists on the vaginal epithelium after standard therapy with oral metronidazole. *Am J Obstet Gynecol* **198**:e91–e96 (2008).
170. Taha TE, Hoover DR, Dallabetta GA, Kumwenda NI, Mtimavalye LA, Yang LP, Liomba GN, Broadhead RL, Chiphangwi JD, Miotti PG. Bacterial vaginosis and disturbances of vaginal flora: Association with increased acquisition of HIV. *AIDS* **12**:1699–1706 (1998).
171. Tan SL, Scammell G, Houang E. The midcycle cervical microbial flora as studied by the weighed-swab method, and its possible correlation with results of sperm cervical mucus penetration tests. *Fertil Steril* **47**:941–946 (1987).

172. Taylor-Robinson D, Lamont RF. Mycoplasmas in pregnancy. *Br J Obstet Gynecol* **118**:164–174 (2011).
173. Teisala K. Endometrial microbial flora of hysterectomy specimens. *Eur J Obstet Gynecol Reprod Biol* **26**:151–155 (1987).
174. Thoma ME, Gray RH, Kiwanuka N, Aluma S, Wang MC, Sewankambo N, Wawer MJ. The short-term variability of bacterial vaginosis diagnosed by Nugent Gram stain criteria among sexually active women in Rakai, Uganda. *Sex Transm Dis* **38**:111–116 (2011).
175. Thoma ME, Gray RH, Kiwanuka N, Wang MC, Sewankambo N, Wawer MJ. The natural history of bacterial vaginosis diagnosed by gram stain among women in Rakai, Uganda. *Sex Transm Dis* **38**:1040–1045 (2011).
176. Thomason JL, Schreckenberger PC, Spellacy WN, Riff LJ, LeBeau LJ. Clinical and microbiological characterization of patients with nonspecific vaginosis associated with motile, curved anaerobic rods. *J Infect Dis* **149**:801–809 (1984).
177. Vallor AC, Antonio MA, Hawes SE, Hillier SL. Factors associated with acquisition of, or persistent colonization by, vaginal lactobacilli: Role of hydrogen peroxide production. *J Infect Dis* **184**:1431–1436 (2001).
178. Vasquez A, Jakobsson T, Ahrne S, Forsum U, Molin G. Vaginal lactobacillus flora of healthy Swedish women. *J Clin Microbiol* **40**:2746–2749 (2002).
179. Verhelst R, Verstraelen H, Claeys G, Verschraegen G, Delanghe J, Van Simaey L, De Ganck C, Temmerman M, Vaneechoutte M. Cloning of 16S rRNA genes amplified from normal and disturbed vaginal microflora suggests a strong association between Atopobium vaginae, Gardnerella vaginalis and bacterial vaginosis. *BMC Microbiol* **4**:16 (2004).
180. Verstraelen H, Verhelst R. Bacterial vaginosis: An update on diagnosis and treatment. *Expert Rev Anti Infect Ther* **7**:1109–1124 (2009).
181. Wagner G, Levin R. Oxygen tension of the vaginal surface during sexual stimulation in the human. *Fertil Steril* **30**:50–53 (1978).
182. Wagner G, Ottesen B. Vaginal physiology during menstruation. *Ann Intern Med* **96**:921–923 (1982).
183. Watts DH, Krohn MA, Hillier SL, Eschenbach DA. The association of occult amniotic fluid infection with gestational age and neonatal outcome among women in preterm labor. *Obstet Gynecol* **79**:351–357 (1992).
184. Weinstein L. The bacterial flora of the human vagina. *Yale J Biol Med* **10**:247–260 (1938).
185. Wertz J, Isaacs-Cosgrove N, Holzman C, Marsh TL. Temporal shifts in microbial communities in nonpregnant African-American women with and without Bacterial vaginosis. *Interdiscipl Perspect Infect Dis* **2008**:181253 (2008).
186. Wiesenfeld HC, Hillier SL, Krohn MA, Amortegui AJ, Heine RP, Landers DV, Sweet RL. Lower genital tract infection and endometritis: insight into subclinical pelvic inflammatory disease. *Obstet Gynecol* **100**:456–463 (2002).
187. Wiesenfeld HC, Hillier SL, Krohn MA, Landers DV, Sweet RL. Bacterial vaginosis is a strong predictor of Neisseria gonorrhoeae and Chlamydia trachomatis infection. *Clin Infect Dis* **36**:663–668 (2003).
188. Wiesenfeld HC, Macio I. The infrequent use of office-based diagnostic tests for vaginitis. *Am J Obstet Gynecol* **181**:39–41 (1999).
189. Wilks M, Wiggins R, Whiley A, Hennessy E, Warwick S, Porter H, Corfield A, Millar M. Identification and H(2)O(2) production of vaginal lactobacilli from pregnant women at high risk of preterm birth and relation with outcome. *J Clin Microbiol* **42**:713–717 (2004).

190. Wilson JD, Ralph SG, Rutherford AJ. Rates of bacterial vaginosis in women undergoing in vitro fertilisation for different types of infertility. *Br J Obstet Gynecol* **109**:714–717 (2002).

191. Wylie JG, Henderson A. Identity and glycogen-fermenting ability of lactobacilli isolated from the vagina of pregnant women. *J Med Microbiol* **2**:363–366 (1969).

192. Yeoman CJ, Yildirim S, Thomas SM, Durkin AS, Torralba M, Sutton G, Buhay CJ, Ding Y, Dugan-Rocha SP, Muzny DM, et al. Comparative genomics of Gardnerella vaginalis strains reveals substantial differences in metabolic and virulence potential. *PLoS ONE* **5**:e12411 (2010).

193. Zervomanolakis I, Ott HW, Hadziomerovic D, Mattle V, Seeber BE, Virgolini I, Heute D, Kissler S, Leyendecker G, Wildt L. Physiology of upward transport in the human female genital tract. *Ann NY Acad Sci* **1101**:1–20 (2007).

194. Zhou X, Bent SJ, Schneider MG, Davis CC, Islam MR, Forney LJ. Characterization of vaginal microbial communities in adult healthy women using cultivation-independent methods. *Microbiology* **150**:2565–2573 (2004).

195. Zhou X, Brotman RM, Gajer P, Abdo Z, Schuette U, Ma S, Ravel J, Forney LJ. Recent advances in understanding the microbiology of the female reproductive tract and the causes of premature birth. *Infect Dis Obstet Gynecol* **2010**:737425 (2010).

196. Zozaya-Hinchliffe M, Lillis R, Martin DH, Ferris MJ. Quantitative PCR assessments of bacterial species in women with and without bacterial vaginosis. *J Clin Microbiol* **48**:1812–1819 (2010).

197. Zozaya-Hinchliffe M, Martin DH, Ferris MJ. Prevalence and abundance of uncultivated Megasphaera-like bacteria in the human vaginal environment. *Appl Environ Microbiol* **74**:1656–1659 (2008).

9

FUNCTIONAL STRUCTURE OF INTESTINAL MICROBIOTA IN HEALTH AND DISEASE

ALEXANDER SWIDSINSKI and VERA LOENING-BAUCKE

The University Hospital Charité of the Humboldt University at Berlin, Berlin, Germany

9.1. INTRODUCTION

The intestine serves as an interface between macroorganisms and the naturally polymicrobial environment. While ingesting food, the organism encounters diverse bacteria. Intestinal microbiota are therefore basically polymicrobial. Different from monocultures, the polymicrobial community contains members that are not easily cultivated in the laboratory in isolation, yet thrive in association with other microbes. The properties of isolated microorganisms do not explain how the polymicrobial community functions or why it can grow under conditions that may be deadly for each constituent. To understand polymicrobial communities, the composite structures as they grow and respond to challenges need to be monitored. One method used to visualize single bacterial species within complex communities involes ribosomal RNA fluorescence *in situ* hybridization (FISH). Each bacterium possesses 10^{3-5} ribosomes. Each ribosome includes an RNA copy. Some of the regions of the ribosomal RNA are strain-specific; others are universal for groups, domains, or even kingdoms. Synthetically produced oligonucleotides that are complementary to sequences of interest can be labeled with fluorescent dye and added to samples containing bacteria. These oligonucleotides, which are called *FISH probes*, hybridize with RNA of bacterial ribosomes. Bacteria can be visualized with the microscope directly without additional enhancement because of the high number of ribosomes within each bacterium [1].

The Human Microbiota: How Microbial Communities Affect Health and Disease, First Edition. Edited by David N. Fredricks.

The following presentation is based largely on data obtained from intestinal microbiota using FISH. The names of the FISH probes are listed according to abbreviations of probeBase online resources for rRNA-targeted oligonucleotid probes (http://www.microbial-ecology.net/probebase/credits.asp) [2].

9.2. INTESTINAL MICROBIOTA

The intestinal pulp is rich in nutrients and provides a desirable environment for many bacteria. The organism therefore develops multiple mechanisms to control bacteria either through suppression (e.g., with lysozyme of the saliva, gastric acid of the stomach, secretion of defensins) or through separation of bacteria from the intestinal wall with a mucus barrier. Most often both suppression and separation are implicated but differently balanced. Polymicrobial communities as compared to monocultures are extremely recalcitrant. They respond in coordination to environmental challenges, resist antibiotic treatment and immune responses, and are able to persist under extreme conditions. The control of bacterial growth by the host is therefore never absolute, and the intestine is never sterile. The occurrence, composition, and organization of intestinal microbiota in each gut segment depend on whether suppression or separation dominates. In gut regions with active suppression of microbiota, the bacteria are occasional and of variable composition and concentration. A complete separation of bacteria from the mucosa and low levels of suppression lead to the development of an intestinal reservoir in which bacteria can grow and reach high concentrations. Bacteria are indigenous here. The balance of suppression and separation mechanisms depends on the evolutionary impact of bacteria on health benefits and hazards. For example, none of the eukaryotic organisms can digest cellulose. Plant-feeding animals use microorganisms for this task. Bacteria, which digest cellulose, are therefore indigenous in the rumen of these animals. Absorption of nutrients takes place in the small intestine. Bacteria are clear competitors here and are suppressed in all (including ruminal) mammals. The presence of single bacterial groups is transient, and concentrations of bacteria are low in the small intestine. The function of the large intestine is resorption of water and electrolytes and reduction of the fecal mass. The intestinal content that leaves the small intestine contains many non-digestible substances. These nondigestible substances can still be degraded by bacteria in the large intestine. Our presentation will be restricted mostly to microbiota of the human intestinal tract. Description of processes in the intestinal tract of rodents will be presented only if their description is necessary for understanding processes in the human intestine.

9.2.1. Bacteria in the Upper Gastrointestinal Tract

The Mouth

The mouth is the host region that has initial contact with bacteria of the outer world. Environmental diversity is high and bacterial diversity is high as long as samples of saliva are evaluated. In contrast, FISH-based investigations of samples taken from the oral cavity demonstrate that the stratified epithelium of the mouth in healthy persons is free of bacteria, despite high concentrations of bacteria in saliva (Figure 9.1).

No bacteria can be found also in saliva taken directly from the salivary duct. In contrast, bacteria can be found in high concentrations on food remnants

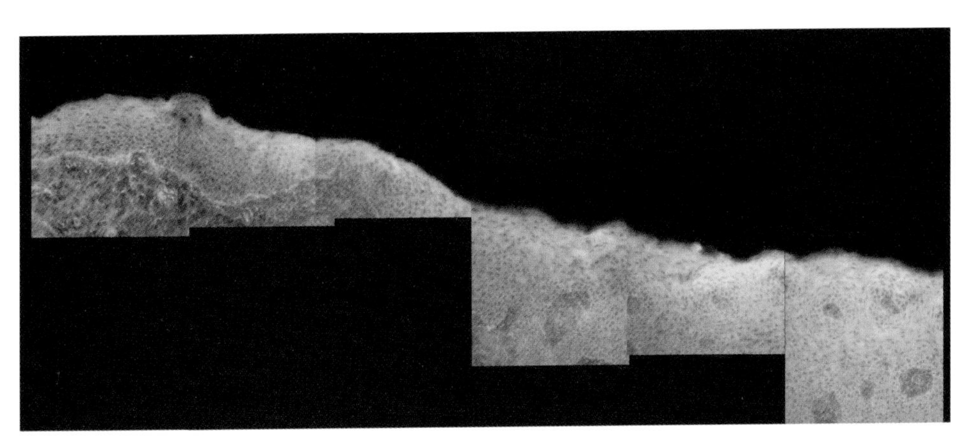

Figure 9.1. Healthy human mouth epithelium hybridized with the universal Eub338 Cy3 bacterial probe. The figure is composed of six consecutively captured microphotographs at 400× magnification. No bacterial signals can be seen.

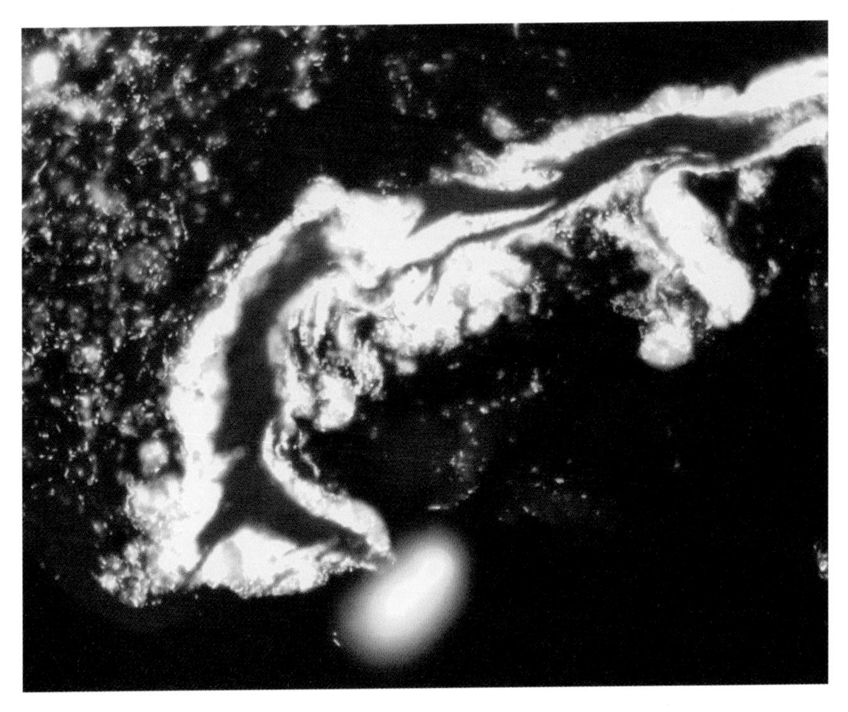

Figure 9.2. Massive bacterial biofilm attached to food remnants in the mouth (universal Eub338 Cy3 probe, yellow fluorescence 400×). Unlike the healthy mucosal surface, food remnants can be covered with prolific bacterial biofilm.

(Figure 9.2), the dental surface, and within or attached to desquamated epithelial cells suspended in oral secretions (Figure 9.3).

The stratified epithelium of the oral cavity and the epithelium of the salivary glands possess efficient mechanisms to suppress bacterial adhesion and growth but are unable to control the bacterial growth on the surfaces of teeth, in food remnants, or on desquamated epithelial cells.

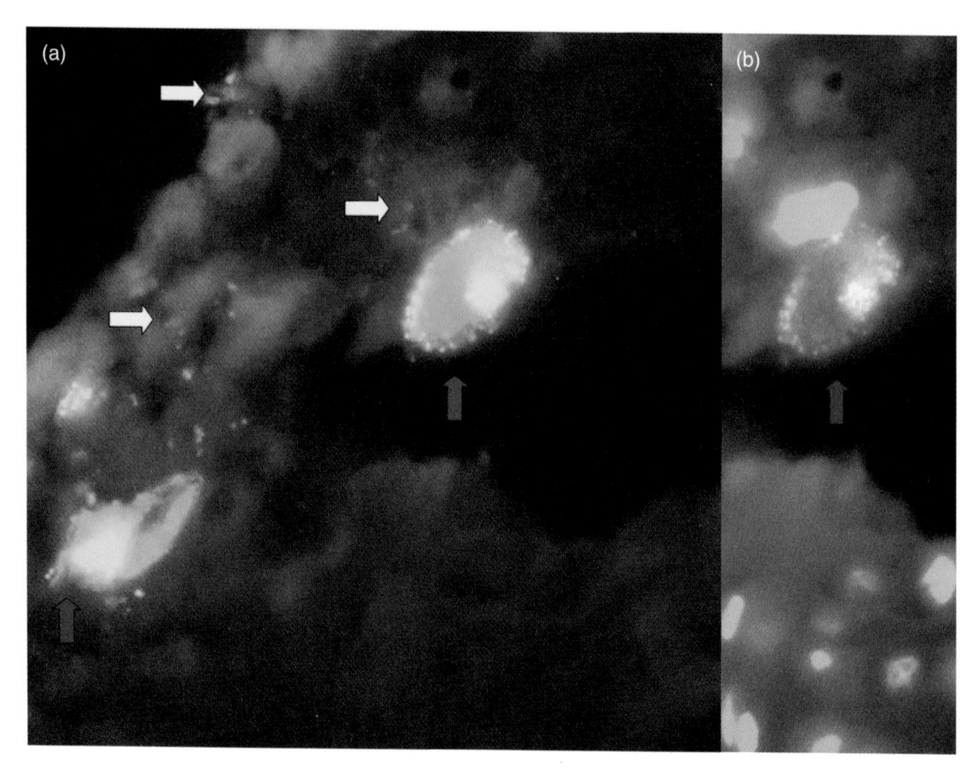

Figure 9.3. (a) Isolated island of bacteria attached to desquamated epithelial cells in saliva (universal bacterial probe Eub338 FITC, green fluorescence) and *Burkholderia* (Burkho Cy3 probe, orange fluorescence). (b) DAPI stain (blue fluorescence of all DNA-rich structures) is overlayed with Burkholderia fluorescence at 1000′. The large blue fluorescence spots on the right are nuclei of desquamated epithelial cells within saliva. The attached bacteria are either irregularly scattered (green fluorescence, white arrows) or organized to oval structures with exact arrangement of single bacterial groups (red arrows).

Tonsils

Similar to the situation in the mouth, no constant pattern of colonization can be found on the surface of the tonsilar epithelium [3]. Most of the epithelial surface of tonsils is free of bacteria, even in tonsillectomy material after chronic tonsillitis (Figure 9.4). When bacteria were found, they were localized either to circumscribed regions of diffuse infiltration (Figure 9.5), within macrophages (Figures 9.6a,b), superficial infiltrates (Figure 9.7), singular purulent fissures (Figure 9.8), or abscesses (Figures 9.9 and 9.10).

The composition of bacteria within infectious tonsilar foci (Table 9.1) is individual and often differs even between different regions of the same tonsil, indicating that bacteria are not indigenous here but represent remnants of incompletely cured purulent processes.

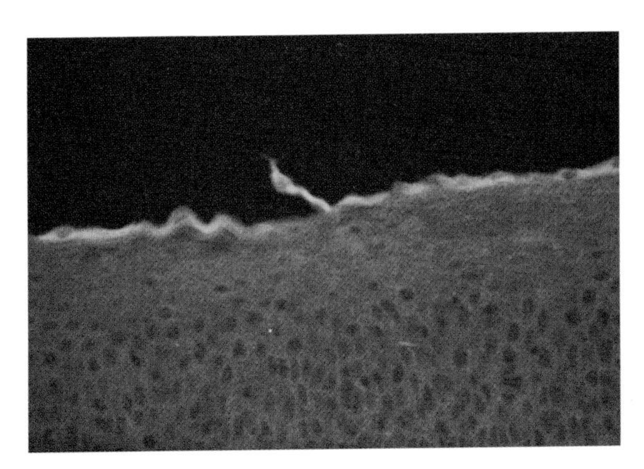

Figure 9.4. Healthy tonsillar epithelium (universal Eub338 Cy3 probe, 400×) free of bacteria.

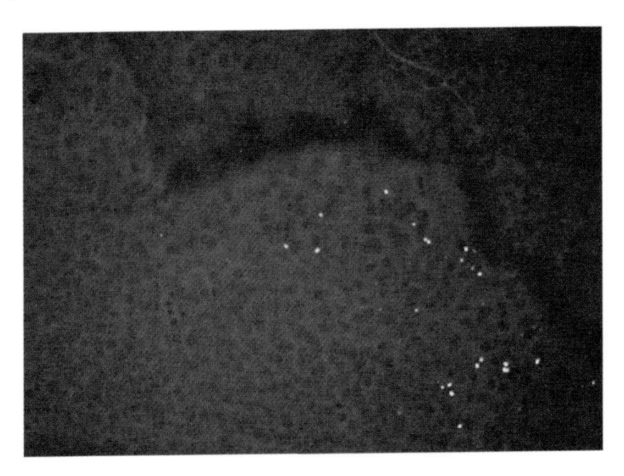

Figure 9.5. Diffuse local infiltration of a tonsil with *Haemophilus influenzae* (Haeinf Cy3 probe, orange fluorescence 400×). Despite diffuse distribution, bacteria are localized only in parts of the tonsil. The remainder of the tonsillar tissue is free of bacteria.

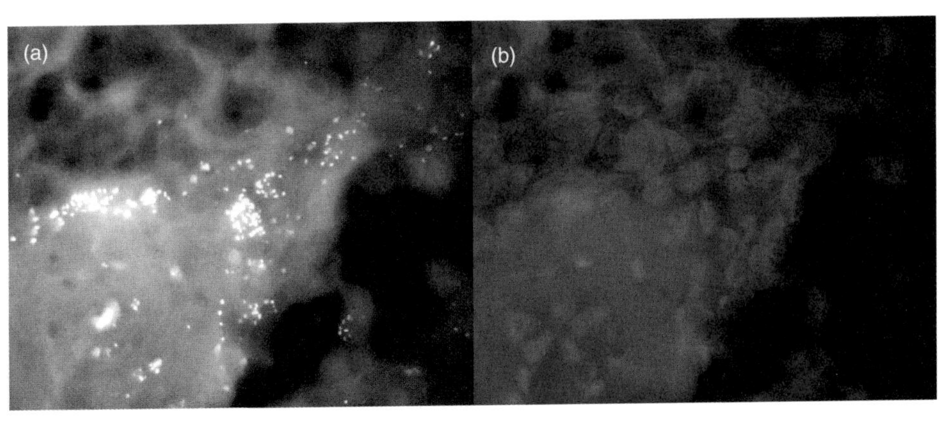

Figure 9.6. (a) Diffuse infiltration with *Streptococcus pyogenes* (Strpyo Cy3 probe, orange fluorescence) and DAPI stain (all DNA structures) overlaid. (b) Isolated DAPI stain of the same microscopic field is presented on the right side to better outline the silhouettes of the eukaryotic cells. Note that bacteria are ordered around the nuclei of eukaryotic cells and probably phagocytized by macrophages.

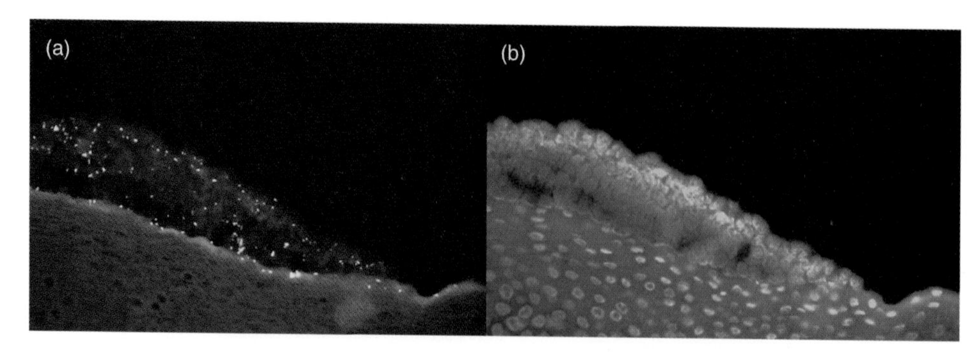

Figure 9.7. (a) Superficial adherence of bacteria to the tonsillar surface (universal Eub338 Cy3 probe, orange florescence). Bacteria are attached to the surface and mixed with an inflammatory infiltrate covering them. (b) DAPI stain of the same microscopic field reveals the DNA structures. Bacteria are enveloped by a lymphatic infiltrate (large blue nuclei of leukocytes).

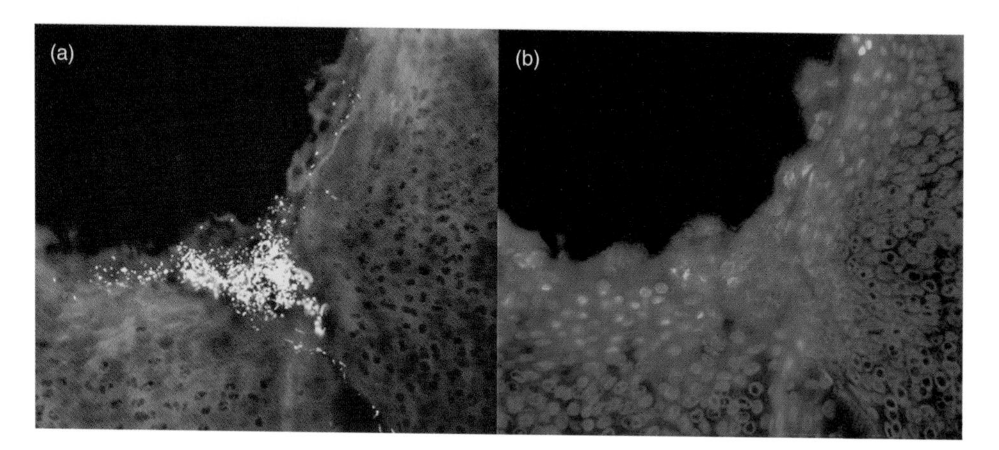

Figure 9.8. (a) Fissure filled with bacteria; (b) the DAPI stain of the same microscopic fields demonstrates that bacteria are surrounded by inflammatory cells indicating an active immunologic response to bacterial adhesion and invasion (Eub338 Cy3 probe, orange fluorescence).

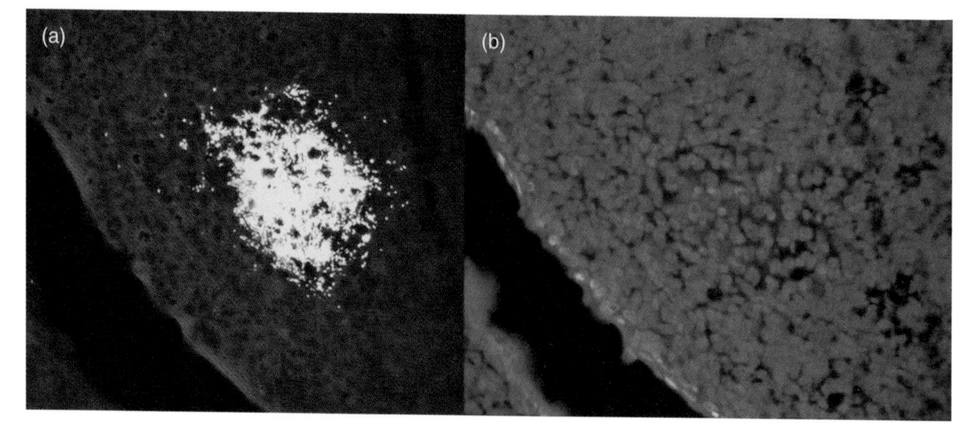

Figure 9.9. Example of microabscess (a) (at 400× magnification) Eub338 Cy3; (b) DAPI stain of the same microscopic field.

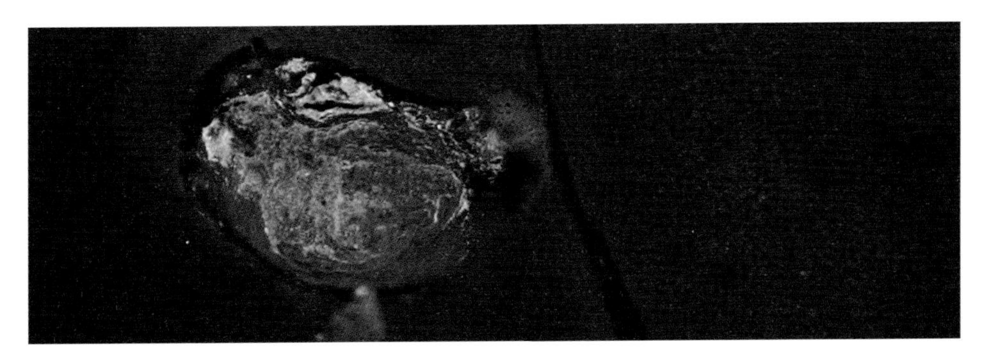

Figure 9.10. Macroabscess (Eub338 Cy3, at 100× magnification) within tonsillar tissue.

TABLE 9.1. Occurrence of Different Bacterial Groups within Local Tonsillar Lesions such as Fissures and Diffuse Infiltrates

Superficial Infiltration and Fissures[a]	%	Diffuse Infiltration[a]	%
Fusobacteria spp. (Fuso)	36	*Firmicutes* (LGC)	74
Pseudomonas (Ps, Pseaer A, Pseaer B)	34	*Streptococcus* (Strc493)	74
Beta-Proteobacteria inclusive; *Neisseria* (Bet42a)	33	*Haemophilus influenzae* (Haeinf)	66
Burkholderia (Burcep, Burkho)	30	*Actinobacteria* (HGC)	50
Lactobacillus and *Enterococcus* (Lab)	24	*Bacteroides/Prevotella* (Bac303)	39
Veillonella group inclusive	23	*Cytophaga-Flavobacteria* (CF319)	34
Veillonella parvula (Veil, Vepa)	20	*Streptococcus pyogenes* (Strpyo)	11
Roseburia (Erec)	11	*Atopobium* and others (Ato291)	6
Staphylococcus aureus (Staaur)	10		
Prevotella intermedia (Prin)	7		
Ruminococcus bromii, R. flavefaciens (Rbro, Rfla)	6		
Coriobacterium group (Cor653)	6		
Listeria,Brochothrix (Lis637,1255)	4		

[a]Bacteria that were observed within regions with diffuse infiltration were also found in fissures but not vice versa.

Stomach and Duodenum

Depending on the digested food, the culture of the gastric or duodenal juices contains bacterial concentrations of 10^3–10^4/mL. For comparison, a 10-µL suspension of bacteria with a concentration of 10^7 cells/mL applied to a glass surface in a 1-cm circle results in 40 cells per average microscopic field at 1000× magnification [3]. No biofilms can be formed at such concentrations.

Fluorescence *in situ* hybridization (FISH) demonstrates that bacteria in the stomach and duodenum are localized strictly within the lumen and separated from the mucosa by a mucus layer. Their composition is heterogeneous, reflecting the heterogeneous composition of the ingested flora. In patients with *Helicobacter pylori* infection, the mucosal surface of the stomach is covered with a bacterial biofilm in which *Helicobacter pylori* is predominant. The data on *H. pylori* biofilms

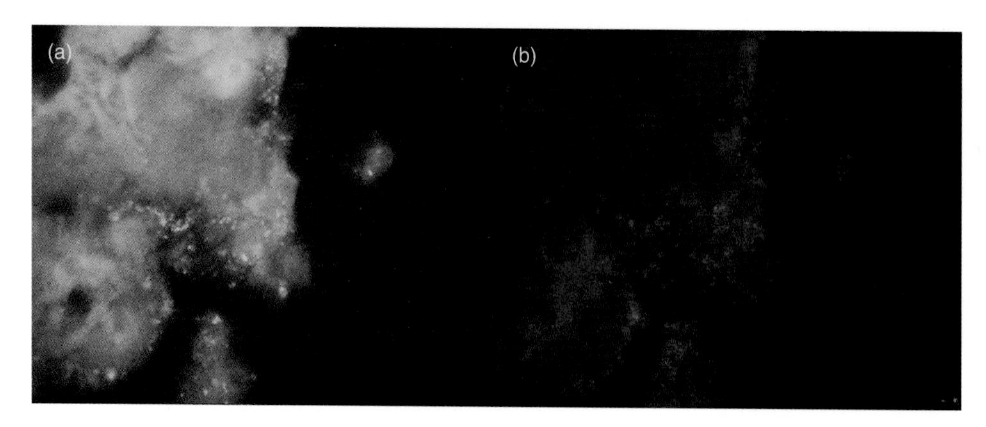

Figure 9.11. Biofilm in a calcified pancreatic duct simultaneously hybridized with (a) group-specific and (b) universal (Eub338 Cy5, red fluorescence) probes at 400× magnification, demonstrating Enterobacteriaceae (Ebac Cy3, orange fluorescence).

are abundant, extensively studied, and reviewed in the literature and will not be specifically referred to here. It is, however, important to mention, that besides *H. pylori* biofilms, other bacterial groups, including *E. coli*, can adhere in a confluent manner to the mucosa, for example, in patients with polyposis of the stomach.

Pancreatic Tract

Healthy pancreatic tissue is not available for investigation. Biopsies taken during endoscopic retrograde cholangiopancreatography (ERCP) are restricted mainly to patients with benign or malignant obstruction. The FISH analysis of microbiota within these samples demonstrates islands of bacterial adhesion in ~70% of the biopsies (Figure 9.11) [4].

The anatomically normal pancreatic duct epithelium is free of bacteria. Bacterial islands are located in regions of disturbed duct anatomy. Bacteria found in the pancreatic duct are diverse, represented mainly by environmental groups, indicating an exogenous origin, and the composition of the epithelium is highly individual. The stability of these findings cannot be verified, because repeated biopsies from these regions were not obtained.

Biliary Tract

Biopsies from the bile ducts are rare. Material of gallbladder resections, electively performed without preoperative antibiotics, is free of bacteria, indicating that the bile duct epithelium is normally uncolonized. The situation is different in the presence of foreign bodies such as biliary stents (Figure 9.12). Colonization of the biliary stents by polymicrobial biofilm can be documented as soon as one week after implantation. The microbial colonization begins with the distal end of the biliary stent and advances proximally. Bacteria are located mainly on the inner surface of the plastic stent. The surface of biliary stents facing normal epithelium is rarely colonized, indicating that the healthy epithelial layer efficiently resists microbial

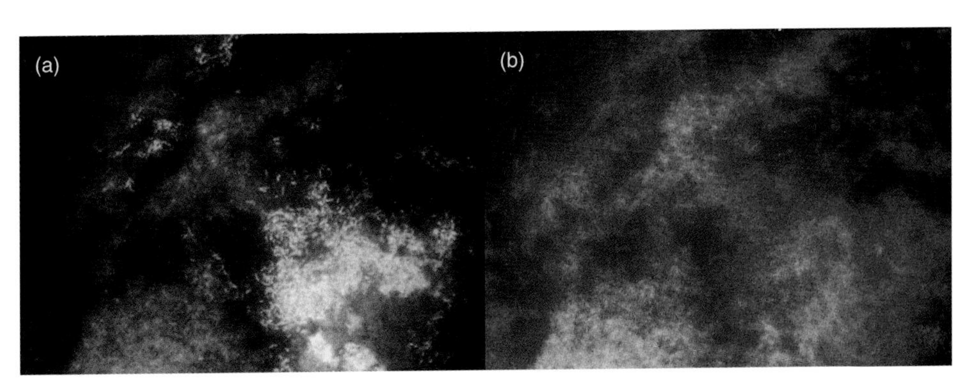

Figure 9.12. Cytophaga *Flavobacterium* group (a, CF319a Cy3, orange fluorescence). Bacteria are of varying composition and have no characteristic form of spatial organization. (b, Eub 338 Cy5 all bacteria, red fluorescence.)

colonization even on contact with the foreign body [4]. The rapid development of biofilms at the inner surface of the biliary stents indicates that the biliary and pancreatic secretions alone are unable to prevent the development of bacterial biofilms on foreign bodies. Both aerobic and anaerobic bacteria, which are commonly found in the intestine, can be identified. Interestingly, the bacterial biofilm disappears as soon as the stent is occluded by sludge (Figure 9.13).

Gallstones

A similar trend of the bacterial biofilm vanishing after local deposition of organic substances can be observed in gallstones. As foreign bodies, gallstones should be permanently colonized. Bacteria can be found in loose brown pigment stones and sludge, which represent an initial stage in the formation of gallstones. The natural history of the gallstone is a procession from brown to composite, and then to cholesterol gallstone. The cholesterol stones may reach considerable sizes and persist in the human body over many decades. However, despite such long history and multiple episodes of acute bacterial cholecystitis, the cholesterol gallstones obtained after cholecystectomy are mostly sterile. Obviously, the sedimentation of cholesterol and sludge within the bacterial biofilm is an integral part of some kind of protective mechanism against otherwise extremely recalcitrant infections [4]. This mechanism appears to be extremely efficient in suppressing bacterial biofilms on foreign bodies when compared to the complete ineffectiveness of presently available antibiotics.

9.2.2. Small Intestine

The epithelial surface of the small intestine in healthy humans is not colonized (Figure 9.14). The bacteria are represented by occasional groups and can be found in low concentrations of 10^5 or less within the lumen. Bacteria do not form conglomerates and spatial structures, and the luminal contents are separated from the mucosa by a mucus layer. The situation is similar in mice but not in rats. In most groups of wildtype rats, segmented filamentous bacteria (SFB) can be observed tightly attached to the epithelial surface and located between villi

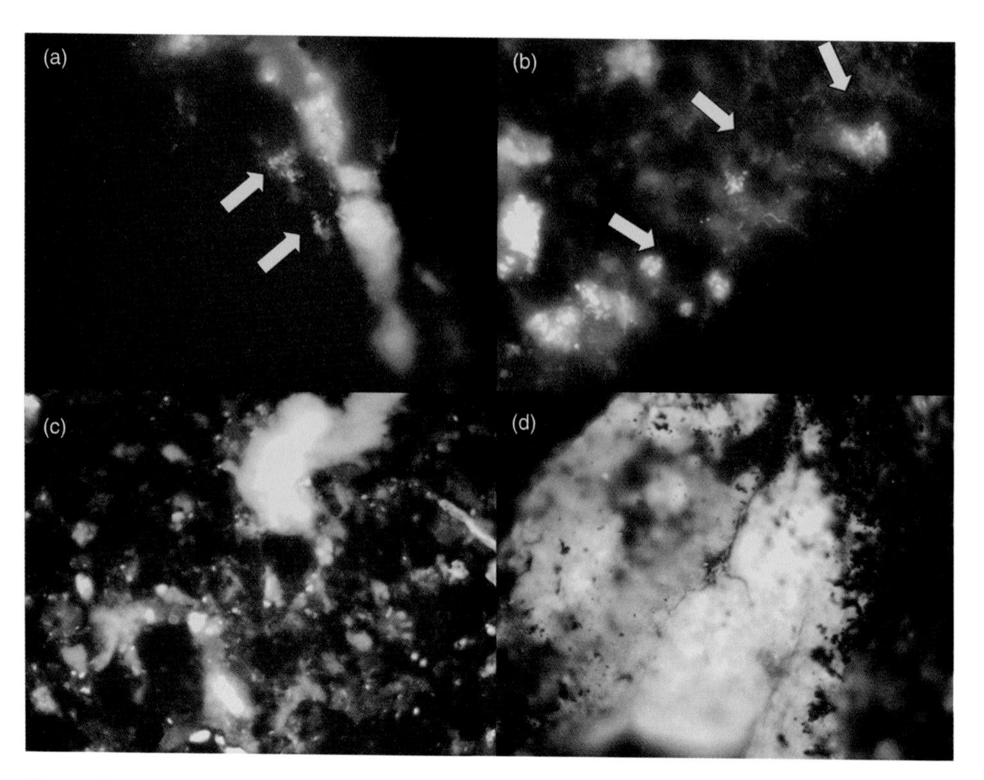

Figure 9.13. Sections of whole biliary stents hybridized with the universal Eub338 Cy3 probe. Bacteria (orange signal can be clearly identified as punctuate signals within the yellowish sludge covering the inner surface of stents (a,b) in partially occluded stents. Bacteria disappear from the stent lumen with increasing occlusion of the bile duct stent [(c) stent center; (d) stent wall]; only large yellowish masses of cholesterol and other sludge can be seen.

throughout the small intestine (Figure 9.15). The SFB adherence in rats is not accompanied by an increase in concentrations of other bacterial groups or leukocytes. It is difficult to ascertain whether segmented filamentous bacteria are pathogens such as *Helicobacter* species, saprophytic or symbiotic bacteria with an unknown role, because of the very high prevalence of adherent SFB in the small intestines of wildtype rats. In humans, the SFB adherence could not be observed in either health or disease.

Pathologic conditions with altered microbiota in the small intestine are acute and chronic infections, bacterial overgrowth, and inflammatory bowel disease (IBD). Typical for all of them are a constituent mucus barrier, loss of bacterial separation between mucosa and lumen, bacterial adherence, invasion, and translocation (see Section 9.3.1) [5].

9.2.3. Bacteria in the Large Intestine

In the colon, bacteria reach concentrations of 10^{11-12}/mL and constitute up to 90% of the fecal mass. Thus high bacterial concentrations can be achieved only under active facilitation of bacterial growth. *Roseburia* spp., *Faecalibacterium prausnitzii*,

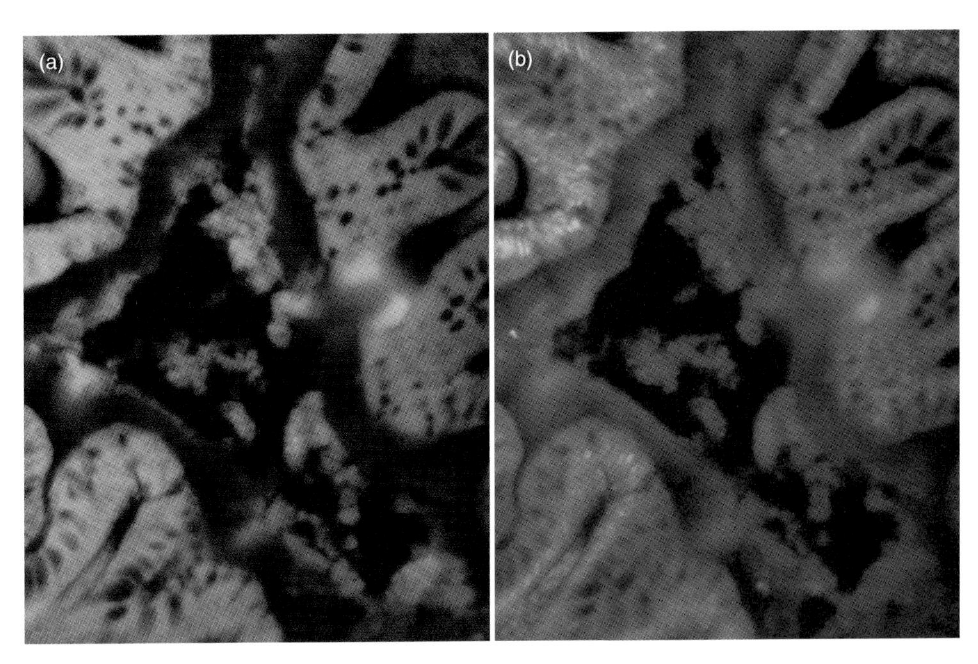

Figure 9.14. Ileum of a healthy person. (a) Hybridization with the universal Eub338 Cy3 probe at 400× magnification. (b) The same microscopic field demonstrating DAPI fluorescence of all DNA structures at 400×. No bacteria can be seen between crypts. The lumen is completely separated from the mucosa by a mucus layer, which is free of bacteria.

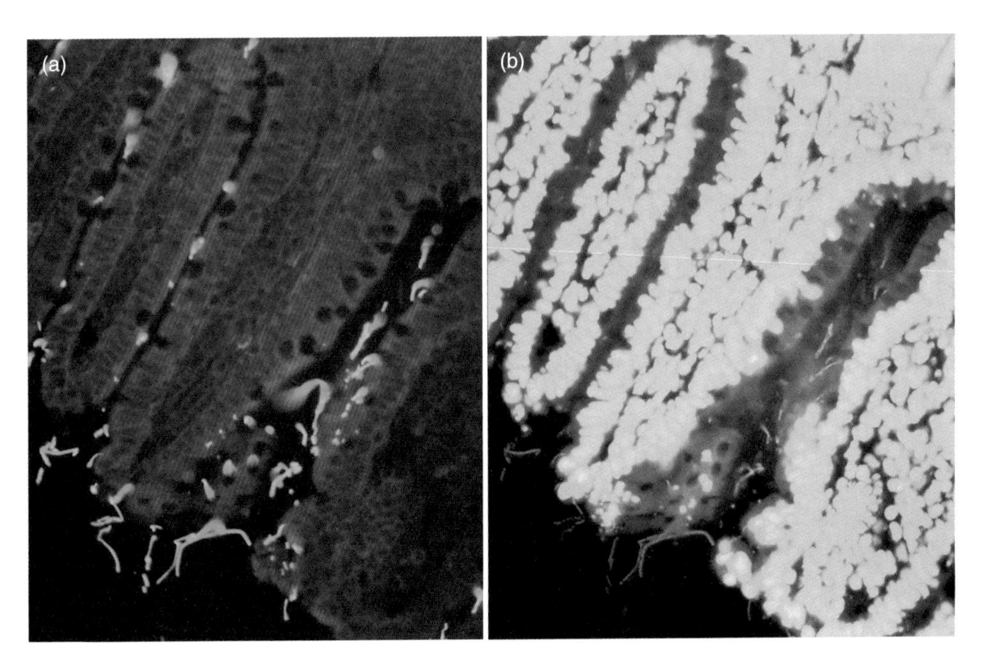

Figure 9.15. Rat small intestine. (a) Hybridization with the SFB FITC probe (segmented filamentous bacteria). (b) The same microscopic field in DAPI fluorescence. Long curly bacteria adhere to the villi and enter deep into the crypts. No other bacterial groups are seen.

and *Bacteroides* groups each represent 10–30% and cumulative 70% of the total microbiota in humans [6,7].

9.2.4. The Role of Microbiota in Colonic Function

The large intestine is a biofermenter, in which the host employs bacteria to degrade undigested leftovers. Bacteria produce valuable substances such as vitamins and short fatty acids by degrading waste products. To recover these nutrients in the small intestine, many species adopt fecophagy. The considerable differences in the size and anatomy of the large intestine observed between mammals indicate that the processes of utilization may differ and that other, nonfecophagic, mechanisms may be available in some species, which allow them to directly absorb and utilize beneficial products of bacterial metabolism.

The global impact of single bacterial groups for the biofermentation process is not known. We can, however, assume that numerically predominant and obligatory present bacteria are indispensable for the biochemical processes that occur in the colon. *Roseburia* spp., *Faecalibacterium prausnitzii*, and *Bacteroides* groups each represent 10–30% and cumulative 70% of the total microbiota in humans [6,7]. Obviously these groups are important for the function of the colonic biofermenter. All other bacterial groups are present only in subgroups of patients or parts of the colon.

Although the fecal flora is one of the most well-characterized microbiota with culture and molecular–genetic methods, many of the bacterial species that inhabit the large intestine are still unknown. Strict anaerobic species are predominant. The diversity of bacteria is high, consisting of approximately 3000 to 5000 species [8].

Peristalsis extensively mixes bacteria with fibers and maintains optimal viscosity, and temperature is one element of this process. The lack of any perceptible suppression by the host is, however, the most astonishing feature. It was previously assumed that the enormous masses of intestinal bacteria directly contact the intestinal wall. The non-pathogenic bacteria are tolerated, while the pathogenic bacteria are responded to. The immune response determines which bacteria should be normally present in the colon and which not. In reality, the residents of the large bowel cannot be this clearly divided into good and evil. Many indigenous bacteria are pathogenic: *Escherichia coli* cause sepsis, *Bacteroides* cause abscesses, *Enterococci* cause endocarditis, and *Clostridium perfringens* cause gas gangrene. These bacterial groups are assumed to be normal inhabitants of the human colon since they can be found in every healthy person. However, these bacteria are not "healthy" in any way. The host is able to distinguish between nonpathogenic and saprophytic bacteria, and the pathways of such recognition are well explored. Mechanisms that could eliminate single bacterial groups from the highly concentrated mass of fecal microbiota without affecting "beneficial" bacteria have yet to be demonstrated. FISH analysis of the mucosal flora clearly demonstrates that the host does not tolerate the indigenous microbiota; rather, it separates them from contact with the mucosa.

9.2.5. Mucus Barrier

Biopsies from healthy human subjects show that the walls are covered with mucus, which is free of bacteria throughout the colon (Figures 9.16, 9.17, and 9.18) and ileum

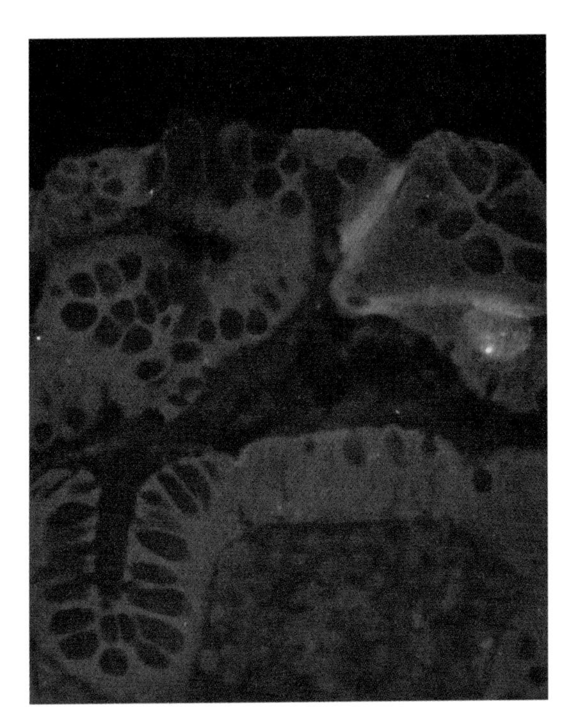

Figure 9.16. Biopsy of ascending colon of a healthy person hybridized with the universal Eub338 Cy3 probe at 400× magnification. The surface of the mucosa is covered with mucus that completely omits bacteria.

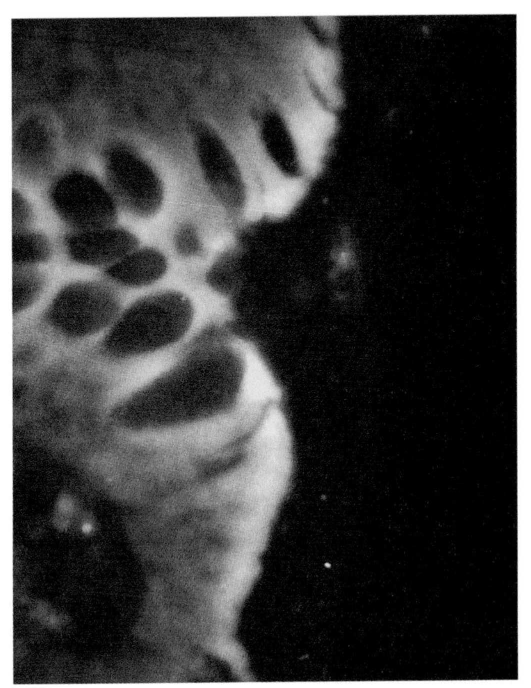

Figure 9.17. Biopsy of descending colon of same subject and under same conditions as in Figure 9.16 but at 1000× magnification.

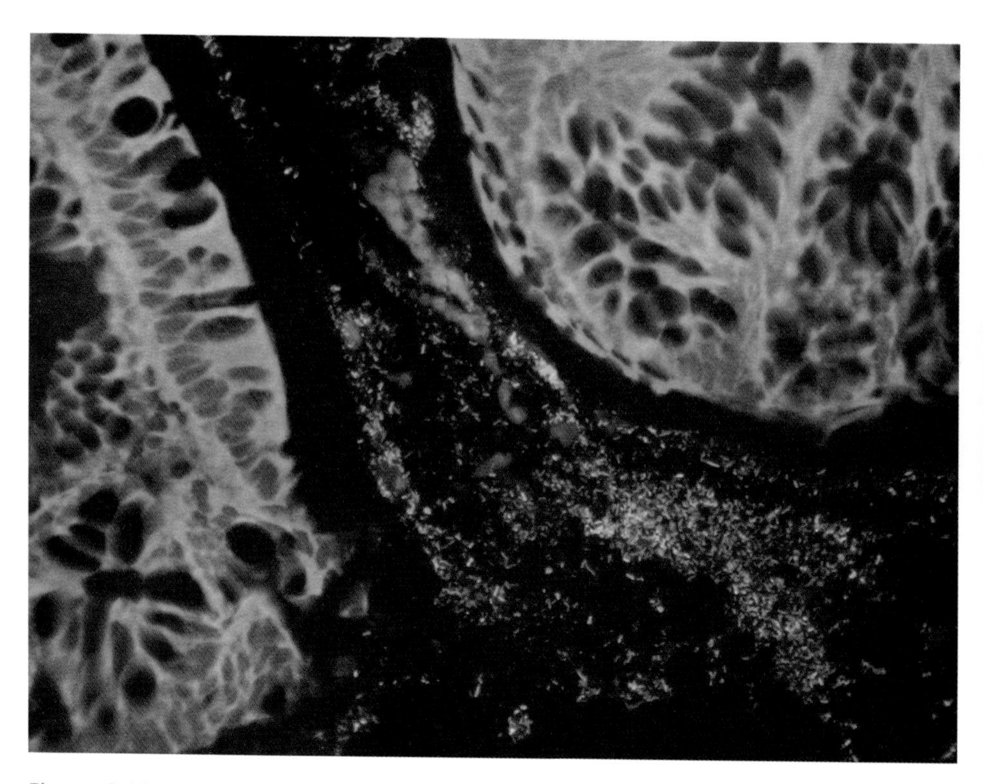

Figure 9.18. Cecal biopsy taken from a patient who was inadequately purged. Remnants of feces remained in the colon. No bacteria adhere to or contact the mucosa despite considerable amounts of intraluminal bacteria (multicolor FISH) (at 400× magnification). *Bacteroides* is shown in orange, *Roseburia*, green; and *Faecalibacterium prausnitzii*, red.)

(Figure 9.14). The separation of fecal bacteria from the mucosa by mucus can be seen in sections of normal appendices, resected for suspected acute appendicitis but found to be normal (Figure 9.19).

A similar separation of colonic bacteria from the mucosa can be observed in the distal colon of rodent animals (Figure 9.20), where the intestine is filled with a fecal mass that can be investigated in total.

Viscosity of the Mucus Barrier

Different from humans, in the proximal colon of mice and rats, bacteria contact the colonic wall (Figure 9.21) and enter crypts in high concentrations [9]. However, contact of bacteria with the mucosa in the proximal colon of mice is selective. Long curly rods of *Roseburia* contact the mucosa and enter crypts in large numbers, while short coccoid rods of *Bacteroides* are separated from the colonic wall. The differences in arrangement of bacterial groups are especially obvious in multicolor FISH that shows simultaneously different bacterial species (Figures 9.22 and 9.23).

Analysis of bacterial groups in the murine proximal colon reveals that only long rods with a curly form contact the mucosa (*Roseburia*, *Bifidobacteriaceae*,

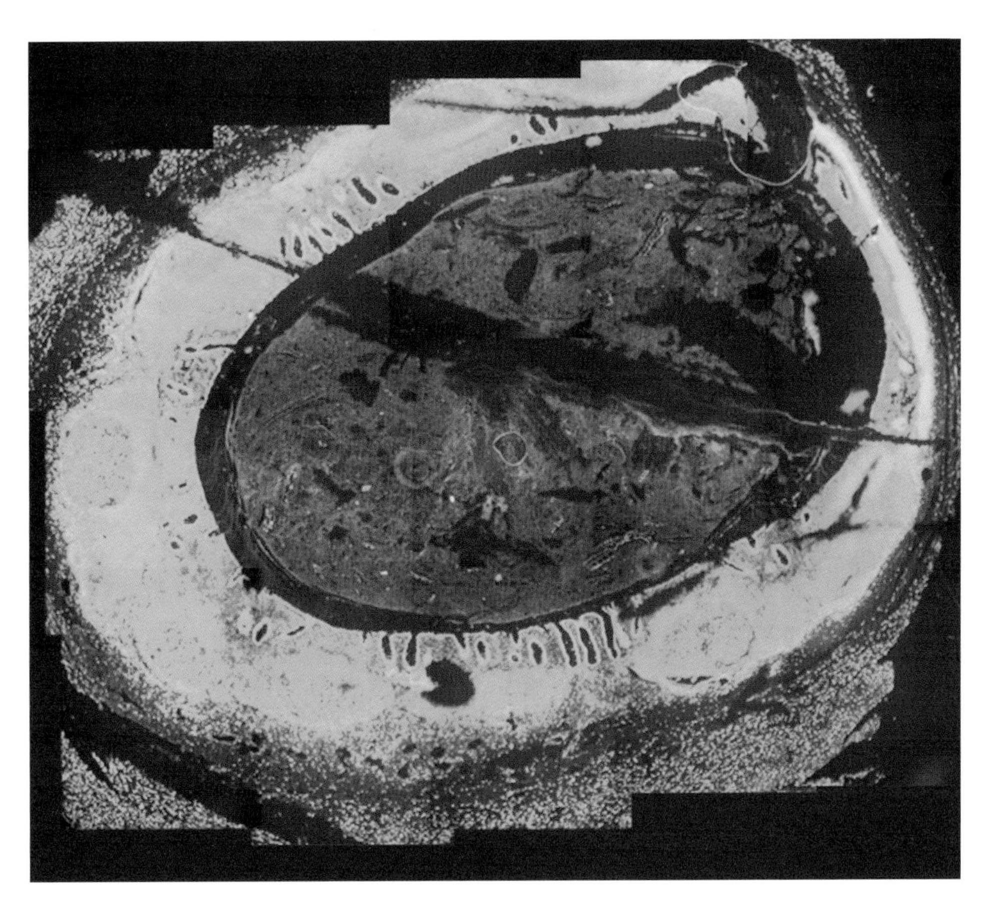

Figure 9.19. DAPI stain of a transverse appendix section taken at 100× magnification. No leukocytes can be seen within the appendix, confirming that the appendix is not inflamed. The bacteria within the lumen are homogeneously distributed over the appendix lumen and clearly demarcated from the mucosa by a mucus layer. The figure is a composite of multiple photographs taken at low magnification.

Lactobacillius). Short rods and coccoid bacteria such as *Bacteroides*, *Enterobacteriaceae*, *Clostridium difficile*, *Veillonella*, and other groups are separated from the mucosa. The difference in the spatial distribution of differently shaped bacteria in the proximal colon of mice indicates that the mucus layer in the proximal colon of rodents is also present; however, because of a lower viscosity it is penetrable for long corkscrew-shaped bacteria but not for short coccoid rods.

The bacterial shape is important for bacterial movement. Short rods are equipped with multiple pili. Pili enable movements in a watery environment but not in slime. Short rods may have flagella, which, through propeller-like action, move them through slime. Long curly rods use complex body movements that penetrate in screw-like motion through gels of high viscosity, but are immobile in water.

Investigations of the velocity of differently shaped bacteria in simulated mucus with variable viscosity indicate that the coccoid *Bacteroides* species have the highest

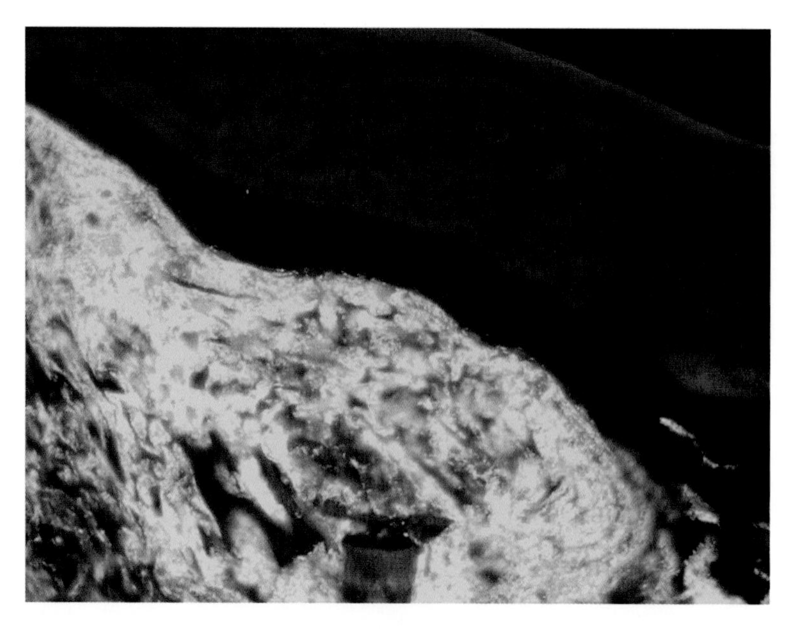

Figure 9.20. Distal colon of a healthy wildtype mouse, universal Eub338 Cy3 probe, at 400×. Bacteria (orange fluorescence) are clearly separated from the colonic wall by a mucus layer. Bacterial concentrations within the colon are high and homogeneously distributed between central regions of the intestinal lumen and mucosa adjacent regions. The water from the fecal stream softens the mucus layer. Pieces of the maturated mucus are carried away by the fecal stream. However, the mucus layer serves as an impenetrable wall between mucosa and feces.

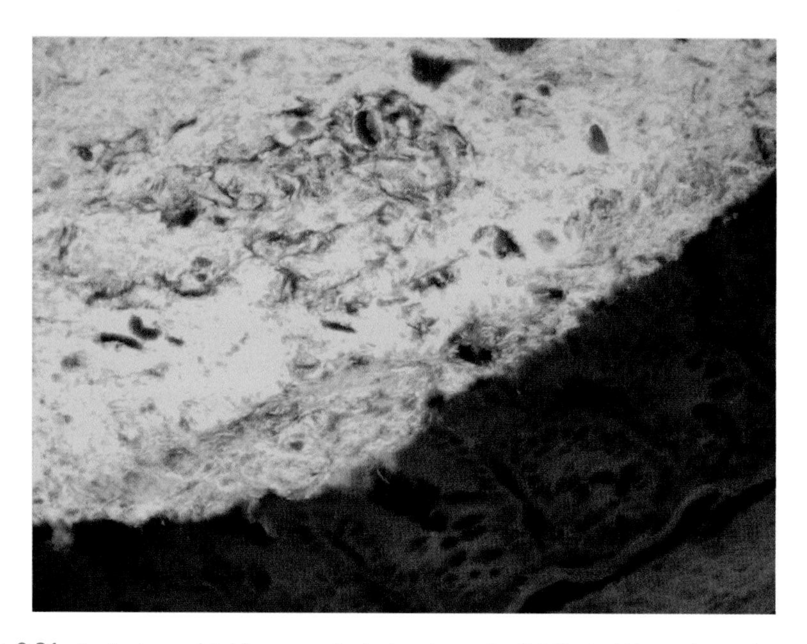

Figure 9.21. Bacteria are highly concentrated and evenly distributed throughout the proximal colon in a healthy mouse (Eub338 Cy3, all bacteria, orange). Bacteria contact the colonic wall and enter crypts in the proximal colon of mice; this pattern is different in the distal colon. There is no reduction in fluorescence of the bacteria contacting the colonic wall, indicating an absence of any suppressive substances produced by the colonic wall.

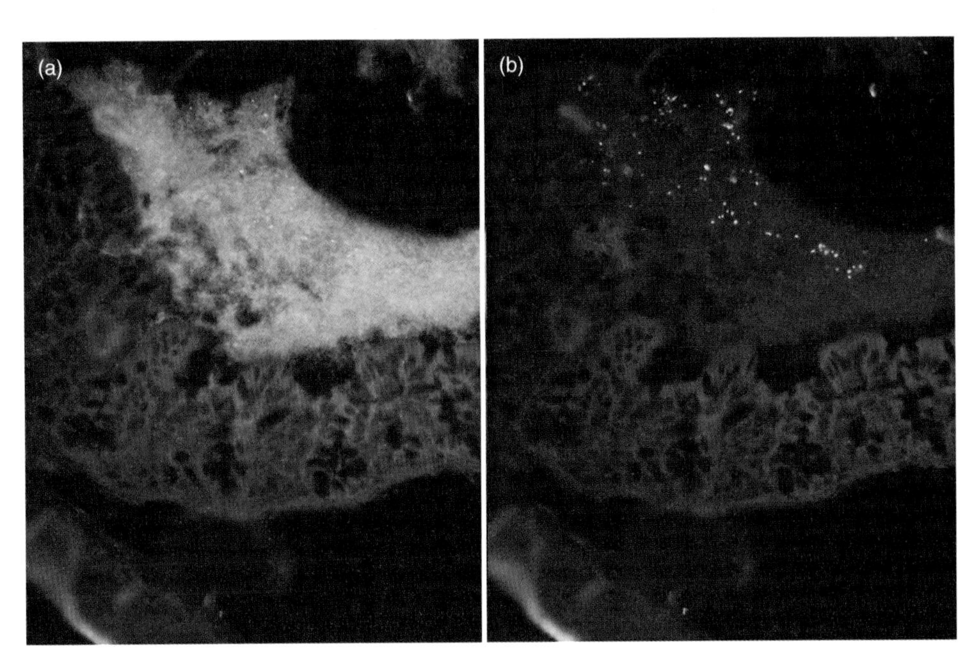

Figure 9.22. Proximal mouse colon. (a) All groups of bacteria are overlaid (*Roseburia* is red, *Bacteroides* is orange, all other bacteria are green). (b) Only *Bacteroides* within the same microscopic field.

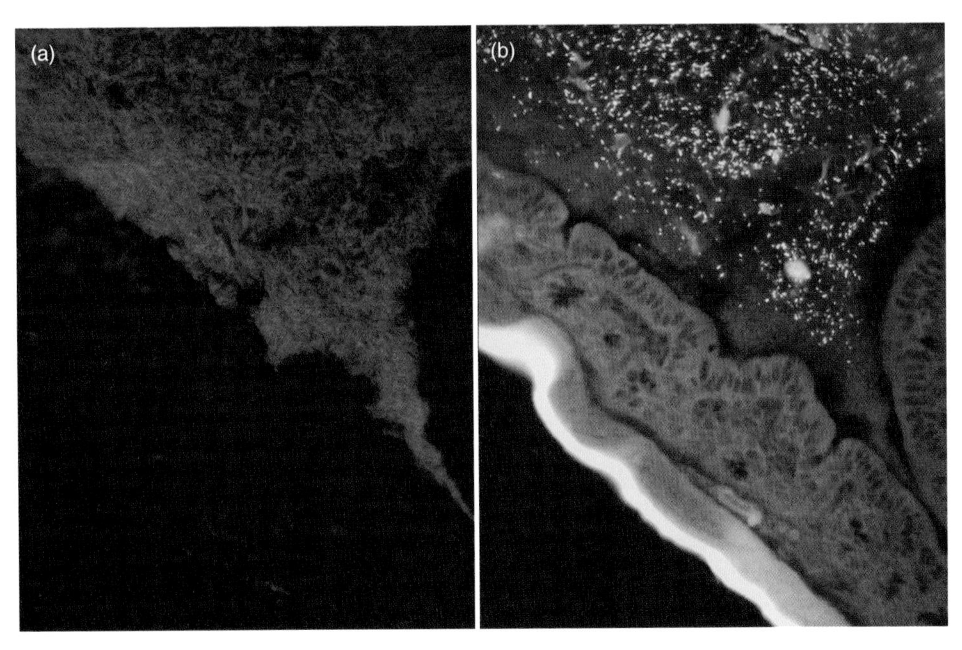

Figure 9.23. Proximal mouse colon. (a) *Roseburia*, (b) *Bacteroides* within the same microscopic field. While long rod-shaped bacteria including the *Roseburia* group contact the colonic wall and enter crypts, the short coccoid rods of *Bacteroides* are completely separated from the mucosa in the same location and do not enter crypts.

velocity at viscosity corresponding to 0.2% agarose and are immobilized at 0.4% agarose, while the long curly rods of the *Roseburia* group have the highest velocity at viscosity of 0.5% agarose. All bacterial movements stop at viscosity of 0.7% agarose [10].

The distribution of long curly bacteria or short coccoid rods within mucus can be used to mark the areas of changing viscosity. In healthy humans, the separation of bacteria from the mucosa is equally perfect in the proximal and distal colon, and bacteria are never found in crypts. In healthy rodents the viscosity of the mucus in the proximal colon is markedly lower than in the distal colon, allowing bacteria with a long curly rod shape to reach and contact the mucosa. The reason for the variation in mucus-layer viscosity of the proximal colon in humans and rodents is unclear, and the adherence of selected bacterial groups here may have some evolutionary advantages. Another possibility is that the region between crypts in the proximal colon of rodents builds the germinal zone for the colonic biofermentor (see Section 9.5). The lumen of the human large intestine is much larger, and the germinal zone is located intraluminally.

The presence of the mucus barrier in the proximal colon of mice can be clearly demonstrated in germ-free mice mono-associated with *Enterobacter cloacae*—a bacterium with a short coccoid form. The distinct mucus layer and separation of bacteria from the colonic wall can be observed in both the distal and proximal colon segments of these mice. Bacteria are perfectly separated in the distal colon (Figure 9.24); however, in the proximal colon some bacteria can be found inside isolated vacuoles of the goblet cells, especially at the bottom of crypts (Figure 9.25).

The undifferentiated epithelial cells at the base of the crypts are primarily mucus-secreting cells, whereas differentiated cells of the columnar epithelium are

Figure 9.24. Distal colon of a mouse monoassociated with *Enterobacter cloacae*, a short coccoid rod. Bacteria are clearly separated from the colonic wall by a mucus layer.

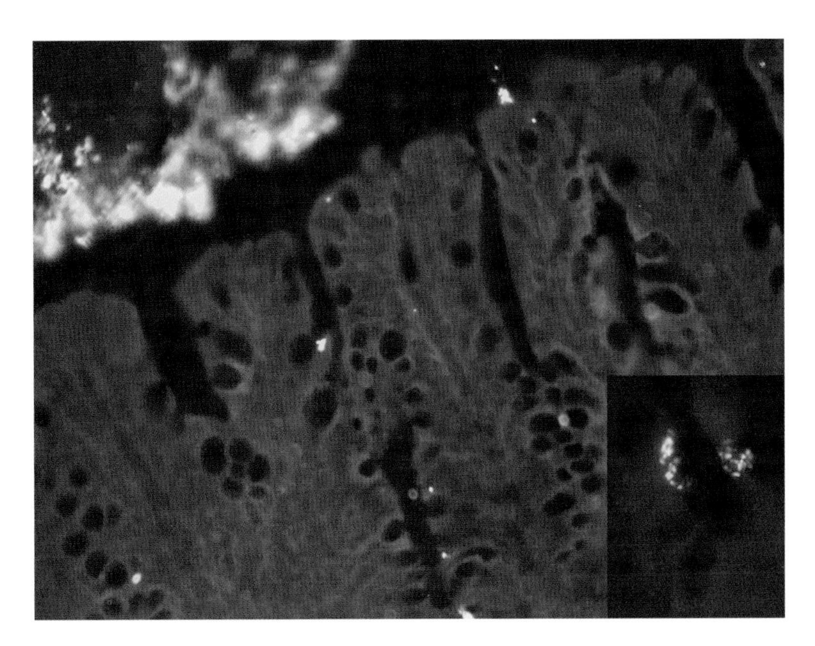

Figure 9.25. Proximal colon of a mouse monoassociated with *Enterobacter cloacae* (orange fluorescence 1000×). The mucus layer is clearly visible. Bacteria can be found in regions of lower mucus viscosity at the bottom of crypts and within vacuoles of goblet cells. The inset shows bacteria in a crypt.

mainly absorptive cells, removing water and electrolytes from the mucus [10]. The epithelial stem cells at the crypt base proliferate and replace surface cells within 4–8 days. The dissemination of *E. cloacae* in crypt bases and goblet cells outlines zones of lower viscosity and confirms independently that during the journey from the crypt base toward the surface epithelium, crypt cells become increasingly differentiated and absorptive.

The adsorptive cells of the crypt neck and of the epithelial cells of the columnar epithelium dehydrate the mucus layer. Dehydration renders the mucus layer solid and impenetrable for bacteria and protects sites of mucus production and the mucosa from encounters with potential pathogens. The lower viscosity of the mucus at the crypt base promotes emptying of crypts and prevents obstruction, but as a drawback it may render these types of cells more vulnerable to invasion by potential pathogens. Indeed, invasion of epithelial cells by *E. cloacae* was observed exclusively at the crypt bottom, whereas no *E. cloacae*-containing cells were observed within the cytoplasm of the columnar epithelial cells in monoassociated mice.

Interestingly, crypt abscesses, which are typical histomorphologic findings in human self-limiting colitis and IBD, are also more abundant in crypt bases.

The changes in viscosity of the mucus layer are controlled by two opposing processes: (1) water resorption by the columnar epithelial cells solidifies the mucus at the mucosal site, and (2) the diffusion of water from the fecal stream thins (dilutes) the mucus from the luminal side. The fecal stream carries away the maturated mucus and incorporates it into the fecal mass (Figure 9.20). The mucus layer

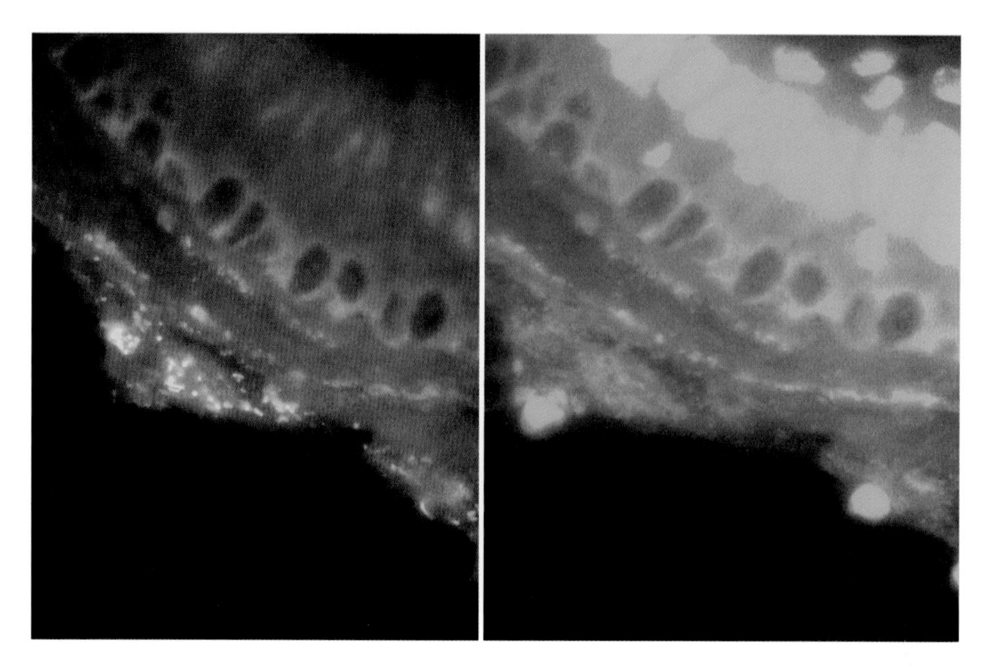

Figure 9.26. Cecal biopsy from a patient with irritable bowel syndrome. Bacteria are hybridized with the universal probe Eub338 Cy3, orange (on left) and counterstained with DAPI (on right). The waves of mucus with incorporated and immobilized bacteria are moving from the mucosa to the luminal side. Goblet cells discharge their contents below these coats of solidified mucus, leading to an onion-like structure of mucus.

is continuously renewed. The viscosity of the mucus secreted by the goblet cells is significantly lower than the viscosity of the dehydrated mucus film, which is attached to the columnar epithelium. The secreted mucus cannot merge with the dehydrated mucus because of differences in consistency. Instead, it heaves up the semisolid dehydrated mucus and spreads below it. This mode of replacement protects freshly secreted mucus from bacterial penetration until it is, in turn, solidified by water resorption and can serve as an impenetrable cover. The waves of secretions, displacement, and solidification lead to an onion-like stratification of the mucus. This stratification can be seen in alcian stain, or even better with FISH in patients with irritable bowel syndrome (IBS), where the zones of massive secretion and imperfect solidification are visualized by bacteria that are entrapped within consecutively displaced and alternating mucus films of moderate and high viscosity (Figure 9.26).

9.2.6. Fecal Microbiota

Biostructure of Fecal Microbiota

The native structure of fecal microbiota within intact intestine cannot be directly investigated in humans, which is not the case in animal experiments. Intestine filled with feces cannot be obtained from healthy persons, except on rare occasions, from

Figure 9.27. Alcian stain of the fecal cylinder from a healthy person. The mucus layer covers the spontaneously defecated formed stool. Note that no bacteria are in the mucus.

appendectomy and elective surgical resections ideally performed without prior use of antibiotics. An alternative approach is to study normal stool samples, since the outer regions of feces represent the luminal surface of the mucus layer and since the structural organization of defecated feces does not differ from that of feces located within the intestine. In analogy to core boring used for investigation of geologic formations, the spatial structure of fecal microbiota can be investigated on sections of punched-out fecal cylinders, which are then fixated and embedded in paraffin [11,12].

In healthy humans, the surface of formed stools is covered with a mucus layer that is similar to the mucus covering the mucosal surface of biopsies (Figure 9.27). Fecal microbiota in healthy humans can be divided into habitual bacterial groups present in all subjects (Figure 9.28) and occasional bacteria, which are present only in subgroups of subjects, either diffusely or locally distributed (Figure 9.29).

Investigation of 86 different bacterial groups demonstrated that *Roseburia* spp., *Bacteroides*, and *Faecalibacterium prausnitzii* groups are habitual and constitute each 20–50% of the fecal flora and together ≥70% of all bacteria present in feces. All other bacterial groups occur only in a subset of patients. The exact incidence and concentrations have been previously reported [11,13]. With regard to the mucus layer, bacteria can be divided in fecomucous, mucophob, and mucotrop (Figures 9.30 and 9.31). All habitual bacteria are fecomucous. Their highest concentrations are within feces; however, they also enter mucus. Their concentrations diminish with increasing distance from the fecal surface. The mucophobe bacteria, a typical representative of which is the *Bifidobacteriaceae* group, avoid mucus. Mucotrope bacteria such as Enterobacteriaceae and Verucomicriaceae are located on the border between feces and mucus.

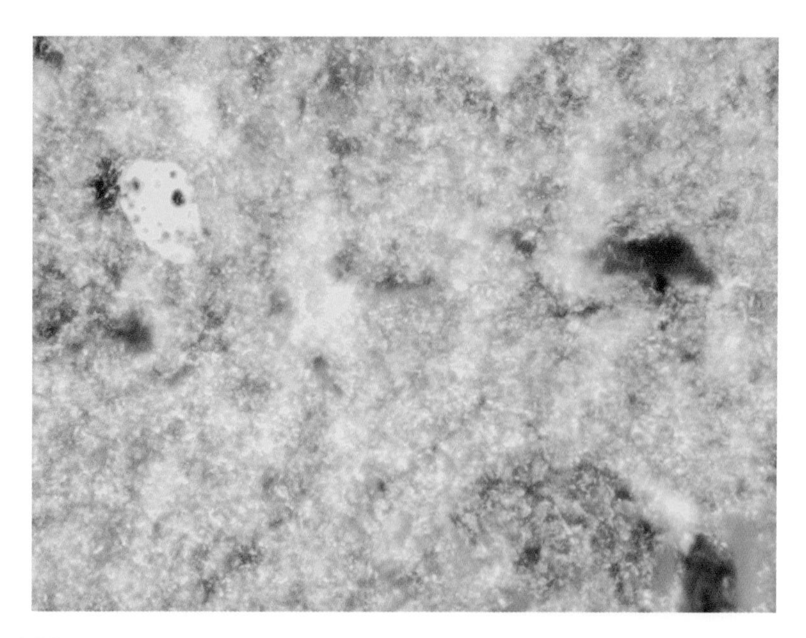

Figure 9.28. Habitual bacterial groups are normally present in each healthy person, homogenously distributed throughout the fecal cylinder and each represent 20–50% of the bacterial biomass. Habitual bacterial groups are represented in humans by *Roseburia* (orange fluorescence, at 400×), *Bacteroides*, and *Faecalibacterium prausnitzii*.

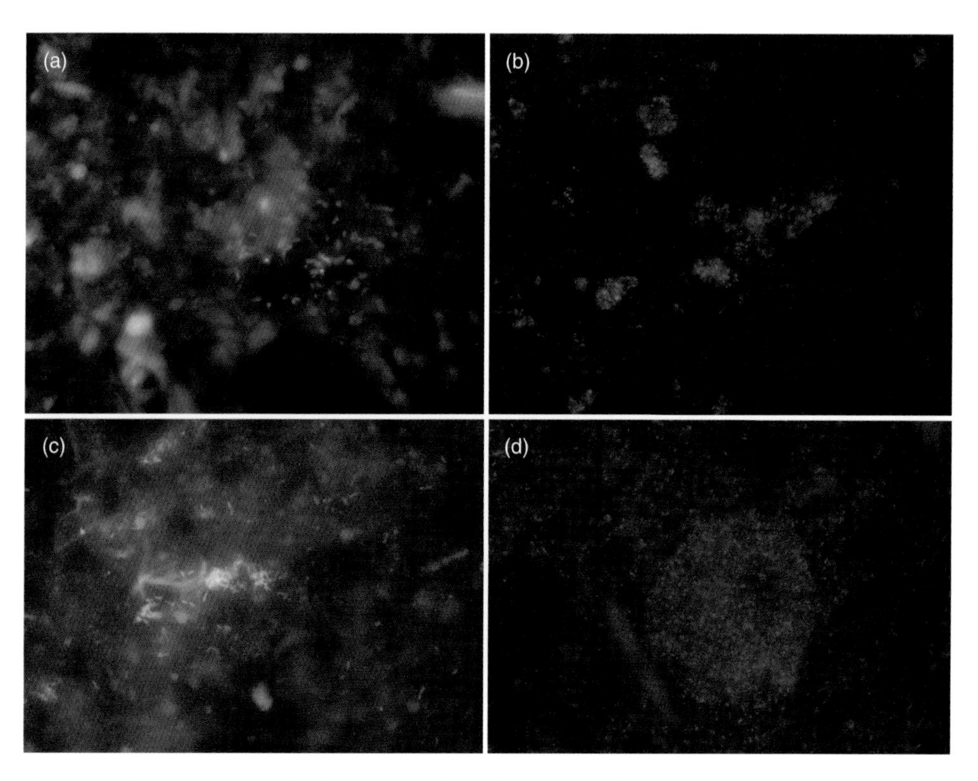

Figure 9.29. Examples of distribution (which can be either diffuse or local of occasional bacteria. Occasional bacterial groups can be detected by FISH only in some persons: (a) *Clostridium histolyticum* (orange, at 1000′); (b) Bifidobacteriacae (red fluorescence, at 400×); (c) *Eubacterium hallii* (orange, at 400×); (d) Enterobacteriaceae (red, at 400×).

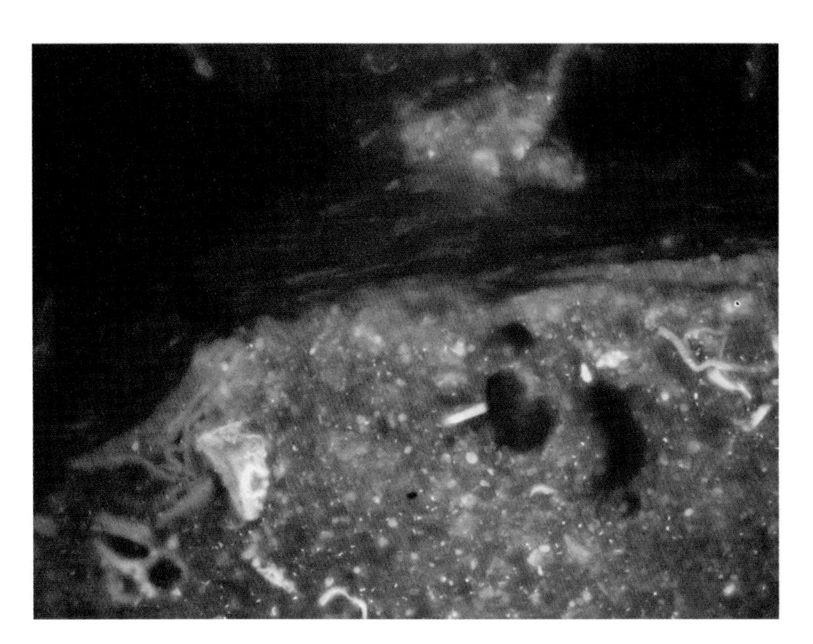

Figure 9.30. Example of mucophobic bacteria. Bifidobacteriaceae are shown avoiding mucus (orange fluorescence, at 400×).

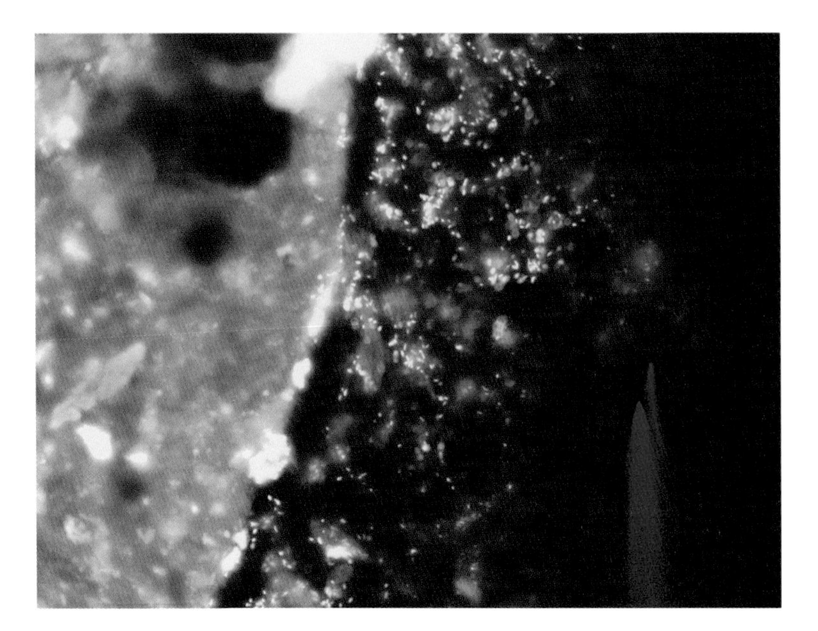

Figure 9.31. Example of mucotropic bacterial groups. Enterobacteriaceae (orange fluorescence, at 400×). Mucotropic bacterial groups including Verucobacteriaceae are preferably located in the region between feces and mucus. All mucotropic bacteria occur only occasionally. Verucobacteriaceae appear to be associated with diarrhea, but can also be found in asymptomatic healthy persons.

The biostructure and composition of bacterial groups is individual in each case. Daily investigation of stools demonstrated a relative stability of the fecal microbiota. However, weekly investigations continued over >6 months have shown that the individual composition of the microbiota change significantly with time.

9.3. CHANGES OF THE COLONIC MICROBIOTA IN DISEASE

9.3.1. Break of the Mucus Barrier in Inflammatory Bowel Disease

The most prominent feature of intestinal inflammation is a break of the mucus barrier (Figures 9.32 and 9.33) with subsequent migration of intestinal bacteria towards the mucosa, adhesion, and cytopathologic effects (Figure 9.34). Bacteria adhere to epithelial cells and build dense adherent layers. Despite massive adhesion, the epithelial barrier holds up against the bacterial invasion in most cases. Finding bacteria in epithelial cells or within submucosal regions is an exception. Even in severe inflammation, multiple sections of the same biopsy have to be investigated before single intraepithelial bacterial inclusions can be detected (Figure 9.35). They are located mainly at the bottom of the crypts, which often remain devoid of bacteria. They are not present in the columnar epithelium, which has direct contact with

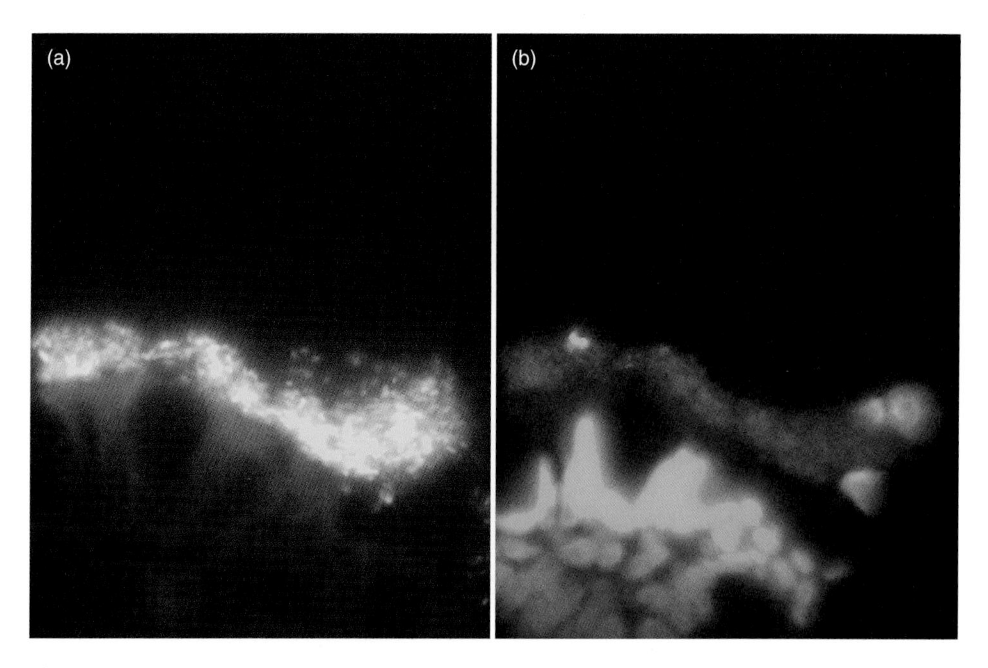

Figure 9.32. Prolific bacterial biofilm covering the colonic mucosa in a patient with Crohn's disease. (a) Multicolor FISH. (b) DAPI stain of DNA structures. *Bacteroides* is shown in orange; *Roseburia* and *Faecalibacterium prausnitzii*, both in red; and all other groups, in green. Despite close attachment to the mucosa, no bacteria can be found intracellularly in the columnar epithelium or submucosa. In DAPI stain only a few leukocytes are seen in the biofilm. The patient was treated with azathioprine. Prolific biofilms without leukocyte response are typical for patients with inflammatory bowel disease (IBD) treated with azathioprine.

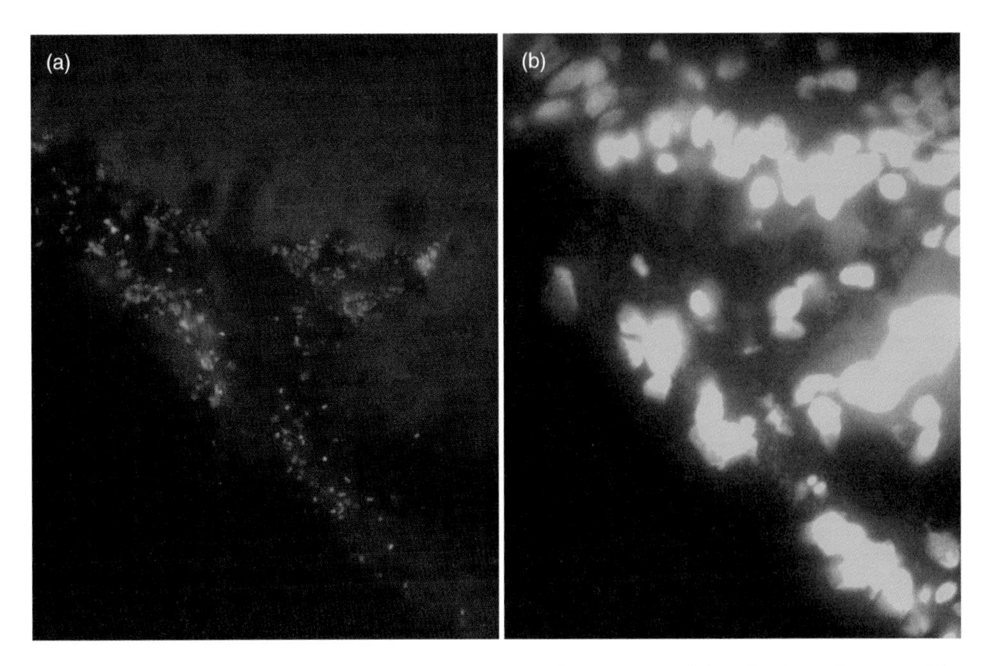

Figure 9.33. Ascending colon of a patient with Crohn's disease. (a) Adhesion of *Bacteroides* (orange) to the epithelial surface and the entrance of crypts. (b) DAPI stain demonstrates the response of the leukocytes (large blue nuclei) that migrate into the mucus of crypts and the surface.

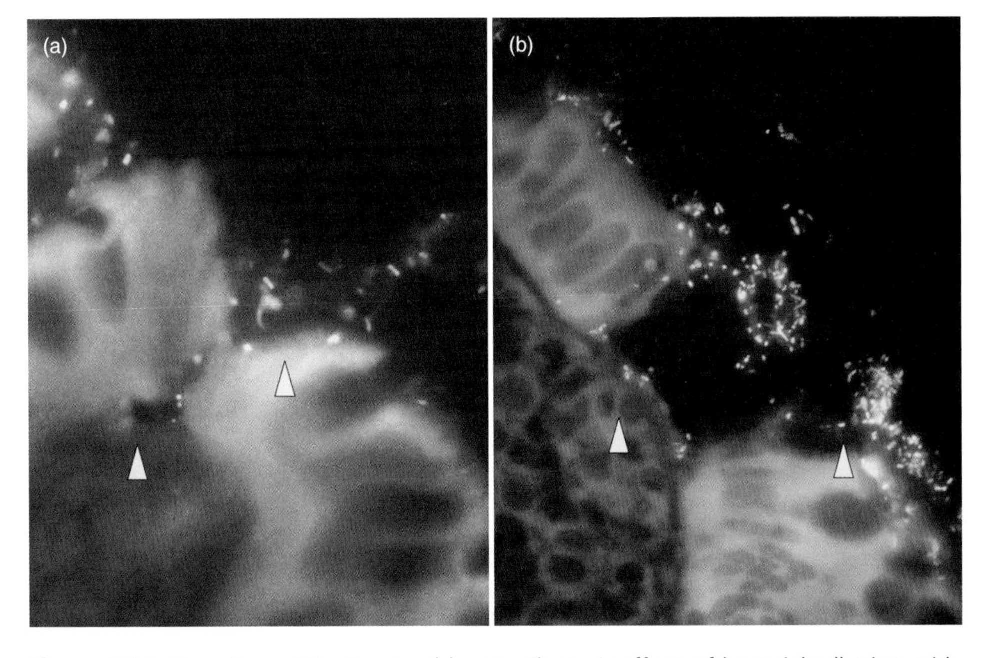

Figure 9.34. Ulcerative colitis, showing (a) cytopathogenic effects of bacterial adhesion, with (b) defects (ulcerations indicated by arrows) of the epithelial layer and migration of bacteria into the submucosa. Bacteria are hybridized with the universal probe (Eub338 Cy3, orange fluorescence, at 1000′). Despite the severe inflammation, the number of bacteria is reduced compared to regions without marked inflammation as shown in Figure 9.32.

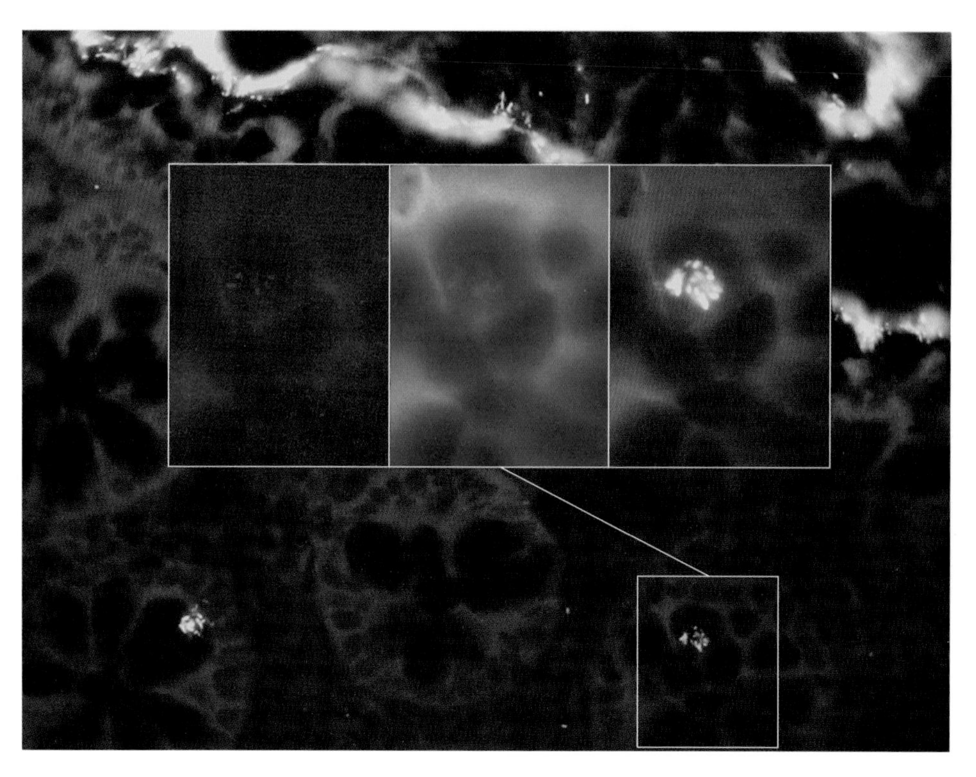

Figure 9.35. Bacterial inclusions in the vacuoles of goblet cells in a patient with Crohn's disease, shown at 400× magnification (inserts at 1000×). All bacteria are orange, *Roseburia* is red. The inset figures (DAPI) demonstrate that the individual fluorescence signals with specific bacterial probes also have signals corresponding to DNA fluorescence. The intracellular inclusions are located in crypts, which are free of bacteria.

the dense masses of bacteria (see discussion of mucus barrier viscosity in Section 9.2.5). Bacteria can often be seen in regions of the biopsy, which are mechanically damaged [5]. Reports of finding submucosal bacteria should be cautiously interpreted as long as they do not exactly define how far away from the biopsy edge the observations were made.

Although adherent bacterial layers are present in nearly all (94% of) patients with IBD not previously treated with antibiotics, the highest concentrations of mucosal bacteria are found, not in the highly inflamed regions of the intestine, but in less macroscopically noninflamed regions. This indicates that the break of the mucus barrier is primary to the inflammation. In inflamed regions, the bacterial concentrations are reduced (Figures 9.34 and 9.35), while leukocytes appear in the mucus in large numbers often arraying the outer regions of the mucus (Figure 9.36). Leukocytes within mucus are absent in biopsies from healthy persons. Bacteria reach the intestinal wall and lead to development of ulcers, fissures, and crypt abscesses, despite high concentrations of leukocytes and reduced numbers of bacteria in the mucus of inflamed gut segments.

The break in the mucus barrier and bacterial adherence to the mucosa is not IBD-specific. Bacterial concentrations of 10^9 bacteria/ml or higher can be found

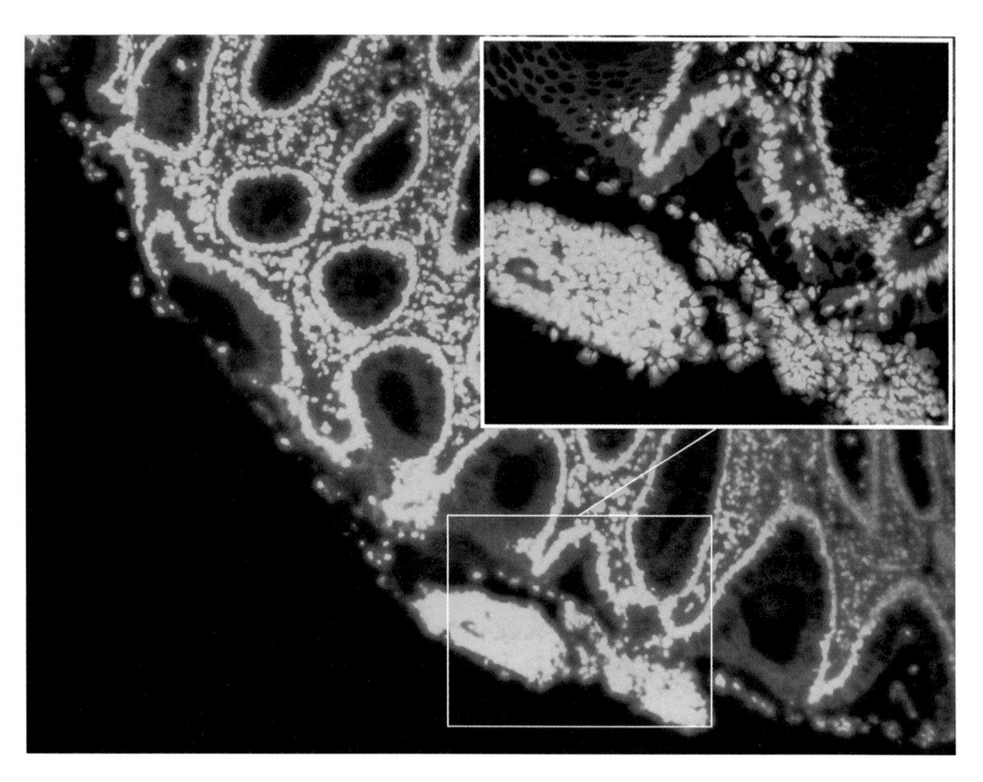

Figure 9.36. Sigmoid colon of a patient with ulcerative colitis (DAPI at 100× and inset at 400×). Leukocytes (large blue nuclei) array the outer regions of the mucus.

within mucus in nearly all patients with IBD and also in patients with celiac disease, 60% of patients with acute diarrhea, 52% of patients with diverticulosis, 45% of patients with carcinoma or polyps, and 38% of patients with irritable bowel syndrome (IBS). A 90% representation of bacteria found in mucus of patients with these diseases is observed in only three groups: *Bacteroides*, *Roseburia*, and *Faecalibacterium prausnitzii* (Figure 9.32). The mean density of the mucosal bacteria is significantly lower in non-IBD disease, and the composition of the biofilm differs. Bacteria of the *Bacteroides fragilis* group are responsible for >50% of the biofilm mass in IBD. In contrast, bacteria that positively hybridize with the Erec (*Roseburia*) and Fprau (*Faecalibacterium prausnitzii*) probes account for >50% of the biofilm in IBS patients, but for only <30% of the biofilm in IBD. The range of individual findings is, however, high, and the differentiation between Crohn's disease, ulcerative colitis, IBS, diverticulosis, or colonic cancer based solely on the FISH analysis of the mucosal flora is at present impossible. Even *Faecalibacterium prausnitzii*, which is completely depleted in feces of Crohn's disease patients (see Section 9.5), cannot be used as a diagnostic criterion. It can be detected in mucus of colonoscopic biopsies from >50% of patients with Crohn's disease but is, for example, absent in most of the biopsies from healthy persons.

The situation is different in cases of self-limiting colitis or specific infections such as *Serpulina*, *Fusobacterium necrophorum* (*nucleatum*), or Wipple's disease. Typical for a specific break of the mucus barrier is an increase of bacterial groups

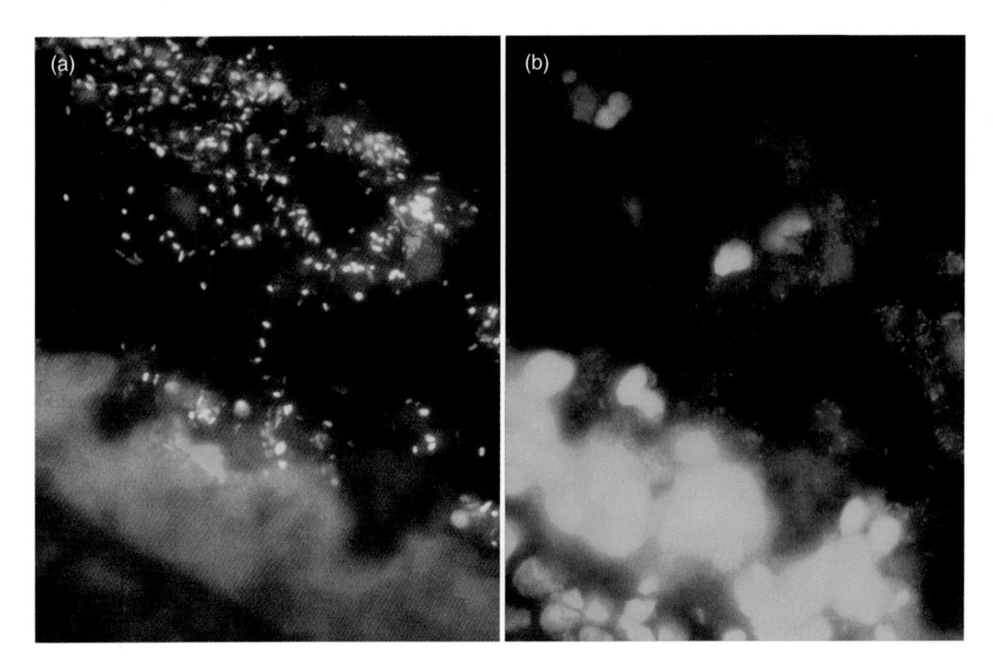

Figure 9.37. Self-limiting colitis in ascending colon. (a) *Bacteroides* is shown in yellow; *Roseburia* and *Faecalibacterium prausnitzii*, both in red, all other bacteria, in green. (b) DAPI stain of the same microscopic field demonstrates leukocyte migration into the mucosa. The proportion of bacteria other than *Bacteroides+Roseburia+Faecalibacterium prausnitzii* exceeds 10% in patients with self-limiting colitis or specific infection. This pattern is different from that in IBD patients. In IBD, these three bacterial groups together constitute 90% of the mucosal bacteria.

other than *Roseburia* + *Faecalibacterium prausnitzii* + *Bacteroides* to 10–70% (Figure 9.37).

9.3.2. Factors Potentially Constitute the Mucus Barrier

Since the beginning of the twentieth century, there has been a steady increase in reported cases of both Crohn's disease and ulcerative colitis, and the peak has obviously not been reached. This increase in IBD affects mainly the developed world, especially populations with high living standards in urban areas. Statistically, the frequency of the disease correlates with introduction of tap water, soap, and improvement in living conditions. The hygiene hypothesis argues that improved hygiene and a lack of exposure to microorganisms of various types have sensitized human immune systems, leading to excessive reactions to harmless bacteria in the environment. Out of these speculations have come recommendations to allow young children a "reasonable" amount of contact with dirt, pets, and other potential sources of infection. A positive effect of such lifestyle changes has not been demonstrated.

9.3.3. The Role of Facultative Pathogens

The statement that exposure to microbes in city dwellers is low is basically wrong and reflects only enteral infections with marked clinical symptoms. The number and diversity of bacteria in the large intestine of the urban population is definitively as high as of those living in rural areas, if not higher. The vegetables and fruits imported from Greece, Portugal, New Zealand, South Africa, and Australia bring a vast variety of microorganisms to people on every continent, providing opportunities for new exposure. The mobility of modern society has led to a profound and rapid exchange of bacteria worldwide that was never encountered in geographically constrained rural populations of the world. The exposure to facultative pathogens has definitively increased since the 1910s; however, their spectrum shifted from *Cholera*, *Salmonella*, *Shigella* and *Yersinia* to less spectacular, clinically "noiseless" pathogens. It is only decades after the first description that *Helicobacter pylori* was accepted as pathogen and moved into the center of research interest. *Helicobacter* is, however, not the only pathogen capable of forming biofilms. *Serpulina*, adhesive Enterobacteriaceae, or *Gardnerella*, which were also shown to form adherent biofilms on the epithelial surface, are still even now largely unnoticed by the medical community. These biofilms are often completely asymptomatic, such as those found with *Helicobacter pylori* infections. Their discovery was in most cases accidental; the clinical relevance remains unknown and implications controversial. Common for all of these "silent infections" is, however, their ability to compromise the mucus barrier and provide niches for growth of other bacteria that normally have no access to the mucosa (Figures 9.38 and 9.39). The exact data on occurrence and incidence of such "silent infections" in the human population still need to be evaluated.

9.3.4. Substances Reducing the Viscosity of the Mucus Barrier

The stated statistical correlation of the hygiene hypothesis between increased incidence of IBD and increased cleanliness of the modern society may have a completely

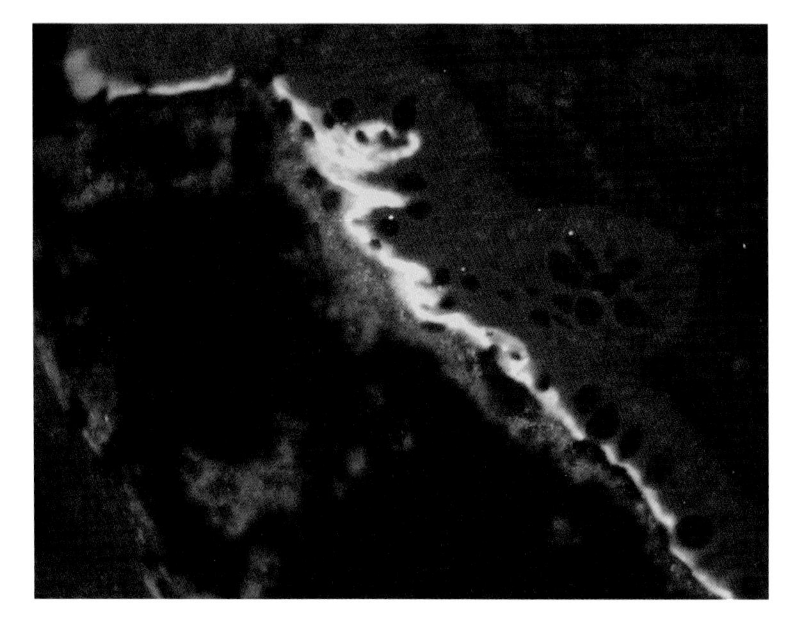

Figure 9.38. A prolific *Serpulina* biofilm adherent to the mucosa (orange serpentine covering the mucosa) opens access for other colonic bacteria (red fluorescence, Eub338 Cy5, at 400×).

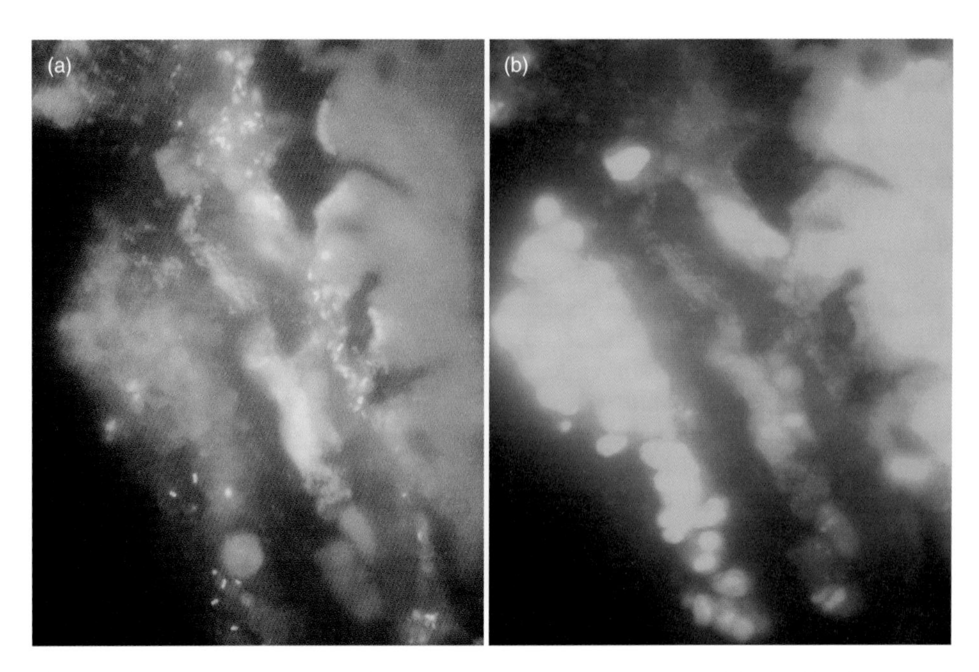

Figure 9.39. (a) Sigmoid colon of a patient with Crohn's disease. Prolific biofilm by AIEC of *Escherichia coli* (*E. coli* is shown in green; *Bacteroides*, in yellow; *Roseburia*, in red). The adherence of *E. coli* opens access for colonic microbiota to the mucosa. Despite an extensive leukocyte response, the number of bacteria present is very high and bacteria are located below the leukocytes that array the outer mucus regions. (b) DAPI stain of the same microscopic field.

different and more disturbing explanation. Detergents clean objects; they do not sterilize them. We know astonishing little about their effects on the intestinal mucosa and the mucus barrier, despite the longtime use of detergents in households. *In vitro* and *in vivo* evidence, however, suggests that substances that reduce the mucus viscosity may contribute to bacterial proliferation on the intestinal mucosa.

Detergents

Addition of detergents such as dextran sodium sulfate (DSS) to an *in vitro* model of stimulated mucus enables migration of bacteria through gels with a viscosity corresponding to agarose concentrations of 0.9%. *Bacteroides* migration could be seen at concentrations of 0.6% and migration of the *Roseburia* group, up to 0.9%. Without detergents, *Bacteroides* is immobilized at viscosity corresponding to agarose concentration of 0.4%, and the migration of all bacterial groups stops at viscosity corresponding to agarose concentrations of 0.7% [10].

Although the effects of detergents on the mucosal barrier in human are unknown, in the mouse model, the addition of DSS to food induces acute colitis (Figure 9.40), which becomes chronic after repeated exposure to DSS. The DSS-induced inflammation in mice is restricted to the large intestine, where bacterial concentrations are high, and bypasses the small intestine, where bacterial concentrations are low. Antibiotics relieve the DSS-induced inflammation. Both peculiarities stress the key role of bacteria in the pathogenesis of DSS colitis.

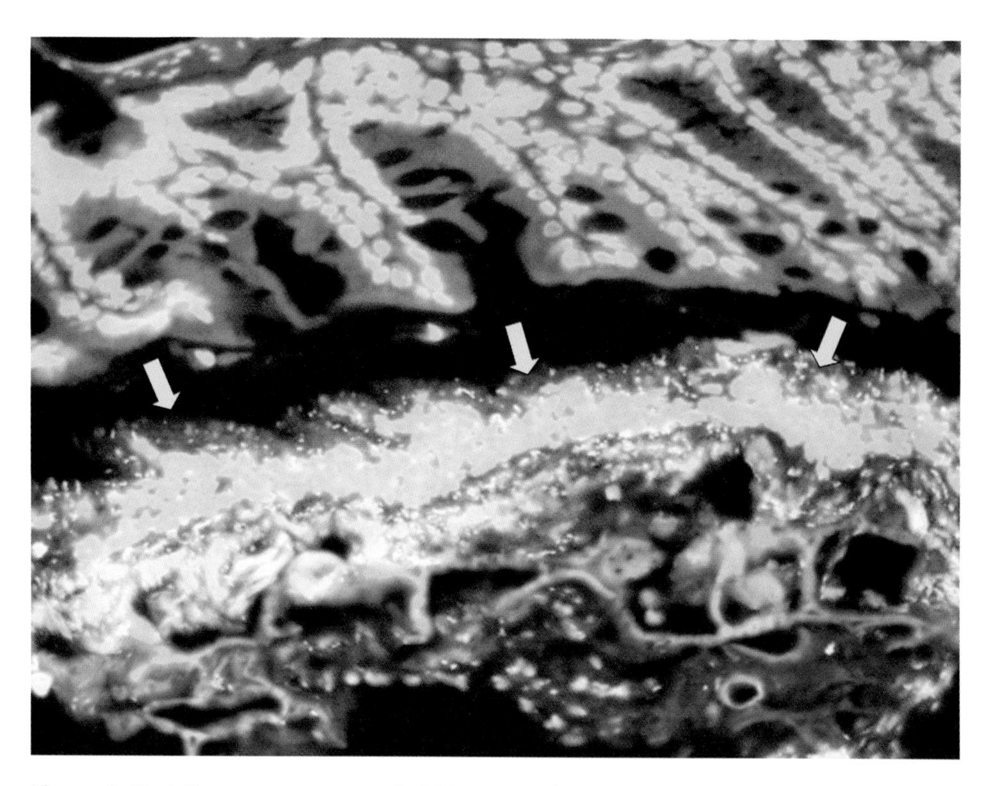

Figure 9.40. Inflammatory response in DSS mouse colitis. *Bacteroides* (Bac303 Cy3, orange) and DAPI stain are overlayed. Leukocytes array the outer regions of the mucus layer (arrows).

Emulsifiers

Emulsifiers are another group of substances that could potentially influence the mucus barrier and have been increasingly used by the food industry since the beginning of the twentieth century.

Data on IL10 gene-deficient mice support the potentially detrimental role of emulsifiers, such as 2% carboxymethyl cellulose (CMC) [14] High bacterial concentrations were found within crypts of Lieberkuhn in the ileum of all CMC-treated IL10 knockout mice (Figure 9.41). The finding resembled visual observations in the ileum of Crohn's disease patients (Figure 9.42).

Bile Acids

Many other factors can influence the mucus barrier. Bile acids, for example, are natural emulsifiers. Normally they are completely reabsorbed in the ileum and do not reach the colon. When the ileum is resected, then the reabsorption is disturbed, bile acids reach the colon and induce diarrhea.

Glutens

Celiac disease is regarded as an allergic response, although the exact structure within the gluten molecule that induces an allergic reaction could not be defined.

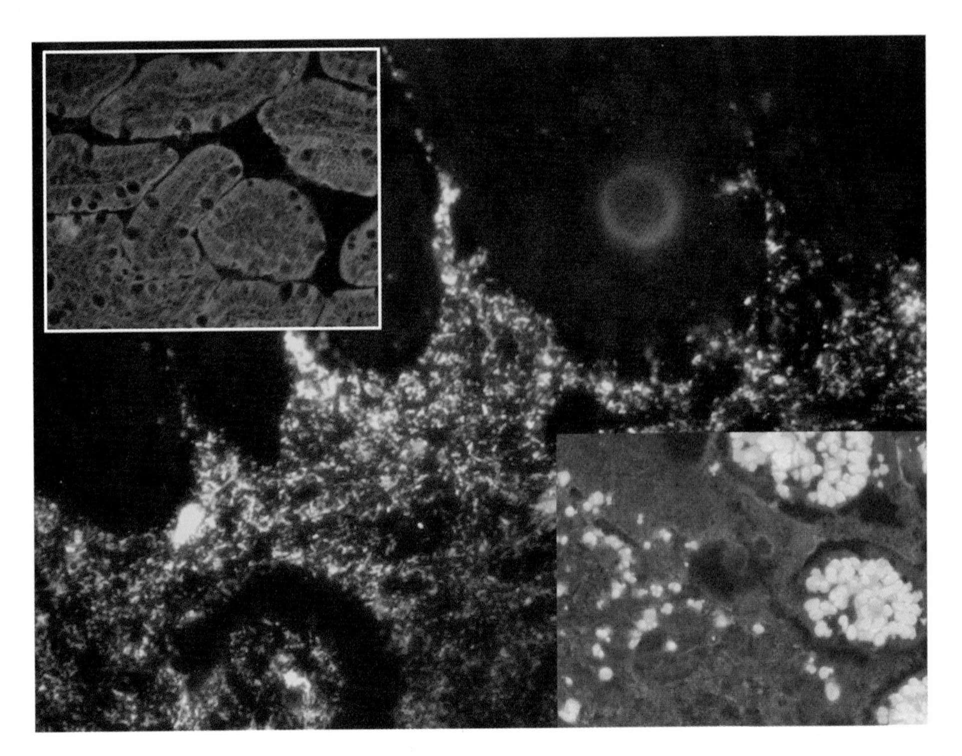

Figure 9.41. Comparison between ileum of the interleukin 10 gene-deficient mouse treated with 2% CMC and control IL10 gene-deficient mouse receiving only water (framed inset). The CMC ingestion is accompanied by a significant increase in intraluminal bacteria in the small intestine. Bacteria enter deep into the crypts of Lieberkuhn, and leukocytes migrate into the lumen of the intestine, shown at 400× magnification (blue inset demonstrates the DAPI fluorescence of the same region).

We do know that symptomatic celiac disease is always ongoing with bacterial overgrowth in the small bowel. The link between bacteria and glutens is poorly understood. Glutens are, however, naturally occurring emulsifiers. It could be that first bacteria make glutens harmful and that progressive destruction of the mucosa in the small intestine leads to decreased suppression and bacterial overgrowth.

Smoking

Smoking stimulates mucus secretion. Epidemiologic studies indicate that tobacco smoke is beneficial in patients with ulcerative colitis (LIC) but detrimental for patients with Crohn's disease. A thicker mucus barrier could, indeed, explain why smoking could be protective in UC patients but has no effect in Crohn's disease, where bacterial suppression is more important than bacterial separation.

Stress

Stress interferes with both mucus production and regulation of the viscosity of the mucus. It is a known fact that in patients with IBD, stress leads to acute exacerbations of the disease.

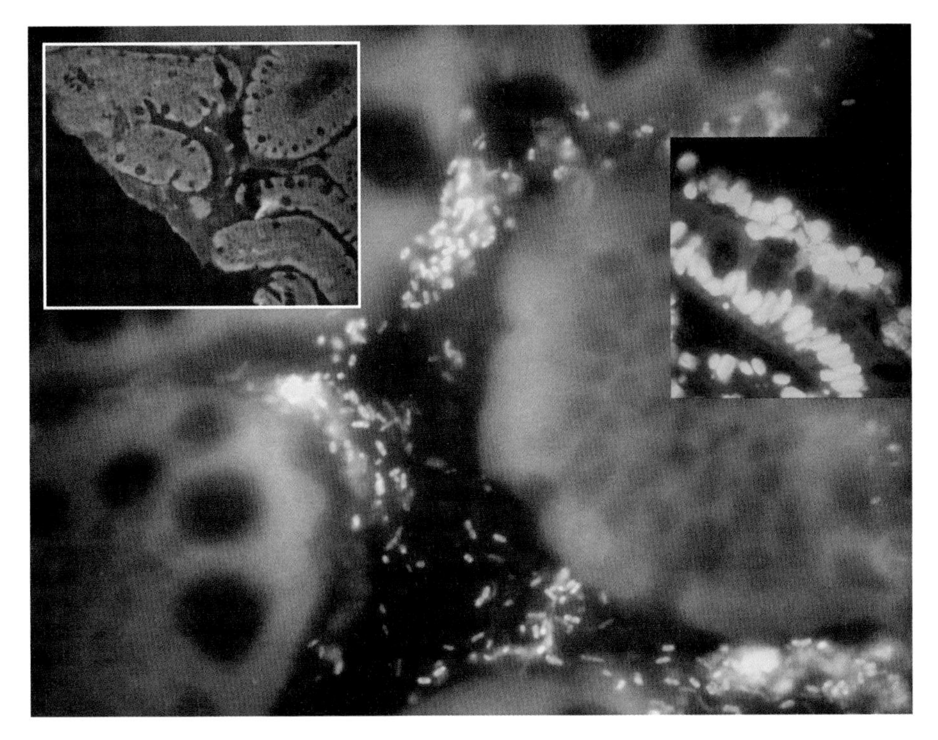

Figure 9.42. Prolific adherent biofilm located between villi in the ileum of a patient with Crohn's disease. *Bacteroides* is shown in orange; *Roseburia*, in red; and all other bacteria, in green, at 1000× magnification. The blue inset demonstrates an intense leukocytic response in the same patient in a neighboring region. The white framed inset shows the ileum of a healthy person.

Multiple other factors, including defensins, probiotics, enteral pathogens, the inflammation itself, and genetic background interfere with the mucus barrier function. As long as the mucus barrier is compromised, a conflict between the organism and the pathogens that inhabit the human colon in large numbers and diversity is inevitable.

9.4. POSSIBLE WAYS TO REMODEL THE MUCUS BARRIER

The following strategies are possible:

- Eradication of occasional pathogens, which compromise the mucus barrier (*Enteroadhesive E. coli*, *Fusobacterium nucleatum*, *Serpulina*)
- Selective control of mucus secretion and dehydration (analogs of cortisol)
- Induction of a higher differentiation of epithelial cells, which leads to a switch from a mainly secretory to an adsorptive function (analogs of anti-TNF-suppressing apoptosis, methotrexate)
- Suppression of adherent bacterial biofilms (a possible effect of 5-aminosalycilic acid (5-asa)

- Reduction of the burden of detergents and emulsifiers in food
- Stimulation of innate immunity (substances such as GM-CSF, probiotics as living vaccines)

Antibiotics can effectively reduce the number of pathogens contacting the mucosa. They have, however, no direct influence on the mucus barrier and cannot sterilize the polymicrobial colonic microbiota. As soon as antibiotics are withdrawn, the situation becomes reversed. In the long term, antibiotics are generally ineffective in IBD because of increasing microbial resistance. The mucus barrier, however, can be compromised not only by environmental or genetic factors but also by specific pathogens such as *Serpulina, Fusobacteria*, Enterobacteriaceae, or *Gardnerella*, which build biofims on the epithelial surface. The identification of adherent biofilms and their eradication could be advantageous.

Prednisolone is a very potent drug. As a glucocorticoid, it stimulates mucus secretion. Its mineralocorticoid activity increases water resorption, thereby increasing the viscosity gradient within the intestinal mucus layer. Development of substances that can selectively control the mucus barrier without the typical side effects of prednisone could be of extreme advantage for IBD treatment.

We have previously mentioned that the columnar epithelial cells are differentiated and mainly resorptive, while crypt cells are immature stem cells and mainly secretors. A balance between both is under TNF control. The cell turnover is increased in inflammation. Anti-TNF agents reduce the apoptosis of differentiated epithelial cells. This may explain why, of many known biological mediators of inflammation, only anti-TNF antibodies have a clinically proven role in the treatment of IBD. The development of drugs with an effect on apoptosis regulation of the epithelial cell turnover should be considered in the future.

Methotrexate is a potent agent for the *in vitro* transformation of immature epithelial cell lines to mucine-producing goblet cells and columnar epithelial cells. Higher concentrations of MTX increase the proportion of columnar to goblet cells.

The role of so-called immune suppressive substances in treatment of IBD remains to be exactly defined.

Mesalazine suppresses bacterial biofilms *in vivo* by mechanisms that are unclear at present. Different from antibiotic therapy, the suppression with mesalazine does not seem to induce bacterial resistance. It is possible that the suppressive effects of mesalazine could be further expanded when the mode of action is clarified.

The importance of reducing the detergent and emulsifier burden in food has been mentioned. We do not know at present which of the substances may reach the colon and accumulate in the human body. This issue needs to be further investigated before any recommendations can be made.

Stimulation of the immune response is an intriguing approach for the treatment of IBD. Previous trials with interferon and GM-CSF were half-hearted and inconclusive. PEG interferon, for example, was not tested at all. The therapeutic potential could be enormous. After all, probiotics may act as living vaccines using attenuated strains to stimulate mucosal immunity. Actually we do not know how probiotics work. However, since the influence of antibiotics on polymicrobial microbiota is limited, the use of microorganisms for the control of indigenous microbiota is intriguing. We must, however, admit that all presently available probiotics use bacterial strains that are present only in small numbers in the human large intestine (<0.01%). They were selected mainly for ease of culture, storage, transport, and stability within food products. The probiotic potential of anaerobes, which constitute the mass of the indigenous flora of the large intestine, has not been studied.

9.5. BIOSTRUCTURE OF FECAL MICROBIOTA IN HEALTH, INFLAMMATORY BOWEL DISEASE, AND OTHER GI DISEASES

The colon is a highly efficient biofermenter. We take the colonic bacterial concentrations of 10^{11-12} bacteria/mL, and diversity of >3000 species, for granted. They are, however, unprecedented for environment and nonreproducible for humans *in vitro*. Any dysfunction of the mucus barrier, irritation of the gut, or inflammatory response could lead to malfunction of the colonic biofermenter and general or specific alterations of microbiota. These alterations could be used for diagnostic purposes. Unfortunately, all previous investigations of the fecal microbiota used homogenized samples of feces and did not reveal any diagnostically valuable features. This may be due to ignorance of the spatial organization of the colonic microbiota.

The situation changed with studies of the microbial biostructure of punched fecal cylinders, which made the investigation of fecal microbiota a reliable, highly reproducible, and easy-to-perform diagnostic assay [11,12] (see also discussion of fecal microbiota biostructure elsewhere in this section).

Functionally, the colonic biofermenter can be divided into three zones:

1. A transparent outer *mucus layer*, which is constantly dehydrated by the columnar epithelium, and therefore highly viscous and impenetrable for bacteria and separates the colonic biofermenter from the mucosa.
2. A *central fermenting zone*, in which bacteria and fibers are stirred and fermented.
3. A transitional *resting zone* between the mucus layer and central zone, in which mucus becomes increasingly diluted by luminal fluids and penetrable for bacteria. The softened mucus is, however, still less versatile for peristalsis and remains attached to the colonic wall for a prolonged period. Bacteria enter these soft portions of the mucus in concentrations inverse to the growing viscosity gradient and become increasingly immobilized. Trapped within the resting zone, bacteria are protected against purging events and can be used for renewed settings of the biofermenter after occasional cleanouts, periods of fasting, or even antibiotic treatment. Bacteria within the resting zone are *germinal stocks of the colonic biofermenter*.

The firm mucus can be easily perceived with alcian stain. In healthy subjects, neither the resting (germinal) zone nor the luminal (fermenting) area can distinguished from one another (Figure 9.43). When the colonic biofermenter functions properly, the composition and density of bacteria found here are similar. In disease, however, the microbial changes in these compartments are different and disease specific (Figures 9.44, 9.45, 9.46, 9.47, and 9.48).

What happens in the laboratory when the biofermenter ceases to perform properly? The laboratory staff puts on the protection suit, discharges the biofermenter,

Figure 9.43. Colonic bacteria in the healthy wildtype mouse are diffusely distributed and have similar high concentrations at the center of feces and in the "germinal" zone.

Figure 9.44. Bacteria are suppressed in a 28-week-old mouse with IL10 deficiency, especially at the center of feces. *Bacteroides* is shown in orange; Roseburia, in red; all other bacteria, in green, at 400× magnification. The suppression of bacteria precedes the actual inflammation. At the time of investigation the animals were a symptomatic.

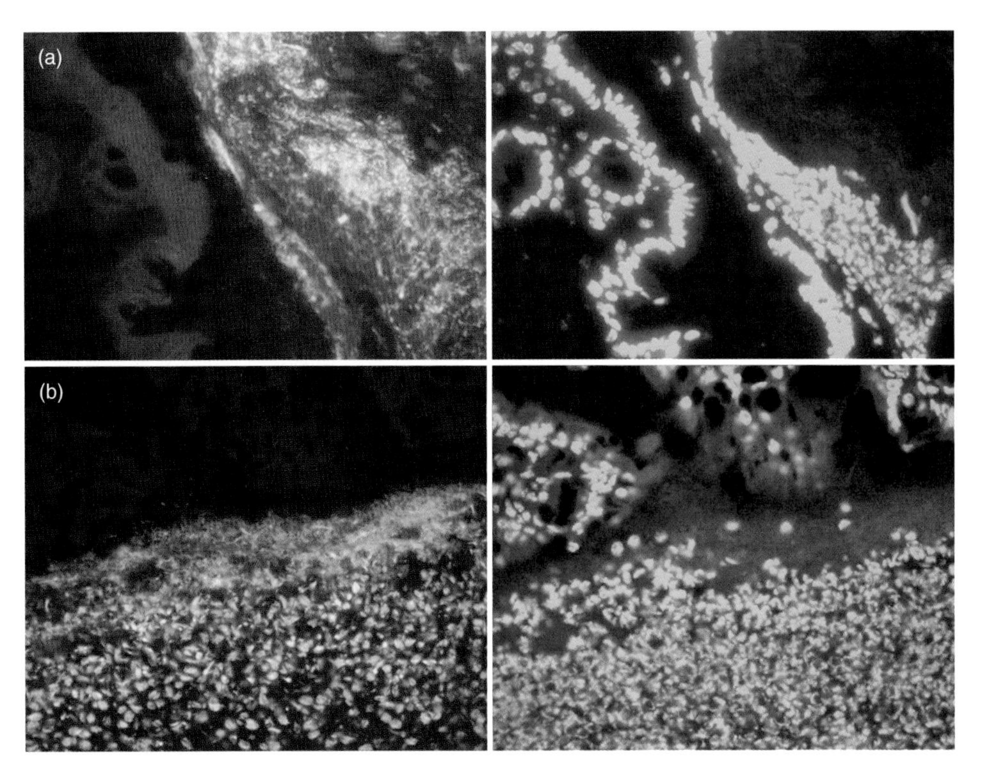

Figure 9.45. (a) A 32-week-old IL10-deficient mouse with colitis; (b) a 20-week-old wildtype mouse with severe dextran sodium sulfate (DSS)-induced colitis. The left panels demonstrate the distribution of bacteria in feces; the right panel demonstrates the DAPI stain of the same microscopy field. Despite massive leukocyte response in both cases presented on the right (the intestinal lumen is completely filled with leukocytes in the DSS mouse) and bacterial reduction at the center of feces (left panels), bacterial concentrations in the transition zone between mucus and feces are completely intact and the fluorescence of bacteria is not inhibited.

decontaminates the contents, and restocks the system. Exactly the same happens in the malfunctioning colonic biofermenter: increased protection of the mucosa, discharge, decontamination, and restocking. Problems with the germinal stock are in the foreground in IBD.

9.6. SITE-DEPENDENT CHANGES OF COLONIC MICROBIAL BIOSTRUCTURE

9.6.1. The Mucus Layer

Diarrhea discharges the colonic biofermenter, reduces the total number of luminal bacteria, and is accompanied by massive increase in mucus production: increased thickness of the mucus layer and growing incorporation of the unstructured mucus within the fecal mass (Figure 9.46).

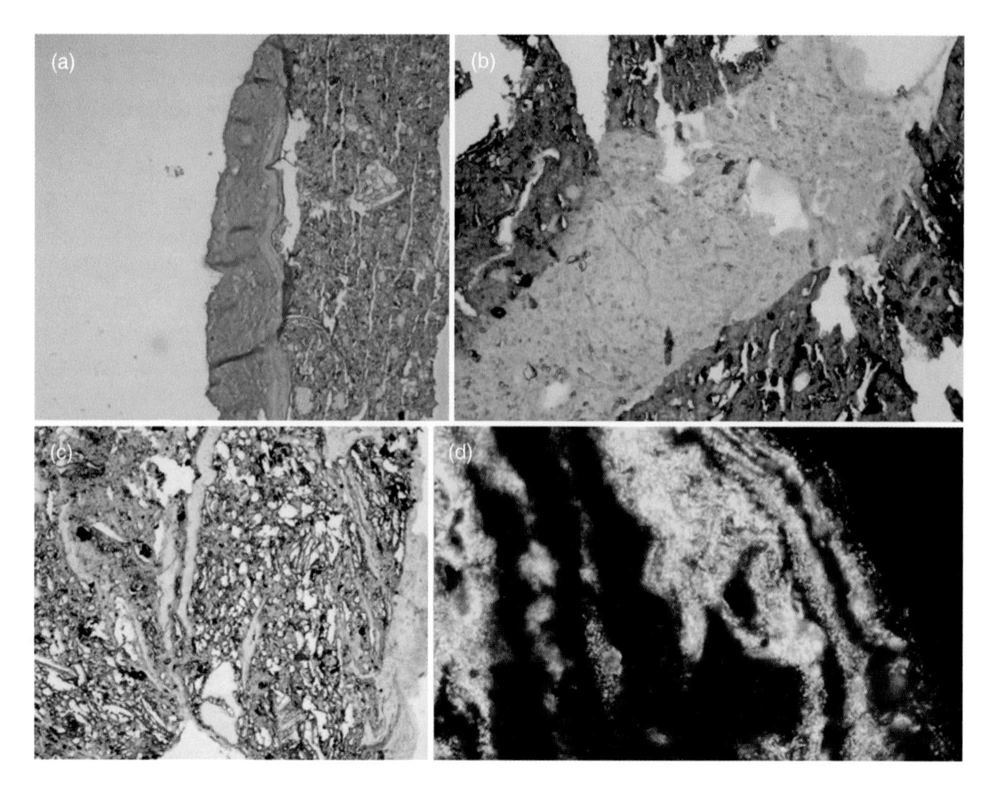

Figure 9.46. Increased mucus production as a sign of irritation in patients with irritable bowel syndrome (a) and diarrhea (b–d). (a)–(c) Mucus can be directly seen with alcian stain. (d) Disruption of the normally homogenous biostructure of the fecal cylinder and development of multiple striae of slime demonstrate the increased mucus production. [Panels (a)–(c) at 100×; panel (d) at 400×.]

Increased mucus production is general for most intestinal disorders. The only exception is ulcerative colitis, where the mucus layer is depleted, probably as a result of exhaustion.

In acute diarrhea, the local concentrations of single bacterial groups and their fluorescence intensity remain unchanged, but the homogeneous structure of healthy feces (Figure 9.28) is interrupted by broad large septa (unformed stools) or multiple thin striae (watery stools; Figure 9.46). In chronic conditions, the bacterial concentrations are in addition markedly reduced, indicating that active decontamination of the biofermenter contents takes place.

9.6.2. The Working (Luminal) Area of the Colonic Biofermenter

The most common general feature of intestinal disturbances is a suppression of bacterial growth and metabolism at the center of the biofermenter by the host. The suppression is general, and is most apparent with respect to the habitual bacteria. Their concentrations in healthy person are especially high, and the distribution throughout the fecal cylinder is normally homogeneous. Initially, only the

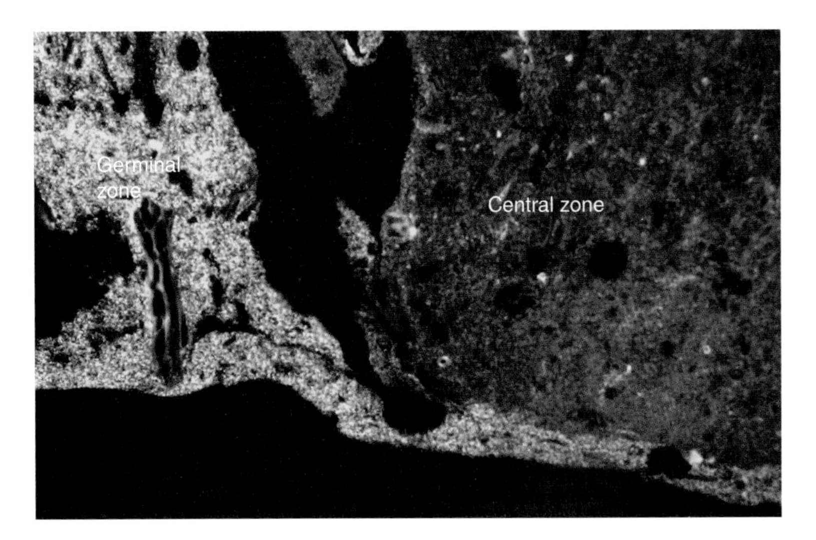

Figure 9.47. Hybridization silence of *Faecalibacterium prausnitzii* in a patient with idiopathic diarrhea (orange fluorescence, at 200×). Bacteria are suppressed at the center of the feces. The concentration and fluorescence of bacteria at the surface of feces are well preserved. Bacterial concentration and fluorescence are excellent in the periphery of feces.

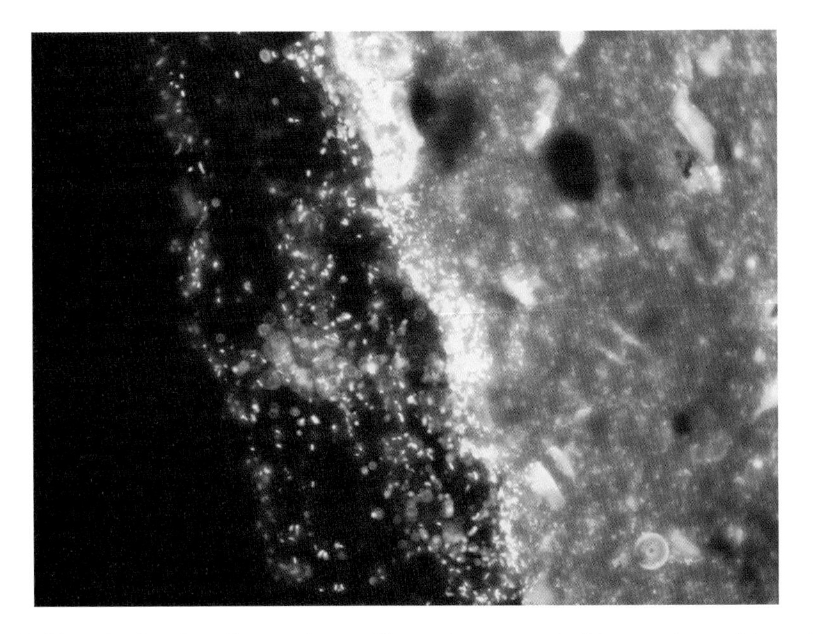

Figure 9.48. Hybridization silence of *Faecalibacterium prausnitzii* in a patient with acute gastroenteritis caused by *Campylobacter jejuni* (yellow fluorescence, at 400×). Bacteria are suppressed throughout the feces, with reduced fluorescence of bacteria (relative hybridization silence). Bacteria are highly concentrated and have excellent fluorescence in the germinal zone, at the transition from feces to mucus, and in the superficial portions of the mucus.

fluorescence signals fade. We call these fading a *hybridization silence* (Figures 9.47 and 9.48), because the number of bacteria remains constant from suppressed to unsuppressed regions and only hybridization signals of bacteria change (relative hybridization silence). With increasing host response the hybridization signals of single bacterial groups may disappear completely (absolute hybridization silence). It is then impossible to discriminate between suppression and physical elimination of bacteria.

The epicenter of suppression is located in the center of feces and with an exception in Crohn's disease, does not involve the superficial "germinal" zone of the fecal cylinder, where the fluorescence of bacteria and bacterial numbers remain high (Figures 9.47 and 9.48). It appears that the production of the suppressive substance takes place in the small intestine or at least upstream from the colon.

The extent to which single habitual bacterial groups are involved is individual and reproducible in repeated investigations of the same patient. It is probable that the involved mechanism of suppression may be different and directed more or less specifically to different bacterial groups. These changes can be exactly quantified on slides of fecal cylinders. Disease progression can be monitored in fecal cylinders obtained at different times.

9.6.3. The Germinal Zone of the Colonic Biofermenter

Changes involving the germinal bacterial zone are characteristic for active inflammatory bowel disease and absent in most other gastrointestinal diseases, with the exceptions of carcinoid of the small bowel and subgroups of patients with celiac disease. The most prominent feature of ulcerative colitis is a replacement of the mucus layer by leukocytes. The leukocytes are located in the germinal zone, which they progressively destroy (Figure 9.49 and 9.50). Active Crohn's disease is characterized by complete depletion of *Faecalibacterium prausnitzii* from the central and germinal zones of feces.

The reproducible detection of these two features in three consecutive fecal cylinders taken at 2-week intervals allow the diagnosis of active CD and UC with a 79/80% sensitivity and 98/100% specificity [12].

Leukocytes within the germinal zone and *Faecalibacterium prausnitzii* depletion are not causes but symptoms of the disease. High-dose prednisolone or anti-TNF therapy is able to reverse these findings within days. The reversal is not permanent. Both *Faecalibacterium prausnitzii* deletion and leukocyte infiltration of the germinal zone return quickly after the depletion of prednisolone or within weeks after Remicade infusion (anti-TNF antibody).

Analysis of occasional bacterial groups within the fecal cylinders of patients with IBD demonstrated astonishing differences in occurrence and concentrations of Enterobacteriaceae, *Bifidobacteria, Atopobium, Eubacterium cylindroides*, and *E. hallii* bacterial groups between Crohn's disease and ulcerative colitis [12]. This indicates that these diseases are distinctly different entities and not merely different expressions of the same inflammatory process. However, because of inconsistent occurrence of occasional bacterial groups in single patients, the IBD-specific changes of these bacterial groups cannot be used for diagnosis in each case.

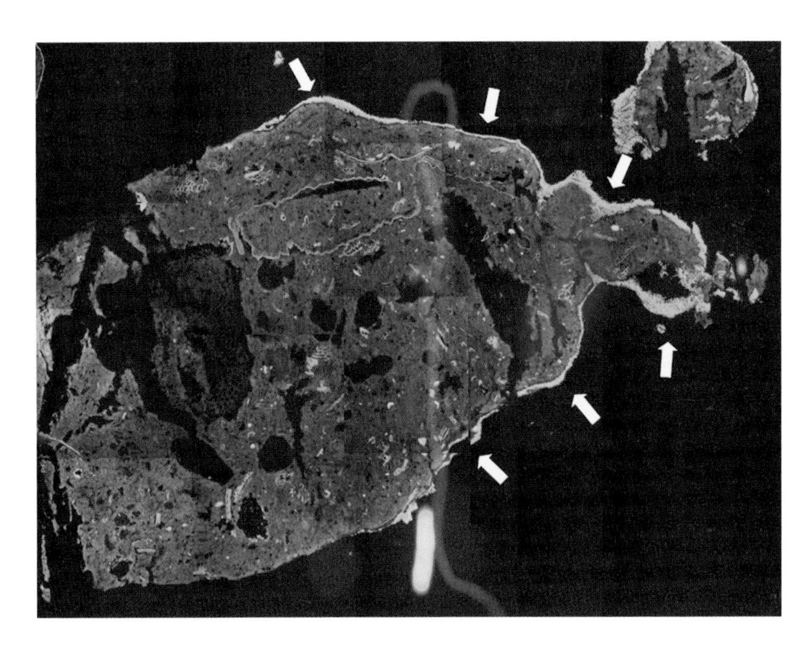

Figure 9.49. Fecal cylinder with DAPI stain. Multiple single microscopic fields photographed at 100× magnification seen in an overview composite. The center of the cylinder contains no leukocytes, and the bacteria are homogeneously distributed. The cylinder has no mucus cover; instead, the germinal zone is infiltrated by many leukocytes (arrows). The findings are typical for ulcerative colitis.

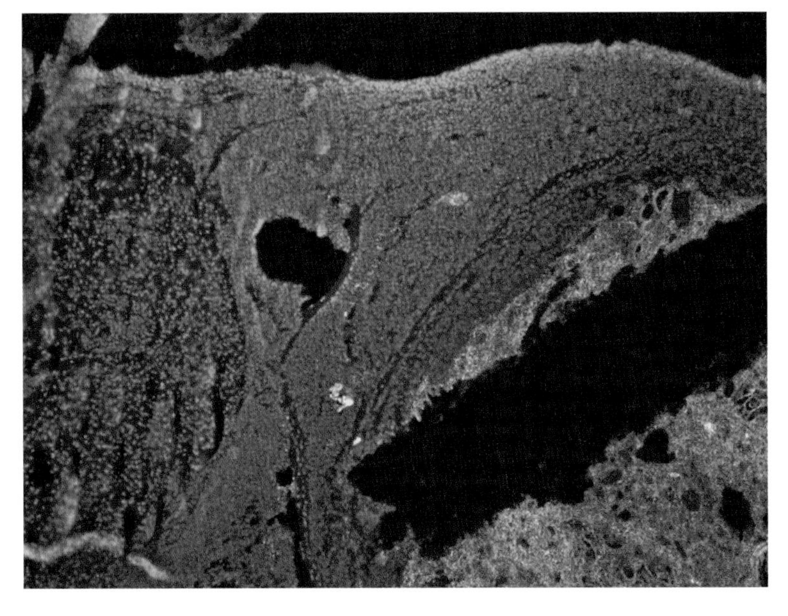

Figure 9.50. Fecal cylinder of a patient with ulcerative colitis. Massive infiltration of the germinal zone by leukocytes (DAPI stain). The fluorescence of *Faecalibacterium prausnitzii* and DAPI are overlayed. The *F. prausnitzii* concentrations are not markedly reduced in ulcerative colitis.

9.7. CONCLUSIONS

The surface of the intestinal tract in healthy humans is free of bacteria in all bowel segments. Adherence of bacteria to epithelial cells is, therefore, a sign of infection. In contrast to the mucosa, the intestinal lumen is never sterile because of the polymicrobial nature of intestinal microbiota. In gut segments such as the stomach or small intestine, where bacteria are actively suppressed, the microbiota are accidental in occurrence, composition, and concentrations. The situation is different in the colon. The colon is a biofermenter that reduces nondigested remnants with bacterial assistance. Here, the bacterial growth is facilitated and the suppression is suspended. Bacterial concentrations and diversity in the colon reach astronomic numbers. Some of these bacteria are indispensable for the function of the colonic biofermenter; others occur occasionally. Many of the indigenous bacteria are potential pathogens: *Bacteroides*, Enterobacteriaceae, *Enterococci*, and *Clostridium histolyticum*. The control of pathogens within the colonic biofermenter is achieved by an impenetrable mucus layer. Although the mucus layer is typical for all mucosal surfaces, the intense resorption of water through the colonic columnar epithelium thickens the colonic mucus to an especially high viscosity, which immobilizes all bacterial movement. As long as the separation is complete, the number and diversity of the colonic bacteria are unimportant. Before bacteria can adhere to and invade the mucosa, they must first traverse the mucus. When pathogens penetrate the mucus and adhere to the epithelial cells, inflammation clears the mucosa from the bacterial contact and the mucus from the bacteria, thus reestablishing the status quo.

The situation is different in the case of prolonged compromise of the mucus barrier. The inflammation cannot clear the mucosa from pathogens that inhabit the colonic lumen and continuously migrate toward the mucosa. The inflammation becomes chronic. Inflammatory bowel disease is a polymicrobial infection that is characterized by a sustained broken mucus barrier, subsequent bacterial migration toward the mucosa, and proliferation of a complex bacterial biofilm on the epithelial surface with resulting invasive and cytopathologic effects. The reasons for malfunction of the mucus barrier could be increased urbanization that leads to increased incidence of silent infections with adherent bacteria that compromise the mucus barrier, excessive use of detergents and emulsifiers that change the viscosity of the mucus, and many other reasons. As long as mucus barrier function is impaired, the inflammatory process cannot successfully clear the bacteria from the mucosal surface, and the inflammation is itself detrimental. Then immunosuppressive therapy remains the main therapeutic option. Other therapeutic measures, including regulation of the mucus secretion and viscosity, suppression of bacterial biofilms, eradication of occasional pathogens, probiotics, and immunostimulation, are possible therapeutic interventions and should be increasingly considered and evaluated in the future.

As a consequence of the inflammatory response, the composition and structure of fecal microbiota is changed. From the biostructure of fecal cylinders, active Crohn's disease and ulcerative colitis can be distinguished from each other and from other gastrointestinal diseases. The specific and noninvasive monitoring of disease activity will enable us to intensify the search for alternative therapeutic strategies aimed at cure of the disease rather than symptom control.

REFERENCES

1. Amann RI, Ludwig W, Schleifer K-H. Phylogenetic identification and in situ detection of individual microbial cells without cultivation. *Microbiol Rev* **59**:143–169 (1995).
2. Loy A, Maixner F, Wagner M, Horn M. ProbeBase—an online resource for rRNA-targeted oligonucleotide probes: New features 2007. *Nucleic Acids Res* **35**:800–804 (2007).
3. Swidsinski A, Göktas Ö, Bessler C, Loening-Baucke V, Hale LP, Andree H, Weizenegger M, Hölzl M, Scherer H, Lochs H. Spatial organization of microbiotica in quiescent adenoiditis and tonsillitis. *J Clin Pathol* **60**:253–260 (2007).
4. Swidsinski A, Schlien P, Pernthaler A, Gottschalk U, Bärlehner E, Decker G, Swidsinski S, Strassburg J, Loening-Baucke V, Hoffmann U, Seehofer D, Hale LP, Lochs H. Bacterial biofilm within diseased pancreatic and biliary tracts. *Gut* **54**:338–395 (2005).
5. Swidsinski A, Weber J, Loening-Baucke V, et al. Spatial organization and composition of the mucosal flora in patients with inflammatory bowel disease. *J Clin Microbiol* **43**:3380–3389 (2005).
6. Franks AH, Harmsen HJ, Raangs GC, Jansen GJ, Schut F, Welling GW. Variations of bacterial populations in human feces measured by fluorescent in situ hybridization with group-specific 16S rRNA targeted oligonucleotide probes. *Appl Environ Microbiol* **64**:3336–3345 (1998).
7. Harmsen HJ, Raangs GC, He T, Degener JE, Welling GW. Extensive set of 16S rRNA-based probes for detection of bacteria in human feces. *Appl Environ Microbiol* **68**:2982–2990 (2002).
8. Dethlefsen L, Huse S, Sogin ML, Relman DA. The pervasive effects of an antibiotic on the human gut microbiota, as revealed by deep 16S rRNA sequencing. *PLoS Biol* **6**(11):e280 (2008).
9. Swidsinski A, Loening-Baucke V, Lochs H, Lochs H, Hale LP. Spatial organization of bacterial flora in normal and inflamed intestine: A fluorescence in situ hybridization study in mice. *World J Gastroenterol* **8**:1131–1140 (2005).
10. Swidsinski A, Sydora BC, Doerffel Y, Loening-Baucke V, Vaneechoutte M, Lupicki M, Scholze J, Lochs H, Dieleman LA. Viscosity gradient within the mucus layer determines the mucosal barrier function and the spatial organization of the intestinal microbiota. *Inflamm Bowel Dis* **13**:963–970 (2007).
11. Swidsinski A, Loening-Baucke V, Verstraelen H, Osowska S, Doerffel Y. Biostructure of fecal microbiota in healthy subjects and patients with chronic idiopathic diarrhea. *Gastroenterology* **135**:568–579 (2008).
12. Swidsinski A, Loening-Baucke V, Vaneechoutte M, Doerffel Y. Active Crohn's disease and ulcerative colitis can be specifically diagnosed and monitored based on the biostructure of the fecal flora. *Inflamm Bowel Dis* **14**:147–161 (2008).
13. Kunzelmann K, Mall M. Electrolyte transport in the mammalian colon: Mechanisms and implications for disease. *Physiol Rev* **82**:245–289 (2002).
14. Swidsinski A, Ung V, Sydora BC, Loening-Baucke V, Doerffel Y, Verstraelen H, Fedorak RN. Bacterial overgrowth and inflammation of small intestine after carboxymethylcellulose ingestion in genetically susceptible mice. *Inflamm Bowel Dis* **15**:359–364 (2009).

10

FROM FLY TO HUMAN: UNDERSTANDING HOW COMMENSAL MICROORGANISMS INFLUENCE HOST IMMUNITY AND HEALTH

JUNE L. ROUND

Department of Pathology, Division of Microbiology and Immunology, University of Utah, Salt Lake City, Utah

10.1. INTRODUCTION

A long-forgotten resident that persists on the bodies of most animals is the vast consortium of commensal bacteria that are collectively referred to as the *microbiota*. Most animals form intimate, unknown relationships with bacteria, and researchers are beginning to appreciate their profound impacts on host development [1]. These impacts range from inducing immune cells to governing intestinal architecture, nutrition, and even mating preference. Thus, the microbiota has become central to the health and wellbeing of multiple organisms. This chapter focuses on what can be learned about commensal–host relationships through multiple animal models including the fruit fly, zebrafish, and mouse. Analyses of these models will be based on comparisons of the models to ascertain commonalities and differences with respect to the influence of the commensal microbiota on host immunity and how information acquired from these models can be translated to treat human disease.

10.2. MICROBIAL DIVERSITY IN HUMANS AND ANIMAL MODELS

During development within the womb (or egg), animals are completely sterile; however, during birth, these once "clean" animals are rapidly colonized by the microbes within the environment. From that point forward, animals are in constant

The Human Microbiota: How Microbial Communities Affect Health and Disease, First Edition. Edited by David N. Fredricks.

contact with bacteria, some commensal and some pathogenic; others are transient while some are stable. These resident commensal bacteria outnumber host cells by an order of magnitude and are present on almost all environmentally exposed surfaces of the body. The greatest diversity and density of bacteria are found within the gastrointestinal tract. The human microbiota collectively contains >100 times more genetic material than our own, and therefore mammals should be considered "super-organisms", and studied in the context of both their own genome and their microbiome [2]. Of the known 70 bacterial divisions and the 13 divisions of Archaea that have been described to date, only two phyla predominate within the mammalian intestine: the Bacteroidetes (16.3%) and the Firmicutes (65.7%) [3,4]. Of the Firmicutes in the human intestine, most of the phylotypes identified belonged to the class Clostridia, including Clostridia cluster XIV and Feacalibacteria. Organisms belonging to this phylum that have been the topic of extensive research include the segmented filamentous bacteria (SFBs). A single Archaea was identified within the human intestine: *Methanobrevibacter smithii*. Other phyla found within the human gut include Actinobacteria (4.7%), Proteobacteria (8.8%), Verrucomicrobia (2.2%), Fusobacteria (0.7%) Spirochaetes (0.46%), DSS1 (0.35%), Fibrobacteres (0.13%) Cyanobacteria (0.10%), and Planctomycetes (0.08%). The fact that the bacteria within the human intestine represent such a small fraction of the total bacteria found on the planet suggests that these bacteria are highly evolved and specialized to live within the mammalian intestine [2].

How does the diversity of commensal organisms of a human compare to the diversity that is seen in other animal models commonly used in research? A summary of this comparison is provided in Figure 10.1. While organisms in the phyla Proteobacteria and Actinobacteria represent <1% of the organisms within the mouse colon, 60–80% of the organisms are a part of the Firmicutes and 20–40% of the organisms fall into the Bacteriodetes phylum [5]. Most of the organisms found within the Firmicutes are in the *Clostridium* cluster XIVa, a common clade in humans that contains butyrate producers. Therefore, mice represent a viable animal model for understanding how the microbiota may be related to human disease.

A more recent study analyzed the microbial diversity of the fruit fly (*Drosophila*) [6]. This model allows for rapid genetic screens and large population sizes. Additionally, *Drosphilia* can be made germ-free (devoid of any colonizing organism), and is thus well suited to understand host–commensal relationships. They used 14 strains of *Drosophilia* that were collected in the wild and compared the diversity to isolated lab strains of *Drosophilia*. There was no clear core microbiome (a set of organisms that was present in every strain tested); however, there were similarities among the strains. The Proteobacteria was the most abundant of the organisms present, with ~70% of the bacteria belonging to this phylum. The orders that were represented within this phylum included 60% of Enterobacteriaceae and 9% Acetobacteraceae. Several of the Enterobacteriaceae isolated from the fly gut are closely related to the species found within the animal intestine, including *Providencia* and *Serratia*. These organisms are used as model pathogens in the fly and also include many obligate endosymbiotic organisms such as *Wolbachia*. While 70% of the bacteria within the fly gut belonged to the Proteobacteria, 21% of these bacteria belonged to the Firmicutes. The orders Lactobacillales and Enterococcaceae were the primary representative of this phylum. The primary genus found in this group

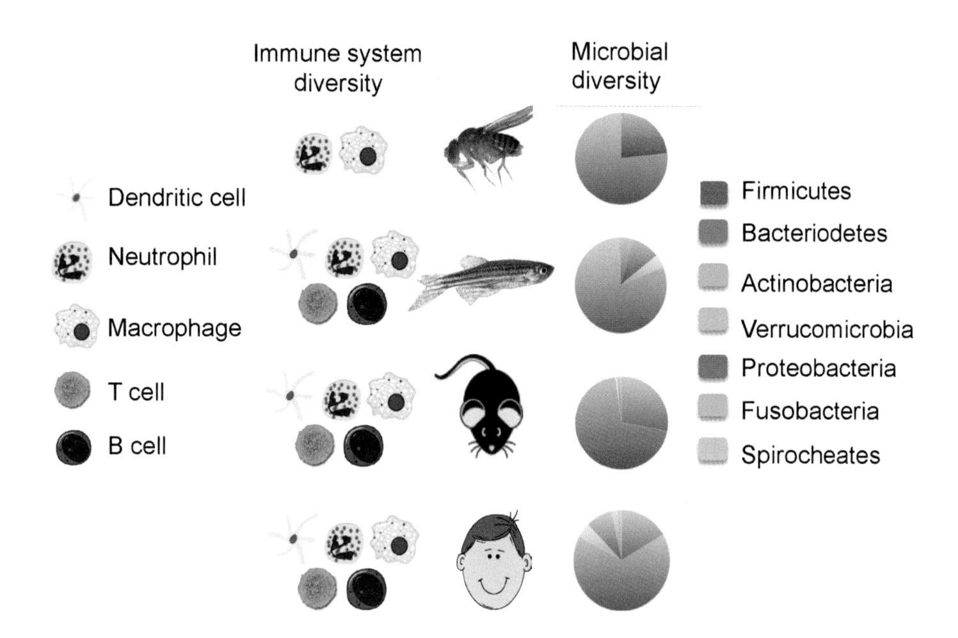

Figure 10.1. Microbial and immune system diversity in animal models of commensal–host interactions. The fly, zebrafish, and mouse represent the most widely used animal models to study commensal bacterial interactions and host immunity. Here, both microbial diversity and immune system diversity are shown to highlight similarities and differences of each model organism in humans. The model organism is illustrated in the middle, the immune system diversity is shown on the left and the microbial diversity within the intestinal tract of the model organism is represented on the right. Schematic representation of microbial diversity is based on information for the fly found in Chandler et al. [6]; zebrafish, in Roeselers et al. [8]; mouse, in Friswell et al. [5]; human, in Ley et al. [4].

was Lactobacillus, which are Gram-positive acidophilic bacteria found on nutrient-rich resources. *Lactobacillus plantarum* is a common species found within the intestine of mammals, and two species isolated from the fly gut had >99% homology to this organism. Fourteen other orders of bacteria represented the last 10% of organisms found in *Drosophilia*. When compared to the organisms found within the mammalian gastrointestinal tract, many differences emerge. While humans and mice are colonized predominantly by bacteria that belong to the phyla Bacteriodetes and Firmicutes, the fly is most abundantly associated with organisms belonging to Proteobacteria. *Drosophilia* have a large representation of organisms belonging to the Firmicutes; however, <1% of the organisms cataloged belong to the Bacteriodetes, indicating differences between the microbiota of flies and mammals. Similarities do emerge, however, as humans are also colonized by *Lactobacillus* and *Enterococcus* species.

Zebrafish (*Danio rerio*) is another animal model that is rapidly being utilized in the laboratory because it is amenable to *in vivo* imaging, genetic analysis, and germ-free studies [7]. This model, has been used in the past to study pathogenic organisms such as *Pseudomonas aeruginosa*, *Staphylococcus aureus*, or

Mycobacterium marinum but has more recently been utilized to study commensal organisms. More recent studies have identified a core microbiome within zebrafish [8]. Rawls and colleagues compared zebrafish reared in the laboratory versus zebrafish that were recently caught. These studies revealed that, unlike *Drosophilia*, there appeared to be a close similarity between the microbiotas, suggesting a strong selection by the host for the bacteria that reside on its body. Much like *Drosophilia*, the zebrafish microbiota is dominated by the bacterial phylum, Proteobacteria. However, the phyla Firmicutes and Fusobacteria are also prevalent during larval and adult stages, respectively. Also representing a small proportion of the zebrafish microbiota are representatives from the phyla Actinobacteria and Bacteriodetes. Thus, the composition of microbiota from zebrafish differs greatly from that of mice or humans in that, instead of members of the phyla Firmicutes and Bacteridetes, Proteobacteria dominate (76–82%) within the zebrafish gut. Despite the differences bacterial populations, much can be learned from these models as rapid genetic screens and large population sizes cannot be done with mice.

10.3. COMPARATIVE IMMUNITY: IMMUNE SYSTEM DEVELOPMENT IN ANIMALS

It seems that all forms have life have evolved methods to defend themselves from other organisms, and therefore possess a functional immune system. Even bacteria and some archaea have evolved mechanisms to fend off phages and invading nucleic acids in the form of clustered regularly interspaced short palindromic repeats (CRISPRS) [9]. These sequences are found in distinct loci and provide acquired immunity by targeting specific sequences from invading elements. Therefore, active immunity is an ancient evolutionary mechanism of survival. While many components of the immune system are conserved from fly to humans, many elements are unique to specific model systems, and thus each organism responds to the commensal microbiota in distinct ways. Figure 10.1 summarizes the cellular immune cells that are found in each model organism. Humans, mice, and zebrafish [7] have both innate and adaptive immunity, which includes the use of cells such as dendritic cells (DCs), phagocytic macrophages, antibody producing B cells, and antigen-specific T cells. However, there is no evidence that there is an adaptive immune system in *Drosophila* [10]. Indeed, no specific antibody response can be detected, and gene rearrangement has not been shown to occur within these animals. Thus, *Drosophila* has emerged as an animal model for innate immunity. Cellular components of the fly immune system include plasmatocytes, phagocytic macrophage-like cells, crystal cells, and lamellocytes. The plasmatocytes and phagocytic cells represent ~90% of the blood population. Crystal cells act as a source of enzymes that aid in the melanization reaction (a process involing deposition of melanin during a wounding response). Lamellocytes are flattened cell types that that are able to differentiate and encapsulate parasite eggs (reviewed in Ref. 10). Perhaps one of the most famous immune system discoveries made in *Drosophila* was the identification of the transmembrane protein toll [11,12]. *Drosophila* has nine different toll receptors that activate inflammatory cascades that lead to activation of NFkB-like proteins. Therefore, the fruit fly has provided a powerful animal model in the identification of multiple components of innate inflammatory signaling.

The zebrafish is a relatively newer model for studying the effects of the immune system. The zebrafish contains both innate and adaptive immunity, and therefore more closely resembles the human immune system. Innate immunity develops quickly with macrophage-like cells appearing ~25 h postfertilization; B and T cells are also present within zebrafish [13]. Hematopoiesis within the zebrafish occurs very similarly to mammals and gives rise to monocytes, macrophages, neutrophils, and eosinophils [14]. Additionally, mast cells and cells with cytotoxic properties much like those of mammalian NK (natural killer) cells have been shown to exist in zebrafish [14]. After ~4 days postfertilization T and B lymphocyte progenitors begin undergoing *rag*-dependent rearrangement within the thymus [15] and pancreas [16]. The structure of the immunologlobulin (Ig) M gene in zebrafish and evidence for heavy-chain variable sequences has also been reported [7]. Additionally, genes that resemble TCRα and MHC have been identified, indicating that zebrafish serve as a good model system for immune system development and hematopoiesis. Even subtypes of T lymphocytes have been identified. An antiinflammatory T cell type, T regulatory cells (Treg), are marked by expression of the transcription factor Foxp3 and function to control inflammation. In the absence of Tregs, humans and mice succumb to overt autoimmunity and are severely immune-deficient. The functional homolog of Foxp3 has been identified in zebrafish and contains many of the same regulatory subunits within its promoter as humans and mice [17]. More importantly, it seems to be restricted to lymphocyte subpopulations as in humans and mice, suggesting it may have similar functions. Another very prominent inflammatory T lineage, Th17 cells, are cells that produce copious amounts of the cytokine IL17. More recently, five genes were identified and cloned in the zebrafish that encode for this cytokine [18], while it is unclear whether the mature T cells in the fish express this cytokine and thus resemble a bona fide Th17 cell, this suggests the possibility that zebrafish serve as a viable model for understanding human immune system development and disease. Infectious disease such as *Mycobacterium*, *Aeromonas*, and even fungal pathogens such as *Lecythophora*, have been well modeled in zebrafish [19]. Therefore, the zebrafish is a highly versatile model for understanding development and infectious disease; however, other human diseases such as autoimmunity have yet to be developed, although they are beginning to emerge. Indeed, a single study explored the use of a chemically induced model of enterocolitis in zebrafish [20]. Using trinitrobenzene sulfonic acid (TNBS) as the inducing agent they found that this model recapitulated key aspects of human IBD, including induction of proinflammatory pathways and degradative enzymes and leukocytosis around the intestine. Additionally, they found both genetic and bacterial susceptibilities to induction of disease that were ameliorated by administration of antibiotics. Rodents, including mice and rats, have long served as model organisms for multiple diseases. All these organisms have the same cellular immune cells that are present in humans, and many of the molecular pathways are conserved. The strength of the mouse model currently is that many models of autoimmunity and infectious disease have been developed. These include models of inflammatory bowel disease (IBD), multiple sclerosis (MS), asthma, allergies, diabetes, and arthritis. Much has been learned about these diseases using mouse animal models and are discussed later in the chapter.

10.4. MAINTAINING INTESTINAL HOMEOSTASIS

The maintenance of intestinal homeostasis is imperative to achieving health. Preserving a balance between inflammatory and antiinflammatory immune responses is a critical component in attaining homeostasis. Therefore, mechanisms by which the host achieves this balance remain important discoveries. More recent studies have demonstrated that the microbiota governs intestinal immune system responses in multiple animal models. The strongest piece of evidence that the microbiota is important for the development of the host immune system is demonstrated when animals are born and reared in a sterile environment. *Drosophilia* have a decreased lifespan when raised axenically; germ-free zebrafish suffer from severe gut deterioration [21] and fail to reach adulthood, and germ-free mice have perturbed intestinal development [22], decreased immune cell function and diversity, and require more food to maintain healthy body weight. Thus, animals seem to have a strong dependence on the presence of the microbiota, which suggests that bacteria play an important role in maintaining homeostasis. However, both pathogenic and commensal bacteria share similar structural features, such as lipopolysaccharide and peptidoglycan, which might make it difficult for the immune system to distinguish between them. Therefore, the immune system is posed with a problem of mounting efficient immune responses toward pathogens while maintaining tolerance to commensals. How this distinction is made remains enigmatic, and the answer likely holds the key to maintaining intestinal health.

As bacteria and their hosts have evolved together overtime, it is not surprising that both sides have devised approaches to preserve health. Therefore, a dynamic balance has been made that includes microbe-initiated and host-initiated strategies to make the relationship work. Some of these examples are represented in Figure 10.2 and explained in more detail in the following paragraphs. Strategies to enforce tolerance include bacterial expressed molecules that are capable of coordinating host immunity and specialized host receptors and detoxifying enzymes that control an overzealous response to benign microbes while also preventing over-growth of the microbiota. This section focuses on our current knowledge of how commensal bacteria shape immunity and homeostasis and discuss mechanisms by which these organisms are tolerated by the host.

NFκB is a transcription factor that is evolutionarily conserved and expressed in multiple immune cell types. Classically, NFκB proteins are induced by pathogenic bacteria and are instrumental in initiating inflammatory cascades that result in clearance of the pathogen. A study in 2008 provided a new role for NFκB (called Relish or Rel in the fly) in maintaining tolerance in the fly gut [23]. It is well documented that NFκB/Rel is constitutively imported into the nucleus in the cells of the fly midgut; however, when flies are reared in a germ-free environment, NFκB/Rel was no longer localized to the nucleus and unable to transactivate target genes. Interestingly, on infection with the pathogen *Erwinia*, the nuclear translocation of NFκB/Rel was even more pronounced when compared to steady state levels. When they analyzed the differences in NFκB/Rel target gene expression during pathogen or commensal infection, they found profound differences in functional gene output. When flies were colonized with only commensal bacteria, they observed that, despite chronic NFκB/Rel translocation to the nucleas, antimicrobial peptides (AMPs) were not induced. However, flies infected with the pathogen *Erwinia* enhanced NFκB/

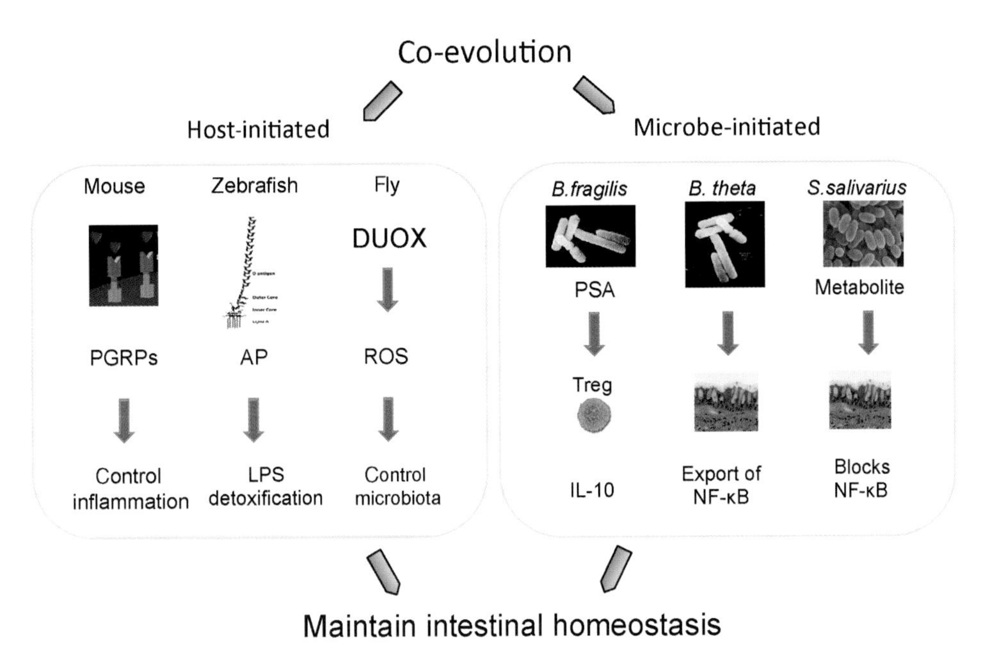

Figure 10.2. Mechanisms from bacteria and host to maintain tolerance. Bacteria and host have evolved together with one another and devised strategies to enforce mucosal tolerance. Some of the known mechanisms of maintaining intestinal homeostasis are derived from the host, and some are derived from the bacteria themselves and have thus been classified as host-initiated or bacteria-initiated pathways, respectively. Loss of any one of these mechanisms has been experimentally shown to result in the loss of the delicate balance that is forged between microbe and host. Many of the microbe-initiated pathways involve manipulating host immunity to enforce tolerance while the host pathways involve downregulation of inflammatory pathways that are typically turned on by pathogens. Therefore, both host and microbe work together to ensure a symbiotic relationship. (Abbreviations: AP—alkaline phosphatase; DUOX—dual oxidase, PGRPs—peptidoglycan recognition proteins, ROS—reactive oxygen species).

Rel activation and coordinately induced expression of AMPs. They further demonstrated that the gene *Caudal*, a transcription factor that binds to the promoter regions of certain AMPs, was important for repression of AMP genes during commensal bacteria colonization. Therefore, to maintain tolerance to the microbiota, flies control levels of AMPs. How the fly immune system is able to accomplish this is thought to be through the detection of peptidoglycan (PG) levels (reviewed in Ref. 24). In the absence of infection, small quantities of peptidoglycan derived by commensal bacteria drive expression of NFκB/Rel. Negative regulators of the pathway (including peptidoglycan cleaving molecules (PGRP-SC1) and peptidoglycan recognition protein LC interacting inhibitor of IMD signaling (PIMS) [25] are also under the transcriptional control of NFκB/Rel. Therefore, during commensal colonization the negative regulators constantly impose a brake on gene expression and provide a buffered threshold of signaling. However, during infection, large amounts of peptidoglycan transiently increase the signaling pathway and override the negative regulators to trigger AMP expression to clear infection. Therefore, detection of PG and regulation of the NFκB/Rel signaling pathway is a mechanism by which the host distinguishes commensal from pathogen in the fly. Interestingly, a distinct pathway exists in the fly to control commensal microbial outgrowth [26]. Reactive oxygen species (ROS) is the second mechanism utilized in the fly gut to maintain homeostatic balance and is regulated in a manner distinct from AMP. ROS are constitutively produced in the gut through membrane-associated dual oxidase (DUOX). When the microbiota interacts with host epithelial cells, it increases expression of DUOX through molecules that are distinct from peptidoglycan (although their identity remains unknown). When the gut bacterial burden increases, so does DUOX expression, thereby inducing ROS production and controlling commensal growth. Interestingly, DUOX and ROS expression are independent of NFκB/Rel signaling, and instead rely on PLC-B and p38 induction. Thus, in the *Drosophila* gut the innate immune responses rely on a finely tuned balance between positive and negative regulatory inputs controlled by specific pathways. Given the conservation of the PGRPs, NFκB, PLC-B, and p38 pathways in mammals, these studies might define conserved mechanisms employed by the host to maintain intestinal homeostasis.

Expression of NFκB in response to microbial challenge is evolutionarily well conserved. Indeed, in response to commensal colonization, the induction of NFκB is seen in flies (discussed above), mammals, and zebrafish. To study the NFκB pathway in zebrafish, Rawls and colleagues fused an eGFP reporter to NFκB and tracked expression in the whole animal during development. Indeed, when germ-free fish were colonized with commensal bacteria, induction of NFκB and its target genes was seen in both intestinal and extra-intestinal tissues [21], suggesting that this pathway also plays a role in maintaining intestinal homeostasis during commensal colonization; however, this mechanism in zebrafish requires further investigation. Guillemin and colleagues have uncovered a novel system of maintaining intestinal tolerance in the zebrafish [27]. The enzyme alkaline phosphatase (AP) modifies LPS by dephosphorylating its lipid A moiety, which accounts for the toxicity of LPS. Vertebrates have broad expression of APs within the intestine; however, their physiological substrates remain unclear. Guillemin and colleagues demonstrated that within the zebrafish intestine, alkaline phosphatase expression increased as the intestine was colonized by commensal bacteria. This increase in expression is lost when animals are maintained in a germ-free environment. They tested the hypothesis that AP activity was important for detoxifying LPS encountered by the intestinal epithelium. Much like mammals, zebrafish are highly sensitive to high doses of LPS; however, when the LPS is dephosphorylated by alkaline phosphatase, it induces less inflammation and the fish can tolerate higher LPS doses. They next depleted alkaline phosphatase from the zebrafish and saw that animals had an increased sensitivity to LPS challenge. Moreover, germ-free fish (that have a reduced expression of alkaline phosphatase activity) are also more sensitive to LPS challenge. Finally, they demonstrated that by detoxifying LPS, neutrophil recruitment is prevented in response to the microbiota. Therefore, the host has evolved a mechanism to detect commensal bacteria and make them more tolerable by modifying a bacterial motif to elicit inflammation. While it remains unclear how the expression of this enzyme changes during pathogenic infection, it highlights an example of a host-initiated pathway of maintaining intestinal tolerance. Interestingly, alkaline phosphatases are conserved throughout mammals, suggesting this as a potential mechanism by which the mammalian host can maintain intestinal homeostasis.

The previous paragraphs have focused on host-initiated mechanisms of maintaining intestinal homeostasis found in fish and flies. Interestingly, similar mechanisms have also been uncovered in mice as well. Earlier we discussed that peptidoglycan levels in the fly gut are sensed by the host and can initiate differential signaling outcomes when commensals or pathogens are present. Similarly, in mice, deficiency in peptidoglycan recognition proteins (PGRPs) leads to a disruption in intestinal homeostasis [28]. Specifically, in the absence of these proteins, a more "inflammatory" microbiota is able to exist within the intestine of these mice. Transfer of the microbiota from PGRP deficient mice into WT germ-free hosts increases their susceptibility to inflammation, indicating that PGRPs in mice are important for the maintenance of intestinal homeostasis. Therefore, these evolutionary conserved host mechanisms allow the host to distinguish pathogen from commensal. But are their mechanisms employed by bacteria to induce tolerance? Extensive studies performed with pathogenic bacteria have identified a multitude of mechanisms by which the bacteria manipulate host immunity to cause disease. It seems reasonable that since commensal bacteria have been present through each stage of animal evolution that they have devised elegant strategies for subverting host immunity to preserve their intestinal niche. The following paragraphs will discuss some examples of studies performed in mammalian model systems that uncovered how particular commensals downregulate the host immune response to maintain a peaceful coexistence with their host.

The commensal *Streptococcus salivarius* is one of the early colonizers of oral mucosal surfaces and remains prevalent in the oral cavity and digestive tract throughout mammalian life. More recently, *S. salivarius* strain K12 was shown to attenuate NFκB, demonstrating a role in suppressing host inflammation [29]. To determine whether products from *S. salivarius* were able to affect the intestinal inflammatory responses, supernatants were collected from growing cultures of *S. salivarius*. Supernatants were then incubated with human intestinal epithelial cells and NFκB activation was measured. Cells that were incubated with the supernatant from *S. salivarius* had greatly reduced NFκB, identifying a secreted bacterial molecule that could directly inhibit inflammation within the intestine. While the identity of the molecule remains enigmatic, it suggests that this bacterium has a dedicated machinery to downregulate intestinal inflammation while it colonizes its host. Further experimentation will be required to define whether this mechanism has evolved to define the bacterial niche within the intestine.

Bacteriodes fragilis is a Gram-negative anaerobic bacterium that predominates in the human colon. These bacteria specialize in the synthesis of up to eight different polysaccharides that are expressed on the cell surface. The ability to express a variety of these polysaccharides is important for competitive colonization within the mammalian intestine [30]. One of these polysaccharides, polysaccharide A (PSA), is the most predominantly expressed sugar and possesses a potent immunomodulatory capacity [31]. Within the mammalian intestine most commensal bacteria are found within the lumen and are unable to gain access to the host tissue. In most cases, when commensals come into contact with the tissue, an inflammatory response ensues and the bacteria are cleared. However, more recent studies have uncovered a novel mechanism by which *B. fragilis* manipulates host immunity to maintain a small tissue associated population of bacteria [32]. T helper cells are separated into four main populations, including proinflammatory IFNγ producing $CD4^+$ T cells (Th1 cells), IL17-producing $CD4^+$ T cells (Th17 cells), IL4-producing $CD4^+$ T cells (Th2 cells), and anti-inflammatory T regulatory cells (Tregs). Both Th1 and Th17 cells are important for the clearance of extracellular pathogens, while Th2 cells are required for the response against helminthes and parasites. Tregs ensure control of inflammatory immunity and prevent immune responses against self- (host) tissue. Tregs have been shown to be important in preventing inflammation that leads to autoimmune diseases such as multiple sclerosis (MS), diabetes, and inflammatory bowel disease (IBD). Germ-free animals that are monoassociated with *B. fragilis* had an increased percentage of IL10-producing Tregs and a complete lack of inflammatory Th17 cells. Induction of Tregs was reliant on PSA as animals monoassociated with *B. fragilis* ΔPSA had little to no $IL10^+$ Tregs. Interestingly, when animals are associated with *B. fragilis* ΔPSA, a potent Th17 response is elicited, suggesting that PSA actively suppresses inflammatory responses within the intestine. Indeed, if Tregs are specifically ablated in animals that are monoassociated with *B. fragilis*, Th17 development is induced in response to bacterial colonization. While inflammation does not occur in response to *B. fragilis* colonization, a small but significant population of *B. fragilis* are found to be closely associated with the mouse tissue within the crypt of the intestinal epithelial cells. Interestingly, *B. fragilis* ΔPSA is unable to associate with host tissue because it induces an inflammatory response. Supporting this, when a neutralizing antibody to IL17A is administered to animals colonized with *B. fragilis* ΔPSA, the organism can now associate with host tissue. Finally, ablation of T regulatory cells abolishes the ability of wild type *B. fragilis* to associate with host tissue. This indicates that *B. fragilis* coordinates host T regulatory cells to specifically suppress immune responses. The host receptor that is responsible for mediating this signal was found to be toll-like receptor 2 (TLR2). TLR2 is known to recognize peptidoglycans and lipoproteins found on the surface of bacteria. It is classically thought to evoke an inflammatory response when triggered on cells of the innate immune system. However, PSA seems to selectively target TLR2 on T cells and triggers an antiinflammatory IL10 response. Therefore, PSA is the incipient member of a class of molecules termed *symbiosis-associated molecular patterns* (SAMPs). Unlike pathogen associated molecular patterns (PAMPs) that induce inflammation and clearance of the bacteria, SAMPs trigger responses that enforce symbiosis. Taken together, these provide examples of how bacteria have evolved specialized molecules to manipulate host immunity to allow for their persistence on host mucosal surfaces. Certainly, other commensal bacteria have evolved similar strategies to coordinate host immunity for their benefit. Identification of these mechanisms will not only allow for a more in-depth understanding of how tolerance is maintained but also may identify novel therapeutic strategies that can be used to enforce homeostasis.

10.5. THE ROLE OF COMMENSAL BACTERIA IN HOST HEALTH

While much remains unclear, more recent studies have solidified that commensal bacteria can have profound influences on the development of the immune system, but how do these development consequences affect host health? A wealth of recent data have explored the effects of commensal bacteria on multiple autoimmune diseases and infections. Most of these studies have been performed in mouse models of disease, as no established models of autoimmune disease exist in fly and zebrafish. The rest of this chapter is dedicated to presenting what has been learned about the role of commensal bacteria in autoimmunity from the use of mouse models.

Autoimmunity is a term used to describe when the immune system begins to mount an inflammatory response directed at "self"-molecules. In healthy individuals, these molecules are tolerated, and autoreactive cells that initiate these responses are deleted early in their development. Autoimmune diseases include multiple sclerosis, diabetes, inflammatory bowel disease, allergies, asthma, and rheumatoid arthritis. More recent epidemiological and clinical reports have described a rapid and marked increase in many of these diseases in Westernized populations, including the United States, Japan, and Europe. Since these increases in incidence have occurred so quickly, it is unlikely that genetics is the sole reason for the rise in incidence and suggests that an environmental factor is to blame. The hygiene/microbiota hypothesis was proposed to explain the rapid increase in autoimmunity. This hypothesis states that perturbations in the gastrointestinal populations of bacteria due to changes in antibiotic use, vaccination, and diet have disrupted mechanisms that are involved in the development of immunological tolerance [33]. Indeed, a correlation between increased antibiotic use and allergies and asthma has been documented and animals reared in germ-free conditions have exhibited severe immunological developmental defects. More recent studies have identified specific organisms that predispose or protect from diseases such as MS, diabetes, and IBD, demonstrating that the microbiota can have profound influences on human health. Details on how the microbiota influences disease progression follow.

10.5.1. Inflammatory Bowel Disease

Inflammatory bowel disease encompasses diseases that involve chronic inflammation of the gastrointestinal tract. Human diseases that fall into this category include Crohn's disease and ulcerative colitis. In the United States alone, there are an estimated 1.4 million people with IBD and in Europe this number nearly doubles to 2.2 million people; 10% of the people suffering from IBD are children, and there is currently very few treatments and no cure. While it remains unclear how IBD develops, it seems that it is a multifactoral disease that involves an environmental stimulus with a genetic predisposition. Indeed, genomewide association studies have pinpointed a few genes that are commonly found to be dysregulated in people with IBD. IBD risk loci include a large number of genes involved in T cell inflammatory responses, recognition of microorganisms, and autophagy pathways. These genes include *IL23R*, *ATG16L1*, *CCR6*, *STAT3*, and *IL12B* [34]. The fact that both microbial recognition pathways and host immune responses are risk alleles for IBD points to a critical role in maintaining these pathways for intestinal health. Supporting this, multiple mouse models of IBD include immune deficiency such as lack of Tregs or the cytokine IL10 [35].

A number of commensal bacteria have been implicated in protection from IBD as shown in mouse models, and this has been reviewed elsewhere [36,37]. Perhaps the most promising results in humans has been a cocktail of bacteria that is referred to as VSL3. VSL3 contains eight different strains of bacteria that include three

Bifidobacterial species, four *Lactobacillus* species, and one *Streptococcus* species. Administration of the probiotic to patients with ulcerative colitis and pouchitis seemed to ameliorate disease. In animal models, preliminary evidence has shown that VSL3 induces antiinflammatory Tregs and the cytokine IL10. This suggests that the of VSL3 cocktail induces an antiinflammatory response within the colon, and this may be the mechanism by which VSL3 protects from colitis. However, what molecules from the bacteria are responsible for inducing this activity and whether this is the mechanism of protection remain to be elucidated. Other organisms that are associated with protection from IBD in humans include *E. coli Nissle* 1917 [38] and *Faecalibacterium prausnitzii* [39]; however, much work remains to be done to determine whether they can protect from human disease. Another organism that has been shown to be protective in mouse models of inflammatory bowel disease is the commensal bacteria, *B. fragilis* [40]. Colonization of animals with *B. fragilis* potently protects in two experimental models of colitis. Protection is dependent on production of the surface polysaccharide, PSA (discussed earlier), as animals that are colonized with a mutant bacteria that do not express PSA (*B. fragilis* ΔPSA) are no longer protected from disease. Interestingly, the host recognizes PSA through TLR2-signaling. Indeed, TLR2 deficient animals are no longer protected from colitis even when treated with purified PSA [41]. The mechanism by which PSA is able to protect from colitis through TLR2 is the induction of IL10-producing Tregs. Tregs are known to play an important role in maintaining intestinal homeostasis and suppression of inflammation. Thus, these studies demonstrate that *B. fragilis* is able to actively maintain intestinal homeostasis through production of PSA, and loss of these mechanisms results in disease. Conversely, commensal organisms have been associated with induction of disease. *Helicobacter hepaticus* is a common commensal bacterial organism found in many strains of animals. Immunocompetent mice peacefully coexist with *H. hepaticus* with no signs of disease; however, animals with immunodeficiencies, such as lack of IL10 production, suffer from chronic colitis induced by the organism [42]. Taken together, these data suggest that microbial composition can be a very important determinant in progression of disease and highlights the important therapeutic potential of commensal organisms.

10.5.2. Multiple Sclerosis

Multiple sclerosis (MS) is an autoimmune disease that affects the central nervous system (CNS). People afflicted with this disease experience a slow deterioration in their ability to move with the most extreme cases resulting in complete paralysis. There is a high rate of discordance in monozygotic twins indicating that disease progression involves a strong environmental factor. Experimental autoimmune encephalomyelitis (EAE) is a mouse model of MS that reproduces many features of the disease seen in humans. In this model, an antigen-specific immune response is induced toward myelin basic protein (MBP) that is found on the outer covering of the neuron. Interestingly, germ-free mice have a greatly reduced incidence of disease that correlates with reduced Th1 and Th17 responses in the CNS and spleen of these animals. Remarkably, colonization of germ-free animals with a single organism, segmented filamentous bacteria (SFB), induces disease [43]. While germ-free animals are completely devoid of Th17 cells within the lamina propria of the small intestine, colonization by SFB induces these cell populations [44]. It is well established that induction of EAE depends on Th17 and Th1 cellular responses. Therefore, a potential mechanism by which SFBs induce EAE in animals is through their ability to induce these populations. Supporting this, dendritic cells (DCs) isolated from GF mice are deficient in inducing Th17 and Th1 cells *in vitro*, while DCs isolated from animals that are mono-associated with SFBs are capable of inducing these cellular responses. Therefore, colonization of animals with SFBs predisposes the animal to the development of EAE. In contrast, PSA isolated from the commensal organism, *B. fragilis*, has been demonstrated to be protective in models of EAE [45]. Animals orally administered the polysaccharide do not develop as severe a disease when compared to mock-treated animals. This protection correlated with induction of IL10-producing Treg cells. Therefore, the contribution of the microbiota to this disease is complex, and further studies will be necessary to define the role of individual bacteria in disease induction and more importantly, identify the molecules formed by the bacteria to induce these responses.

10.5.3. Diabetes

Diabetes is a chronic metabolic disease that is characterized by increased blood sugar due to either the lack of insulin production or lack of response to insulin. In the year 2000, 171 million people suffered from diabetes. The incidence is rapidly increasing, and it is estimated that by the year 2030, this number will double. Currently, there is no cure for diabetes, but the disease can be managed by changes in lifestyle and medication. Much like IBD, diabetes is a multifactorial disease that involves both environmental and genetic influences. A recent study demonstrated a role for the microbiota in induction of disease [46]. The NOD mouse spontaneously develops a form of diabetes as it ages and is used as a mouse model of diabetes. When this mouse strain is crossed to an animal that lacks MyD88 (the adapter molecule that governs the signaling pathway for most TLRs and IL1 receptors), they find that the incidence of diabetes decreases. When these animals are given antibiotics or rederived germ-free, the incidence of diabetes increases. Giving back MyD88$_{-/-}$ animals a commensal flora restores protection from diabetes, indicating that commensal bacteria play a protective role in the development of diabetes. Along these lines, another group found that the presence of SFB correlated with disease protection in female animals. Consistently, they find the presence of Th17 cells in animals colonized with SFBs, consistent with the commensal microbiota playing a protective role in diabetes [47]. Defining the role of other commensal bacteria in this disease will be of utmost importance. As in IBD, certain bacteria might play a negative role in diabetes induction. Additionally, identifying the mechanisms by which segmented filamentous bacteria (SFB) induce protection may provide insight into novel therapies for this disease. These studies highlight that we have much to learn about the role of commensal bacteria in disease development. Figure 10.3 illustrates how individual bacteria are able to influence host immunity. Maintaining intestinal homeostasis and maintaining health are intimately tied to one another. While some bacteria induce regulatory or antiinflammatory immune responses (such as *B. fragilis*), other commensal bacteria induce inflammatory pathways (such as SFB). Both pathways are required for health. Indeed, if only regulatory pathways were induced, then the host would become susceptible to infection, while in the presence of excess inflammation, autoimmunity would ensue. Therefore,

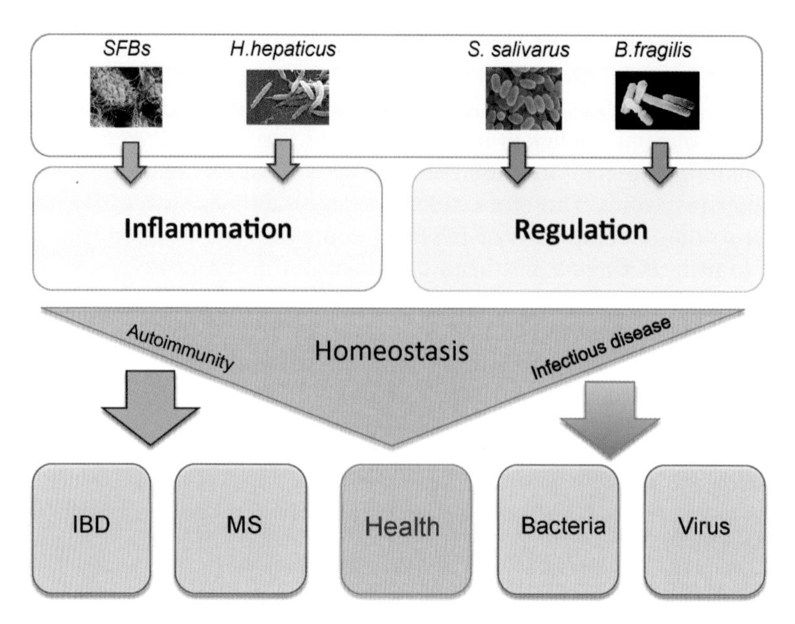

Figure 10.3. Finding the right mix of commensal bacteria. Maintenance of intestinal homeostasis is imperative to health. Researchers are finding that specific bacteria are capable of directing the development of multiple host pathways, including both antiinflammatory and inflammatory cellular responses. Therefore, microbial composition is likely very important in maintaining a homeostatic environment. A host that is colonized with too many microbes that elicit a regulatory or antiinflammatory response (blue box) may become susceptible to infectious pathogens. In contrast, if the host becomes colonized with bacteria that only induce inflammatory pathways (red box), the balance is tipped toward the development of autoimmunity. Therefore, a healthy intestinal immune response relies on the commensal microbial population.

the maintenance of intestinal homeostasis and host health may rely on the composition of the microbiota. Having an equal mix of bacteria that can induce these opposing pathways can ensure the health of an individual.

10.6. FINDING THE RIGHT MIX OF BACTERIA: HARNESSING OUR UNDERSTANDING OF COMMENSAL HOST RELATIONSHIPS FOR THERAPEUTIC BENEFIT

Our understanding of how commensal bacteria influence the development of the immune response is only in its infancy. There are 1000s of different species living within the gastrointestinal tract, in the oral cavity, and on the skin. Identifying these organisms and how they influence host health will be important future scientific endeavors. If bacterial organisms have evolved molecules to coordinate host health, then these molecules can be purified from the bacteria and used therapeutically to treat multiple diseases for which there is currently no cure. Additional studies should focus on bacterial relationships within the intestine as well. While it is imperative to understand how individual commensal bacteria influence the host, it is also important to understand how organisms influence the biology of one another. Understanding interbacterial relationships within the environment have been one of the ways that we have discovered molecules such as antibiotics, as bacteria and fungi utilize these to defend their niches from one another. Since the gastro-intestinal tract represents an amalgam of species of bacteria, there is competition for nutrients and space. During this competition, there is a possibility that novel molecules are being utilized to define the bacterial niche within the intestine. Therefore, studying communities of bacteria instead of bacteria in isolation may lead to the identification of novel therapeutic molecules.

Commensal bacteria not only represents a place to harvest therapeutic molecules that can be given to patients retrospectively as a therapy; if we can uncover the unique role that these bacteria play in host health and how these organisms interact with one another, we can find the right mix of bacteria to give to people before disease ensues. While it remains unclear, it is likely that many of the effects of the microbiota have temporal facets that we have yet to appreciate. Therefore, if a commensal organism is able to colonize the host at a young age while the immune system is still taking shape, that organism may be able to influence host immunity more profoundly and ultimately even prevent disease from occurring. Therefore, identifying a collection of commensal species that can be given at birth may provide the host immune system with the boost that it needs to ward off infection during an early age and protect against the development of autoimmune diseases later in life.

REFERENCES

1. McFall-Ngai MJ. Unseen forces: The influence of bacteria on animal development. *Dev Biol* **242**:1 (2002).
2. Gill SR et al. Metagenomic analysis of the human distal gut microbiome. *Science* **312**:1355 (2006).
3. Eckburg PB et al. Diversity of the human intestinal microbial flora. *Science* **308**:1635 (2005).
4. Ley RE et al. Evolution of mammals and their gut microbes. *Science* **320**:1647 (2008).
5. Friswell MK et al. Site and strain-specific variation in gut microbiota profiles and metabolism in experimental mice. *PLoS ONE* **5**:e8584 (2010).
6. Chandler JA, Morgan Lang J, Bhatnagar S, Eisen JA, Kopp A. Bacterial communities of diverse Drosophila species: Ecological context of a host-microbe model system. *PLoS Genet* **7**:e1002272 (2011).
7. Yoder JA, Nielsen ME, Amemiya CT, Litman GW. Zebrafish as an immunological model system. *Microbes Infect* **4**:1469 (2002).
8. Roeselers G et al. Evidence for a core gut microbiota in the zebrafish. *Isme J* **5**:1595 (2011).
9. Horvath P, Barrangou R. CRISPR/Cas, the immune system of bacteria and archaea. *Science* **327**:167 (2010).
10. Brennan CA, Anderson KV. Drosophila: The genetics of innate immune recognition and response. *Annu Rev Immunol* **22**:457 (2004).
11. Lemaitre B, Nicolas E, Michaut L, Reichhart JM, Hoffmann JA. The dorsoventral regulatory gene cassette spatzle/Toll/cactus controls the potent antifungal response in Drosophila adults. *Cell* **86**:973 (1996).

12. Poltorak A et al. Defective LPS signaling in C3H/HeJ and C57BL/10ScCr mice: Mutations in Tlr4 gene. *Science* **282**:2085 (1998).
13. Herbomel P, Thisse B, Thisse C. Ontogeny and behaviour of early macrophages in the zebrafish embryo. *Development* **126**:3735 (1999).
14. Kanther M, Rawls JF. Host-microbe interactions in the developing zebrafish. *Curr Opin Immunol* **22**:10 (2010).
15. Willett CE, Cherry JJ, Steiner LA. Characterization and expression of the recombination activating genes (rag1 and rag2) of zebrafish. *Immunogenetics* **45**:394 (1997).
16. Danilova N, Steiner LA. B cells develop in the zebrafish pancreas. *Proc Natl Acad Sci USA* **99**:13711 (2002).
17. Mitra S, Alnabulsi A, Secombes CJ, Bird S. Identification and characterization of the transcription factors involved in T-cell development, t-bet, stat6 and foxp3, within the zebrafish, Danio rerio. *FEBS J* **277**:128 (2010).
18. Gunimaladevi I, Savan R, Sakai M. Identification, cloning and characterization of interleukin-17 and its family from zebrafish. *Fish Shellfish Immunol* **21**:393 (2006).
19. Sullivan C, Kim CH. Zebrafish as a model for infectious disease and immune function. *Fish Shellfish Immunol* **25**:341 (2008).
20. Oehlers SH et al. A chemical enterocolitis model in zebrafish larvae that is dependent on microbiota and responsive to pharmacological agents. *Dev Dynam* **240**:288 (2011).
21. Kanther M et al. Microbial colonization induces dynamic temporal and spatial patterns of NF-kappaB activation in the zebrafish digestive tract. *Gastroenterology* **141**:197 (2011).
22. Smith K, McCoy KD, Macpherson AJ. Use of axenic animals in studying the adaptation of mammals to their commensal intestinal microbiota. *Semin Immunol* **19**:59 (2007).
23. Ryu JH et al. Innate immune homeostasis by the homeobox gene caudal and commensal-gut mutualism in Drosophila. *Science* **319**:777 (2008).
24. Muyskens JB, Guillemin K. Bugs inside bugs: What the fruit fly can teach us about immune and microbial balance in the gut. *Cell Host Microbe* **3**:117 (2008).
25. Lhocine N et al. PIMS modulates immune tolerance by negatively regulating Drosophila innate immune signaling. *Cell Host Microbe* **4**:147 (2008).
26. Ha EM et al. Coordination of multiple dual oxidase-regulatory pathways in responses to commensal and infectious microbes in drosophila gut. *Nat Immunol* **10**:949 (2009).
27. Bates JM, Akerlund J, Mittge E, Guillemin K. Intestinal alkaline phosphatase detoxifies lipopolysaccharide and prevents inflammation in zebrafish in response to the gut microbiota. *Cell Host Microbe* **2**:371 (2007).
28. Saha S et al. Peptidoglycan recognition proteins protect mice from experimental colitis by promoting normal gut flora and preventing induction of interferon-gamma. *Cell Host Microbe* **8**:147 (2010).
29. Kaci G et al. Inhibition of the NF-kappaB pathway in human intestinal epithelial cells by commensal Streptococcus salivarius. *Appl Environ Microbiol* **77**:4681 (2011).
30. Liu CH, Lee MS, Vanlare JM, Kasper DL Mazmanian SK. Regulation of surface architecture by symbiotic bacteria mediates host colonization. *Proc Natl Acad Sci USA* **105**:3951 (2008).
31. Mazmanian SK, Liu CH, Tzianabos AO, Kasper DL. An immunomodulatory molecule of symbiotic bacteria directs maturation of the host immune system. *Cell* **122**:107 (2005).
32. Round JL et al. The Toll-like receptor 2 pathway establishes colonization by a commensal of the human microbiota. *Science* **332**:974 (2011).
33. Noverr MC, Huffnagle GB. Does the microbiota regulate immune responses outside the gut? *Trends Microbiol* **12**:562 (2004).

34. Barrett JC et al. Genome-wide association defines more than 30 distinct susceptibility loci for Crohn's disease. *Nat Genet* **40**:955 (2008).
35. Kullberg MC et al. Helicobacter hepaticus-induced colitis in interleukin-10-deficient mice: cytokine requirements for the induction and maintenance of intestinal inflammation. *Infect Immun* **69**:4232 (2001).
36. Round JL, Mazmanian SK. The gut microbiota shapes intestinal immune responses during health and disease. *Nat Rev Immunol* **9**:313 (2009).
37. Round JL, O'Connell RM, Mazmanian SK. Coordination of tolerogenic immune responses by the commensal microbiota. *J Autoimmun* **34**:J220 (2010).
38. Grabig A et al. Escherichia coli strain Nissle 1917 ameliorates experimental colitis via toll-like receptor 2- and toll-like receptor 4-dependent pathways. *Infect Immun* **74**:4075 (2006).
39. Sokol H et al. Faecalibacterium prausnitzii is an anti-inflammatory commensal bacterium identified by gut microbiota analysis of Crohn disease patients. *Proc Natl Acad Sci USA* **105**:16731 (2008).
40. Mazmanian SK, Round JL, Kasper DL. A microbial symbiosis factor prevents intestinal inflammatory disease. *Nature* **453**:620 (2008).
41. Round JL, Mazmanian SK. Inducible Foxp3+ regulatory T-cell development by a commensal bacterium of the intestinal microbiota. *Proc Natl Acad Sci USA* **107**:12204 (2010).
42. Kullberg MC et al. Helicobacter hepaticus triggers colitis in specific-pathogen-free interleukin-10 (IL-10)-deficient mice through an IL-12- and gamma interferon-dependent mechanism. *Infect Immun* **66**:5157 (1998).
43. Lee YK, Menezes JS, Umesaki Y, Mazmanian SK. Proinflammatory T-cell responses to gut microbiota promote experimental autoimmune encephalomyelitis. *Proc Natl Acad Sci USA* **108**(Suppl 1):4615 (2011).
44. Ivanov II et al. Induction of intestinal Th17 cells by segmented filamentous bacteria. *Cell* **139**:485 (2009).
45. Ochoa-Reparaz J et al. A polysaccharide from the human commensal Bacteroides fragilis protects against CNS demyelinating disease. *Mucosal Immunol* **3**:487 (2010).
46. Wen L et al. Innate immunity and intestinal microbiota in the development of type 1 diabetes. *Nature* **455**:1109 (2008).
47. Kriegel MA et al. Naturally transmitted segmented filamentous bacteria segregate with diabetes protection in nonobese diabetic mice. *Proc Natl Acad Sci USA* **108**:11548 (2011).

11

INSIGHTS INTO THE HUMAN MICROBIOME FROM ANIMAL MODELS

BETHANY A. RADER

Department of Molecular and Cell Biology, University of Connecticut, Storrs, Connecticut

KAREN GUILLEMIN

Institute of Molecular Biology, University of Oregon, Eugene, Oregon

11.1. INTRODUCTION

All humans are ecosystems, home to hundreds of trillions of microorganisms. Among the medical community there is a growing appreciation for the importance of these tiny residents in human health and disease. Fueled by initiatives such as the National Institutes of Health-supported Human Microbiome Project and enabled by next-generation sequencing technologies, we have gained a vast amount of information about the species composition (microbiota) and gene content (microbiome) of human-associated microbial communities. However, there are many questions about host-associated microbial communities that cannot be answered from these burgeoning datasets. For example, how do these communities assemble in space and time, and what are the relative contributions of host genetics, environmental factors such as diet, and stochastic sampling, to their assembly? What functions do these communities contribute to normal physiological processes in their hosts? How and under what circumstances do these communities promote pathogenic processes? Precise mechanistic answers to these questions are essential for translating our new knowledge of our microbial residents into clinical practices to prevent or reverse detrimental conditions associated with perturbed microbe–host interactions.

The Human Microbiota: How Microbial Communities Affect Health and Disease, First Edition. Edited by David N. Fredricks.

Essential to the research mission of human microbiome studies are animal models. Below we discuss several animal models that are making important contributions to our understanding of our coexistence with microorganisms. We emphasize that no single system is suitable for modeling all aspects of human–microbe interactions and answering all that we would like to know about the human microbiome, and we point out the various strengths of individual animal models in this research endeavor.

11.2. ANIMAL MODEL SYSTEMS

All of animal life evolved in a microbial world, and thus all animals possess strategies to maintain intimate interactions with microbes. Here we will use the original definition of "symbiosis" coined by Heinrich Anton de Bary in 1879 to mean "the living together of unlike organisms" and encompassing the spectrum of interactions from pathogenesis (in which one partner is harmed) to mutualism (in which both partners benefit). Several models of microbie–host mutualisms have made major contributions to our understanding of symbiosis. Some of these microbe–host models, such as the bobtail squid, were developed specifically to study mutualism, whereas others grew out of long research histories of the model animals, such as fruit flies, mice, and zebrafish. A key feature of most of these models is that the animal host can be grown in the absence of its associated microbes (axenic or germ-free). For all of these models, at least some representative microbial partners can be cultured. Thus the partnerships can be manipulated in the laboratory, and in the case of animals with naturally complex consortia, they can be maintained with a completely known complement of microbes (gnotobiotic). For some of these models, the bacteria or the host can be manipulated genetically to test the function of individual genes and genetic processes in the establishment and maintenance of the symbiosis.

We distinguish these models by the complexity of the host-associated microbial consortia. In general, invertebrates harbor simple microbial communities and provide natural examples of simple binary associations (Figure 11.1) such as the monoculture of luminescent bacteria that colonizes the squid light organ. Even multimember consortia tend to be low complexity in invertebrates (Figure 11.2), such as the two-member community in the leech crop. A notable exception is the

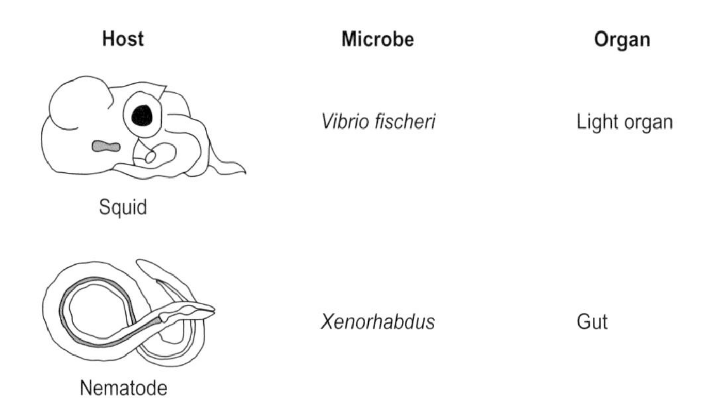

Figure 11.1. Animal models of binary host–microbe associations.

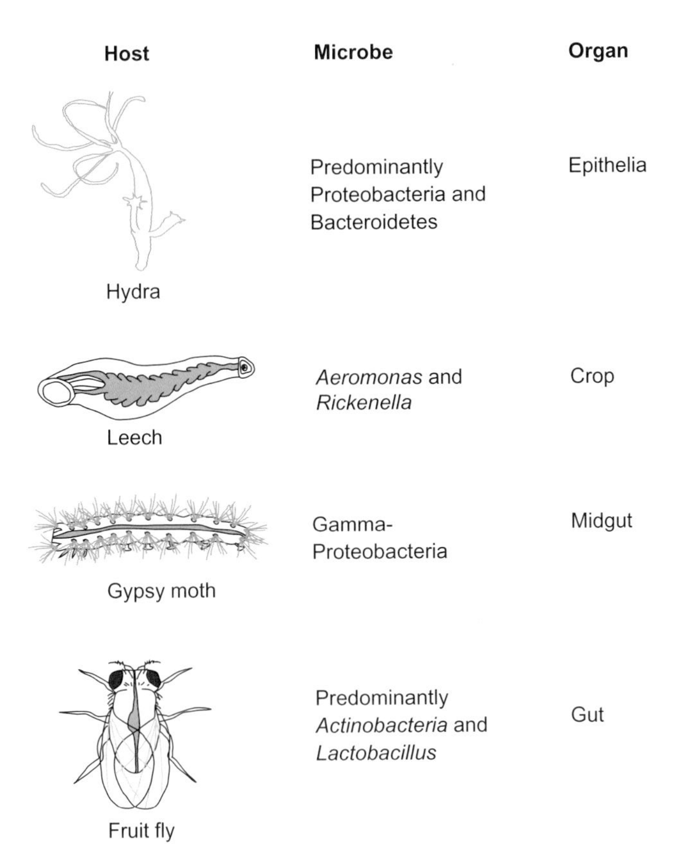

Figure 11.2. Animal models of host associations with low-complexity microbial communities.

complex microbial community of the termite midgut that harbors archea and microscopic eukaryotes as well as bacteria that collectively contribute to the digestion of a cellulose-based diet. Vertebrates, in contrast, almost exclusively harbor complex microbial communities (Figure 11.3), where low complexity communities are characteristic of disease, as in the case of pathogenic overgrowth of *Clostridium difficile* in the mammalian intestine. Indeed, the adaptive immune system of vertebrates is hypothesized to have evolved for the maintenance of complex host-associated microbial consortia [1].

Below we describe briefly the important animal model hosts for studying symbiosis illustrated in Figures 11.1, 11.2, and 11.3. In the following sections we describe how these different model systems have contributed to our understanding of basic processes of host–microbiota interactions relevant to humans.

11.2.1. Squid

A powerful model for studying bacterial–animal mutualisms is the squid–*Vibrio* system, a binary association between the Gram-negative bioluminescent bacteria *Vibrio fischeri* and the Hawaiian bobtail squid, *Euprymna scolopes* (reviewed in Ref. 2). This system has been used to study the establishment, development, and

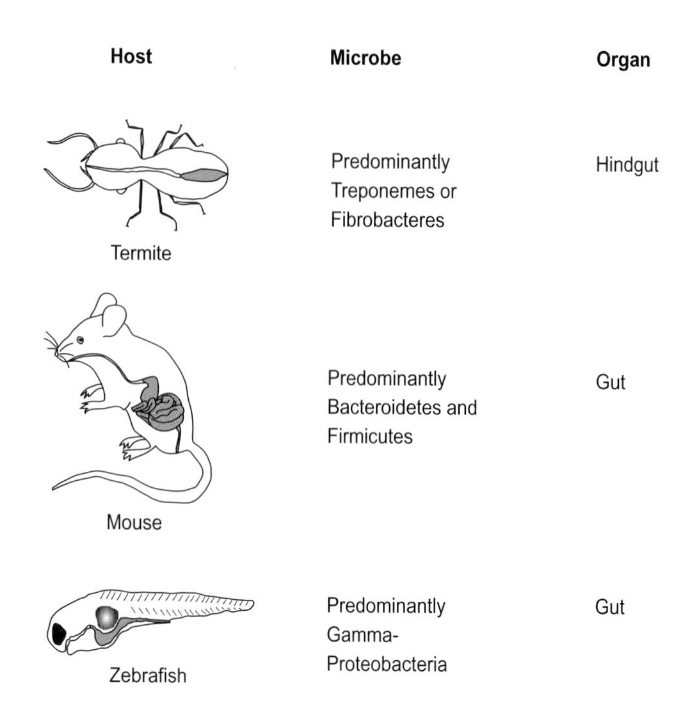

Figure 11.3. Animal models of host associations with high-complexity microbial communities.

maintenance of beneficial symbioses. Through a complex process involving reciprocal communication between host and symbiont, *V. fischeri* colonizes a specialized tissue structure called the light organ within the animal's mantle cavity. While the host gives the symbiont a protected and nutrient-rich place to live, it utilizes the light produced by the bacteria for counterillumination to avoid predators during its nocturnal hunting behavior. The simplicity of this one host–one symbiont interaction allows for the evaluation of the exact contribution of each to the symbiosis. The strength of the system lies in the amenability of the bacterial symbiont, *V. fischeri*, to genetic manipulations such as gene knockout, overexpression of target genes, and mutational screening. Using these methods, an exceptionally detailed account of the events of colonization has been developed.

11.2.2. Nematode

A second binary bacterial–animal mutualism is the association between the enteric bacteria *Xenorhabdus* and *Photorabdus* and their respective nematode hosts, nematodes of the families Steinernematidiae and Heterorhabditiae (reviewed in Refs. 3 and 4). This association has the added complexity of the pathogenic interaction between the nematode and an insect host. During this pathogenic phase the nematode regurgitates the bacteria into the insect host, where it will proliferate and eventually cause the death of the insect, at which time it will reassociate with the nematode host, which has also gone through several generations. The bacterial and nematode partners are amenable to genetic manipulation, and the mutualist–pathogen–mutualist cycle can be carried out in the laboratory. Investigations into these associations has yielded insight into the complex regulation of the lifecycle events of both partners with emphasis on both how the bacteria transitions from mutualist to pathogen and back, and comparisons of the mutualistic and pathogenic states. Furthermore, the bacterial partners are classified in the family Enterobactericeae (enteric bacteria), which also includes the mammalian pathogens *Escherichia coli*, *Salmonella*, and *Yersinia* spp., making it possible to compare the mechanisms involved in this symbiosis to the human enteric symbiosis.

11.2.3. Hydra

The Cnidarian Hydra is one of the most basal groups in the Kingdom Animalia. It contains a simple body plan consisting of two epithelial layers (endoderm and ectoderm) forming a tube-like structure, a single body axis with a head, gastric region, and foot, and has previously been used as a model in developmental biology. More recently, Hydra has become a model for studying ancestral associations between microbes and epithelial tissues (reviewed in Ref. 5) since the epithelia of Hydra is colonized by a relatively simple microbial community and has an effective innate immune system with which it interacts with the bacteria at the epithelial surface, similar to the human intestine. Also, manipulations of the host's epithelia have revealed that changes in epithelial homeostasis can cause changes in the microbial community, implying a direct interaction between epithelia and the microbiota [6]. Thus the Hydra symbiosis may help us understand the evolutionary impact of a microbe-dependent lifestyle.

11.2.4. Leech

The medicinal leech *Hirudo verbana* represents a relatively new model system for the study of digestive tract symbioses. The largest compartment of the leech digestive tract, the crop, stores the ingested blood meal and houses a simple two-member bacterial community composed of *Aeromonas veronii*, which can colonize a variety of host digestive tracts including humans, and a Rikenella-like bacteria that until recently [7] was unculturable.The leech–bacteria symbiosis is thought to be beneficial, and the bacteria are hypothesized to aid in the digestion of the blood meal and provide essential nutrients that are lacking in the leeches' restricted blood diet. The association with *A. veronii* is thought to prevent colonization by other potentially detrimental microorganisms and provide a microenvironment for Rikenella. As in the squid–*Vibrio* symbiosis, the symbiont *A. veronii* is amenable to genetic manipulation [8]. Further distinguishing the leech tripartite symbiosis is the use of symbiont genetic manipulation to understand the role of symbiont–symbiont interactions in the establishment and function of the symbiosis. Also, since *A. veronii* is a human pathogen that can be transmitted by leech bites, this model system can be used to identify factors that allow bacteria to switch from a mutualistic to a pathogenic relationship with animals.

11.2.5. Gypsy Moth

Lepidopterans harbor bacterial communities in their midguts that have been hypothesized to provide essential nutrients to the host or assist in the function of the midgut. The composition and regulation of these communities are of interest because of the challenging environment of the midgut resulting from the ingestion of a broad range of chemically diverse plant species in their diet. The gypsy moth *Lymantria dispar* is a lepidopteran insect and an invasive defoliator of deciduous trees in North America. A relatively simple midgut bacterial consortial community was identified and consists largely of γ-Proteobacteria, and contains *Enterococcus* and *Enterobacter* species, bacteria that are also members of the human gut microbiota [9]. More recent work on this model system has shown that the midgut microbiota promotes toxicity of the insect pathogen *Bacillus thuringiensis* [10], due in part to stimulation of an immune response to the pathogen, which requires the presence of the midgut bacterial community [11]. Although not a human pathogen, *B. thuringiensis* causes disease using a mechanism also employed by a number of human pathogens. Thus the gypsy moth serves as a model for studying a cooperative relationship between pathogen bacteria and the indigenous gut microbiota.

11.2.6. Fruit Fly

The fruit fly *Drosophila melanogaster* was among the first eukaryotes used for genetic analysis, and has one of the most sophisticated sets of genetic tools available in any experimental system. Microbial pathogenesis studies in *Drosophila* helped uncover the conserved components of innate immune signaling [12]. More recently, *Drosophila* has emerged as a powerful system for studying microbial–host mutualisms. The *Drosophila* gut microbial community is composed largely of *Actinobacteria* and *Lactobacilli*, with some variation across different wild-caught and laboratory populations, and a typical complexity of about five dominant members [5, 13–17]. More recent work has demonstrated the importance of the *Drosophila*-associated microbiota in regulating host metabolic homeostasis. Genetic analysis of a bacterial resident, *Actinobacter pomorum*, revealed that this member of the microbiota regulates host insulin signaling through acetic acid production [18]. The fact that microbial regulation of host innate immune and metabolic signaling is conserved between *Drosophila* and mammals suggests that this highly genetically tractable system will be important for future symbiosis studies.

11.2.7. Termite

Termites, famed in popular culture for eating wood and other plant materials, house a microbial community that aids in the digestion of lignocellulose. This community in most termites is a combination of both eukaryotic protists and prokaryotic bacteria that are either free-living or associated within or attached to the outside of the protists, housed within a distinct portion of the hindgut. Although the termite host produces digestive enzymes that begin the process of breaking down lignocellulose, the complete breakdown of lignocellulose and the acquisition of carbon and energy by the host is dependent on its microbiota (reviewed in Ref. 19). The majority of gut microbes of termites still remains uncultured, and the detailed mechanism of this process is unclear; however, more recent transcriptomic, metagenomic, and metaproteomic studies have demonstrated that the breakdown of plant cellulose fibers is a complex and coordinated effort that involves the fermentation of cellulose to produce acetate, which is utilized by the host (reviewed in Ref. 20). These studies also implicate the microbiota in utilization of hydrogen, which is an end product of fermentation and can be inhibitory to the fermentation process, as well as nitrogen fixation. Although far from simple, the termite gut microbiota is an excellent model for studying both the interactions of gut microbes with each other and the coordination of microbial metabolisms within a host.

11.2.8. Mouse

Vertebrate model systems are best suited for studying aspects of human–microbiota interactions specific to vertebrates, such as the assembly of highly complex microbial communities and interactions with the adaptive immune system. Mouse researchers have sophisticated tools for genetic manipulation including, spatiotemporal control of gene function [21]. The mouse is a well-established research model for studies of infectious disease and immunology. It also has a long history of use in gnotobiology dating back to the 1950s, with a more recent renaissance in gnotobiotic mouse experimentation (reviewed in Ref. 22). Much of what we infer about the function of the human microbiota comes from studies in gnotobiotic mice.

11.2.9. Zebrafish

A newer vertebrate model for studying host–microbe interactions is the zebrafish [23]. Zebrafish have traditionally been used to study fundamental processes of embryonic development because of their transparency, rapid early development, and high fecundity [24]. These same attributes, along with a growing toolbox for genetic manipulation, make zebrafish well suited for studies of postembryonic interactions with microbes and a powerful complement to mouse models. The animals' optical transparency even at larval stages allows visualization of bacterial colonization of the digestive tract [25]. Because 1000s of germ-free zebrafish embryos can be derived at a time through *in vitro* fertilization and surface sterilization of the embryos [26], large-scale experiments can be performed with gnotobiotic zebrafish that would not be feasible with mice.

11.3. ESTABLISHMENT OF HOST-ASSOCIATED MICROBIAL COMMUNITIES

The human microbiota is acquired horizontally, meaning that each new generation acquires its microbial partners from the external environment. An individual's microbiota is thus a product of a colonization process that includes initial contact of the host with the potentially 10s of 1000s of microbial species present in the environment and selection of those species that will successfully adapt to a new microenvironment to promote a stable and functional association. An ideal experimental system for studying the process of microbiota establishment would allow for visualization of the colonization process in real time. To study the mechanisms that underlie the colonization process, one would want to be able to genetically manipulate both the host and the microbial associates. The synthesis of live imaging observation with genetic dissection of the colonization process has started to outline a complex "conversation" that takes place between host and symbiont to facilitate proper association.

11.3.1. Visualizing Colonization and Identifying Symbiosis-Defective Bacterial Mutants

The simplicity of the squid–*Vibrio* association has allowed a detailed microscopic analysis of the colonization process in real time. Development of fluorescently labeled *V. fischeri* has allowed researchers to visualize and document the important early events of colonization. This process begins with the host capturing and aggregating the symbiont in mucus secreted from the epithelia of the light organ. Bacterial aggregates are readily visible at high resolution using laser scanning confocal microscopy. Once they aggregate in the mucus, they can be visualized migrating to the pores, the entry point to the light organ, through the light organ as they reach and colonize the internal crypt spaces [27]. Along with unprecedented documentation of colonization, these studies made it possible to further dissect the mechanisms of colonization through bacterial genetics.

With this knowledge of the colonization process in hand, researchers were able to categorize a rich collection of mutant *V. fischeri*, isolated during 25 years of experimentation on this model system, that are defective for colonization of the host (reviewed in Ref. 28). Initiation mutants either are delayed or fail to colonize the light organ and include motility mutants (regardless of flagellation state) [29] and gene regulation mutants such as RscS [30, 31]. Accommodation mutants colonize the light organ, but to a lower level and include amino acid synthesis mutants [32] and lipopolysaccharide (cell wall) mutants [33]. Persistence mutants colonize the light organ to levels similar to those of wildtype but fail to maintain the association over time and include luminescence mutants [34], and bacterial cell communication or quorum-sensing mutants [35]. These mutant studies identified not only three stages that promote the development of symbiosis, but also the concept that successful colonization requires particular phenotypes from both host and symbiont. It will be interesting to see whether these same concepts apply to colonization of the vertebrate intestine by complex microbial communities.

Colonization of the mammalian digestive tract initiates at birth. Although it is not possible to visualize this process in real time, the genetic requirements for colonization by a dominant resident, *Bacteroides thetaiotaomicron*, have been dissected [36], using a negative selection strategy traditionally employed to characterize bacterial pathogenesis [37]. In this study, the authors created a high complexity library of transposon insertions in *B. thetaiotaomicron*, which they used to inoculate germ-free mice. After 2 weeks of colonization, they harvested the output bacterial community and compared its membership to the input community, thereby identifying genes that, when disrupted, rendered the bacteria at a selective disadvantage for colonization. They further explored the host and microbial contexts in which these genes were required. Remarkably, when they colonized either wildtype or immunocompromised germ-free mice, there was very little difference in output community of *B. thetaiotaomicron* transposon insertions harvested from these host environments. In contrast, when they simultaneously colonized the mice with defined bacterial consortia, the different preexisting microbial communities dramatically influenced the suite of genes required by *B. thetaiotaomicron* to survive in the mouse digestive tract, suggesting that microbe–microbe interactions profoundly shape host-associated microbial communities. The initial colonization of the mammalian intestine appears to be a relatively stochastic sampling of environmental microbes [38, 39]. The finding that microbial context strongly determines colonization success of new members suggests that stochastic early colonization events could have longlasting effects on host-associated microbial community structures. Studies in vertebrate systems that allow direct visualization of the colonization process, such as zebrafish [40], will help elucidate the mechanisms underlying how succession events influence microbiota assembly.

Another important determinant of microbiota assembly within animal digestive tracts is the availability of nutrients from the host diet. A number of studies have demonstrated that genetically identical animals raised on different diets have different microbiota (reviewed in Ref. 41). For example, mice on a Western diet high in fat and refined sugar exhibit a bloom of a particular Mollicutes class of the Firmicutes as compared to littermates on a standard mouse chow diet [42]. A more recent study by Sonnenburg and colleagues dissected the bacterial genetic requirements for utilization of specific sugars ingested by the host, again using the model mammalian gut symbiont *B. thetaiotaomicron*. They identified a genetic locus of *B. thetaiotaomicron* that functions in the utilization of the polysaccharide inulin [43]. This locus is conserved but varies genetically and functionally across *Bacteroides* species and allowed the authors to predict changes in bacterial abundance when pairs of *Bacteroides* species were inoculated in germ-free mice in competition with each other and administered an inulin-rich diet. This study demonstrates the feasibility of using dietary manipulations to engineer gut microbiota, although much more analysis will be required before we understand the assembly rules of complex microbial communities well enough to rationally design prebiotics for promoting healthy microbiota in humans.

11.3.2. Identifying Host Genetic Determinants of Microbiota Composition

There is currently much debate about the relative contribution of host genetics versus environmental factors, such as microbial exposures or diet, to microbiota composition [44]. On one hand, a comparison of the microbiota composition of stool samples of monozygotic and dizygotic twin pairs revealed no significant difference between the two groups, arguing against a strong role for host genetics in the assembly of gut microbial communities [45]. On the other hand, several studies of genetically distinct mice have uncovered marked differences in their gut microbiota. For example, disruption of the *obese* gene encoding the appetite-regulating hormone leptin results in a gut microbiota with an overrepresentation of the phylum *Firmicutes* relative to *Bacteroidetes* [46]. Other examples are the shift toward a more proinflammatory gut microbiota seen in mice deficient for the immune regulator Tbet [47, 48], and the shift toward a diabetes-promoting microbial community in *tlr5* deficient mice [49]. Importantly, in all of these mouse models, not only were the host genotype-specific microbiota different in their composition, as assayed by the complement of 16*S* rRNA genes amplified from the communities; they were also shown to be functionally different in their ability to promote disease symptoms when transplanted into germ-free wildtype mice [48–50].

By what mechanisms might host genes regulate microbiota structure? One elegant example comes from *Drosophila*. Lee and colleagues discovered that the transcription factor caudal, originally identified for its role in patterning the embryonic anteroposterior body axis, has a postembryonic role in regulating intestinal epithelial expression of antimicrobial peptides (AMPs) [16]. In the absence of intestinal caudal function, AMPs were inappropriately over-expressed, resulting in a pathogenic shift in the composition of the microbial community. Because of the simplicity of this community, composed of just five major bacterial members, the authors were able to show that the overexpression of AMPs led to loss of a protective member of the community, an Acetobacteraceae species, which allowed the overgrowth of an opportunistic pathogen *Gluconobacter* species that was normally a minor member kept in check by the Acetobacteraceae species. This simple *Drosophila* model is likely representative of host AMP regulation of more complex microbiota in mammals [51]. Disregulation of AMP expression, resulting in microbiota dysbiosis, is suspected as an etiology of some cases of inflammatory bowel disease (IBD) [52].

In an effort to directly identify host genetic factors that structure the bacterial community in the mouse intestine, Benson and colleagues used a quantitaive trait locus (QTL) mapping strategy to assocaite abundance of fecal bacteria with genetic loci across a panel of intercross mouse lines [53]. Their study revealed 18 loci that varied with microbiota membership; however, their analysis was complicated by the strong influence of cohousing on microbiota composition, which could not be easily disassociated from genetic effects without a cross-fostering experimental design. Future QTL mapping studies will continue to refine our understanding of host genetic factors that shape intestinal microbiota composition. Fish models are particularly amenable to such studies because of their large family sizes and the ability to cohouse progeny from different families in shared aquaria with a common microbial environment.

11.4. THE FUNCTION OF RESIDENT MICROBIAL COMMUNITIES IN HOST BIOLOGY

Perhaps the most important question that researchers wish to ask about host-associated microbial communities is also the question that is the hardest to answer: What functions do microbiota provide to the host? We can begin to answer this question by conducting a simple yet elegant experiment of raising animals in the absence of their associated microbes and compare them to conventionally reared animals in which the microbiota has not been perturbed. In some cases, such as the squid–*Vibrio* symbiosis, the animals can be raised in their normal microbial-rich environment as long as it is devoid of the particular symbiont. For studying the function of complex and less selective microbial associations, such as those in the vertebrate gut, animals are raised under germ-free conditions. For studying the activities of immune cells that require the microbiota for their normal development, another experimental approach is to attempt to sterilize colonized animals with high doses of broad-spectrum antibiotics, but this approach is unlikely to entirely eliminate all microbial associates.

11.4.1. Microbial Influences on Epithelial Tissues

A common theme that emerges from many of the germ-free or aposymbiotic animal models is the influence of microbial associates on epithelial tissues with which they interact. One of the most dramatic examples of this is seen in the squid–*Vibrio* symbiosis. Normally, on colonization by their symbiont, juvenile squid undergo a tissue remodeling of the external epithelial cell layer of the light organ, involving apoptosis beginning at roughly 12 h after first contact with the bacteria [54] and ending with full regression of the epithelial appendages by 96 h. Regression does not occur in squid reared in the absence of *V. fischeri*. It was further shown that although bacterial cell wall molecules stimulate a low level of epithelial cell regression, only actively secreted signals from the symbiont could induce full epithelial regression [55].

Microbial influences on the epithelial tissues are likely to be very ancient, as demonstrated by the example of Hydra, the freshwater invertebrate whose body plan consists of a simple epithelium of only two cell layers that is colonized by microbes. An early study of axenic Hydra found that they are unable to undergo asexual budding, a process that could be rescued by addition of bacteria [56]. To determine whether epithelial homeostasis and microbial community are linked, Bosch and colleagues created a transgenic Hydra that lacked one of three epithelial stem cell lineages. Removal of the interstital stem cell lineage resulted in significant changes in the microbial population associated with the epithelia, implying a direct interaction between the microbiota and the epithelia [6].

Regulation of epithelial cell proliferation by associated microbes is observed in a broad range of animal models. The intestinal epithelia of germ-free *Drosophila*, mice, and zebrafish all have reduced rates of cell proliferation [57–61]. In the zebrafish model, the microbial signal that promotes intestinal epithelial cell proliferation is transduced through the innate immune signaling molecule Myd88, a common adaptor of the toll-like receptors. Myd88 is also required for normal cell proliferation in a chemically injured mouse intestine [62, 63]. In contrast to the injury model, the normal developmental response to the microbiota in zebrafish does not require tumor necrosis factor cytokine signaling [58]. This difference suggest that there are two distinct roles for the microbiota in epithelial homeostasis: one that regulates normal cell proliferation, and one involved in a protective response to injury.

11.4.2. Microbial Influences on the Immune System

A second theme that emerges from germ-free animal models is the importance of host-associated microbes in shaping the immune system. This function is observed in both invertebrates, possessing solely an innate immune system, and vertebrates with both innate and adaptive immune systems. In newly hatched squid, the hemocytes, macrophage-like blood cells, start to traffic to the external epithelia of the light organ 2 h after exposure to bacteria, where they remain until the regression of the epithelial fields is complete (reviewed in Ref. 2). Hemocytes in hatchling squid reared in the absence of *V. fischeri* do not traffic to the light organ. In symbiont-colonized juveniles, the number of hemocytes in the internal space that houses *V. fischeri* increases after 36 h. A similar recruitment of neutrophils occurs in the zebrafish intestine as it is colonized by bacteria [64]. This process is absent in germ-free zebrafish and can be mimicked by addition of the Gram-negative bacterial cell wall constituent, lipopolysaccharide. Neutrophil recruitment in response to microbiota in zebrafish requires both Myd88-dependent innate immune signaling and tumor necrosis factor cytokine signaling. In the mouse intestine, maturation of isolated lymphoid follicles also requires colonization by the microbiota and fails to occur in germ-free animals [65]. In this case, the signal that stimulates the immune system maturation process is a Gram-negative bacteria-specific monomer of peptidoglycan, and requires the innate immune receptor NOD1 and the chemokine receptor CCR6.

In the early stage of the squid–*Vibrio* symbiosis, the hemocytes are often observed with phagocytosed bacteria in their vacuoles. In symbiotic animals, hemocytes continue to traffic to the symbiont-containing tissues of adult animals, but bacteria are never found within theses cells. This suggests that the presence of the symbiont "educates" the hemocytes, modulating their response to the symbiont [66]. In the mammalian intestine, the microbiota play an important function in educating the adaptive immune system [67]. For example, germ-free mice have a paucity of $CD4^+$ T cells and a disproportionate expression of Th2 versus Th1 cytokines. These deficiencies can be corrected by addition of a single capsule polysaccharide, PSA, from the mouse intestinal resident *Bacteroides fragilis* [68]. *B. fragilis* is also important for promoting antiinflammatory regulatory T cell differentiation [69], as are several other *Clostridium* members of the microbiota [70], whereas another *Clostridium* member, *Segmented Filamentous Bacteria*, promotes the differentiation of proinflammatory Th17 cells [71]. These emerging studies on the immunomodulatory functions of individual members of the mammalian microbiota provide strong support for the idea that diseases of gastrointestinal immune dysfunction, such as inflammatory bowel diseases, could be due to unhealthy microbiota membership.

11.5. CONCLUDING REMARKS

Although we are still a long way from fully delineating the rules that govern the assembly of the human microbiota or understanding the effects of resident microbes on their hosts, animal models provide a starting point for the rational design of therapies to promote human health through manipulation of the microbiota.

One strategy would be to promote the assembly of a healthy microbiota or to correct a perturbed microbial community. This could be achieved with probiotics, or microbial dietary supplements designed to promote health. Studies of the processes of animal colonization, such as those on the colonization determinants of *B. thetaiotamicron* [36], will help us design better probiotics that are capable of colonizing established microbial communities. Another promising avenue is the use of prebiotics, or nonliving dietary supplements designed to promote the growth of beneficial bacteria. Continued analysis of the nutritional strategies of microbiota members, such as *Bacteroides* species utilization of inulin [43], should provide information needed for the designing of better prebiotics. In this regard, human milk, among its many health-promoting properties, appears to be an exemplar naturally designed prebiotic, containing oligosaccharides that are not absorbed by the nursing infant but serve as a nutrient source for beneficial intestinal bacteria [72].

Designing probiotics and prebiotics will require a better understanding of the effects of microbial residents on host tissues and the nature of the signaling involved. The ability of mucosa-associated bacteria to promote epithelial cell proliferation could be harnessed for treatment of intestinal injury or inhibited as a strategy to prevent gastrointestinal cancers. The immunomodulatory effects of microbiota members offer exciting new opportunities to treat immune dysfunction. Some currently available probiotics have antiinflammatory properties that hold clinical promise. However, this first generation of probiotics was selected for therapeutic testing with limited understanding of the mechanisms of their effects. In the future, animal models will be indispensable for acquiring new basic knowledge about host–microbiota interactions and for applying this knowledge to the design of new microbe-based therapies for promoting human health.

REFERENCES

1. McFall-Ngai M. Adaptive immunity: care for the community. *Nature* **445**:153 (2007).
2. McFall-Ngai M, Nyholm SV, Castillo MG. The role of the immune system in the initiation and persistence of the Euprymna scolopes—Vibrio fischeri symbiosis. *Semin Immunol* **22**:48–53 (2004).
3. Clarke DJ. Photorhabdus: A model for the analysis of pathogenicity and mutualism. *Cell Microbiol* **10**:2159–2167 (2008).
4. Richards GR, Goodrich-Blair H. Masters of conquest and pillage: Xenorhabdus nematophila global regulators control transitions from virulence to nutrient acquisition. *Cell Microbiol* **11**:1025–1033 (2009).
5. Augustin R, Fraune S, Franzenburg S, Bosch TC. Where simplicity meets complexity: Hydra, a model for host-microbe interactions. *Adv Exp Med Biol* **710**:71–81 (2012).
6. Fraune S, Abe Y, Bosch TC. Disturbing epithelial homeostasis in the metazoan Hydra leads to drastic changes in associated microbiota. *Environ Microbiol* **11**:2361–2369 (2009).
7. Bomar L, Maltz M, Colston S, Graf J. Directed culturing of microorganisms using metatranscriptomics. *mBio* **2**:e00012–11 (2011).
8. Rio RV, Anderegg M, Graf J. Characterization of a catalase gene from Aeromonas veronii, the digestive-tract symbiont of the medicinal leech. *Microbiology* **153**:1897–1906 (2007).
9. Broderick NA, Raffa KF, Goodman RM, Handelsman J. Census of the bacterial community of the gypsy moth larval midgut by using culturing and culture-independent methods. *Appl Environ Microbiol* **70**:293–300 (2004).
10. Broderick NA, Raffa KF, Handelsman J. Midgut bacteria required for Bacillus thuringiensis insecticidal activity. *Proc Natl Acad Sci USA* **103**:15196–15199 (2006).
11. Broderick NA, Raffa KF, Handelsman J. Chemical modulators of the innate immune response alter gypsy moth larval susceptibility to Bacillus thuringiensis. *BMC Microbiol* **10**:129 (2010).
12. Lemaitre B, Hoffmann J. The host defense of Drosophila melanogaster. *Annu Rev Immunol* **25**:697–743 (2007).
13. Corby-Harris V, Pontaroli AC, Shimkets LJ, Bennetzen JL, Habel KE, Promislow DE. Geographical distribution and diversity of bacteria associated with natural populations of Drosophila melanogaster. *Appl Environ Microbiol* **73**:3470–3479 (2007).
14. Cox CR, Gilmore MS. Native microbial colonization of Drosophila melanogaster and its use as a model of Enterococcus faecalis pathogenesis. *Infect Immun* **75**:1565–1576 (2007).

15. Ren C, Webster P, Finkel SE, Tower J. Increased internal and external bacterial load during Drosophila aging without life-span trade-off. *Cell Metab* **6**:144–152 (2007).
16. Ryu JH, Nam KB, Oh CT, Nam HJ, Kim SH, Yoon JH, Seong JK, Yoo MA, Jang IH, Brey PT, et al. The homeobox gene Caudal regulates constitutive local expression of antimicrobial peptide genes in Drosophila epithelia. *Mol Cell Biol* **24**:172–185 (2004).
17. Wong CN, Ng P, Douglas AE. Low-diversity bacterial community in the gut of the fruitfly Drosophila melanogaster. *Environ Microbiol* **13**:1889–1900 (2011).
18. Shin SC, Kim SH, You H, Kim B, Kim AC, Lee KA, Yoon JH, Ryu JH, Lee WJ. Drosophila microbiome modulates host developmental and metabolic homeostasis via insulin signaling. *Science* **334**:670–674 (2011).
19. Ohkuma M. Symbioses of flagellates and prokaryotes in the gut of lower termites. *Trends Microbiol* **16**:345–352 (2008).
20. Hongoh Y. Toward the functional analysis of uncultivable, symbiotic microorganisms in the termite gut. *Cell Mol Life Sci* **68**:1311–1325 (2011).
21. Lewandoski M. Conditional control of gene expression in the mouse. *Nat Rev Genet* **2**:743–755 (2001).
22. Macpherson AJ, Harris NL. Interactions between commensal intestinal bacteria and the immune system. *Nat Rev Immunol* **4**:478–485 (2004).
23. Kanther M, Rawls JF. Host-microbe interactions in the developing zebrafish. *Curr Opin Immunol* **22**:10–19 (2010).
24. Grunwald DJ, Eisen JS. Headwaters of the zebrafish—emergence of a new model vertebrate. *Nat Rev Genet* **3**:717–724 (2002).
25. Rawls JF, Mahowald MA, Goodman AL, Trent CM, Gordon JI. In vivo imaging and genetic analysis link bacterial motility and symbiosis in the zebrafish gut. *Proc Natl Acad Sci USA* **104**:7622–7627 (2007).
26. Milligan-Myhre K, Charette JR, Phennicie RT, Stephens WZ, Rawls JF, Guillemin K, Kim CH. Study of host-microbe interactions in zebrafish. *Meth Cell Biol* **105**:87–116 (2011).
27. Nyholm SV, Stabb EV, Ruby EG, McFall-Ngai MJ. Establishment of an animal-bacterial association: Recruiting symbiotic vibrios from the environment. *Proc Natl Acad Sci USA* **97**:10231–10235 (2000).
28. Nyholm SV, McFall-Ngai MJ. The winnowing: Establishing the squid-vibrio symbiosis. *Nat Rev Microbiol* **2**:632–642 (2004).
29. Graf J, Dunlap PV, Ruby EG. Effect of transposon-induced motility mutations on colonization of the host light organ by Vibrio fischeri. *J Bacteriol* **176**:6986–6991 (1994).
30. Mandel MJ, Wollenberg MS, Stabb EV, Visick KL, Ruby EG. A single regulatory gene is sufficient to alter bacterial host range. *Nature* **458**:215–218 (2009).
31. Visick KL, Skoufos LM. Two-component sensor required for normal symbiotic colonization of euprymna scolopes by Vibrio fischeri. *J Bacteriol* **183**:835–842 (2001).
32. Graf J, Ruby EG. Host-derived amino acids support the proliferation of symbiotic bacteria. *Proc Natl Acad Sci USA* **95**:1818–1822 (1998).
33. DeLoney CR, Bartley TM, Visick KL. Role for phosphoglucomutase in Vibrio fischeri-Euprymna scolopes symbiosis. *J Bacteriol* **184**:5121–5129 (2002).
34. Visick KL, Foster J, Doino J, McFall-Ngai M, Ruby EG. Vibrio fischeri lux genes play an important role in colonization and development of the host light organ. *J Bacteriol* **182**:4578–4586 (2000).
35. Lupp C, Urbanowski M, Greenberg EP, Ruby EG. The Vibrio fischeri quorum-sensing systems ain and lux sequentially induce luminescence gene expression and are important for persistence in the squid host. *Mol Microbiol* **50**:319–331 (2003).

36. Goodman AL, McNulty NP, Zhao Y, Leip D, Mitra RD, Lozupone CA, Knight R, Gordon JI. Identifying genetic determinants needed to establish a human gut symbiont in its habitat. *Cell Host Microbe* **6**:279–289 (2009).
37. Saenz HL, Dehio C. Signature-tagged mutagenesis: technical advances in a negative selection method for virulence gene identification. *Curr Opin Microbiol* **8**:612–619 (2005).
38. Koenig JE, Spor A, Scalfone N, Fricker AD, Stombaugh J, Knight R, Angenent LT, Ley RE. Succession of microbial consortia in the developing infant gut microbiome. *Proc Natl Acad Sci USA* **108**(Suppl 1):4578–4585 (2011).
39. Palmer C, Bik EM, DiGiulio DB, Relman DA, Brown PO. Development of the human infant intestinal microbiota. *PLoS Biol* **5**:e177 (2007).
40. Taormina MJ, Jemielita M, Stephens WZ, Burns AR, Troll JV, Parthasarathy R, Guillemin K. Investigating bacterial-animal symbioses with light sheet microscopy. *Biol Bull* **223**:7–20 (2012).
41. Ley RE. Obesity and the human microbiome. *Curr Opin Gastroenterol* **26**:5–11 (2010).
42. Turnbaugh PJ, Backhed F, Fulton L, Gordon JI. Diet-induced obesity is linked to marked but reversible alterations in the mouse distal gut microbiome. *Cell Host Microbe* **3**:213–223 (2008).
43. Sonnenburg ED, Zheng H, Joglekar P, Higginbottom SK, Firbank SJ, Bolam DN, Sonnenburg JL. Specificity of polysaccharide use in intestinal bacteroides species determines diet-induced microbiota alterations. *Cell* **141**:1241–1252 (2010).
44. Spor A, Koren O, Ley R. Unravelling the effects of the environment and host genotype on the gut microbiome. *Nat Rev Microbiol* **9**:279–290 (2011).
45. Turnbaugh PJ, Hamady M, Yatsunenko T, Cantarel BL, Duncan A, Ley RE, Sogin ML, Jones WJ, Roe BA, Affourtit JP, et al. A core gut microbiome in obese and lean twins. *Nature* **457**:480–484 (2009).
46. Ley RE, Backhed F, Turnbaugh P, Lozupone CA, Knight RD, Gordon JI. Obesity alters gut microbial ecology. *Proc Natl Acad Sci USA* **102**:11070–11075 (2005).
47. Garrett WS, Gallini CA, Yatsunenko T, Michaud M, DuBois A, Delaney ML, Punit S, Karlsson M, Bry L, Glickman JN, et al. Enterobacteriaceae act in concert with the gut microbiota to induce spontaneous and maternally transmitted colitis. *Cell Host Microbe* **8**:292–300 (2010).
48. Garrett WS, Lord GM, Punit S, Lugo-Villarino G, Mazmanian SK, Ito S, Glickman JN, Glimcher LH. Communicable ulcerative colitis induced by T-bet deficiency in the innate immune system. *Cell* **131**:33–45 (2007).
49. Vijay-Kumar M, Aitken JD, Carvalho FA, Cullender TC, Mwangi S, Srinivasan S, Sitaraman SV, Knight R, Ley RE, Gewirtz AT. Metabolic syndrome and altered gut microbiota in mice lacking Toll-like receptor 5. *Science* **328**:228–231 (2010).
50. Turnbaugh PJ, Ley RE, Mahowald MA, Magrini V, Mardis ER, Gordon JI. An obesity-associated gut microbiome with increased capacity for energy harvest. *Nature* **444**: 1027–1031 (2006).
51. Bevins CL, Salzman NH. Paneth cells, antimicrobial peptides and maintenance of intestinal homeostasis. *Nat Rev Microbiol* **9**:356–368 (2011).
52. Wehkamp J, Schmid M, Stange EF. Defensins and other antimicrobial peptides in inflammatory bowel disease. *Curr Opin Gastroenterol* **23**:370–378 (2007).
53. Benson AK, Kelly SA, Legge R, Ma F, Low SJ, Kim J, Zhang M, Oh PL, Nehrenberg D, Hua K, et al. Individuality in gut microbiota composition is a complex polygenic trait shaped by multiple environmental and host genetic factors. *Proc Natl Acad Sci USA* **107**:18933–18938 (2010).

54. Doino J, McFall-Ngai MJ. A transient exposure to symbiosis-competent bacteria induces light organ morphogenesis in the host squid. *Biol Bull* **189**:347–355 (1995).
55. Koropatnick TA, Engle JT, Apicella MA, Stabb EV, Goldman WE, McFall-Ngai MJ. Microbial factor-mediated development in a host-bacterial mutualism. *Science* **306**:1186–1188 (2004).
56. Rahat M, Dimentman C. Cultivation of bacteria-free Hydra viridis: Missing budding factor in nonsymbiotic hydra. *Science* **216**:67–68 (1982).
57. Abrams GD, Bauer H, and Sprinz H. Influence of the normal flora on mucosal morphology and cellular renewal in the ileum. A comparison of germ-free and conventional mice. *Lab Invest* **12**:355–364 (1963).
58. Cheesman SE, Neal JT, Mittge E, Seredick BM, Guillemin K. Epithelial cell proliferation in the developing zebrafish intestine is regulated by the Wnt pathway and microbial signaling via Myd88. *Proc Natl Acad Sci USA* **108**(Suppl 1):4570–4577 (2011).
59. Lee WJ. Bacterial-modulated host immunity and stem cell activation for gut homeostasis. *Genes Dev* **23**:2260–2265 (2009).
60. Rawls JF, Mahowald MA, Ley RE, Gordon JI. Reciprocal gut microbiota transplants from zebrafish and mice to germ-free recipients reveal host habitat selection. *Cell* **127**:423–433 (2006).
61. Rawls JF, Samuel BS, Gordon JI. Gnotobiotic zebrafish reveal evolutionarily conserved responses to the gut microbiota. *Proc Natl Acad Sci USA* **101**:4596–4601 (2004).
62. Pull SL, Doherty JM, Mills JC, Gordon JI, Stappenbeck TS. Activated macrophages are an adaptive element of the colonic epithelial progenitor niche necessary for regenerative responses to injury. *Proc Natl Acad Sci USA* **102**:99–104 (2005).
63. Rakoff-Nahoum S, Paglino J, Eslami-Varzaneh F, Edberg S, Medzhitov R. Recognition of commensal microflora by toll-like receptors is required for intestinal homeostasis. *Cell* **118**:229–241 (2004).
64. Bates JM, Akerlund J, Mittge E, Guillemin K. Intestinal alkaline phosphatase detoxifies lipopolysaccharide and prevents inflammation in zebrafish in response to the gut microbiota. *Cell Host Microbe* **2**:371–382 (2007).
65. Bouskra D, Brezillon C, Berard M, Werts C, Varona R, Boneca I.G, Eberl G. Lymphoid tissue genesis induced by commensals through NOD1 regulates intestinal homeostasis. *Nature* **456**:507–510 (2008).
66. Nyholm SV, Stewart JJ, Ruby EG, McFall-Ngai MJ. Recognition between symbiotic Vibrio fischeri and the haemocytes of Euprymna scolopes. *Environ Microbiol* **11**:483–493 (2009).
67. Lee YK, Mazmanian SK. Has the microbiota played a critical role in the evolution of the adaptive immune system? *Science* **330**:1768–1773 (2010).
68. Mazmanian SK, Liu CH, Tzianabos AO, Kasper DL. An immunomodulatory molecule of symbiotic bacteria directs maturation of the host immune system. *Cell* **122**:107–118 (2005).
69. Round JL, Lee SM, Li J, Tran G, Jabri B, Chatila TA, Mazmanian SK. The Toll-like receptor 2 pathway establishes colonization by a commensal of the human microbiota. *Science* **332**:974–977 (2011).
70. Atarashi K, Tanoue T, Shima T, Imaoka A, Kuwahara T, Momose Y, Cheng G, Yamasaki S, Saito T, Ohba Y, et al. Induction of colonic regulatory T cells by indigenous Clostridium species. *Science* **331**:337–341 (2011).
71. Ivanov, II, Frutos Rde L, Manel N, Yoshinaga K, Rifkin DB, Sartor RB, Finlay BB, Littman DR. Specific microbiota direct the differentiation of IL-17-producing T-helper cells in the mucosa of the small intestine. *Cell Host Microbe* **4**:337–349 (2008).
72. Zivkovic AM, German JB, Lebrilla CB, Mills DA. Human milk glycobiome and its impact on the infant gastrointestinal microbiota. *Proc Natl Acad Sci USA* **108**(Suppl 1):4653–4658 (2010).

12

TO GROW OR NOT TO GROW: ISOLATION AND CULTIVATION PROCEDURES IN THE GENOMIC AGE

KARSTEN ZENGLER

Department of Bioengineering, University of California, San Diego, La Jolla, California

12.1. INTRODUCTION

Estimates regarding microbial diversity have changed drastically since the early 2000s, mainly because of advances in detection, the application of new computational modeling methods and algorithms, and the changing concept of what defines a microbial species [1–5]. For example, numbers of different species or operational taxonomic units [6] vary from <500 [7] to 2000 [8], 10,000 [9], 21,000 [10], 500,000 [11], to even $\leq 10^7$ cells within a given soil sample [12,13]. This disparity can be due to the diversity of the samples themselves [10,14,15], as well as the varying approaches used to estimate the diversity [5,16]. However, it is clear that the vast majority of the microbial diversity in any given environment, including the human body, has not yet been isolated and cultured and therefore hamper in-depth studies of effects on human health and disease [17,18]. Many bacterial divisions recognized to date consist entirely of as yet uncultured bacteria that have been described solely by their 16S rRNA gene sequences [19,20]. The picture gets even blurrier if we look at the quantitative distribution of isolates across the phylogenetic tree (Figure 12.1). Over 90% of all cultures in our culture collections fall within four major phyla, namely proteobacteria, firmicutes, bacteriodetes, and actinobacteria. Even current efforts to obtain sequence information from underrepresented phyla [21] won't be able to breach the tremendous knowledge gap that results from the lack of available isolates. Furthermore, only a small fraction of organisms that can be cultured have

The Human Microbiota: How Microbial Communities Affect Health and Disease, First Edition. Edited by David N. Fredricks.

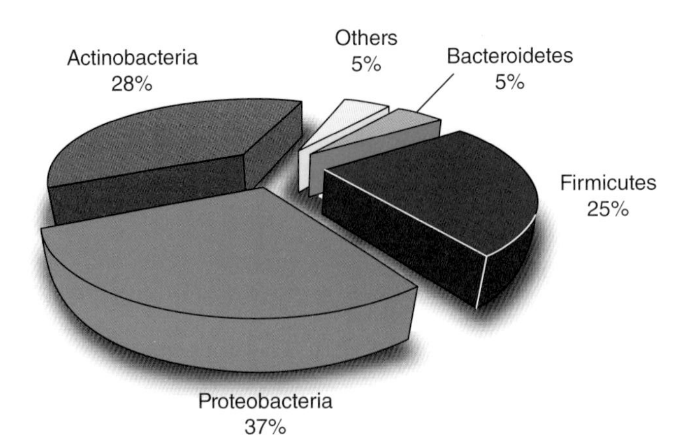

Figure 12.1. Percentage of bacterial phyla with cultivated representatives (from Keller and Zengler [19]).

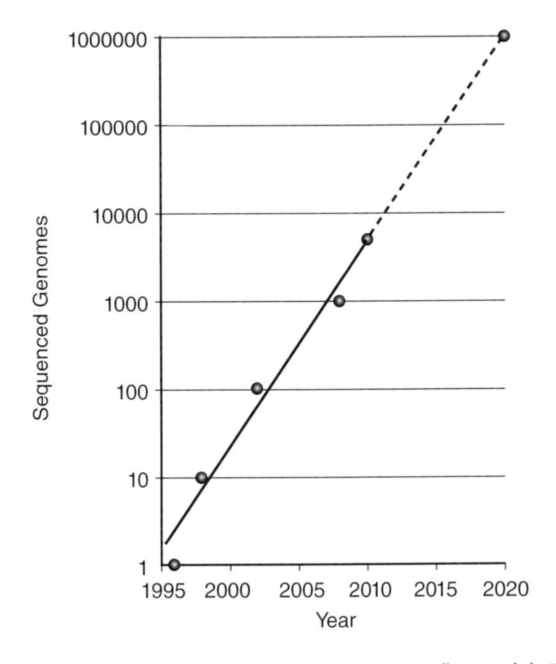

Figure 12.2. Number of genome sequences available per year (log scale). The dotted line represents an extrapolation.

been at least partially characterized, and in most cases in-depth knowledge about general metabolism and physiology is still missing [22,23]. Almost 5000 bacterial genomes have been fully sequenced as of 2010 (Figure 12.2). If we assume that sequencing technology will continue its current trend and will become even more effective and affordable at the same time, we can estimate that by the year 2020, around 1 million bacterial and archaeal genomes could be sequenced (Figure 12.2). However, the two largest culture collections (ATCC and DSMZ) currently hold less than 40,000 different strains combined (many of them existing in both collections). This means that with all the sequencing power available today, it is likely that we will run out of cultures to sequence within the next decade. At the same time, it is envisioned that the number of sequences obtained by metagenomics or single cell sequencing efforts will grow steadily [24,25].

It is of uttermost importance, however, to keep in mind that genome sequences themselves (partial or full) cannot provide any quantitative data about metabolism, physiology, or pathogenicity [26]. The genome sequence can be viewed as a foundation for microbial physiology. How this foundation is organized on a structural and organizational level and how it is utilized to define a phenotype cannot be elucidated from the genome sequence alone, but requires experimental methods [27], which, again, calls for making microorganisms available in the laboratory [28]. Several advances in isolation and cultivation techniques have been made over the last few decades and could provide the cultures needed to broaden our knowledge about microorganisms in great detail. The term *microorganism* used throughout this chapter is not restricted to members of the domains Bacteria and Archaea but expands to the whole microbial community, including members of the Eukarya such as fungi and protists. The kingdoms of protists and fungi contain not only environmentally relevant organisms but also large numbers of human pathogens. Protists, a highly diverse taxa of single-celled eukaryotic organisms, were described by Hooke and van Leeuwenhoek in the seventeenth century [29,30]. While estimates of protist diversity vary in the literature from as few as 20,000 [31] to >300,000 [32], it is fair to say that like bacteria and archaea, only a tiny fraction can be maintained and studied in the laboratory. The same is true for the fungi. While mycologists have been collecting and growing fungi for thousands of years, the ~80,000 fungal species described so far represent only ~5% of the estimated total diversity [33]. Since fungi are well known producers of toxins, antibiotics, and other secondary metabolites, they are commercially used to produce not only antibiotics but also other valuable products such as vitamins and anticancer drugs. However, because of engineering difficulties related to most fungal species, recombinant production of viable products are restricted to a few genetically tractable species [34,35]. Thus, access to the fungus with necessary metabolic traits often remains the only option to obtain and generate the desired product.

12.2. ISOLATION, GROWTH, AND CULTIVATION

Enrichment, isolation, growth, and cultivation techniques were first developed by researchers such as Cohn, Beijerinck, Koch, and Winogradsky [36–39] in the late nineteenth century. Isolation of an organism (or multiple organisms) describes the process by which individual cells are physically separated from each other and/or from matrix material, such as eukaryotic tissue. It is of critical importance to keep in mind that the selection of the microorganism targeted for cultivation determines all of the necessary isolation, growth, and cultivation steps. Therefore the choice of separation and isolation, potentially critical cell-to-cell, cell-to-host, or cell-to-substrate contact, and the composition of the medium have to be carefully evaluated before the experimental work begins. The term *isolation* describes the process of obtaining and establishing a defined coculture in the laboratory and is the most critical step in establishing a pure culture. This separation of individual cells (or group of cells) is essential to cultivation efforts and is normally achieved by physical separation; however, other methods of separation are also possible, including the use of antibiotics to inhibit growth of bacteria in order to facilitate isolation of archaea [40]. The isolation step can take place before or after cells are grown [41].

12.2.1. Isolation

The most common method used to physically separate cells is to spread them onto solid media, such as agar plates. While multiple advances have been made in isolating microorganisms on solidified media since Robert Koch introduced this method over a century ago [38], the principle has been unchanged [42]. The separation process follows the isolation process itself, whereby cells or colonies are isolated and propagated, such as by using a loop or toothpick. While this process appears to be straightforward, it actually represents the most daunting and essential task—having cells divide and grow in the first place. Harold Conn and Heinrich Winterberg realized this almost 100 years ago when they reported that the majority of cells observed under the microscope will not form colonies on solid media [43,44]. The same phenomenon was termed "the great plate count anomaly" half a century later by James Staley and Alan Konopka [45]. Many theories have been subsequently formulated to explain this phenomenon.

Since the beginning of the century it has become clear that we have been underestimating the complexity of microbial niches and requirements. It now seems preposterous to have expected the plethora of microorganisms in nature to feel at home on a handful (at most) of media (Luria-Bertani and ZoBell are the most common ones among them). This is akin to feeding all the animals at the zoo the same diet and expecting them not only to survive but also to thrive and reproduce. The notion that microorganisms, because of their size and simplicity, must have equally simple growth requirements likely originated from an animal (human)-centric view of biology. We now know that the metabolic capabilities of bacteria (and therefore their growth needs) outnumber those in the Eukarya world by far—a simple look at the number of electron donors and acceptors utilized by bacteria underlines this.

Another lesson learned more recently is that the microbial community itself has a much larger role in growing cells in isolation than previously thought. Cell-to-cell communication plays a critical role in cell growth [46]. Efforts have been made to simulate this kind of communication in the laboratory by adding signal compounds [47–51] or by keeping the microbial community as a whole intact [41,52]. This intercelluar communication is widespread within microorganisms and represents the foundation for several aspects of growth and physiology, cell cycling, molecular clocks, and oscillation. These signals can silence competitors [53], induce growth [54], or cause death [55].

Obviously, factors that induce cell growth are the most important for isolation and cultivation efforts [54,56]. Unfortunately, none of these factors are universal; on the contrary, they seem to promote growth of only a very small subset of bacteria. In addition, elucidation of the structure and mechanism of action of these growth promoting factors can be quite tedious. As stated above, isolation can also take place before growth with methods that include the use of flow cytometry [57,58], microfluidics [59,60], or micromanipulation using focused laser beams (optical tweezers)

[61,62]. These methods detect and subsequently separate individual cells. An approach that does not require single-cell detection for separation is the isolation of bacteria by microencapsulation [41,63]. A more commonly used technique is the isolation of bacteria via liquid serial dilution [64–67]. This technique can also be used if bacteria do not form colonies on solid surfaces.

12.2.2. Growth

Growth of microorganisms implies cell division and therefore duplication in cell number. Observing and measuring microbial growth, especially *in situ*, can be surprisingly challenging. Since not all cells will form large colonies or dense cultures, it is often not possible to use standard procedures to detect growth in the laboratory [68,69]. On the other hand, some microbes in the lab are known to grow very slowly [70], and certain organisms in nature have estimated doubling times of several 100s to 1000s of years [71]. It should be noted that rapid growth, desired by the researcher, is not the main objective of most environmental organisms. In fact, the opposite is likely the case, for instance, to avoid lysis by phages [72]. Slow growth sometimes also requires the use of a microscope to observe colony formation [52,73,74]. It is well known that several species will not form larger colonies or never grow to high cell densities even after prolonged incubation [42,75–77]. The combination of isolating cells by encapsulation in microcapsules and sorting by flow cytometry is a more high-throughput method for detection of growth, which at the same time allows for higher sensitivity [63]. Instead of using a flow cytometer, this encapsulation technique can also be combined with a microfluidic approach to monitor division of cells [78], or cells can be observed directly.

12.2.3. Cultivation

Cultivation describes the growth and maintenance of cells (usually millions or billions of them) in the laboratory. To maintain and grow cells in the laboratory, one needs to provide them with the right nutrients in the medium. The choice of medium is yet another task that can be quite challenging. Every medium component can have inhibitory effects on cell growth [28]. Even the quality of the water and glass- or plasticware is critical for cultivation success [42,79]. Accounting for all these inhibitory effects is a daunting task, and the potential number of media to test can range in the billions [28]. Accounting for variations in carbon, nitrogen, and sulfur source, pH, salinity, temperature, and atmospheric pressure of key gases such as CO_2 and O_2 makes it nearly impossible to find the optimal condition for a single organism, let alone *all* the members of a microbial community in which we are interested. Cultivation success is probably not hampered exclusively by *what* nutrients are offered to the microorganisms for growth but more likely *how much* of these nutrients are provided [28]. High concentrations of nutrients often have inhibitory effects on cell growth, especially for organisms adapted to low nutrient environments [80–82]. Therefore, several methods have been developed that employ low-nutrient media [41,52,73,75,83–86]. Using these media often results in lower biomass, which requires modified techniques to determine growth (see above) or increased culture volumes [75]. Another solution is the use of flow-through setups, such as a chemostat, to obtain higher cell densities.

As outlined above, the steps involved in obtaining a microbial culture can be challenging. So what are the best choices if we only have limited time and a limited budget with which to obtain cultures of interest? To maximize the cultivation success, the simulation of the natural environmental conditions should be considered the most critical—this includes providing the right "partners" of the microbial community. Unfortunately, our knowledge of the natural environment at levels of resolution (spatiotemporal) relevant to microorganisms is very limited. However, only such high-resolution measurement of varying environmental parameters, and therefore, definition of niches at the microscale [87,88], can provide us with the knowledge needed to improve our culture condition and thereby increase our success rate.

12.3. CHANGES IN A CHANGING WORLD

After a culture has been established in the laboratory, two questions arise: (1) whether this culture is similar to the culture targeted for isolation, and (2) if so, whether it is identical to the culture at the beginning of the isolation and cultivation process? The first question is not easy to answer even if we reduce it to the genome sequence (which would require a complete genome sequence from the environmental clone as well as from the isolate). It is known that genomes of different strains belonging to the same species can vary substantially [89–93]. For example, the genome of the virulent *E. coli* O157:H7 strain is ~25% larger than that of the laboratory strain K-12 and encodes for 1632 proteins and 20 tRNAs that are not present in K-12 [94]. Commonly used molecular biology tools that are based on the 16*S* rRNA gene cannot decipher these fundamental molecular differences, not to mention their phenotypic manifestations. An exact match of environmental clone and isolate (down to the base pair resolution) is not needed to pursue most research questions. The second question about how the organism has changed during the cultivation process itself can be addressed in part by genome sequencing. It is often observed during initial isolation and cultivation efforts and year-long propagation in the laboratory that the cultures eventually begin to grow progressively better in the new environment provided [95]. Adaptation to laboratory conditions, for example, can include colony formation on plates [76,96]. Researchers are normally pleased by adaptations that result in better growth since it means shorter incubation times, and therefore, experiment times, and often more robust cultures. However, these adaptations are an indication that the existing strain in the laboratory has evolved, and (often) minuscule changes in the genome are responsible for these changes in phenotype [95,97–103]. It has been demonstrated that organisms undergo genetic changes over time when cultivated under the same conditions, resulting in genetically diverse subpopulations [98,104–106]. However, linking changes in the genome sequence to the phenotype is still not a straightforward process. Not all mutations in the genome will lead to a beneficial effect for the microorganism. Determining the causality of mutations by gene knockins or knockouts and reconstituting the wildtype can be challenging and requires a genetic system for the organism [102,107].

Comprehensive knowledge of quantitative processes at the cellular level requires the integration of various datasets and approaches. Microbial interactions and the mutual influence of microbial cells with their biotic and abiotic environment, such as, the human body, can be elucidated only by combining several approaches, which can be grouped into bottom–up and top–down approaches [28]. Top–down approaches, for example, include biodiversity assessments, rate measurements, and isotopic signatures, as well as biomarker and all kinds of metaomics studies. Bottom–up approaches consist of isolation and cultivation methods, single-cell techniques, and all sorts of omics studies. Because of their relatively high coverage, top–down approaches are advantageous for studies at the community level, such as by metaomics methods. In depth knowledge about phenotype, metabolism, and pathogenicity can thus far be inferred only by these methods [28]. Bottom–up approaches, on the other hand, utilize direct measurements performed on the cellular level. These approaches include isolation and cultivation as well as single-cell techniques, such as single cell genome sequencing [108–111], various *in situ* hybridization methods [112,113], secondary ion mass spectrometry [114,115], and Raman [116] and Fourier transform infrared spectroscopy [117,118]. Since many microorganisms are currently recalcitrant to our cultivation approaches, integration of bottom–up and top–down approaches seems advantageous in order to understand and predict functions of complex biological systems. Therefore, isolation and cultivation methods are still most valuable in the genomic age and represent a cornerstone in the study of the human microbiome, especially when we move beyond descriptive assessments. The most direct measurements possible today are still experiments with defined pure or mixed cultures. Combined with genetic manipulation techniques, these lay the foundation for proving mutualism and pathogenicity.

REFERENCES

1. Huse SM, Dethlefsen L, Huber JA, Welch DM, Relman DA, Sogin ML. Exploring microbial diversity and taxonomy using SSU rRNA hypervariable tag sequencing. *PLoS Genet* **4**(11):e1000255 (2008).
2. Sogin ML, Morrison HG, Huber JA, et al. Microbial diversity in the deep sea and the underexplored "rare biosphere." *Proc Natl Acad Sci USA* **103**(32):12115–12120 (2006).
3. Vetsigian K, Goldenfeld N. Global divergence of microbial genome sequences mediated by propagating fronts. *Proc Natl Acad Sci USA* **102**(20):7332–7337 (2005).
4. Lozupone CA, Knight R. Species divergence and the measurement of microbial diversity. *FEMS Microbiol Rev* **32**(4):557–578 (2008).
5. Quince C, Lanzen A, Curtis TP, et al. Accurate determination of microbial diversity from 454 pyrosequencing data. *Nat Meth* **6**(9):639–641 (2009).
6. Rosselló-Mora R, López-López A. The least common denominator: Species or operational taxonomic units? In Zengler K, ed., *Accessing Uncultivated Microorganisms. From the Environment to Organisms and Genomes and Back*, ASM Press, Washington, DC, 2008, pp. 117–130.
7. Hughes JB, Hellmann JJ, Ricketts TH, Bohannan BJ. Counting the uncountable: Statistical approaches to estimating microbial diversity. *Appl Environ Microbiol* **67**(10):4399–4406 (2001).
8. Schloss PD, Handelsman J. Toward a census of bacteria in soil. *PLoS Comput Biol* **2**(7):e92 (2006).
9. Torsvik V, Daae FL, Sandaa RA, Øvreås L. Novel techniques for analysing microbial diversity in natural and perturbed environments. *J Biotechnol* **64**(1):53–62 (1998).

10. Roesch LF, Fulthorpe RR, Riva A, et al. Pyrosequencing enumerates and contrasts soil microbial diversity. *ISME J* **1**(4):283–290 (2007).
11. Dykhuizen DE. Santa Rosalia revisited: Why are there so many species of bacteria? *A van Leeuwenhoek* **73**:25–33 (1998).
12. Gans J, Wolinsky M, Dunbar J. Computational improvements reveal great bacterial diversity and high metal toxicity in soil. *Science* **309**(5739):1387–1390 (2005).
13. Bunge J, Epstein SS, Peterson DG. Comment on "Computational improvements reveal great bacterial diversity and high metal toxicity in soil." *Science* **313**(5789):918; author reply 918 (2006).
14. Fulthorpe RR, Roesch LF, Riva A, Triplett EW. Distantly sampled soils carry few species in common. *ISME J* **2**(9):901–910 (2008).
15. Lamarche J, Bradley RL, Hooper E, Shipley B, Simao Beaunoir AM, Beaulieu C. Forest floor bacterial community composition and catabolic profiles in relation to landscape features in Quebec's southern boreal forest. *Microbial Ecol* **54**(1):10–20 (2007).
16. Quince C, Curtis TP, Sloan WT. The rational exploration of microbial diversity. *ISME J* **2**(10):997–1006 (2008).
17. Amann RI, Ludwig W, Schleifer K-H. Phylogenetic identification and in situ detection of individual microbial cells without cultivation. *Microbiol Rev* **59**(1):143–169 (1995).
18. Dunbar J, Barns SM, Ticknor LO, Kuske CR. Empirical and theoretical bacterial diversity in four Arizona soils. *Appl Environ Microbiol* **68**(6):3035–3045 (2002).
19. Keller M, Zengler K. Tapping into microbial diversity. *Nat Rev Microbiol* **2**(2):141–150 (2004).
20. Pace NR. Mapping the tree of life: progress and prospects. *Microbiol Mol Biol Rev* **73**(4):565–576 (2009).
21. Wu D, Hugenholtz P, Mavromatis K, et al. A phylogeny-driven genomic encyclopaedia of Bacteria and Archaea. *Nature* **462**(7276):1056–1060 (2009).
22. Rosselló-Mora R, Amann R. The species concept for prokaryotes. *FEMS Microbiol Rev* **25**(1):39–67 (2001).
23. Zengler K. Does cultivation still matter? In Zengler K, ed., *Accessing Uncultivated Microorganisms From the Environment to Organisms and Genomes and Back*, ASM Press, Washington, DC, 2008, pp. 3–10.
24. Dinsdale EA, Edwards RA, Hall D, Angly F, Breitbart M, Brulc JM, Furlan M, Desnues C, Haynes M, Li L, et al. Functional metagenomic profiling of nine biomes. *Nature* **452**(7187):629–632 (2008).
25. Podar M, Abulencia CB, Walcher M, Hutchison D, Zengler K, Garcia JA, Holland T, Cotton D, Hauser L, Keller M. Targeted access to the genomes of low-abundance organisms in complex microbial communities. *Appl Environ Microbiol* **73**(10):3205–3214 (2007).
26. Oremland RS, Capone DG, Stolz JF, Fuhrman J. Whither or wither geomicrobiology in the era of "community metagenomics." *Nat Rev Microbiol* **3**(7):572–578 (2005).
27. Qiu Y, Cho BK, Park YS, Lovley D, Palsson BO, Zengler K. Structural and operational complexity of the *Geobacter sulfurreducens* genome. *Genome Res* **20**:1304–1311 (2010).
28. Zengler K. Central role of the cell in microbial ecology. *Microbiol Mol Biol Rev* **73**(4):712–729 (2009).
29. Caron DA, Gast RJ. The diversity of free-living protists seen and unseen, cultured and uncultured. In Zengler K, ed. *Accessing Uncultivated Microorganisms: From the Environment to Organisms and Genomes and Back*, ASM Press, Washington, DC, 2008, pp. 67–93.

30. Jeon S, Bunge J, Leslin C, Stoeck T, Hong S, Epstein SS. Environmental rRNA inventories miss over half of protistan diversity. *BMC Microbiol* **8**:222 (2008).
31. Finlay BJ. Global dispersal of free-living microbial eukaryote species. *Science* **296**(5570):1061–1063 (2002).
32. Foissner W. Protist diversity and distribution: Some basic considerations. *Biodivers Conserv* **17**:235–242 (2008).
33. Hawksworth DL, Rossman AY. Where are all the undescribed fungi? *Phytopathology* **87**(9):888–891 (1997).
34. Fincham JR. Transformation in fungi. *Microbiol Rev* **53**(1):148–170 (1989).
35. Hawkins KM, Smolke CD. Production of benzylisoquinoline alkaloids in *Saccharomyces cerevisiae*. *Nat Chem Biol* **4**(9):564–573 (2008).
36. Cohn F. Untersuchungen über Bakterien. *Beitr Biol Pflanz* **1**:127–224 (1875).
37. Beijerinck WM. Ueber *Spirillum desulfuricans* als Ursache von Sulfatreduktion. *Zentralblatt Bakteriol* **1**:1–9, 49–59, 104–114 (1895).
38. Koch R. Untersuchungen über Bakterien VI. Verfahren zur Untersuchung, zum Conservieren und Photographieren. *Beitr Biol Pflanz* **2**:399–434 (1877).
39. Winogradsky S. Ueber Schwefelbacterien. *Botanische Zeitung* **45**(31–36):489–507, 513–523, 529–539, 545–559, 569–576, 585–594, 606–610 (1887).
40. Konneke M, Bernhard AE, de la Torre JR, Walker CB, Waterbury JB, Stahl DA. Isolation of an autotrophic ammonia-oxidizing marine archaeon. *Nature* **437**(7058):543–546 (2005).
41. Zengler K, Toledo G, Rappe M, Elkins J, Mathur EJ, Short JM, Keller M. Cultivating the uncultured. *Proc Natl Acad Sci USA* **99**(24):15681–15686 (2002).
42. Janssen PH. New cultivation strategies for terrestrial microorganisms. In Zengler K ed., *Accessing Uncultivated Microorganisms: From the Environment to Organisms and Genomes and Back*, ASM Press, Washington, DC, 2008, pp. 173–192.
43. Winterberg H. Zur Methodik der Bakterienzaehlung. *Zeitschr Hyg* **29**:75–93 (1898).
44. Conn HJ. The microscopic study of bacteria and fungi in soil. *NY Agric Exp Stn Tech Bull* **64**:3–20 (1918).
45. Staley JT, Konopka A. Measurement of in situ activities of nonphotosynthetic microorganisms in aquatic and terrestrial habitats. *Annu Rev Microbiol* **39**:321–346 (1985).
46. Wintermute EH, Silver PA. Emergent cooperation in microbial metabolism. *Mol Syst Biol* **6**:407 (2010).
47. Bruns A, Cypionka H, Overmann J. Cyclic AMP and acyl homoserine lactones increase the cultivation efficiency of heterotrophic bacteria from the Central Baltic Sea. *Appl Environ Microbiol* **68**(8):3978–3987 (2002).
48. Bruns A, Nübel U, Cypionka H, Overmann J. Effect of signal compounds and incubation conditions on the culturability of freshwater bacterioplankton. *Appl Environ Microbiol* **69**(4):1980–1989 (2003).
49. Guan LL, Kamino K. Bacterial response to siderophore and quorum-sensing chemical signals in the seawater microbial community. *BMC Microbiol* **1**(1):27 (2001).
50. Guan LL, Onuki H, Kamino K. Bacterial growth stimulation with exogenous siderophore and synthetic *N*-acyl homoserine lactone autoinducers under iron-limited and low-nutrient conditions. *Appl Environ Microbiol* **66**(7):2797–2803 (2000).
51. Bussmann I, Philipp B, Schink B. Factors influencing the cultivability of lake water bacteria. *J Microbiol Meth* **47**(1):41–50 (2001).
52. Kaeberlein T, Lewis K, Epstein SS. Isolating "uncultivable" microorganisms in pure culture in a simulated natural environment. *Science* **296**(5570):1127–1129 (2002).

53. Leadbetter JR, Greenberg EP. Metabolism of acyl-homoserine lactone quorum-sensing signals by *Variovorax paradoxus*. *J Bacteriol* **182**(24):6921–6926 (2000).
54. Nichols D, Lewis K, Orjala J, Mo S, Ortenberg R, O'Connor P, Zhao C, Vouros P, Kaeberlein T, Epstein SS. Short peptide induces an "uncultivable" microorganism to grow in vitro. *Appl Environ Microbiol* **74**(15):4889–4897 (2008).
55. Mashburn LM, Whiteley M. Membrane vesicles traffic signals and facilitate group activities in a prokaryote. *Nature* **437**(7057):422–425 (2005).
56. D'Onofrio A, Crawford JM, Stewart EJ, Witt K, Gavrish E, Epstein S, Clardy J, Lewis K. Siderophores from neighboring organisms promote the growth of uncultured bacteria. *Chem Biol* **17**(3):254–264 (2010).
57. Porter J, Edwards C, Morgan JA, Pickup RW. Rapid, automated separation of specific bacteria from lake water and sewage by flow cytometry and cell sorting. *Appl Environ Microbiol* **59**(10):3327–3333 (1993).
58. Ferrari BC, Oregaard G, Sorensen SJ. Recovery of GFP-labeled bacteria for culturing and molecular analysis after cell sorting using a benchtop flow cytometer. *Microbial Ecol* **48**(2):239–245 (2004).
59. Taylor RJ, Falconnet D, Niemisto A, Ramsey SA, Prinz S, Shmulevich I, Galitski T, Hansen CL. Dynamic analysis of MAPK signaling using a high-throughput microfluidic single-cell imaging platform. *Proc Natl Acad Sci USA* **106**(10):3758–3763 (2009).
60. Marcy Y, Ouverney C, Bik EM, Losekann T, Ivanova N, Martin HG, Szeto E, Platt D, Hugenholtz P, Relman DA, et al. Dissecting biological "dark matter" with single-cell genetic analysis of rare and uncultivated TM7 microbes from the human mouth. *Proc Natl Acad Sci USA* **104**(29):11889–11894 (2007).
61. Fröhlich J, König H. Rapid isolation of single microbial cells from mixed natural and laboratory populations with the aid of a micromanipulator. *Syst Appl Microbiol* **22**(2):249–257 (1999).
62. Ishoy T, Kvist T, Westermann P, Ahring BK. An improved method for single cell isolation of prokaryotes from meso-, thermo- and hyperthermophilic environments using micromanipulation. *Appl Microbiol Biotechnol* **69**(5):510–514 (2006).
63. Zengler K, Walcher M, Clark G, Haller I, Toledo G, Holland T, Mathur EJ, Woodnutt G, Short JM, Keller M. High-throughput cultivation of microorganisms using microcapsules. *Meth Enzymol* **397**:124–130 (2005).
64. Button DK, Schut F, Quang P, Martin R, Roberston BR. Viability and isolation of marine bacteria by dilution culture: theory, procedures, and initial results. *Appl Environ Microbiol* **59**:881–891 (1993).
65. Schoenborn L, Yates PS, Grinton BE, Hugenholtz P, Janssen PH. Liquid serial dilution is inferior to solid media for isolation of cultures representative of the phylum-level diversity of soil bacteria. *Appl Environ Microbiol* **70**(7):4363–4366 (2004).
66. Connon SA, Giovannoni SJ. High-throughput methods for culturing microorganisms in very-low-nutrient media yield diverse new marine isolates. *Appl Environ Microbiol* **68**(8):3878–3885 (2002).
67. de Man JC. The probability of most probable numbers. *Eur J Appl Microbiol* **1**:67–78 (1975).
68. Ishikuri S, Hattori T. Formation of bacterial colonies in successive time intervals. *Appl Environ Microbiol* **49**(4):870–873 (1985).
69. Winding A, Binnerup SJ, Sørensen J. Viability of indigenous soil bacteria assayed by respiratory activity and growth. *Appl Environ Microbiol* **60**(8):2869–2875 (1994).
70. Zengler K, Richnow HH, Rossello-Mora R, Michaelis W, Widdel F. Methane formation from long-chain alkanes by anaerobic microorganisms. *Nature* **401**(6750):266–269 (1999).

71. Biddle JF, Lipp JS, Lever MA, Lloyd KG, Sorensen KB, Anderson R, Fredricks HF, Elvert M, Kelly TJ, Schrag DP, et al. Heterotrophic Archaea dominate sedimentary subsurface ecosystems off Peru. *Proc Natl Acad Sci USA* **103**(10):3846–3851 (2006).

72. Thingstad F. Elements of a theory for the mechanisms controlling abundance, diversity, and biogeochemical role of lytic bacterial viruses in aquatic systems. *Limnol Oceanogr* **45**:1320–1328 (2000).

73. Joseph SJ, Hugenholtz P, Sangwan P, Osborne CA, Janssen PH. Laboratory cultivation of widespread and previously uncultured soil bacteria. *Appl Environ Microbiol* **69**(12): 7210–7215 (2003).

74. Ferrari BC, Gillings MR. Cultivation of fastidious bacteria by viability staining and micromanipulation in a soil substrate membrane system. *Appl Environ Microbiol* **75**(10):3352–3354 (2009).

75. Rappé MS, Connon SA, Vergin KL, Giovannoni SJ. Cultivation of the ubiquitous SAR11 marine bacterioplankton clade. *Nature* **418**(6898):630–633 (2002).

76. Cho JC, Giovannoni SJ. Cultivation and growth characteristics of a diverse group of oligotrophic marine *Gammaproteobacteria*. *Appl Environ Microbiol* **70**(1):432–440 (2004).

77. Watve M, Shejval V, Sonawane C, Rahalkar M, Matapurkar A, Shouche Y, Patole M, Phadnis N, Champhenkar A, Damle K, et al. The "K" selected oligotrophic bacteria: A key to uncultured diversity? *Curr Sci* **78**:1535–1542 (2000).

78. Koster S, Angile FE, Duan H, Agresti JJ, Wintner A, Schmitz C, Rowat AC, Merten CA, Pisignano D, Griffiths AD, et al. Drop-based microfluidic devices for encapsulation of single cells. *Lab Chip* **8**(7):1110–1115 (2008).

79. McDonald GR, Hudson AL, Dunn SM, You H, Baker GB, Whittal RM, Martin JW, Jha A, Edmondson DE, Holt A. Bioactive contaminants leach from disposable laboratory plasticware. *Science* **322**(5903):917 (2008).

80. Olsen RA, Bakken LR. Viability of soil bacteria: optimization of plate-counting techniques and comparisins between total counts and plate counts within different size groups. *Microbial Ecol* **13**:59–74 (1987).

81. Aagot N, Nybroe O, Nielsen P, Johnsen K. An altered *Pseudomonas* diversity is recovered from soil by using nutrient-poor *Pseudomonas*-selective soil extract media. *Appl Environ Microbiol* **67**(11):5233–5239 (2001).

82. Andrews JH, Harris RF. *r*- and *K*-selection and microbial ecology. *Adv Microbial Ecol* **9**:99–147 (1986).

83. Hahn MW, Stadler P, Wu QL, Pockl M. The filtration-acclimatization method for isolation of an important fraction of the not readily cultivable bacteria. *J Microbiol Methods* **57**(3):379–390 (2004).

84. Ferrari BC, Binnerup SJ, Gillings M. Microcolony cultivation on a soil substrate membrane system selects for previously uncultured soil bacteria. *Appl Environ Microbiol* **71**(12):8714–8720 (2005).

85. Aoi Y, Kinoshita T, Hata T, Ohta H, Obokata H, Tsuneda S. Hollow fiber membrane chamber as a device for in situ environmental cultivation. *Appl Environ Microbiol* **75**(11):3826–3833 (2009).

86. Ingham CJ, van den Ende M, Pijnenburg D, Wever PC, Schneeberger PM. Growth and multiplexed analysis of microorganisms on a subdivided, highly porous, inorganic chip manufactured from anopore. *Appl Environ Microbiol* **71**(12):8978–8981 (2005).

87. Bertics VJ, Ziebis W. Biodiversity of benthic microbial communities in bioturbated coastal sediments is controlled by geochemical microniches. *ISME J* **3**(11):1269–1285 (2009).

88. Bertics VJ, Ziebis W. Bioturbation and the role of microniches for sulfate reduction in coastal marine sediments. *Environ Microbiol* **12**(11):3022–3034 (2010).
89. Perna NT, Plunkett G 3rd, Burland V, Mau B, Glasner JD, Rose DJ, Mayhew GF, Evans PS, Gregor J, Kirkpatrick HA, et al. Genome sequence of enterohaemorrhagic *Escherichia coli* O157:H7. *Nature* **409**(6819):529–533 (2001).
90. Coleman ML, Sullivan MB, Martiny AC, Steglich C, Barry K, Delong EF, Chisholm SW. Genomic islands and the ecology and evolution of *Prochlorococcus. Science* **311**(5768):1768–1770 (2006).
91. Médigue C, Rouxel T, Vigier P, Hénaut A, Danchin A. Evidence for horizontal gene transfer in *Escherichia coli* speciation. *J Mol Biol* **222**(4):851–856 (1991).
92. Welch RA, Burland V, Plunkett G, III, Redford P, Roesch P, Rasko D, Buckles EL, Liou SR, Boutin A, Hackett J, et al. Extensive mosaic structure revealed by the complete genome sequence of uropathogenic *Escherichia coli. Proc Natl Acad Sci USA* **99**(26):17020–17024 (2002).
93. Alm RA, Ling LS, Moir DT, King BL, Brown ED, Doig PC, Smith DR, Noonan B, Guild BC, deJonge BL, et al. Genomic-sequence comparison of two unrelated isolates of the human gastric pathogen *Helicobacter pylori. Nature* **397**(6715):176–180 (1999).
94. Hayashi T, Makino K, Ohnishi M, Kurokawa K, Ishii K, Yokoyama K, Han CG, Ohtsubo E, Nakayama K, Murata T, et al. Complete genome sequence of enterohemorrhagic *Escherichia coli* O157:H7 and genomic comparison with a laboratory strain K-12. *DNA Res* **8**(1):11–22 (2001).
95. Cooper TF, Rozen DE, Lenski RE. Parallel changes in gene expression after 20,000 generations of evolution in *Escherichiacoli. Proc Natl Acad Sci USA* **100**(3):1072–1077 (2003).
96. Davis KE, Joseph SJ, Janssen PH. Effects of growth medium, inoculum size, and incubation time on culturability and isolation of soil bacteria. *Appl Environ Microbiol* **71**(2):826–834 (2005).
97. Babu MM, Aravind L. Adaptive evolution by optimizing expression levels in different environments. *Trends Microbiol* **14**(1):11–14 (2006).
98. Rainey PB, Rainey K. Evolution of cooperation and conflict in experimental bacterial populations. *Nature* **425**(6953):72–74 (2003).
99. Ibarra RU, Edwards JS, Palsson BO. *Escherichia coli* K-12 undergoes adaptive evolution to achieve in silico predicted optimal growth. *Nature* **420**(6912):186–189 (2002).
100. Fong SS, Nanchen A, Palsson BO, Sauer U. Latent pathway activation and increased pathway capacity enable *Escherichia coli* adaptation to loss of key metabolic enzymes. *J Biol Chem* **281**(12):8024–8033 (2006).
101. Dekel E, Alon U. Optimality and evolutionary tuning of the expression level of a protein. *Nature* **436**(7050):588–592 (2005).
102. Herring CD, Raghunathan A, Honisch C, Patel T, Applebee MK, Joyce AR, Albert TJ, Blattner FR, van den Boom D, Cantor CR, et al. Comparative genome sequencing of *Escherichia coli* allows observation of bacterial evolution on a laboratory timescale. *Nat Genet* **38**(12):1406–1412 (2006).
103. Applebee MK, Herrgard MJ, Palsson BO. Impact of individual mutations on increased fitness in adaptively evolved strains of *Escherichia coli. J Bacteriol* **190**(14):5087–5094 (2008).
104. Woods R, Schneider D, Winkworth CL, Riley MA, Lenski RE. Tests of parallel molecular evolution in a long-term experiment with *Escherichia coli. Proc Natl Acad Sci USA* **103**(24):9107–9112 (2006).
105. Finkel SE. Long-term survival during stationary phase: Evolution and the GASP phenotype. *Nat Rev Microbiol* **4**(2):113–120 (2006).

106. Rozen DE, Philippe N, Arjan de Visser J, Lenski RE, Schneider D. Death and cannibalism in a seasonal environment facilitate bacterial coexistence. *Ecol Lett* **12**(1):34–44 (2009).
107. Tremblay PL, Summers ZM, Glaven RH, Nevin KP, Zengler K, Barrett CL, Qiu Y, Palsson BO, Lovley DR. A c-type cytochrome and a transcriptional regulator responsible for enhanced extracellular electron transfer in *Geobacter sulfurreducens* revealed by adaptive evolution. *Environ Microbiol* **13**(1):13–23 (2010).
108. Podar M, Abulencia CB, Walcher M, Hutchison D, Zengler K, Garcia JA, Holland T, Cotton D, Hauser L, Keller M. Targeted access to the genomes of low-abundance organisms in complex microbial communities. *Appl Environ Microbiol* **73**(10):3205–3214 (2007).
109. Zhang L, Cui X, Schmitt K, Hubert R, Navidi W, Arnheim N. Whole genome amplification from a single cell: implications for genetic analysis. *Proc Natl Acad Sci USA* **89**(13):5847–5851 (1992).
110. Zhang K, Martiny AC, Reppas NB, Barry KW, Malek J, Chisholm SW, Church GM. Sequencing genomes from single cells by polymerase cloning. *Nat Biotechnol* **24**(6):680–686 (2006).
111. Mussmann M, Hu FZ, Richter M, de Beer D, Preisler A, Jørgensen BB, Huntemann M, Glockner FO, Amann R, Koopman WJ, et al. Insights into the genome of large sulfur bacteria revealed by analysis of single filaments. *PLoS Biol* **5**(9):e230 (2007).
112. Lee N, Nielsen PH, Andreasen KH, Juretschko S, Nielsen JL, Schleifer K-H, Wagner M. Combination of fluorescent in situ hybridization and microautoradiography—a new tool for structure-function analyses in microbial ecology. *Appl Environ Microbiol* **65**(3):1289–1297 (1999).
113. Ouverney CC, Fuhrman JA. Combined microautoradiography-16S rRNA probe technique for determination of radioisotope uptake by specific microbial cell types in situ. *Appl Environ Microbiol* **65**:1746–1752 (1999).
114. Halm H, Musat N, Lam P, Langlois R, Musat F, Peduzzi S, Lavik G, Schubert CJ, Singha B, Laroche J, et al. Co-occurrence of denitrification and nitrogen fixation in a meromictic lake, Lake Cadagno (Switzerland). *Environ Microbiol* 2009(**11**):1945–1958.
115. Musat N, Halm H, Winterholler B, Hoppe P, Peduzzi S, Hillion F, Horreard F, Amann R, Jorgensen BB, Kuypers MM. A single-cell view on the ecophysiology of anaerobic phototrophic bacteria. *Proc Natl Acad Sci USA* **105**(46):17861–17866 (2008).
116. Huang WE, Stoecker K, Griffiths R, Newbold L, Daims H, Whiteley AS, Wagner M. Raman-FISH: combining stable-isotope Raman spectroscopy and fluorescence in situ hybridization for the single cell analysis of identity and function. *Environ Microbiol* **9**(8):1878–1889 (2007).
117. Naumann D, Helm D, Labischinski H. Microbiological characterizations by FT-IR spectroscopy. *Nature* **351**:81–82 (1991).
118. Zhao H, Kassama Y, Young M, Kell DB, Goodacre R. Differentiation of *Micromonospora* isolates from a coastal sediment in Wales on the basis of Fourier transform infrared spectroscopy, 16S rRNA sequence analysis, and the amplified fragment length polymorphism technique. *Appl Environ Microbiol* **70**(11):6619–6627 (2004).

13

NEW APPROACHES TO CULTIVATION OF HUMAN MICROBIOTA

SLAVA S. EPSTEIN, MARIA SIZOVA, and AMANDA HAZEN

Department of Biology, Northeastern University, Boston Massachusetts

13.1. INTRODUCTION

One of the most interesting observations in microbiology is that most microorganisms refuse to grow in the laboratory [1,2]. This was first noted well over a century ago as a gross mismatch between the number of cells in an environmental sample compared to the number of colonies they produced on artificial media [3]. This observation became especially intriguing when, almost a century later, the rRNA approach to microbial detection and identification revealed that it was not just an issue of inoculated cells versus. colony-forming unit (CFU) number. It became abundantly clear that it was also a species diversity phenomenon; the richness of microbial species in nature greatly exceeded the number of cultivated species [4–11]. Today, this phenomenon is known as the "great plate count anomaly" [12]. Despite the obvious importance of this phenomenon for progress in microbiology and biotechnology [13,14], its nature remains largely unresolved.

The anomaly is observed in the human microbiota as well. The size of the cultivated versus uncultivated pools of species from the human microbiome has not been fully determined, and estimates vary significantly among the microbial sites in the human body. It seems that the cultivable fraction is fairly small in the gut [15], relatively large (≤90%) on the skin [16], and has intermediate values at other sites. For example, the current estimate of the uncultivated species number in the human oral cavity ranges between 30% and 50% [17–19]. This is calculated from the number

The Human Microbiota: How Microbial Communities Affect Health and Disease,
First Edition. Edited by David N. Fredricks.

of cultivated species and estimates of the total richness (~700 [20] phylotypes). The latter may actually be orders of magnitude higher [21], which may change the estimate of the proportion of uncultivated species. Regardless of the exactness of the current estimates, the "missing" species are interesting because they likely play important roles in the function of microbial communities in the human body, and some may be implicated in a variety of systemic diseases [22–24].

13.2. MICROBIOTA CULTIVATION METHODOLOGY

Microbial culture remains a principal tool for the study of microbes, and the lack of access to cultures of many of them prompted a more recent resurgence of cultivation efforts. Some of these efforts represented a substantial departure from conventional techniques, whereby the researchers adopted single-cell and high-throughput strategies [25–27], developed new tools to better mimic the natural milieu [28–31], and increased the length of incubation and lowered the concentration of nutrients [32]. Our research group contributed to the effort by designing several new cultivation approaches and modifying some conventional methodologies [33–35]. The initial applications of these developments were in environmental settings and covered a range of habitats, such as marine sediments, freshwater ponds, soils, and groundwater [35–37]. The overall idea behind our effort is to replicate as closely as possible the natural conditions, in some cases even moving the actual cultivation process into the environment. This idea is as applicable to environmental species as it is to the human microbiota, and recently we have started adapting relevant approaches for use in the human body [38]. This chapter outlines the principles of these (but by no means all) methodologies and summarizes the experiences we gained from their recent applications to the environmental and human microbiota.

Method 1: Diffusion Chamber

It is self-evident that microbial species that do not grow in the lab can and do grow *in situ*. The difference between the two is that nature supplies such species with growth components necessary and sufficient for their growth, whereas artificial media do not. We therefore considered the possibility of cultivating microbial species inside diffusion chambers incubated in the natural environment of these species. In principle, a membrane bound device could serve two purposes: containment for microbial growth and chemical exchange via diffusion. The expectation is that once the chamber, together with cells enclosed inside, is introduced into the environment, diffusion will automatically provide natural conditions for these cells, allowing for their cultivation with no prior knowledge of their growth requirements. Applying this idea, we designed and tested a simple diffusion chamber built from a metal washer and two 0.02-μm pore-size polycarbonate membranes (Figure 13.1). The intial test of the idea employed marine sediment bacteria, and showed a 300-fold increase in microbial recovery compared to the parallel trials with standard petri dishes and artificial media [33]. This success provided significant support for the notion that the diffusion chamber-based approach to microbial cultivation may allow isolation of some previously uncultivated species responsible for the great plate count anomoly. In the follow-up research, we used similar devices to recover

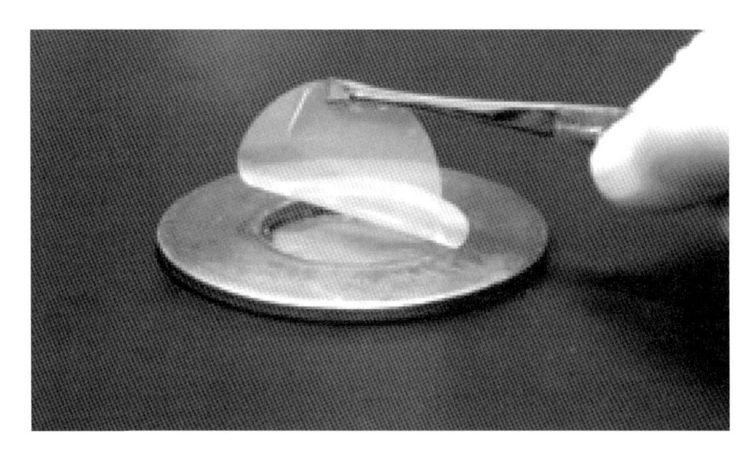

Figure 13.1. Diffusion chamber for *in situ* cultivation of microorganisms. A membrane was glued to the bottom of a metal disk with central orifice; the second membrane is being glued to the upper part of the disk, sealing the inner space. This space can be filled with agar mixed with test cells, and the assembly is then incubated in the natural environment of the cells.

microorganisms from a freshwater pond [36] and contaminated groundwater [37]. In the first study, the overlap between species isolated via the diffusion-chamber-based approach versus standard cultivation on petri dishes was low (7%), and the first led to isolation and propagation of representatives of rarely cultivated phyla Verrucomicrobia and Acidobacteria. The second study was conducted in an area where many rRNA gene surveys had previously taken place (Field Research Center of Oak Ridge National Laboratory, TN). Microbial culture collections are typically very different from the species lists produced by molecular surveys. In that study, we investigated whether diffusion-chamber-derived collections were different from those obtained by conventional approaches, possibly providing a better match to the molecular inventories. Indeed, 30% of the isolates obtained by the new culture method were closely related to species known only from molecular surveys conducted in the same area [e.g., a previous inventory of the depth reported 29 bacterial operational taxonomic units (OTUs)] [39]. We were able to cultivate six of them, from three different phyla, including multiple isolates matching strains that molecular studies reported from the area multiple times [39,40]. In addition, we reasoned that since microbial growth inside diffusion chambers occurred in the presence of the environmental stress factors, the raised isolates must be environmentally relevant, that is, tolerant of these factors (in the present case, high concentrations of heavy metals, nitrate, organic contaminants, extremes of pH, etc.). We investigated the physiology of 14 isolates and showed that the majority were, indeed, capable of growth under low pH and/or high concentrations of U, Co, Ni, and nitrate. One diffusion-chamber-grown isolate, closely related to *Microbacterium laevaniformans*, displayed intensive growth at the concentrations of many stress factors even higher than the range observed *in situ*.

A separate, and very important, question is whether it is possible to domesticate the diffusion-chamber-reared isolates for growth *ex situ*. During our experimentation with such isolates, we serendipitously noticed that a fraction of them could, in fact, grow on artificial media, and this fraction increased as the chamber-grown material was moved from diffusion chamber to diffusion chamber for repetitive *in situ* cultivation. We therefore attempted to determine whether several rounds of cultivation in a series of generations of diffusion chambers facilitated domestication of the grown strains. The results demonstrated a positive correlation between the number of cultivation rounds *in situ* and the probability of obtaining a variant capable of growth *ex situ* [41]. Of the 23 strains tested, approximately a quarter acquired the ability to grow on standard media after just one round, and this proportion steadily grew to 70% after four cultivation rounds. The nature of the domestication process remains unclear, but the empirical observations suggest that prolonged *in situ* cultivation adapted a significant number of otherwise "uncultivable" strains to growth under the conventional conditions of standard petri dishes.

The general conclusion from these data is that *in situ*–based cultivation leads to the isolation of species that are environmentally relevant, are active in nature, are on the list of uncultivated species as detected by molecular surveys, and could be domesticated for growth on artificial media.

Method 2: Ichip

The diffusion chamber design shown in Figure 13.1 has an important limitation: relatively low throughput. In a typical application, it is inoculated with a mixture of different species. This leads to a mixed culture, necessitating purification and isolation efforts, often substantial. To address this limitation, we developed the *isolation chip*, or ichip for short (Figure 13.2) [35]. It is based on the same principle of microbial cultivation *in situ* but is designed differently and consists of dozens or hundreds of miniature diffusion chambers, each loaded with an average of one cell per chamber. This combines microbial growth and isolation into a single step.

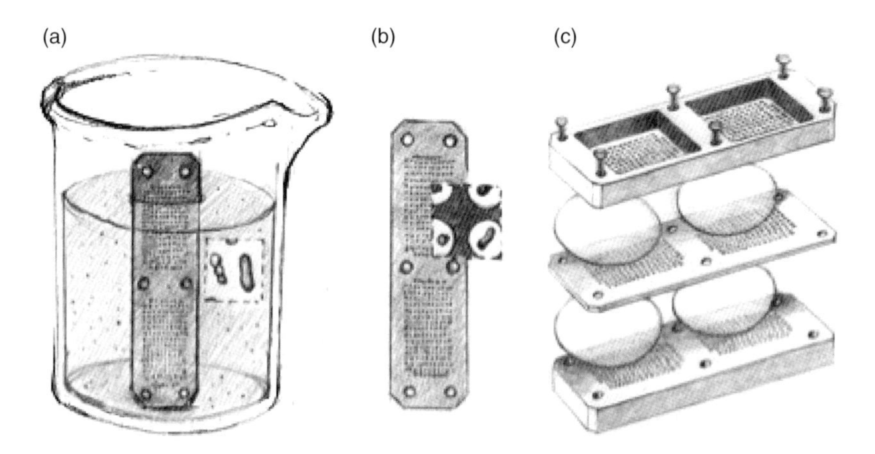

Figure 13.2. Isolation chip, or ichip, for high-throughput microbial cultivation *in situ*: (a) dipping a plate with multiple through-holes into a suspension of cells leads to capturing (on average) single cell (b); (c) ichip assembly; membranes cover arrays of through-holes from each side: Upper and bottom plates with matching holes press the membranes against the central (loaded) plate. Screws provide sufficient pressure to seal the content of individual through-holes, each becoming a miniature diffusion chamber containing (on average) a single cell. (Reproduced from Nichols et al. [35]) with permission.)

We tested the ichip concept using seawater and soil environments and made three important observations: (1) it performed as well or better than the original diffusion chamber, allowing for ≤50% of inoculated cells to form colonies; (2) the ichip produced a unique collection of microorganisms that shared only one species with that obtained by conventional methods; and (3) among the ichip-grown microorganisms, 70–90% represented novel species and genera, which was 2–3 times better than the degree of microbial discovery afforded by parallel conventional petri dishes. We concluded that the ichip is a useful method for studies aimed at large-scale isolation of novel microorganisms.

After its repeated application to various environments ranging from aquatic to soil to subterranean, we discovered a singular disadvantage of the design presented in Figure 13.2—its plastic parts easily wore out, compromising the seal. To compensate for this, we designed, built, and are presently testing a new variant of the ichip. This variant will have two advantages over the one originally proposed: it will be durable and also compatible with standard laboratory high throughput devices such as microtiter plates. This newest variant is presented in Figure 13.3, with explanations.

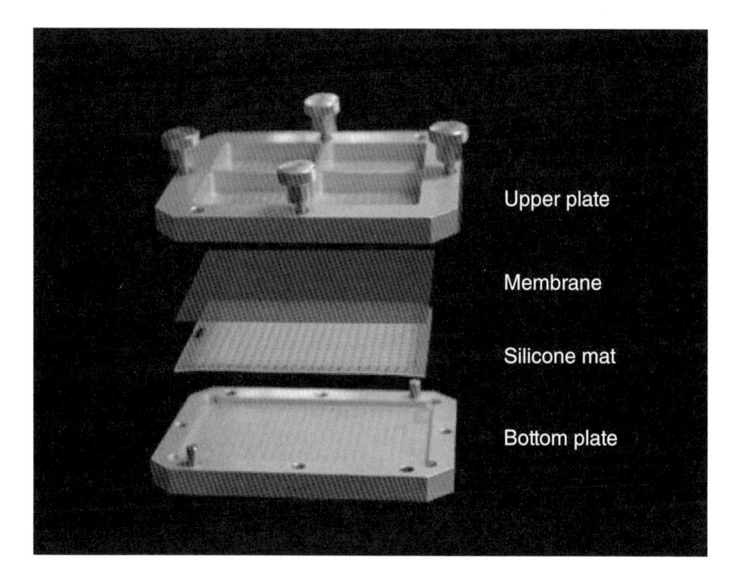

Figure 13.3. A new variant of the ichip. Upper and bottom plates are made from surgical steel, and each has 384 registered through holes that also match wells of a standard silicone sealing mat for PCR plates. Microbial cells are inoculated into wells of the mat by cell sorting or dilution such that each well receives (on average) a single cell. The mat is covered by a precut 0.02-μm pore-size polycarbonate membrane, and is sandwiched between the upper and bottom plates. Screws on the upper plate provide compression to seal the contents of individual wells. The assembled ichip is then incubated in the environment of the inoculated cells. On incubation, grown material can be subsampled by a syringe, or a cassette of syringes, through the holes in the bottom plate. Subsamples can be transferred to standard 384-well plates preloaded with nutrient medium, or injected into a preassembled device of similar design for the second round on *in situ* incubation. The silicone mat is self-healing and can withstand multiple punctures without leaking.

Method 3: Trap

In our search for approaches allowing for *in situ* enrichment for actinomycetes and other specific microorganisms, we realized that this could be achieved by using the above mentioned devices but in a reverse fashion. For example, actinomycetes and fungi grow by forming filaments capable of penetrating soft substrates, and can pass through pores as small as 0.2 µm. If so, a diffusion chamber or ichip loaded with sterile agar and with membrane pores of >0.2 µm, could become a trap for filamentous, chain forming, and possibly actively moving cells. Our test study supported these expectations [34]. In this study, we built traps following the design presented in Figure 13.1, loaded it with sterile agar, and replaced its bottom membrane with a standard 0.2–0.6-µm pore-size polycarbonate filter. The traps were placed on top of moist garden soil, ensuring that the bottom filter with larger pores was in contact with the substrate. The small (0.02-µm) pore size of the upper membrane prevented contamination from the air. After incubation the membranes were peeled off, the slab of solid agar was removed and examined for growth; visible microcolonies were sampled, purified on artificial media, and identified. Unlike the situation in parallel conventional petri dishes, the majority of organisms grown in the traps proved to be actinomycetes, some of which represented rare and unusual species from the genera *Dactilosporangium*, *Catellatospora*, *Catenulispora*, *Lentzea*, and *Streptacidiphilus*. We concluded that the trap method was a useful addition to our suite of *in situ* cultivation methodologies, leading to a selective capture of filamentous actinomycetes enriched for new and rare species. Note that ichip variants (Figures 13.2 and 13.3) can work as traps as well as the original diffusion chamber (Figure 13.1).

13.3. APPLICATION TO HUMAN MICROBIOTA

Encouraged by the apparent success of the approaches that involve *in situ* cultivation as the first step in the overall isolation process, we attempted to tailor these approaches to the cultivation of human microbiota [38]. Because of the ease of access, the most straightforward application was in the human oral cavity. This required scaling down the above mentioned devices, and we started from miniaturizing the trap, hereafter referred to as the *minitrap* (Figure 13.4). Its design is analogous to shown in Figure 13.2, but it is built from surgical steel (we note that steel is not sufficiently compressible to provide the necessary seal for its use as the ichip). The minitrap, loaded with sterile agar, was inserted into a window precut into a palatal appliance molded from the upper maxilla of the subject and affixed with superglue. The subject was allowed to eat, drink, and perform normal oral hygiene. During its 48-h incubation in the subject's mouth, various microorganisms penetrated the membranes and established colonies within the minitrap's compartments. After incubation, the appliance was removed and placed in an anaerobic glovebox. The minitrap was separated from the appliance, and disassembled, and the grown material was pour-plated and anaerobically incubated.

After isolation into pure culture, microorganisms were identified, and the species list was compared to that obtained by conventional enrichment/isolation techniques using subgingival samples from the same subject. Two observations deserve mention:

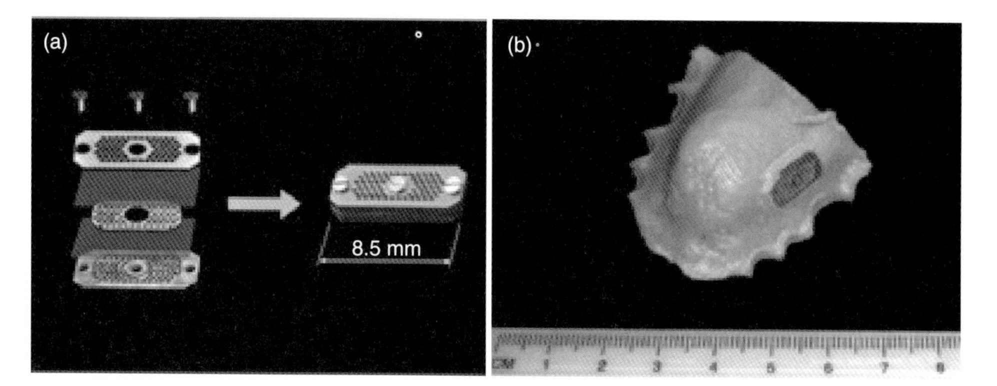

Figure 13.4. Minitrap used for *in situ* cultivation of oral microorganisms. (a) basic design (explanations provided in the text); (b) general view of the subject's dental mold, with the minitrap glued to a window cut in the mold. (Reprinted from Sizova et al. [38], with permission.)

1. Of the cells grown in the minitrap, 11% formed colonies on an artificial medium. This uncharacteristically high rate of microbial recovery is in line with our earlier observations of a high domestication rate of *in situ*–grown environmental cells [36,41].
2. The species lists obtained by the two approaches shared only one species in common. The likeliest explanation is that the differences between the culture collections are due to the respective biases of the cultivation techniques used. Indeed, the minitrap method selects for species active in the mouth at the time of incubation, as only actively growing species would be expected to colonize the space within the minitrap. Conventional enrichment/plating, on the other hand, selects for species most competitive *ex situ*. Therefore, the resulting culture collections are unlikely to be inclusive of each other. This nonredundancy of the cultivation methods confirms our earlier observations made during environmental applications of these methods [35–37,42].

13.4. LOOKING AHEAD

A general—and, in retrospect, predictable—conclusion that we can draw from these experiments is that an ensemble of novel and traditional cultivation techniques is a promising tool with which to close the gap between microorganisms available in culture and those present in the human microbiota. The *in situ* incubation steps clearly enrich for organisms that are not available or are too rare on conventional petri dishes. Clearly, the devices available to conduct such steps have not yet been perfected. One area of improvement will be to ensure a good seal of the individual compartments of the ichip to afford true single-cell incubations. Such a seal appears difficult to achieve with all-metal devices, making them suitable mostly as traps. A possible solution is to use a flexible middle part, perhaps analogous to the silicone mats shown in Figure 13.3. We are currently exploring this idea using the ichip modification presented in Figure 13.4. Additionally, human microbiota applications demand true miniaturization of the devices. While this does not necessarily pose

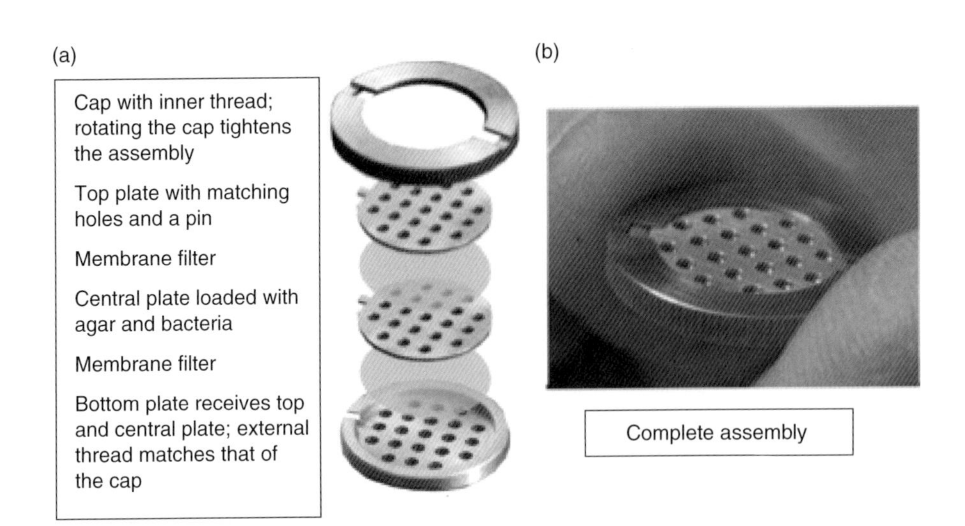

Figure 13.5. Miniature ichip for *in vivo* microbial cultivation in the oral cavity. (a) assembly process; (b) completed assembly.

design/production difficulties, it presents logistical challenges during experimentation. Many target microbiome species are obligate or facultative anaerobes, and maintenance of anaerobic conditions is a must in cultivating some of these species. Working in an anaerobic glovebox imposes its own specific restrictions, and renders the assembly of very small devices (e.g., Figure 13.4) difficult or even impractical. Simplifying the way in which such devices are sandwiched and tightened is one way to fine-tune the ichip approach for use with the human microbiota. In collaboration with Dr. D. Fredricks (Fred Hutchinson Cancer Research Center, Seattle WA), we are currently investigating whether the ichip variant presented in Figure 13.5 has advantages over the previously developed units for cultivation of novel vaginal microorganisms.

Conventional cultivation also offers room for improvement. As is the case with *in situ* incubation devices, maintenance of strict anaerobic conditions is often necessary in more standard applications, starting from sampling to strain purification in petri dish cultivation. This is because oxygen in the mouth (and other body cavities) is rapidly consumed by aerobic, early bacterial colonizers (e.g., *Neisseria* spp.) or facultative anaerobes (e.g., *Streptococcus* and *Actinomyces* spp.). This creates within the biofilm oxic and redox conditions suitable for a large number of obligate anaerobes [43]. Tanner et al. [44] concluded that strict anaerobic cultivation may bring into culture as wide a diversity of species present in early childhood caries as could be detected by cloning approaches. Importantly, Moore and Moore [45] presented evidence that the overall proportion of anaerobes increases during periodontal disease progression and, conversely, that of aerobe and facultative species decreases. However, while arguments in favor of stricter oxygen regimes during cultivation have been made repeatedly since the introduction of modern anaerobic gloveboxes [46,47], this practice has not become a universally accepted norm.

The composition of nutrient media for microbial isolation is also worth experimentation. For example, the majority of microbial culture media used to date have been nutrient-rich. These conditions may favor the growth of faster-growing bacteria at the expense of slower-growing species, some of which thrive in nutrient-poor environments [25,48,49] and may even be inhibited by substrate-rich conventional media. Consequently, the use of dilute nutrient media has led to the successful cultivation of previously uncultivated bacteria from various natural terrestrial habitats [25–27,50]. In our experience, using sugar-free media may lead to the isolation of species from the human oral cavity that have not been seen growing on more traditional media formulations [38].

To summarize, over the past decade, a number of innovations have been introduced into the art of microbial cultivation. These range from moving the cultivation process into the natural habitat of the target microorganisms to the use of dilute and otherwise modified media. The resulting approaches appear to have different biases for different species, and thus can work synergistically toward better microbial discovery. The ensemble of new and time-honored techniques is thus a promising tool in our attempts to resolve the great plate count anomaly.

ACKNOWLEDGMENTS

Research presented here was supported by NIH grants 1RC1DE020707-01 and R21 DE018026-01A1 to SSE and 1R01HG005816-01 to D. Fredricks (Fred Hutchinson Cancer Research Center); NSF grants OCE-0221267 and OCE-0102248 to SSE; and DOE grants DE-FG02-04ER63782, DE-FG02-07ER64507, and DE-FG02-04ER63782 to SSE.

REFERENCES

1. Rappe MS, Giovannoni SJ. The uncultured microbial majority. *Annu Rev Microbiol* **57**:369–394 (2003).
2. Handelsman J. Metagenomics: Application of genomics to uncultured microorganisms. *Microbiol Mol Biol Rev* **68**:669–685 (2004).
3. Winterberg H. Zur Methodik der Bakterienzahlung. *Zeitschr Hyg* **29**:75–93 (1898).
4. Giovannoni SJ, Britschgi TB, Moyer CL, Field KG. Genetic diversity in Sargasso Sea bacterioplankton. *Nature* **345**:60–63 (1990).
5. Ward DM, Weller R, Bateson MM. 16S rRNA sequences reveal numerous uncultured microorganisms in a natural community. *Nature* **345**:63–65 (1990).
6. DeLong EF. Archaea in coastal marine environments. *Proc Natl Acad Sci USA* **89**:5685–5689 (1992).
7. Fuhrman JA, McCallum K, Davis AA. Novel major archaebacterial group from marine plankton. *Nature* **356**:148–149 (1992).
8. Liesack W, Stackebrandt E. Occurrence of novel groups of the domain Bacteria as revealed by analysis of genetic material isolated from an Australian terrestrial environment. *J Bacteriol* **174**:5072–5078 (1992).
9. Hugenholtz P, Goebel BM, Pace NR. Impact of culture-independent studies on the emerging phylogenetic view of bacterial diversity. *J Bacteriol* **180**:4765–4774 (1998).
10. Ravenschlag K, Sahm K, Pernthaler J, Amann R. High bacterial diversity in permanently cold marine sediments. *Appl Environ Microbiol* **65**:3982–3989 (1999).

11. Dojka MA, Harris JK, Pace NR. Expanding the known diversity and environmental distribution of an uncultured phylogenetic division of bacteria. *Appl Environ Microbiol* **66**:1617–1621 (2000).
12. Staley JT, Konopka A. Measurement of in situ activities of nonphotosynthetic microorganisms in aquatic and terrestrial habitats. *Annu Rev Microbiol* **39**:321–346 (1985).
13. Osburne MS, Grossman TH, August PR, MacNeil IA. Tapping into microbial diversity for natural products drug discovery. *ASM News* **66**:411–417 (2000).
14. Hurst CJ. Divining the future of microbiology. *ASM News* **71**:262–263 (2005).
15. Frank DN, Pace NR. Gastrointestinal microbiology enters the metagenomics era. *Curr Opin Gastroenterol* **24**:4–10 (2008).
16. Gao Z, Tseng CH, Pei Z, Blaser MJ. Molecular analysis of human forearm superficial skin bacterial biota. *Pro Nat Acad Sci USA* **104**:2927–2932 (2007).
17. Paster BJ, Boches SK, Galvin JL, Ericson RE, Lau CN, Levanos VA, Sahasrabudhe A, Dewhirst FE. Bacterial diversity in human subgingival plaque. *J Bacteriol* **183**:3770–3783 (2001).
18. Paster BJ, Olsen I, Aas JA, Dewhirst FE. The breadth of bacterial diversity in the human periodontal pocket and other oral sites. *Periodontology 2000* **42**:80–87 (2006).
19. Dewhirst FE, Chen T, Izard J, Paster BJ, Tanner AC, Yu WH, Lakshmanan A, Wade WG. The human oral microbiome. *J Bacteriol* **192**:5002–5017 (2010).
20. Aas JA, Paster BJ, Stokes LN, Olsen I, Dewhirst FE. Defining the normal bacterial flora of the oral cavity. *J Clin Microbiol* **43**:5721–5732 (2005).
21. Keijser BJ, Zaura E, Huse SM, van der Vossen JM, Schuren FH, Montijn RC, ten Cate JM, Crielaard W. Pyrosequencing analysis of the oral microflora of healthy adults. *J Dent Res* **87**:1016–1020 (2008).
22. Beck J, Garcia R, Heiss G, Vokonas PS, Offenbacher S. Periodontal disease and cardiovascular disease. *J Periodontol* **67**:1123–1137 (1996).
23. Scannapieco FA. Role of oral bacteria in respiratory infection. *J Periodontol* **70**:793–802 (1999).
24. Dodman T, Robson J, Pincus D. Kingella kingae infections in children. *J Paediatr Child Health* **36**:87–90 (2000).
25. Connon SA, Giovannoni SJ. High-throughput methods for culturing microorganisms in very-low-nutrient media yield diverse new marine isolates. *Appl Environ Microbiol* **68**:3878–3885 (2002).
26. Rappe MS, Connon SA, Vergin KL, Giovannoni SJ. Cultivation of the ubiquitous SAR11 marine bacterioplankton clade. *Nature* **418**:630–633 (2002).
27. Zengler K, Toledo G, Rappe M, Elkins J, Mathur EJ, Short JM, Keller M. Cultivating the uncultured. *Proc Natl Acad Sci USA* **99**:15681–15686 (2002).
28. Bruns A, Cypionka H, Overmann J. Cyclic AMP and acyl homoserine lactones increase the cultivation efficiency of heterotrophic bacteria from the central Baltic Sea. *Appl Environ Microbiol* **68**:3978–3987 (2002).
29. Stevenson BS, Eichorst SA, Wertz JT, Schmidt TM, Breznak JA. New strategies for cultivation and detection of previously uncultured microbes. *Appl Environ Microbiol* **70**:4748–4755 (2004).
30. Ferrari BC, Binnerup SJ, Gillings M. Microcolony cultivation on a soil substrate membrane system selects for previously uncultured soil bacteria. *Appl Environ Microbiol* **71**:8714–8720 (2005).
31. Aoi Y, Kinoshita T, Hata T, Ohta H, Obokata H, Tsuneda S. Hollow-fiber membrane chamber as a device for in situ environmental cultivation. *Appl Environ Microbiol* **75**:3826–3833 (2009).

32. Davis KE, Joseph SJ, Janssen PH. Effects of growth medium, inoculum size, and incubation time on culturability and isolation of soil bacteria. *Appl Environ Microbiol* **71**:826–834 (2005).
33. Kaeberlein T, Epstein SS, Lewis K. Isolating "uncultivable" microorganisms in pure culture in a simulated natural environment. *Science* **296**:1127–1129 (2002).
34. Gavrish E, Bollmann A, Epstein S, Lewis K. A trap for in situ cultivation of filamentous actinobacteria. *J Microbiol Meth* **72**:257–262 (2008).
35. Nichols D, Cahoon N, Trakhtenberg EM, Pham L, Mehta A, Belanger A, Kanigan T, Lewis K, Epstein SS. Use of Ichip for high-throughput in situ cultivation of "uncultivable" microbial species. *Appl Environ Microbiol* **76**:2445–2450 (2010).
36. Bollmann A, Lewis K, Epstein SS. Incubation of environmental samples in a diffusion chamber increases the diversity of recovered isolates. *Appl Environ Microbiol* **73**:6386–6390 (2007).
37. Bollmann A, Palumbo AV, Lewis K, Epstein SS. Isolation and physiology of bacteria from contaminated subsurface sediments. *Appl Environ Microbiol* **76**:7413–7419 (2010).
38. Sizova MV, Hohmann T, Hazen A, Paster BJ, Halem SR, Murphy CM, Panikov NS, Epstein SS. Cultivability of oral bacteria: New approaches for isolation of previously uncultivated species. *Appl Environ Microbiol* **78**:194–203 (2012).
39. Abulencia CB, Wyborski DL, Garcia JA, Podar M, Chen W, Chang SH, Chang HW, Watson D, Brodie EL, Hazen TC, et al. Environmental whole-genome amplification to access microbial populations in contaminated sediments. *Appl Environ Microbiol* **72**:3291–3301 (2006).
40. Fields MW, Yan T, Rhee SK, Carroll SL, Jardine PM, Watson DB, Criddle CS, Zhou J. Impacts on microbial communities and cultivable isolates from groundwater contaminated with high levels of nitric acid-uranium waste. *FEMS Microbiol Ecol* **53**:417–428 (2005).
41. Nichols D, Lewis K, Orjala J, Mo S, Ortenberg R, O'Connor P, Zhao C, Vouros P, Kaeberlein T, Epstein SS. Short peptide induces an "uncultivable" microorganism to grow in vitro. *Appl Environ Microbiol* **74**:4889–4897 (2008).
42. Epstein SS. General model of microbial uncultivability. In Epstein SS, ed., *Uncultivated Microorganisms*, Spinger, Heidelberg, 2009, pp. 131–150.
43. Marsh PD, Moter A, Devine DA. Dental plaque biofilms: communities, conflict and control. *Periodontol 2000* **55**:16–35 (2011).
44. Tanner AC, Mathney JM, Kent RL, Chalmers NI, Hughes CV, Loo CY, Pradhan N, Kanasi E, Hwang J, Dahlan MA, et al. Cultivable anaerobic microbiota of severe early childhood caries. *J clin microbiol* **49**:1464–1474 (2011).
45. Moore WEC, Moore LVH. The bacteria of periodontal diseases. *Periodontology 2000* **5**:66–77 (1994).
46. Socransky SS, MacDonald JB, Sawyer S. The cultivation of *Treponema microdentium* as surface colonies. *Arch Oral Biol* **1**:171–172 (1959).
47. Rosebury T, Reynolds JB. Continuous anaerobiosis for cultivation of spirochetes. *Proc Soc Exptl Biol Med* **117**:813–815 (1964).
48. Koch AL. Microbial physiology and ecology of slow growth. *Microbiol Mol Biol Rev* **61**:305ff. (1997).
49. Vartoukian SR, Palmer RM, Wade WG. Strategies for culture of "unculturable" bacteria. *FEMS Microbiol Lett* **309**:1–7 (2010).
50. Watve M, Shejval V, Sonawane C, Rahalkar M, Matapurkar A, Shouche Y, Patole M, Phadnis N, Champhenkar A, Damle K, et al. The "K" selected oligophilic bacteria: A key to uncultured diversity? *Curr Sci* **78**:1535–1542 (2000).

14

MANIPULATING THE INDIGENOUS MICROBIOTA IN HUMANS: PREBIOTICS, PROBIOTICS, AND SYNBIOTICS

GEORGE T. MACFARLANE and SANDRA MACFARLANE

The University of Dundee, Microbiology and Gut Biology Group, Ninewells Hospital Medical School, Dundee, United Kingdom

14.1. INTRODUCTION

The human large intestine harbors a complex microbiota that comprises 100s, possibly 1000s, of different bacterial species, subspecies, and strains. While many of the structure/function relationships between different components of the microbiota are unclear, this complex and dynamic multicellular entity plays an important role in maintaining homeostasis in the mammalian host [1]. Some idea of the diversity of its interactions with the body can be seen in Figure 14.1. The intricacy of the microbiota arises principally from the multiplicity of different carbon and energy sources that are accessible for bacterial growth. Thus the types and amounts of food that are consumed are major factors that control microbiota structure and metabolism. Additionally, a number of host determinants are important in regulating form and function in intestinal microbial communities, particularly those concerned with gut anatomy and physiology, such as colonic transit time. Other regulatory processes include competition for nutrients and space, as well as cooperative interactions between individual groups of bacteria, such as those involved in the breakdown of complex polymeric substances, or in syntrophic associations involving the production and utilization of hydrogen gas [2,3]. The microbiota is a stable and immensely complex entity, which, like all climax communities, is self-regulating to a considerable degree. Although many microorganisms are able to invade and temporarily colonize the gastrointestinal tract, indigenous species confer protection to the host by acting

The Human Microbiota: How Microbial Communities Affect Health and Disease, First Edition. Edited by David N. Fredricks.

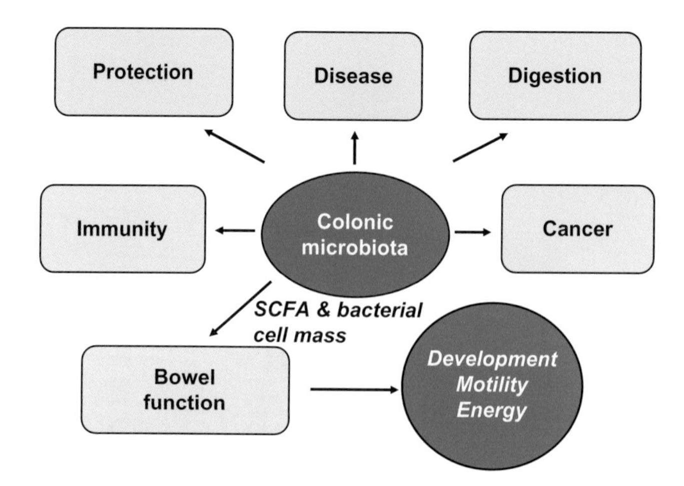

Figure 14.1. Schematic showing major interactions between the human colonic microbiota and host physiology.

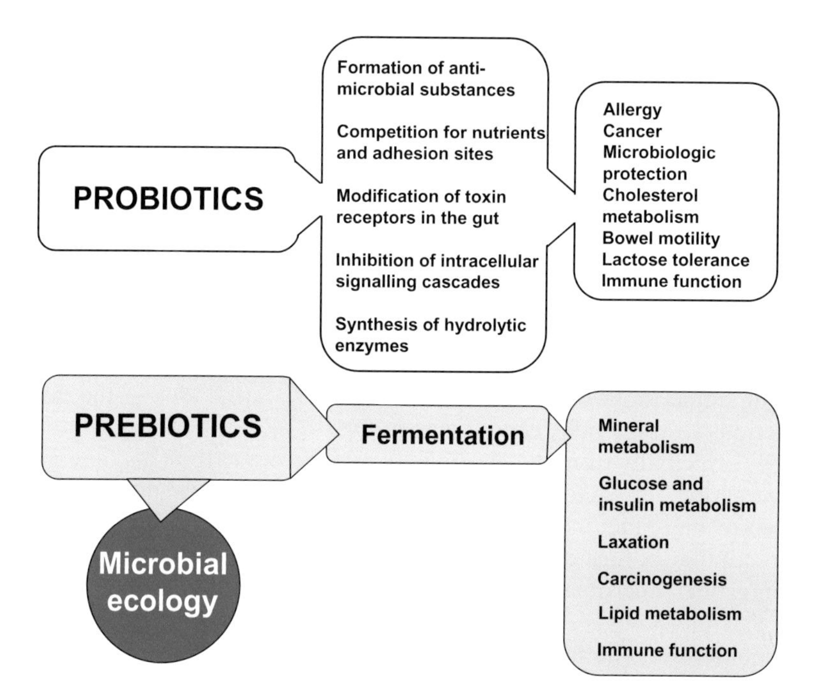

Figure 14.2. Metabolic and physiologic processes mediated by bacteria growing in the large bowel.

as a barrier to invading pathogens, but these activities are often diminished during illness, stress, or aging, or by antibiotic treatment. However, it is now known that the composition of the microbiota and its metabolic potential can be modified through relatively simple changes to the diet, particularly those involving the use of probiotics, prebiotics, and synbiotics. Much has been claimed in terms of the health benefits associated with consumption of these so-called functional foods (see Figure 14.2), with various degrees of scientific or clinical justification. As a consequence, while there is considerable support in the food industry, there is still a significant degree of scepticism concerning the efficacy of their use in scientific and legislative circles.

14.2. PROBIOTICS

As early as 1907, the Russian microbiologist Elie Metchnikoff recognized that consumption of sourmilk or yogurt could benefit human health and wellbeing, which he believed was due to the presence of *Lactobacillus bulgaricus* in the product. The term *probiotic*, referring to bacteria used to treat health disorders, was used later when Parker [4] described probiotics as "organisms and substances, which contribute to intestinal microbial balance." The current widely accepted definition of probiotics is that they are "living microorganisms, which upon ingestion in adequate amounts exert health benefits beyond inherent general nutrition" [5]. Several different microrganisms have been used as probiotics in humans (see Table 14.1); bifidobacteria and lactobacilli are the most commonly investigated in clinical trials. Probiotics must be nontoxigenic or nonpathogenic, and be able to survive gastrontestinal transit. Other desirable traits include temporary colonization of the gastrointestinal tract; secretion of inhibitory substances, including short-chain fatty acids (SCFA) and lactate; and the production of beneficial products such as vitamins [6]. There is also a substantial and growing body of evidence showing that they can inhibit the growth of pathogenic species, and improve host immunity.

TABLE 14.1. Microorganisms Claimed to Have Probiotic Properties in Humans

Gram-Positive Rods	Gram-Positive Cocci	Other Organisms
Bifidobacterium infantis	*Enterococcus faecium*	*Saccharomyces boulardii*
B. lactis	*E. faecalis*	*S. cerevisiae*
B. bifidum	*Lactococcus lactis*	*Escherichia coli*
B. animalis	*L. cremoris*	
B. breve	*Streptococcus thermophilus*	
B. longum	*Leuconostoc mesenteroides*	
B. adolescentis	*Pediococcus pentosaceus*	
Lactobacillus acidophilus		
L. bulgaricus		
L. rhamnosus		
L. casei		
L. paracasei		
L. gasseri		
L. plantarum		
L. reuteri		
Bacillus subtilis		
B. licheniformis		
Propionibacterium freudenreichii		

14.2.1. Health Benefits Associated With Probiotics

Crohn's Disease

Crohn's disease (CD) is a complex disorder of the gastrointestinal tract, and together with ulcerative colitis (UC), it is one of the two principal forms of idiopathic inflammatory bowel disease (IBD) in humans. Patients with CD condition can be asymptomatic, or have a life-threatening presentation. CD is characterized by patchy, transmural zones of inflammation that can affect any part of the gastrointestinal tract, from the oral cavity to the rectum. CD is incurable, and standard treatments are based on maintenance therapies with antiinflammatory drugs, steroids, monoclonal antibody therapy, and surgery. Several different probiotics have been used in CD, with little success. For example, in one 12-month placebo-controlled study involving 38 patients with active CD, *Lactobacillus rhamnosus* GG was used to reduce the rate or severity of recurrence after surgery [7]. No significant differences were found in clinical or endoscopic recurrence rates between the group receiving the probiotic (16.6%) and the placebo group (10.5%), administered maltodextrin and sorbitol. Similarly, no differences were seen using the same probiotic in maintenance of remission, when compared to mesalazine in a 12-month trial involving 35 CD patients [8], and in pediatric CD [9].

Two double-blind randomized controlled trials (DBRCTs) with *L. johnsonii*, aimed at preventing surgical relapse, revealed that the organism was of no significant benefit in preventing endocopic recurrence [10,11]. Malchow et al. [12] used a nonpathogenic *E. coli* strain Nissle 1917 to maintain remission in a placebo-controlled trial involving 28 active CD patients for one year. No differences in relapse rates were detected, although subjects in the placebo group went into remission earlier. The yeast *Saccharomyces boulardii* has also been used to prevent relapse in CD [13]. One gram of the probiotic was given twice per day with mesalazine to 32 patients, and they were compared with individuals receiving mesalazine [1 g 3 times daily (tid)]. No statistical differences in clinical disease activities were detected; however, after 6 months, relapse rates were lower in the group receiving the probiotic and mesalazine (6%), compared to the control mesalazine (38%). VSL3 is a multispecies probiotic preparation containing four lactobacilli (*L. plantarum*, *L. casei*, *L. acidophilus*, *L. delbrueckii* subsp. *bulgaricus*), three bifidobacteria (*B. infantis*, *B. longum*, *B. breve*), and *Streptococcus salivarius* subsp. *thermophilus*. In one trial, patients were randomized to the antibiotic rifaxamin for 3 months, and then VSL3 for 9 months, or were given mesalazine for 12 months [14]. No differences were observed between the groups at the end of the study. In a more recent meta-analysis of the use of probiotics for CD, it was concluded from the eight trials used in the assessment that probiotics conferred no benefits over placebos in maintaining remission, or in preventing endoscopic or clinical relapses in CD [15].

Ulcerative Colitis

Ulcerative colitis is distinct from crohn's disease in several ways, endoscopically, histologically, and in the location of the disease. UC is a chronic, episodic inflammation that affects only the large bowel. It begins distally in the rectum, progressing more proximally to the right colon, and is characterized by bloody diarrhea in the absence of a positive stool culture for bacteria, ova, or parasites. The disease tends to be defined according to its distribution in the gut, using the terms *distal colitis*, *left-sided colitis*, and *pancolitis*. There is some preliminary evidence that probiotics may be effective in UC. The probiotic *E. coli* strain Nissle 1917 has been reported in several trials to be effective as treatment with mesalazine in inducing and maintenance of remission [16–18]. Rembacken et al. [17] compared the ability of *E. coli* Nissle 1917 to prevent UC relapse with mesalazine in a two-phase trial. In the first part, 57 patients were given two capsules of the probiotic daily, containing 2.5×10^{10} bacteria, and 59 patients were given 800 mg mesalazine tid. The patients were monitored for 12 weeks, and those in remission entered the second phase of the trial in which the probiotic was given twice daily (bid), and mesalazine reduced to 400 mg tid. Patients remained in the study for 12 months unless they relapsed. In the first phase, remission occurred in 68% of the probiotic group, compared to 75% of patients receiving mesasalazine, and of these 67% in the probiotic group and 73% in the mesalazine group relapsed in the second phase.

In a non-placebo-controlled trial, 20 patients were given capsules of VSL3, containing 5×10^{11} bacteria, bid for 12 months [19]. In this trial, 15 (75%) were found to be in remission at the end of the study. In a more recent trial of VSL3 in 29 children with mild to moderate UC, remission occurred in 93% of patients in the probiotic group, and 36% of those receiving the placebo. A relapse rate of 21% was recorded in those taking the probiotic, compared to 73% with the placebo [20]. In a small study involving 24 patients with mild to moderate UC, the volunteers were given 250 mg of *S. boulardii* and mesalazine for 4 weeks. The yeast was shown to reduce clinical disease activity scores in 68% of the patients [21]. Ishikawa et al. [22] administered fed 21 UC patients *B. breve* and *B. bifidum* delivered in 100 mL fermented milk for one year. A lower rate of relapse was observed in the probiotic group (3 out of 11) than in the controls with no probiotic (9 out of 10). In a short 2-month study, a mixture of *L. bifidus*, *L acidophilus*, and *Enterococcus* sp. was able to maintain remission in 20% of people receiving the probiotic, with 93% of patients in the placebo group relapsing [23]. However, other probiotic studies in UC patients have not been as effective. In a large open-label study, *L. rhamnosus* GG was given to 187 patients for 12 months with the objective to maintain remission. No differences in remission rates were found compared to mesalamine, or with mesalamine and the probiotic combined [24]. Similarly, no benefits were seen with either *L. salivarius* or *B. infantis* in maintaining remission [25]. A Cochrane review concluded that there was limited evidence that probiotics with standard therapy may provide modest benefits in reducing disease activity indices in patients with mild to moderately severe UC, but not in improving in overall remission rates [26].

Pouchitis

Up to 45% of patients who undergo ileal pouch surgery for UC will suffer from pouchitis, which is a bacteria-mediated inflammation of the ileal pouch, which is quite distinct from the original disease. Although the number of studies is limited, VSL3 has been shown to be the most promising probiotic tested in the treatment of pouchitis [27–29], and has been added to the British Society of Gastroenterology guidelines for management of the disease [30]. Gionchetti et al. [27] carried out a 9-month randomized placebo-controlled trial to maintain remission in pouchitis, in which 40 volunteers in remission received either 3×10^{12} cells of the probiotic daily, or a placebo containing maize starch. The probiotic was shown to be beneficial, with only 15% of the patients relapsing, compared to relapse occurring in all subjects of the placebo group. In another randomized placebo-controlled study by the same researchers, using the same amount of probiotic for 9 months, the probiotic was similarly reported capable of preventing the occurrence of pouchitis in patients postsurgery, with the disease occurring in 15% of the probiotic group, and in 100% of the placebos [28]. Both of these investigations showed high levels of the probiotic organisms in feces of the patients in the probiotic group. In another study of 36 recurrent pouchitis patients induced to remission by 4 weeks of ciprofloxacin and metronidazole, patients were given VSL3 or a placebo for 12 months, or until the subjects relapsed [29]. The probiotic was considered to be successful in maintaining remission, since only 6% of patients receiving it relapsed, compared to 85% in those receiving placebo. *Lactobacillus rhamnosus* GG has also been used in one study, in which it was shown to reduce the risk of pouchitis after colectomy [31].

Irritable Bowel Syndrome (IBS)

Many different probiotics have been tested for their use in IBS. In one of the early studies, administration of *Enterococcus faecium* PR88 for 12 weeks to 28 patients wiith severe diarrhea resulted in a 68% reduction in IBS symptoms [32]. Lactobacilli do not seem to be effective in IBS, and no significant improvements in clinical symptoms in people with untreated IBS were found after the consumption of 6.25×10^9 *L. plantarum* 299V for 4 weeks [33]. Similarly, enterocoated *L. rhamnosus* GG was used in a placebo-controlled study in 19 patients for 8 weeks to reduce disease severity. At the end of the trial, no significant differences were detected in fecal urgency, abdominal pain, or bloating between the probiotic and placebo groups; however, diarrhea did improve in some of the probiotic patients. Two randomized controlled trials also found that *L. rhamnosus* GG was not effective in reducing abdominal pain in children with IBS [34,35]. In contrast, a multispecies combination of *L. rhamnosus* GG, *L. rhamnosus* LC705, *Propionibacterium freudenreichii* subsp. *shermanii* JS, and *B. animalis* subsp. *lactis* Bb12 was shown to be of benefit, and to alleviate IBS symptoms in a clinical trial in which the probiotic preparation was given in capsule form for 6 months [36]. Total IBS symptom scores were reduced in both studies, with an average of 40% reduction for the probiotic compared to <10% for the placebo. No differences in the patients' quality of life were found in the studies. In a systematic review of probiotic trials in IBS, Brenner et al. [37] concluded that there was good evidence from two randomized controlled trials that *B. infantis* 35624 may be effective in IBS. However, the data were found to be insufficient for the authors of the review to comment on the efficacy of other probiotics in IBS.

In one 4-week study, involving >360 individuals living in the community, 10^8 *B. infantis* 35624 administered daily was shown to provide global relief, and to improve all of the primary IBS symptoms [38]. Interestingly, dose may be important, since in the same investigation, reducing the amount of probiotic given to 10^6 *B. infantis* daily elicited no beneficial effects.

Antibiotic-Associated Diarrhea

There have been a number of studies dealing with the effectiveness of probiotics in antibiotic-associated diarrhea (AAD). *Saccharomyces boulardii* has been shown to reduce the incidence of diarrhea in several placebo-controlled investigations in patients with AAD [39,40], however, its effectiveness may be reduced with aging, since in one study, no benefit was found in administering the yeast to 69 elderly patients with AAD [41]. *Saccharomyces boulardii* has also been shown to be effective in inhibiting the recurrence of infection with *Clostridium difficile* [42], and it was the only probiotic to demonstrate benefit for *C. difficile* disease in a review of probiotic trials in AAD [43]. Studies with *L. rhamnosus* GG have yielded mixed results; in one placebo-controlled trial involving 16 patients treated with erythromycin, *L. rhamnosus* GG was shown to decrease the incidence of diarrhea, and to reduce the drug's sideeffects such as bloating and flatulence [44]. This contrasts with another larger placebo-controlled study with 267 hospitalized patients taking antibiotics, who were given *L. rhamnosus* GG or a placebo for 2 weeks, where the incidence of diarrheal symptoms was found to be similar in both groups [45]. In a large placebo-controlled study involving 119 children given antibiotics for respiratory infections, *L. rhamnosus* GG was reported to reduce symptoms of diarrhea by 70%, but not disease severity [46]. In another trial using 202 children, LGG reduced the incidence of diarrhea from 26% in the placebo group to 8% in those receiving the probiotic [47]. From the studies on various probiotics used to treat AAD, the most promising seem to be those involving *S. boulardii*, with *L. rhamnosus* GG observed to be more effective in children. Several meta-analyses have shown that AAD can be significantly reduced if probiotics are given prior to antibiotic therapy [48–50]; the results varied according to the dose and strain used [51].

Acute Gastroenteritis

There is evidence from a number of clinical trials that probiotics are effective in the prevention or treatment of acute diarrhea. The majority of investigations have focused on pediatric gastroenteritis. The most widely studied probiotics have been *L. rhamnosus* GG and *S. boulardii*. In a multicenter placebo-controlled trial, 291 infants aged between one month and three years, with moderate to severe diarrhea, were given *L. rhamnosus* GG together with standard oral hydration therapy [52]. The duration of diarrhea was reduced from 72 h in the placebo group to 58 h in children receiving the probiotic. Individuals receiving LGG also had shorter hospital stays, while the incidence of long-term diarrheal episodes (lasting >7 days) was reduced to 2.7%, compared to 11% in the placebo group. In another large study involving 100 infants, the mean duration of diarrhea was reduced to 3 days with *L. rhamnosus* GG, compared to 6 days in the placebo group [53]. Szajewski et al. [54], in a randomized placebo-controlled trial of 81 of hospitalized infants, demonstrated that *L. rhamnosus* GG could reduce the risk of developing diarrhea, with 7% of patients developing disease symptoms in the probiotic group, compared to 33% in the placebos. A meta-analysis of five RCT trials with *S. boulardii*, involving 619 children, also showed that the probiotic significantly reduced the duration of diarrhea by 1.1 days in children, compared to placebo [55]. The probiotic also reduced the risk of diarrhea on days 3, 6, and 7, as well as diarrheal symptoms that lasted >7 days.

Other probiotics have been investigated for their use as therapeutic agents in diarrheal disease. Saavedra et al. [56] gave *B. bifidum* and *S. thermophilus* to 55 infants aged 5–24 months in a DBRCT. The incidence of diarrhea was reduced to 7% in children receiving the probiotic, compared to 31% in the placebo group. Furthermore, only 3% of patients who were given the probiotic were found to be shedding rotavirus, compared to 39% of the controls. *Lactobacillus* GG, in particular, has been shown to be effective when used in combination with oral hydration therapy in the treatment of rotavirus infection in children. In 49 infants aged between 6–35 months, who were administered LGG bid for 5 days, the probiotic reduced the incidence of diarrhea, after rehydration therapy, and was also shown to enhance immune function by increasing rotavirus-specific serum IgA, and rotavirus-specific antibody secreting cells [57]. Several lactobacili, including LGG, have also been shown to reduce rotavirus excretion in feces, which helps to prevent spread of the virus. Meta-analysis involving the use of single probiotics in children concluded that *L. rhamnosus* GG and *S. boulardii* had beneficial effects in acute gastroenteritis [55,58,59]. On the basis of the available evidence of probiotic use in pediatric diarrhea, a joint statement was issued by the European Society of Paediatric Infectious Diseases, and the European Society for Paediatric Gastroenterology, Hepatology and Nutrition stating that probiotics with proven efficacy (*L. rhamnosus* GG, *S. boulardii*) may be used as an effect adjunct to hydration therapy in the management of diarrhea [60].

Constipation

Several probiotics have been demonstrated to increase stool frequency and consistency. In a DBRCT in children with chronic functional constipation, 8 weeks of treatment with *L. reuteri* DSM 17938 was shown to improve stool frequency, but not to affect stool consistency [61]. *Bifidobacterium animalis* DN173 010 has been reported to improve colonic transit times in constipated [62] and healthy people [63]. Chmielewska and Szajewska [64] carried out a systematic review of five randomized controlled trids (RCTs): three in adults ($N = 266$) and two in children ($N = 111$). Results indicated that in adults, *B. lactis* DN173 010, *L. casei* Shirota, and *E. coli* Nissle 1917 could have a favorable effect on stool consistency and frequency of defecation; however, in children, *L. casei rhamnosus* Lcr35 was beneficial, but *L. rhamnosus* GG was not. There was no difference in the rate of treatment success between the probiotic and placebo groups in pooled results of the trials in children, and the authors concluded that more data were required before probiotics could be recommended for the treatment of constipation.

Allergy-Related Diseases

Probiotics have shown some promise in preventing allergic disorders, mainly atopic diseases. The first investigations were carried out in Finnish infants at high risk for atopic disease. *Lactobacillus rhamnosus* GG or a placebo was given to 159 pregnant women, with a family history of either eczema, rhinitis, or asthma, for 2–4 weeks prior to delivery [65]. The probiotic continued to be given to the breast-feeding mothers after birth, or directly to the children who were not breastfed for 6 months. Probiotic use reduced the frequency of atopic dermatitis by 50% in the children at two years, compared to the placebos. In a follow-up study at 4 years, children who had received the probiotic were found to have had a significantly lower incidence of atopic eczema than those in the placebo group [66]. No differences were detected in circulating IgE, or skin-prick responses, suggesting that IgE-mediated disease was not modulated by probiotics. However, in another randomized controlled trial of 232 families with allergic disease, a significant reduction in IgE-associated eczema was found in the second year of life when *L. reuteri* was given to mothers 4 weeks before birth, and to the infants for one year [67], although there was no difference in eczema incidence between the probiotic and placebo groups. Kopp et al. [68] also administered LGG prenatally, and for 3 months after birth, in a similar manner to that for the Finnish study. At 2 years, the severity or incidence of atopic dermatitis was not reduced in the infants at high risk for allergic disorders. Niers et al. [69] undertook a DBRCT, in which the probiotic combination *B. bifidum* W23, *B. lactis* W52, and *Lactococcus lactis* W58 was administered to 156 families with a history of allergies. The probiotic mixture was given in the last 6 weeks of pregnancy, and to the infants for 12 months after birth. Probiotic use was shown to prevent eczema in the children, with a reduction of 58% in relative risk at 3 months, and 26% at one year. A meta-analysis of six trials for the prevention of pediatric atopic dermatitis indicated that a significant reduction in the incidence of atopic dermatitis could be achieved by prenatal and/or postnatal administration of probiotics to infants or pregnant mothers [70].

Studies have also been carried out on the use of probiotics for the treatment of atopic dermatitis in children. In a small study involving 31 infants with atopic dermatitis, the subjects were randomized to either 5×10^8 *L. rhamnosus* GG daily in hydrolyzed whey formula, or the formula alone [71]. A significantly lower disease activity index was found in the group receiving the probiotic after one month. In another placebo-controlled trial, 27 infants with atopic eczema were fed either *B. lactis* Bb12, *L. rhamnosus* GG, or a placebo. Both probiotics were demonstrated to reduce the severity of atopic eczema, with a significant improvement in skin condition after 2 months [72]. A reduction in the severity of atopic dermatitis was also found in a double-blind crossover study in 41 children, with moderate to severe atopic dermatitis, who were administered *L. rhamnosus* and *L. reuteri* for 6 weeks [73]. However, evidence for the use of probiotics in treating eczema is conflicting; one meta-analysis of four RCTs concluded that there was insufficient data for the use of probiotics for the treatment of pediatric atopic dermatitis [70], while another Cochrane review of 12 RCT also determined that probiotics were not effective for treatment of eczema [74], and that there may be a small risk of adverse effects with probiotic treatment. In contrast, Betsi et al. [75] reviewed 10 trials evaluating probiotics for atopic dermatitis treatment, and found that in half of them, probiotics were able to reduce the severity of atopic dermatitis.

Therefore, while probiotics may be useful in preventing atopic dermatitis, the results are not consistent, and more work needs to be carried out before they can be recommended for the prevention or treatment of the disease.

14.2.2. Aging

There have been few studies on the use of probiotics in the elderly. Feeding 30 healthy elderly subjects with *B. lactis* HNO19 for 3 weeks was shown to affect immune function, and increase levels of activated ($CD25^+$), total helper ($CD4^+$) T lymphocytes, and natural killer (NK) cells in peripheral bloods [76].

14.2.3. Other Uses of Probiotics

Several applications of probiotics in gastrointestinal health have been covered above; however, the field of probiotics is constantly expanding, and other areas of interest include the use of probiotics as a treatment for urogenital infections, lowering cholesterol, and cancer prevention, as shown in Figure 14.2.

14.3. PREBIOTICS

Almost any carbohydrate that reaches the human large intestine will provide a substrate for the commensal microbiota, and will affect its growth and metabolic activities. This has been demonstrated with respect to nonstarch polysaccharides, or dietary fiber, and will occur with other complex macromolecules such as resistant starches, mucopolysaccharides, and mucins, as well as simple sugar alcohols and lactose. However, this stimulation of growth is a nonspecific, generalized effect, which inevitably involves many of the major saccharolytic species in the gut, and associated cross-feeding organisms. The definition of what a prebiotic should be, has been, and currently is, is a matter of some debate. In our view, the designation offered by Pineiro and colleagues [77] should be seriously considered: namely, that a prebiotic is a nonviable food component that confers a health benefit on the host associated with modulation of the microbiota. The following qualifications can be added to this basic definition; thus, a prebiotic is not an organism or drug, but is a substance that can be characterized chemically so that in most cases it will be a food-grade material. Regarding the conferment of health benefits, these should be quantifiable, and should not result from straightforward absorption of the food component into the bloodstream, or simply be a consequence of the prebiotic acting alone (in which case it would probably qualify as a drug). Moreover, it should be demonstrated that the presence of the prebiotic, and the way in which it is being delivered, changes the composition or activities of the host microbiota. The selective properties of prebiotics are commonly supposed to relate to the growth of bifidobacteria and lactobacilli, at the expense of other groups of bacteria in the gut, such as bacteroides, clostridia, eubacteria, enterobacteria, and enterococcis. This has often been reported in *in vitro* and in *in vivo* experiments, but research also shows that such selectivity is variable, and the extent to which changes in the microbiota allow a substance to be called a prebiotic have not really been established. For example, it is highly unlikely that only bifidobacteria and lactobacilli fulfill the criteria of being beneficial bacteria in the gastrointestinal tract. Despite this, more rigorous analysis will need to be done in the near future, for food labeling and health claims legislation purposes.

The majority of research activities involving prebiotic oligosaccharides have been made with inulins and their fructooligosaccharide (FOS) derivatives, together with various types of galacto-oligosaccharides (GOS). Although a number of intestinal bacteria appear to be able to ferment these substances to a certain degree, as mentioned above, most investigations have indicated that the growth of putatively beneficial species such as bifidobacteria, and to a lesser degree, lactobacilli, is particularly favored. One consequence of focusing on these organisms is that we do not know what the global effects of prebiotics are on the structure of the microbiota. Another important factor to bear in mind when using prebiotics to selectively modify the composition of bacterial communities in the large bowel, is that prebiotics on their own can only enhance the growth of bacteria that are already present. So, if for whatever reason, for example, starvation, drug-induced disease, aging, or antibiotic therapy, the target organisms are not present in the microbiota, the prebiotic may be ineffectual.

14.3.1. Health Benefits Associated with Prebiotics

Because of their safety, general stability, organoleptic properties, resistance to digestion in the upper gut, and fermentability in the colon, inulin/FOS/GOS prebiotics are being increasingly incorporated into the Western diet. Inulin-derived oligosaccharides and GOS are mildly laxative, and can result in flatulence if consumed in large amounts, but any deleterious effects on bowel habit are relatively minor [78]. While the literature dealing with the health significance of prebiotics is not as extensive as that on probiotics, considerable evidence has accrued showing that prebiotic consumption can have significant health benefits, particularly in relation to their influence on laxation, mineral absorption, potential anticancer properties, lipid metabolism, and antiinflammatory and other immune effects, including atopic disease (reviewed by Macfarlane et al. [79]). Physiologic processes in the body that are affected by prebiotics result from their breakdown by colonic microorganisms, and the consequent formation of short-chain fatty acids (SCFAs). The principal SCFA produced in the large bowel are acetate, propionate, and butyrate, and the vast majority (>95%) are absorbed from the gut and are metabolized in different ways at diverse body sites [80,81]. The effects of these substances on metabolic processes in the body are summarized in Figures 14.3, 14.4, and 14.5.

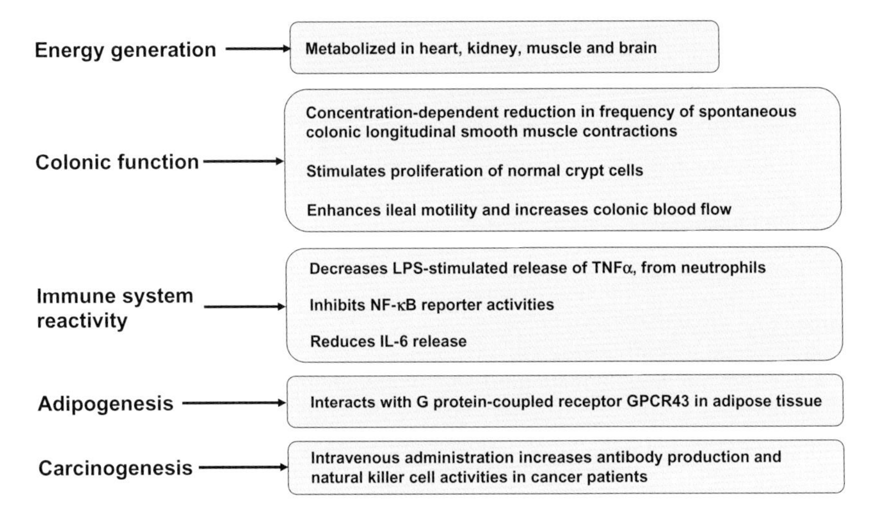

Figure 14.3. Physiologic effects of acetate produced by bacteria colonizing the large gut.

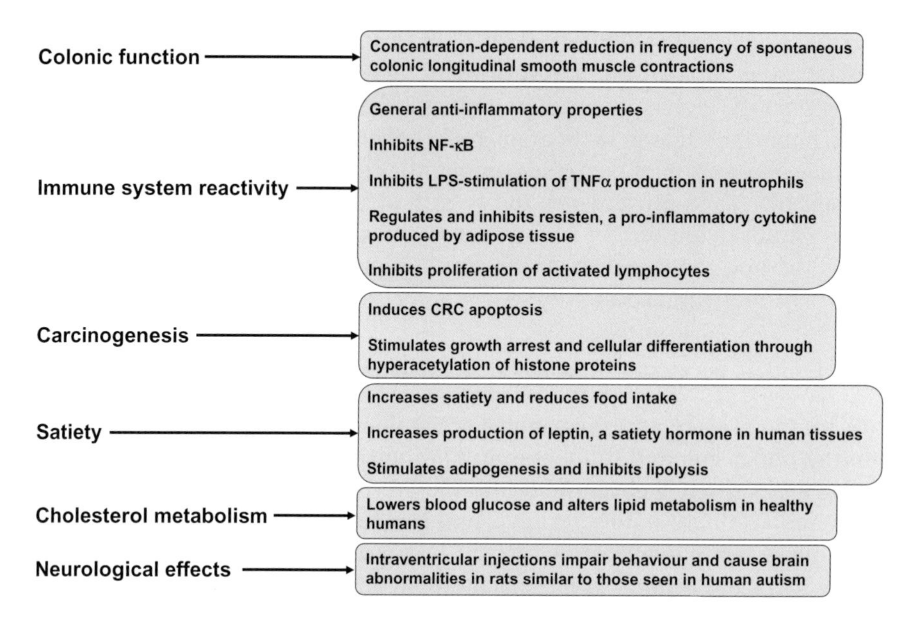

Figure 14.4. Physiologic effects of propionate produced by bacteria colonizing the large gut.

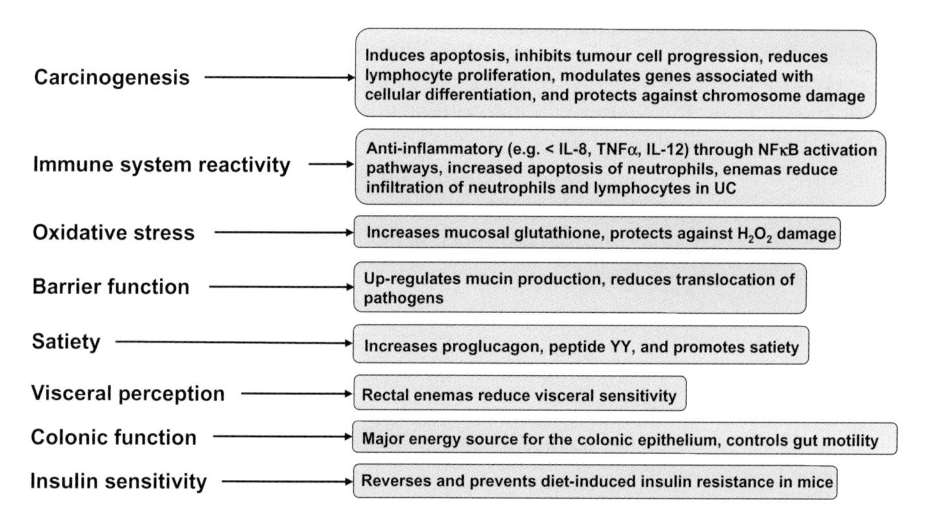

Figure 14.5. Physiologic effects of butyrate produced by bacteria colonizing the large gut.

Inflammatory Bowel Disease

Unlike the situation with with probiotics, or even synbiotics (see Section 14.4), only a small number of studies have been done with prebiotics in relation to IBD (reviewed in Macfarlane et al. [82]), and unfortunately, they were inadequately powered to be more than pilot trials. However, it has been reported that there were improvements in pouchitis when inulin was administered over a relatively short experimental period, while in two small open-label studies of children and adults, significant improvements were seen in clinical scores, with reduced expression of toll-like receptors 2 and 4.

Other Health Benefits Associated with Prebiotics

The effects of relatively small amounts (5–20 g/day) of lactulose, inulin, FOS, and GOS on the composition of the colonic microbiota and its metabolic activities have been documented extensively (reviewed by Macfarlane et al. [79]). Various host and microbial interactions in the digestive tract have been associated with prebiotic use. For example, in the mouth and upper digestive tract prebiotics are believed to be protective against dental caries, improve calcium and magnesium uptake in the small intestine, attach to cellular binding sites for pathogenic bacteria, and exert osmotic effects that can sometimes result in diarrhea. In the large intestine, prebiotic fermentation results in increased production of bacterial cell mass, with improved laxation. In animal studies, mice and rats that were challenged with carcinogens and mutagens showed that prebiotics alone, or in synbiotic combinations, can have strong anticancer effects by minimizing damage to DNA in colonic epithelial cells, reducing numbers of aberrant crypt foci on the mucosal surface and preventing tumor formation [83].

14.4. SYNBIOTICS

Prebiotics can only affect the growth of bacteria that are already present in the large bowel, whereas probiotics are allochthonous, or invading species, that have to compete with other microorganisms indigenous to the gut, that have evolved over millennia to compete under stringent nutritional and environmental growth conditions. Commensal microbes already occupy most of the available ecologic niches, making it difficult for newly introduced probiotics to gain a foothold. This has lead to the development of synbiotics, which are combinations of probiotics and prebiotics. If correct combinations of probiotic and probiotic are employed, the rationale is that the prebiotic should assist the probiotic strain to establish in the gut [83]. Relatively few clinical studies have been done with synbiotics, and perhaps the most interesting results have come from work on IBD patients.

14.4.1. Inflammatory Bowel Disease

There have been two studies by the same group showing that synbiotics may be of therapeutic benefit in IBD. In the first trial, Furrie et al. [84] carried out a DBRCT involving 18 patients with active UC, for a period of one month. The synbiotic consisted of a probiotic *B. longum* isolated from the colon of a healthy volunteer [85], combined with the prebiotic Synergy 1, which is a mixture of chicory long-chain inulin and oligofructose. Sigmoidoscopy scores were reduced in subjects receiving the synbiotic, compared to the placebos, while mRNA levels for inducible human β-defensins, which are produced by colonic epithelial cells, were strongly upregulated during inflammation, and were significantly reduced by the synbiotic. Formation of the proinflammatory cytokines IL1α and TNFα, the main drivers of mucosal inflammation in UC, were also strongly reduced after treatment (Figure 14.6). The colonic mucosa in the synbiotic group was shown to have markedly reduced

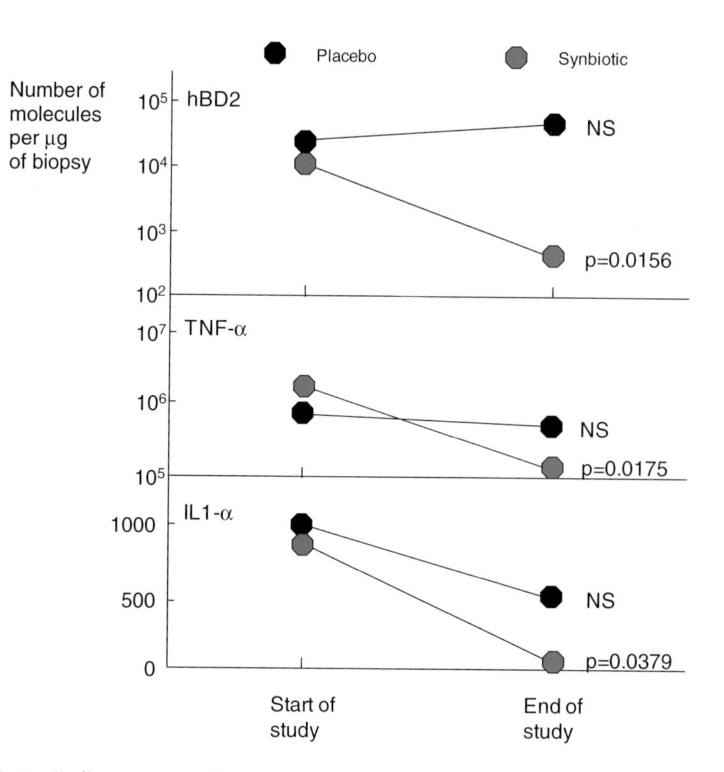

Figure 14.6. Antiinflammatory effects of a synbiotic comprising *Bifidobacterium longum* and Synergy 1 on cytokine formation, and the production of human β-defensin 2 (hBD2) in rectal tissues taken from UC patients during a 4-week feeding trial (NS—not significant). (Data are from Furrie et al. [84].)

inflammatory infiltration, and significant regeneration of healthy epithelial tissue (Figure 14.7). Increased levels of mucosa-associated bifidobacteria and marked improvements in bowel habit (Table 14.2) were also found in the patients administered the synbiotic.

The second investigation involved 35 patients with active CD, in a similar DBRCT, involving the same synbiotic combination, which ran for 6 months [86]. At 3 months, significant reductions in the principal CD proinflammatory cytokine TNFα were detected in mucosal tissues in the synbiotic group, but not in the placebos. Significant improvements in Crohn's disease activity indices, and histology scores were also found in the patients receiving the synbiotic, together with increased mucosal *B. longum*, and other mucosal bifidobacterial populations. The synbiotic was most effective in patients with large bowel Crohn's, and improvements in individuals who had small intestinal involvement were slight.

Other work with synbiotics and IBD has indicated that administration of a combination of *B. breve* strain Yakult, and a galactooligosaccharide prebiotic in 41 patients with mild to moderate UC, improved disease status, as assessed by colonoscopy [87]. Synbiotic 2000 is a mixture containing four probiotics (*L. plantarum* 2362, *L. paracasei* subsp. *paracasei* F19, *Leuconostoc mesenteroides* 77 : 1 and *Pediococcus pentosaceus* 5-33 : 3), the prebiotic inulin, as well as resistant starch,

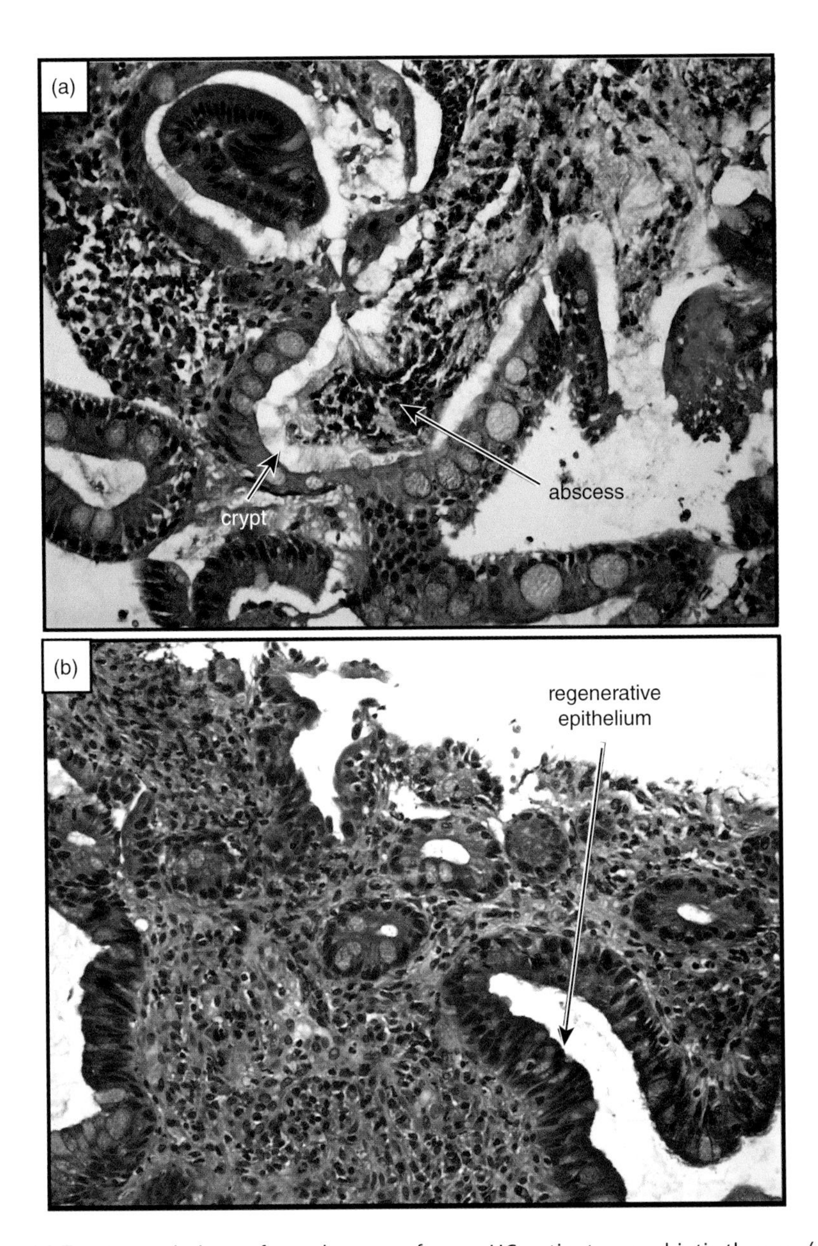

Figure 14.7. Histopathology of rectal mucosa from a UC patient presynbiotic therapy (a), and post-treatment (b), showing reductions in inflammatory infiltrates, reductions in tissue damage, and regenerating healthy epithelium.

β-glucan and pectin. A small trial evaluating the efficacy of Synbiotic 2000 in extending remission following surgery was conducted in 30 CD patients, randomized 2 : 1 into synbiotic or placebo groups [88]. As with many studies with IBD patients, the investigation suffered from a high dropout rate, with 21 subjects leaving during the course of the study. No significant differences were detected in clinical or endoscopic relapses in the small number of patients remaining.

TABLE 14.2. Specimen Bowel Habit Diary from a UC Patient Receiving a Synbiotic Containing *Bifidobacterium longum* and Synergy I during a 4-week Study Period

Details	Week 1	Week 2	Week 3	Week 4
Total stools	64	49	32	35
Loose stools	28	20	7	0
Discomfort	64	31	0	0
Blood present	16	9	14	5
Mucus present	16	9	7	0
Diarrhea	0	0	0	0
Bowel habit index	188	118	60	40

Source: Data are from Furrie et al. [84].

14.4.2. Other Uses of Synbiotics

A total of 111 children with acute diarrhea were treated with oral hydration, and the synbiotic food supplement Probiotical (containing *S. thermophilus, L. rhamnosus, L. acidophilus, B. lactis, B. infantis*, and a fructooligosaccharide) or placebo in a randomized placebo-controlled trial [89]. The duration of diarrhea was reduced by one day in the synbiotic group, compared to the children fed placebo, and normal stool consistency in the synbiotic group was significantly higher on days 2 (21% vs. 2%) and 3 (50% vs. 24%).

A synbiotic comprising *B. longum* BL99 and 90%GOS/10%FOS or control formula was administered to 138 infants aged 14 days old for 7 months in a DBRCT. The synbiotic was shown to be well tolerated by children in the synbiotic group, who had fewer incidences of constipation, and a general trend toward fewer respiratory infections compared to the group on the control formula [90]. In another study involving 66 young males with functional constipation, 4-week consumption of a commercial synbiotic capsule containing four lactobacilli (*L.casei, L. rhamnosus, L acidophilus, L. bulgaricus*), two bifidobacteria (*B. longum, B. breve*), *Streptococcus thermophilus*, and a fructooligosaccharide was demonstrated to increase the frequency and consistency of stools compared to the placebo group [91].

The role of synbiotics in prevention and treatment of atopic dermatitis shows that they may be beneficial in IgE-associated disease. In one placebo-controlled trial, the probiotic *L. rhamnosus* combined with a prebiotic mixture comprising 10% inulin and 90% GOS, was given to 48 children (>2 years old) with atopic dermatitis for 3 months [92]. Both groups were found to have significantly reduced atopic dermatitis scores at the end of the study, with no significant differences observed between those receiving the synbiotic and placebo. In another large study, the preventative effects of the prebiotic GOS combined with a probiotic mixture comprising *L. rhamnosus*, *B. breve*, and *P. freudenreichii* subsp. *shermanii* was evaluated in 925 pregnant mothers at high risk of delivering atopic infants. The women took the synbiotic for 4 weeks prior to delivery, and it was fed to the infants for 6 months after birth. While the trial found that there was no reduction in the cumulative incidence of allergic disease between the study groups, the synbiotic was shown to increase fecal IgA, and to reduce atopic eczema, as well as the incidence of IgE-associated diseases [93]. Children in the synbiotic group were also shown to have less frequent respiratory infections, and to require fewer antibiotic prescriptions during the 2-year follow-up period than the controls [94].

More recently, a synbiotic mixture containing *B. breve* M-16V, and a galactooligosaccharide was assessed as a therapy for atopic dermatitis. The DBRCT was carried out in 90 formula-fed infants <7 months of age who had atopic dermatitis. Comparison of the AD score demonstrated that there was no difference in disease improvement between the groups at the end of the 12-week study period, although in a subgroup of 48 infants with IgE-associated AD, there was a significant improvement compared to the placebos [95]. The combination of *L. salivarius* and fructooligosaccharides has also been shown to be more beneficial than the prebiotic alone, in children with moderate to severe AD [96].

Bartosch et al. [97] carried out a DBRCT with a synbiotic comprising the prebiotic Synergy I, in combination with the probiotics *B. bifidum* BB-02 and *B. lactis* BL-01. The work involved 18 healthy female elderly volunteers. The synbiotic was taken bid, and compared to those given a placebo containing maltoologosaccharide and potato starch. In volunteers given the synbiotic, levels of fecal bifidobacteria increased during the feeding and postfeeding periods, which suggests that synbiotic use was effective in restoring levels of these beneficial bacteria, which are often diminished in older people, and which may lead to improved immune function in the aging population. In a more recent randomized placebo-controlled study, the probiotic *L. acidophilus* NCFM was combined with lactitol, a prebiotic disaccharide polyol derived from lactose, and given in a 2-week intervention to 51 elderly subjects over 65 years of age, who were receiving nonsteroidal antiinflammatory drugs [98]. Volunteers administered the synbiotic had a slightly higher stool frequency, and increased numbers of bifidobacteria at the end of the intervention period, compared to the placebo group.

14.5. CONCLUSION

The increasing use of probiotics, prebiotics, and synbiotics reflects our ability to manipulate the human microbiota through comparatively small dietary changes, thereby impacting host physiology and immune response and potentially prolonging human life [99]. This will open up the possibility of developing new and effective biotherapeutic methods in treatment of IBD and other gut diseases and also in other areas where gut dysfunction can be a source of morbidity, such as aging. The advent of new molecular tools for analyzing bacterial communities in the gastrointestinal tract, particularly those colonizing mucosal surfaces, should enable disease patterns to be more closely linked to specific bacteria, or more likely, groups of organisms in the near future.

REFERENCES

1. Cummings JH, Macfarlane GT. Colonic microflora: Nutrition and health. *Nutrition* **13**:476–478 (1997).
2. Gibson GR, Macfarlane S, Macfarlane GT. Metabolic interactions involving sulphate-reducing and methanogenic bacteria in the human large intestine. *FEMS Microbiol Ecol* **12**:117–125 (1993).

3. Degnan BA, Macfarlane GT. Arabinogalactan utilization in continuous cultures of *Bifidobacterium longum*: Effect of co-culture with *Bacteroides thetaiotaomicron*. *Anaerobe* **1**:103–112 (1995).
4. Parker RB. Probiotics, the other half of the antibiotic story. *Animal Nutr Health* **29**:4–8 (1974).
5. FAO. Guidelines for the evaluation of probiotics in food. Report of a Joint FAO/WHO Working Group, London, Ontario, Canada, 4/30–5/1/02. In *Probiotics in Food: Health and Nutritional Properties and Guidelines for Evaluation*, FAO Food and Nutrition Paper 85, Rome, 2006, pp. 1–56.
6. Macfarlane GT, Cummings JH. Probiotics, infection and immunity. *Curr Opin Infect Dis* **15**:501–506 (2002).
7. Prantera C, Sribano ML, Falasco G, Andreoli A, Luzi C. Ineffectiveness of probiotics in preventing recurrence of after curative resection for Crohn's disease: a randomised controlled trial with *Lactobacillus* GG. *Gut* **51**:405–409 (2002).
8. Zocco MA, Zileri Dal Verne L, Armuzzi A, Nista EC, Papa A, Candelli M. Comparison of *Lactobacillus* GG and mesalazine inmaintaining remission of ulcerative colitis and Crohn's disease. *Gastroenterology* **124**(4 Suppl 1):A201 (2003).
9. Bousvaros A, Guandalini S, Baldassano RN, Botelho C, Evans J, Ferry GD, Goldin B, Hartigan L, Kugathasan S, Levy J, et al. A randomized, double-blind trial of *Lactobacillus* GG versus placebo in addition to standard maintenance therapy for children with Crohn's disease. *Inflamm Bowel Dis* **11**:833–839 (2005).
10. Marteau P, Lehman M, Seksik P, Laharie D, Colombel JF, Bouhnik Y, Cadiot G, Soule JC, Bourreille A, Metman E, et al. Ineffectiveness of *Lactobacillus johnsonii* LA1 for prophylaxis of postoperative recurrence of Crohn's disease: A randomised, double-blind, placebo controlled GETAID trial. *Gut* **55**:842–847 (2006).
11. Van Gossum A, Dewit O, Louis E, de Hergtogh G, Baert F, Fontaine F, Devos M, Enslen M, Paintin M, Franchimont D. Multicenter randomized-controlled clinical trial of probiotics (*Lactobacillus johnsonii*, LA1) on early endoscopic recurrence of Crohn's disease after ileo-caecal resection. *Inflamm Bowel Dis* **13**:135–142 (2007).
12. Malchow HA. Crohn's disease and *Escherichia coli*—a new approach in therapy to maintain remission of colonic Crohn's disease? *J Clin Gastroenterol* **25**:653–858 (1997).
13. Guslandi M, Mezzi G, Sorghi M, Testoni PA. *Saccharomyces boulardii* in maintenance treatment of Crohn's disease. *Digest Dis Sci* **45**:1462–1464 (2000).
14. Campieri M, Rizzello F, Venturi A, Poggili G, Ugolini F, Helwig U. Combination of antibiotic and probiotic treatment is efficacious in prophylaxis of post-operative recurrence of Crohn's disease: A randomized controlled study versus mezalamine. *Gastroenterology* **118**(4 Suppl 2):A781 (2000).
15. Rahimi R, Nikfar S, Rahimi F, Elahi B, Derakhshani S, Vafaie M, Abdollahi M. A meta-analysis on the efficacy of probiotics for maintenance of remission and prevention of clinical and endoscopic relapse in Crohn's disease. *Digest Dis Sci* **53**:2524–2531 (2008).
16. Krius W, Scutz E, Fric P, Judmaier G, Stolte M. Double-blind comparison of an oral *Escherichia coli* preparation and mesasalazine in maintaining remission of ulcerative colitis. *Aliment Pharmacol Ther* **11**:853–858 (1997).
17. Rembacken BJ, Snelling AM, Hawkey PM, Chalmers DM, Axon AT. Non-pathogenic *Escherichia coli* versus mesalazine for the treatment of ulcerative colitis: A randomised trial. *Lancet* **354**:635–639 (1999).
18. Kruis W, Fric P, Pokrotnieks J, Lukas M, Fixa B, Kascak M, Kamm MA,Weismueller J, Beglinger C, Stolte M, et al. Maintaining remission of ulcerative colitis with the probiotic *Escherichia coli* Nissle 1917 is as effective as with standard mesalazine. *Gut* **53**:1617–1623 (2004).

19. Venturi A, Gionchetti P, Rizzello F, Johansson R, Zucconi E, Brigidi P, Matteuzzi D, Campieri M. Impact on the composition of the faecal flora by a new probiotic preparation: preliminary data on maintenance treatment of patients with ulcerative colitis. *Aliment Pharmacol Ther* **13**:1103–1108 (1999).
20. Miele E, Pascarella F, Giannetti E, Quaglietta L, Baldassano RN, Stalano A. Effect of a probiotic preparation (VSL3) on induction and maintenance of remission in children with ulcerative colitis. *Am J Gastroenterol* **104**:437–443 (2009).
21. Guslandi M, Giollo P, Testoni PA. A pilot trial of *Saccharomyces boulardii* in ulcerative colitis. *Eur J Gastroenterol Hepatol* **15**:697–698 (2003).
22. Ishikawa H, Akedo I, Umesaki Y, Tanaka R, Imaoka A, Otani T. Randomized controlled trial of the effect of bifidobacteria-fermented milk on ulcerative colitis. *J Am Coll Nutr* **22**:56–63 (2003).
23. Cui HH, Chen CL,Wang JD, Yang YJ, Cun Y, Wu JB, Liu YH, Dan HL, Jian YT, Chen XQ. Effects of probiotic on intestinal mucosa of patients with ulcerative colitis. *World J Gastroenterol* **10**:1521–1525 (2004).
24. Zocco MA, dal Verme LZ, Cremonini F, Piscaglia AC, Nista EC, Candelli M, Novi M, Rigante D, Cazzato IA, Ojetti V, et al. Efficacy of *Lactobacillus* GG in maintaining remission of ulcerative colitis. *Aliment Pharmacol Ther* **23**:1567–1574 (2006).
25. Shanahan F, Guarner F, von Wright A, Vilpponene-Salmela T, O'Donoghue D, Kiely B. A one year, randomised, double blind, placebo controlled trial of a lactobacillus or a bifidobacterium probiotic for maintenance of steroid-induced remission of ulcerative colitis. *Gastroenterology* **130**(S2):A44 (2006).
26. Mallon P, McKay D, Kirk S, Gardiner K. Probiotics for induction of remission in ulcerative colitis. *Cochrane Database Syst Rev* 4:CD005573 (2007).
27. Gionchetti P, Rizzello F, Venturi A, Brigidi P, Matteuzzi D, Bazzocchi G, Poggioli G, Miglioli M, Campieri M. Oral bacteriotherapy as maintenance treatment in patients with chronic pouchitis: A double-blind, placebo-controlled trial. *Gastroenterology* **119**:305–309 (2000).
28. Gionchetti P, Rizzello F, Helwig U, Venturi A, Lammers KM, Brigandi P, Vitali B, Poggioli G, Miglioli M, Campieri M. Prophylaxis of pouchitis onset with probiotic therapy: a double-blind, placebo-controlled trial. *Gastroenterology* **124**:1202–1209 (2003).
29. Mimura T, Rizzullo F, Helwig U, Poggioli G, Schreiber S, Talbot IC, Nicholls RJ, Gionchetti P, Campieri M, Kamm MA. Once daily high dose probiotic therapy (VSL#3) for maintaining remission in recurrent or refractory pouchitis. *Gut* **53**:108–114 (2004).
30. Carter MJ, Lobo AJ, Travis SPL, on behalf of the IBD section of the British Society of Gastroenterology. Guidelines for the management of inflammatory bowel disease in adults. *Gut* **53**:v1–v16 (2004).
31. Gosselink MP, Schouten WR, Van Lieshout LM, Hop WC, Laman JD, Ruseler-van Embden JG. Delay of the first onset of pouchitis by oral intake of the probiotic strain *Lactobacillus rhanmnosus* GG. *Dis Colon Rectum* **47**:876–884 (2004).
32. Hunter J, Lee A, King T, Barratt M, Linggood M, Blades J. *Enterococcus faecium* strain PR88—an effective probiotic. *Gut* **38**:A62 (1996).
33. Sen S, Mullan MM, Parker TJ, Woolner JT, Tarry SA, Hunter JO. Effect of *Lactobacillus plantarum* 299v on colonic fermentation and symptoms of irritable bowel syndrome. *Digest Dis Sci* **47**:2615–2620 (2002).
34. Baussermann M, Michail S. The use of *Lactobacillus* GG in irritable bowel syndrome in children: A double-blind randomized control trial. *J Pediatr* **147**:197–201 (2005).
35. Gawronska A, Dziechciarz P, Horvath A, Szajewska H. A randomized double-blind placebo-controlled trial of *Lactobacillus* GG for abdominal pain disorders in children. *Aliment Pharmacol Ther* **25**:177–184 (2007).

36. Kajander K, Hatakka K, Poussa T, Farkkila M, Korpela R. A probiotic mixture alleviates symptoms in irritable bowel syndrome patients: a controlled 6-month intervention. *Aliment Pharmacol Ther* **22**:387–394 (2005).
37. Brenner DM, Moeller MJ, Chey WD, Schoenfeld PS. The utility of probiotics in the treatment of irritable bowel syndrome: A systematic review. *Am J Gastroenterol* **104**:1033–1049 (2009).
38. Whorwell PJ, Altringer L, Morel J, Bond Y, Charbonneau D, O'Mahony L, Kiely B, Shanahan F, Quigley EM. Efficacy of an encapsulated probiotic *Bifidobacterium infantis* 35624 in women with irritable bowel syndrome. *Am J Gastroenterol* **101**:1581–1590 (2006).
39. Surawicz CM, Elmer GW, Speelman P, McFarland LV, Chinn J, Van Belle G. Prevention of antibiotic-associated diarrhea by *Saccharomyces boulardii*: A prospective study. *Gastroenterology* **96**:981–988 (1989).
40. McFarland LV, Surawicz CM, Greenberg RN, Elmer GW, Moyer KA, Melcher SA, Bowen KE, Cox JL. Prevention of beta-lactam-associated diarrhea by *Saccharomyces boulardii* compared with placebo. *Am J Gastroenterol* **90**:439–448 (1995).
41. Lewis SJ, Potts LF, Barry RE. The lack of therapeutic effect of *Saccharomyces boulardii* in the prevention of antibiotic-related diarrhoea in elderly patients. *J Infect* **36**:171–174 (1998).
42. Castagliuolo I, Riegler MF, Valenick L, LaMont JT, Pothoulakis C. *Saccharomyces boulardii* protease inhibits the effects of *Clostridium difficile* toxins A and B in human colonic mucosa. **67**:302–307 (1999).
43. McFarland LV. Meta-analysis of probiotics for the prevention of antibiotic associated diarrhea and the treatment of *Clostridium difficile* disease. *Am J Gastroenterol* **101**:812–822 (2006).
44. Siitonin S, Vapaatalo H, Salminen S, Gordin A, Saxelin M, Wikberg R, Kirkkola AL. Effect of *Lactobacillus* GG yogurt in prevention of antibiotic associated diarrhoea. *Ann Med* **22**:57–59 (1990).
45. Thomas MR, Litin SC, Osmon DR, Corr AP, Weaver AL, Lohse CM. Lack of effect of *Lactobacillus* GG on antibiotic-associated diarrhea: a randomized placebo-controlled trail. *Mayo Clin Proc* **76**:883–889 (2001).
46. Arvola T, Laiho K, Torkkeli S, Mykkanen H, Salminen S, Maunula L, Isolauri E. Prophylactic *Lactobacillus* GG reduces antibiotic-associated diarrhea in children with respiratory infections: A randomized study. *Pediatrics* **104**:64–67 (1999).
47. Vanderhoof JA, Whitney DB, Antonson DL, Hanner TL, Lupo JV, Young RJ. *Lactobacillus* GG in the prevention of antibiotic-assocaiated diarrhea in children. *J Pediatr* **135**:564–568 (1999).
48. Cremonini F, Di Caro S, Nista EC, Bartolozzi F, Capelli G, Gasbarrini G. Meta-analysis: The effects of probiotic administration on antibiotic-associated diarrhoea. *Aliment Pharmacol Ther* **16**:1461–1467 (2002).
49. Szajewska H, Ruszczynski M, Radzikowski A. Probiotics in prevention of antibiotic associated diarrhea in children: A meta-analysis of randomized controlled trials. *J Pediatr* **149**:367–372 (2006).
50. Johnston BC, Supina AL, Ospina M, Vohra S. Probiotics for the prevention of pediatric antibiotic-associated diarrhea. *Cochrane Database Syst Rev* **2**:CD004827 (2007).
51. McFarland LV. Evidence-based review of probiotics for antibiotic-associated diarrhea and *Clostridium difficile* infections. *Anaerobe* **15**:274–280 (2009).
52. Guandalini S, Pensabene L, Zikri MA, Dias JA, Casali LG, Hoekstra H, Kolacek S, Massar K, Micetic-Turk D, Papadopoulou A, et al. *Lactobacillus* GG administered in oral

rehydration solution to children with acute diarrhea: A multicenter European trial. *J Pediatr Gastroenterol Nutr* **30**:54–60 (2000).

53. Guarino A, Canani RB, Spagnuolo MI, Albano F, Di Benedetto L. Oral bacterial therapy reduces the duration of symptoms and of viral excretion in children with mild diarrhea. *J Pediatr Gastroenterol Nutr* **25**:516–519 (1997).
54. Szajewska H, Kotowska M, Mrukowicz JZ, Armanska M, Milolajczyk W. Efficacy of *Lactobacillus* GG in prevention of nosocomial diarrhea in infants. *J Pediatr* **138**:361–365 (2001).
55. Szajewska H, Skorka A, Dylag M. Meta-analysis: *Saccharomyces boulardii* for treating acute diarrhoea in children. *Aliment Pharmacol Ther* **25**:257–264 (2007).
56. Saavedra JM, Bauman NA, Oung I, Perman JA, Yolken RH. Feeding of *Bifidobacterium bifidum* and *Streptococcus thermophilus* to infants in hospital for prevention of diarrhoea and shedding of rotavirus. *Lancet* **344**:1046–1049 (1994).
57. Majamaa H, Isolauri E, Saxelin M, Vesikari T. Lactic acid bacteria in the treatment of acute rotavirus gastroenteritis. *J Pediatr Gastroenterol Nutr* **20**:333–338 (1995).
58. Szajewska H, Skorka A, Ruszczynski M, Gieruszczak-Bialek D. Meta-analysis: *Lactobacillus* GG for treating acute diarrhoea in children. *Aliment Pharmacol Ther* **25**:871–881 (2007).
59. Szajewska H, Skorka A. *Saccharomyces boulardii* for treating acute gastroenteritis in children: Updated meta-analysis of randomized controlled trials. *Aliment Pharmacol Ther* **30**:960–961 (2009).
60. Guarino A, Albino F, Ashkenazi S, Gendrel D, Hoekstra JH, Shamir R, Szajewska H. European Society for Pediatric Gastroenterology, Hepatology and Nutrition/European Society for Pediatric Infectious Diseaeses evidence-based guidelines for the management of acute gastroenteritis in children in Europe. *J Pediatr Gastroenterol Nutr* **46**:S81–S122 (2008).
61. Coccorullo P, Strisciuglio C, Martinelli M, Miele E, Greco L, Staiano A. *Lactobacillus reuteri* (DSM 17938) in infants with functional chronic constipation: A double-blind, randomized, placcebo-controlled study. *J Pediatr* **157**:589–602 (2010).
62. Agrawal A, Houghton LA, Morris J, Reilly BD, Guyonnet D, Goupil-Feuillerat N, Schlumberger A, Jakob S, Whorwell PJ. Clinical trial: The effects of a fermented milk product containing *Bifidobacterium lactis* DN-173-010 on abdominal distension and gastrointestinal transit in irritable bowel syndrome with constipation. *Aliment Pharmacol Ther* **29**:104–114 (2008).
63. Picard C, Fioramonti J, Francois A, Robinson T, Neant F, Matuchansky C. Review article: "Bifidobacteria as probiotic agents:" physiological effects and clinical benefits. *Aliment Pharmacol Ther* **22**:495–512 (2005).
64. Chmielewska A, Szajewska H. Systematic review of randomised controlled trials: probiotics for functional constipation. *World J Gastroenterol* **16**:69–75 (2010).
65. Kalliomaki M, Salminen S, Arvilommi H, Kero P, Koskinen P, Isolauri E. Probiotic in primary prevention of atopic disease: a randomised placebo-controlled trial. *Lancet* **357**:1076–1079 (2001).
66. Kalliomaki M, Salminen S, Poussa T, Arvilonmi H, Isolari E. Probiotics and prevention of atopic disease: 4-year follow-up of a randomized placebo-controlled trial. *Lancet* **361**:1869–1871 (2003).
67. Abrahamsson TR, Jakobsson T, Bottcher MF, Fredrikson M, Jenmalm MC, Bjorksten B, Oldaeus G. Probiotics in prevention of IgE-associated eczema: A double-blind, randomized placebo-controlled trial. *J Allergy Clin Immunol* **119**:1174–1180 (2007).

68. Kopp MV, Hennemuth I, Heinzmann A, Urbanek R. Randomized, double-blind, placebo-controlled trial of probiotics for primary preventiomn: No clinical effects of *Lactobacillus* GG supplementation. *Pediatrics* **121**:e850–e856 (2008).
69. Niers L, Martin R, Rijkers G, Sengers F, Timmerman H, van Uden N, Smidt H, Kimpen J, Hoekstra M. The effects of selected probiotic strains on the development of eczema. *Allergy* **64**:1349–1358 (2009).
70. Lee J. Seto D, Biewlory L. Meta-analysis of clinical trials of probiotics for prevention and treatment of pediatric atopic dermatitis. *J Allergy Clin Immunol* **121**:116–121 (2008).
71. Majamaa H, Isolauri E. Probiotics: A novel approach in the management of food allergy. *J Allergy Clin Immunol* **99**:179–185 (1997).
72. Isolauri E, Arvola T, Sutas Y, Moilanaen E, Salminen S. Probiotics in the management of atopic eczema. *Clin Exp Allergy* **30**:1604–1610 (2000).
73. Rosenfeldt V, Benfeldt E, Nielsen SD, Michaelsen KF, Jeppesen DL, Valerius NH, Paerregaard A. Effect of probiotic *Lactobacillus* strains in children with atopic dermatitis. *J Allergy Clin Immunol* **111**:389–395 (2003).
74. Boyle RJ, Bath-Hextall FJ, Leonarsi-Bee J, Murrell DF, Tang MLK. Probiotics for treating eczema. *Cochrane Database Syst Rev* **4**:CD006135 (2008).
75. Betsi GI, Papadavid E, Falagas ME. Probiotics for the prevention and treatment of atopic dermatitis: A review of the evidence from randomized controlled trials. *Am J Clin Dermatol* **9**:93–103 (2008).
76. Gill HS, Rutherfurd KJ, Cross ML, Gopal PK. Enhancement of immunity in the elderly by dietary supplementation with the probiotic *Bifidobacterium lactis* HN019. *Am J Clin Nutr* **74**:833–839 (2001).
77. Pineiro M, Asp NG, Reid G, Macfarlane S, Morelli L, Brunser O, Tuohy K. FAO Technical Meeting on Prebiotics. *J Clin Gastroenterol* **42**:S156–S159 (2008).
78. Macfarlane S, Macfarlane GT, Cummings JH. Prebiotics in the gastrointestinal tract. *Aliment Pharmacol Ther* **24**:701–714 (2006).
79. Macfarlane GT, Steed H, Macfarlane S. Bacterial metabolism and health-related effects of galacto-oligosaccharides and other prebiotics. *J Appl Microbiol* **104**:305–344 (2008).
80. Cummings JH, Pomare EW, Branch WJ, Naylor CPE, Macfarlane GT. Short chain fatty acids in human large intestine, portal, hepatic and venous blood. *Gut* **28**:1221–1227 (1987).
81. Macfarlane GT, Macfarlane S. Fermentation in the human large intestine, its physiological consequences and the potential contribution of prebiotics. *J Clin Gastroenterol* **45**:120–127 (2011).
82. Macfarlane S, Steed H, Macfarlane GT. Intestinal bacteria in inflammatory bowel disease. *Crit Rev* in *Clin Lab Sci* **46**:25–54 (2009).
83. Steed H, Macfarlane GT, Macfarlane S. Prebiotics, synbiotics and inflammatory bowel disease. *Mol Nutr Food Res* **52**:898–205 (2008).
84. Furrie E, Macfarlane S, Kennedy A, Cummings JH, Walsh SV, O'Neil DA, Macfarlane GT. Synbiotic therapy (*Bifidobacterium* longum / Synergy 1) initiates resolution of inflammation in patients with active ulcerative colitis: a randomised controlled pilot trial. *Gut* **54**:242–249 (2005).
85. Macfarlane S, Furrie E, Cummings JH, Macfarlane GT. Chemotaxonomic analysis of bacterial populations colonizing the rectal mucosa in patients with ulcerative colitis. *Clin Infect Dis* **38**:1690–1699 (2004).
86. Steed H, Macfarlane GT, Blackett KL, Bahrami B, Reynolds N, Walsh SV, Cummings J, Macfarlane S. Clinical trial: The microbiological and immunological effects of synbiotic consumption—a randomized double-blind placebo-controlled study in active Crohn's disease. *Aliment Pharmacol Ther* **32**:872–883 (2010).

87. Ishikawa H, Matsumoto S, Ohashi Y, Imaoka A, Setoyama H, Umesaki Y, Tanaka R, Otani T. Beneficial effects of probiotic bifidobacterium and galacto-oligosaccharide in patients with ulcerative colitis: A randomized controlled study. *Digestion* **84**:128–133 (2011).
88. Chermesh I, Tamir A, Reshef R, Chowers Y, Suissa A, Katz D, Gelber M, Halpern Z, Bengmark S, Eliakim R. Failure of Synbiotic 2000 to prevent postoperative recurrence of Crohn's disease. *Dig Dis Sci* **52**:385–389 (2007).
89. Vandenplas Y, De Hert SG; PROBIOTICAL-study group. Randomised clinical trial: The synbiotic food supplement Probiotical vs. placebo for acute gastroenteritis in children. *Aliment Pharmacol Ther* **34**:862–867 (2011).
90. Puccio G, Caiozzo C, Meli F, Rochat F, Grathwohl D, Steenhout P. Clinical evaluation of a new starter formula for infants containing live *Bifidobacterium longum* BL99 and prebiotics. *Nutrition* **23**:1–8 (2007).
91. Fateh R, Iravani S, Frootan M, Rasouli MR, Saadat S. Synbiotic preparation in men suffering from functional constipation: A randomized controlled trial. *Swiss Med Wkly* **30**:141 (2011).
92. Passeron T, Lacour JP, Fontas E, Ortonne JP. Prebiotics and synbiotics: Two promising approaches for the treatment of atopic dermatitis in children above 2 years. *Allergy* **61**:431–437 (2006).
93. Kukkonen K, Savilahti E, Haahtela T, Juntunen-Backman K, Korpela R, Poussa T, Turre T, Kuitunen M. Probiotics and pre-biotic galacto-oligosaccharides in the prevention of allergic diseases: A randomized, placebo-controlled trial. *J Allergy Clin Immunol* **119**: 192–198 (2007).
94. Kukkonen K, Savilahti E, Haahtela T, Juntunen-Backman K, Korpela R, Poussa T, Turre T, Kuitunen M. Long-term safety impact on infection rates of postnatal probiotic and prebiotic(synbiotic) treatment: Randomized, double-blind, placebo-controlled trial. *Pediatrics* **122**:8–12 (2008).
95. Van der Aa LB, Heymans HS, van Aalderen WM, Sillevis Smitt JH, Knol J, Ben Amor K, Goossens DA, Sprikkelman AB; Synbad Study Group. Effect of a new synbiotic mixture on atopic dermatitis in infants: a randomized-controlled trial. *Clin Exp Allergy* **5**:795–804 (2010).
96. Wu KG, Li TH, Peng HJ. *Lactobacillus salivarius* plus fructo-oligosaccharide is superior to fructo-oligosaccharide for treating children with moderate to severe atopic dermatitis: A double-blind randomized clinical trial of efficacy and safety. *Br J Dermatol* **1**:129–136 (2012).
97. Bartosch S, Woodmansey EJ, Paterson JC, McMurdo MEM, Macfarlane GT. Microbiological effects of consuming a synbiotic containing *Bifidobacterium bifidum*, *Bifidobacterium lactis* and oligofructose in elderly patients, determined by real-time polymerase chain reaction and counting of viable bacteria. *Clin Infect Dis* **40**:28–37 (2005).
98. Ouwehand AC, Tiihonen K, Saarinen M, Putalaa H, Rautonen N. Influence of a combination of *Lactobacillus acidophilus* NCFM and lactitol on healthy elderly: Intestinal and immune parameters. *Br J Nutr* **101**:367–375 (2009).
99. Metchnikoff E. *The Prolongation of Life*, Heinemann, London, 1907.

INDEX

Note: Page numbers in *italics* indicate figures; tables are noted with *t*.

AAD. *See* Antibiotic-associated diarrhea
ABI SOLiD, 56
Absolute hybridization silence, 250
Abundance and activities of microbiota, bacterial community diversity and, 81
Accommodation mutants, *V. fischeri*, 280
Acetate, physiologic effects of, produced by bacteria colonizing in large gut, 325, *325*
Acetobacteraceae, in fruit fly, 256
Acinetobacter sp., ventilator-associated pneumonia and, 125
Acquired pellicle, 138, 139
Actinobacillin, 141
Actinobacteria
  Crohn's disease and, 108
  in human gut, 256
  vaginal microbiome and, 178
Actinomyces oris, in oral cavity, 141
Actinomyces sp.
Actinomyces sp., in oral cavity, 136, 140, 310
Actinomycetes, traps and, 308
Active immunity, 258
Adaptive immune system, 3
Adaptive immunity, 112, 258
Adherent invasive *E. coli*, Crohn's disease and, 107
Adhesin-receptor mechanisms, in oral cavity, 140
Aerococcus, in BV-negative women, 183
Aeromonas, modeling in zebrafish, 259
Aeromonas veronii, leech and, 277
Affymetrix, Inc., 77
Agglutinins, oral cavity and, 139
Aggregatibacter actinomycetemcomitans, 141, 142, 152
Aging population
  probiotics and, 323–324
  synbiotics and, 331
AIEC. *See* Adherent invasive *E. coli*
Airway diseases, chronic, polymicrobial communities and, 80
Airway microbiome, human, 119–129
ALF969, 59
Alkaline phosphatase, homeostatic balance in fly gut and, 262
Allergies, 45, 265
  mouse animal models and learning about, 259
  probiotics and, 322–323
Alpha diversity, analyses of, 86
Alpha-amylase, oral cavity and, 139
American Heart Association, on antibiotic prophylaxis prior to dental procedures, 149
American Type Culture Collection, 8
Amnionitis, 171
Amplicons, detecting in qPCR, 59
Amplified ribosomal DNA restriction analysis, 177
AMPs. *See* Antimicrobial peptides
Amsel's criteria, 194, 195
  bacterial vaginosis and, *174*, 175, 176
Anaerobes
  in CF airway samples, 122
  ventilator-associated pneumonia and, 125
Anaerobic bacteria, gum disease sites and, 136
Anaerobic cultivation, strict, issues related to, 310
Anaerococcus, in BV-negative women, 183
Angina pectoris, 151

The Human Microbiota: How Microbial Communities Affect Health and Disease, First Edition. Edited by David N. Fredricks.

Animal models, 274–279
of binary host-microbe associations, *274*
of commensal-host interactions, microbial and immune system diversity in, *257*
fruit fly, 278
gypsy moth, 277–278
of host associations with high-complexity microbial communities, *276*
of host associations with low-complexity communities, *275*
Hydra, 277
leech, 277
microbial diversity in, 255–258
mouse, 279
nematode, 276–277
squid, 275–276
termite, 278–279
zebrafish, 279
Animals, immune system development in: comparative immunity, 258–259
Antagonistic interbacterial interactions, in oral cavity, 141–142
Antibiotic prophylaxis, prior to dental procedures, AHA on, 149
Antibiotic-associated diarrhea, probiotics and, 321
Antibiotics, 269
amelioration of intestinal inflammation and, 107
mucus barrier and, 244
Antimicrobial peptides, 260, 261, 282
AP. *See* Alkaline phosphatase
Appendix section, transverse, DAPI stain of, *225*
Archaea, 256
ARDRA. *See* Amplified ribosomal DNA restriction analysis
ARISA. *See* Automated rRNA intergenic spacer analysis
Arthritis, mouse animal models and learning about, 259
Asthma, 45, 121, 126–127, 259, 265
ATCC. *See* American Type Culture Collection
ATG16L1
Crohn's disease and, 109
inflammatory bowel disease and, 265
phylum-level alterations in enteric microbiota and, 112
ATG16L1 genotype, Crohn's disease and, 111
ATG16L1 loci, Crohn's disease and, 108, 109
ATG16L1 risk alleles, younger age of surgery and, 110
Atg16L1 protein, 109
*ATG16L1*T300A allele, 110
Atheromas
formation of, 151
oral bacteria associated with, 152
Atherosclerosis
oral communities and, *147*, 151–153
pathogen-accelerated, putative mechanisms for, 152–153
predisposing factors for, 151–152
Atopic dermatitis
in children, studies of, 22
synbiotics and, 330, 331
Atopobium
in BV-negative women, 183
infected root canals and, 145
Atopobium parvulum, 145
Atopobium vaginae, 184, 189, 190
AI2, dental plaque formation and, 142
Autism, 45
Autofluorescence, 60
Autoimmune diseases, 265
commensal bacteria and, 267–268
Tregs and, 264
Autoimmunity, 265, *268*
Automated rRNA intergenic spacer analysis, 55

B lymphocytes, in zebrafish, 259
Bacillus thuringiensis, gypsy moth and, 278
Bacteremia, oral communities and, 146, 148
Bacteria
animal intimate, unknown relationships with, 255
in biliary tract, 218–219
colonic, 252
finding right mix of, *268*, 268–269
human oral cavity and communities of, *137*
in human oral cavity and nasopharynx, 135, 136
intestinal homeostasis and, 260–264, *261*
in large intestine, 220, 222
mobility of modern society and worldwide exchange of, 239
in mouth, 212–213
in pancreatic tract, 218
in small intestine, 219–220

in stomach and duodenum, 217–218
in tonsils, 214
in upper gastrointestinal tract, 212–219
Bacterial community composition, profiling, schematic comparison of phylogenetic microarray and sequencing approaches to, *80*
Bacterial community diversity, treatment of *Clostridium difficile* infected mice and, 81
Bacterial cultivars, comparing, developments in, 53
Bacterial genomes, fully sequenced, 290
"Bacterial lobster trap," 63
Bacterial phyla, percentage of, with cultivated representatives, *290*
Bacterial vaginosis, 22, 169, 188, 198
asymptomatic, 176
current gold standard diagnostics for, 173–176, *174*
etiology of, 173
knowledge of natural history dynamics and, 194–195
Lactin-V and, 193–194
relapse, extravaginal reservoirs of BVABs and, 195–196
scientific controversy over, 172–176
treatment for, failure of, 173
Bacteroides, 252
abscesses caused by, 222
in fecal flora, 231, *232*
in mouse with IL10 deficiency, *246*
self-limiting colitis in ascending colon and, *238*
between villi in ileum of patient with Crohn's disease, *243*
viscosity of mucus barrier and, 224, 225
Bacteroides fragilis, 284
biofilm mass in IBD and, 237
Crohn's disease and, 114
intestinal homeostasis and, 263–264
Bacteroides groups, colonic biofermenter and, 222
Bacteroides sp., Crohn's disease and, 108
Bacteroides thetaiotaomiron, 280–281
Bacteroidetes
Crohn's disease and, 108, 114
within mammalian intestine, 256
vaginal microbiome and, 178
Bacteroidetes spp., 154, 173, 281
BAL. *See* Bronchoalveolar lavage
Barrett's esophagus, 21
Bary, Heinrich Anton de, 274
Baylor College of Medicine, 7
Belgian Co-ordinated Collection of Microorganisms (BCCM), 8
Bellerophon, 77
Bergey's taxonomic outline, 77
Beta diversity, phylogenetic measures of, 86
Bias, shotgun metagenomics and, 61
Bifidobacteria, 266, 317, 325
Bifidobacteriacae, 224, 231, *232*
Bifidobacteriales sp., vaginal microbiome studies and, 178
Bifidobacterium animalis DN173 010, 322
Bifidobacterium animalis subsp. *lactis* Bb12, 320
Bifidobacterium bifidum BB-02, 331
Bifidobacterium breve M-16V, 331
Bifidobacterium breve strain Yakult, 328
Bifidobacterium infantis, 320
Bifidobacterium lactis BL-01, 331
Bifidobacterium lactis HNO19, 323
Bifidobacterium longum
antiinflammatory effects of, 327, *328*
specimen bowel habit diary from UC patient receiving Synergy I and, 330*t*
Bifidobacterium longum BL99, 330
Bile acids, intestinal mucosa and, 241
Biliary stents
polymicrobial biofilm and colonization of, 218–219
sections of, hybridized with universal Eub338 Cy3 probe, *220*
Biliary tract
bacteria in, 218–219
cytophaga *Flavobacterium* group, *219*
Biodefense and Emerging Infections Research Resource Repository, 8
Biodiversity assessments, 295
Biofilm matrix, in oral cavity, 143
Bioinformatic tools, 56, 57
Birth cohort studies, 47
Bobtail squid, 274
Bottom-up approaches, 295
BPD. *See* Bronchopulmonary dysplasia
Bray–Curtis measurement, 57, 78
Broad Institute, 7
Bronchoalveolar lavage, 127
Bronchopulmonary dysplasia, 126
Burkholderia cepacia complex, cystic fibrosis and, 122
Burkholderia PhyloChip, 58

Burkholderiaceae, CF genotype, age and, 123*t*, 124
Butyrate
  Crohn's disease and, 114
  physiologic effects of, produced by bacteria colonizing the large gut, 325, *326*
BV. *See* Bacterial vaginosis
BVAB. *See* BV-associated bacteria
BVAB1, 185, 186
BVAB2, 185, 186, 189
BVAB3, 185, 186
BVAB species, 189
BV-associated bacteria, 173, 195–196

Campylobacter jejuni, hybridization silence of *Faecalibacterium prausnitzii* in patient with gastroenteritis caused by, *249*
Candida albicans, 142, 189
Candidatus arthromitus, 79
Capnocytophaga canimorsus, 63–64
CARD-FISH, 59
Cardiovascular disease, atherosclerosis and, 151
Catellatospora, trap method and cultivation of, 308
Caudal gene, 261
CCR6, 284
CCUG. *See* Culture Collection from University of Goteborg
CD. *See* Crohn's disease
cd-hit, 88
Cecal biopsy, from patient who was inadequately purged, *224*
Celiac disease, 45, 237, 241–242
Central fermenting zone, in colonic biofermenter, 245
Central nervous system, 266
Cervix, microbial population of, 169–170
Cesarean birth, 45
Chemostat, 293
Childbirth, 170
Chimeras, 86
Chimeric (artifactual hybrid) sequences, removal of, 57
Chlamydia trachomatis, 171
Chorioamnionitis, 171
Chronic obstructive pulmonary disease, 121, 127–128
Circumcised men, microbiota of uncircumcised men vs. microbiota of, 196–197
Circumcision, penile microbiome and, 22
Clindamycin, for bacterial vaginosis, 173
Clinical care of patients, understanding the microbiome and, 120–121
Clone libraries, 56, 58, 177
Cloning biases, 61
Clostridia, in human intestine, 256
Clostridiales
  bacterial vaginosis and, 185
  novel BVABs and, 185–186
Clostridium cluster XIVa
  adaptive immunity, dysbosis and, 112
  in human intestine, 256
Clostridium difficile, 275, 321
Clostridium histolyticum, *232*, 252
Clostridium perfringens, gas gangrene caused by, 222
Clustered regularly interspaced short palindromic repeats, 258
Cnidarian Hydra, 277
CNS. *See* Central nervous system
Colitis
  crypt abscesses and, 229
  IL10-deficient mouse with, *247*
  self-limiting, in ascending colon, *238*
Colon
  ascending, biopsy of, for healthy person, *223*
  bacteria in, 220
  descending, *223*
  as highly efficient biofermenter, 245
Colonic bacteria, 252
Colonic biofermenter
  control of pathogens within, 252
  diarrhea and, 247, 248
  germinal zone of, 250
  working (luminal) area of, 248, 250
  zones of, 245
Colonic function, role of microbiota in, 222
Colonic microbial biostructure
  site-dependent changes of, 247–251
    germinal zone of colonic biofermenter, 250
    mucus layer, 247–248
    working (luminal) area of colonic biofermenter, 248, 250
Colonic microbiota
  disease-related changes in
    break of mucus barrier in IBD, 234, 236–238
  human, major interactions between host physiology and, *316*
Colonization, visualizing, 280–281

Colony-forming unit (CFU) number, 303
Columnar epithelial cells, intestinal microbiota and, 229, 230, 234, 244
comA, 142
comB, 142
comD, 142
comE, 142
Commensal bacteria
  finding right mix of, *268*
  host health and role of, 267–268
Commensal-host relationships, 255
  diversity of commensal organisms in humans vs. in animal models, 256, *257*
  harnessing our understanding of, for therapeutic benefit, 268–269
  post-birth, 255–256
Community fingerprinting, 54–55
  ARISA, 55
  DGGE/TGGE, 54–55
  T-RFLP, 55
Competence-stimulating peptide, dental plaque formation and, 142
comX, 143
Conception, vaginal microbes and, 169
Conditional analysis of sample coverage
  embedding algorithm, 99–100
  thinning property of Poisson point processes, 97–98
Conn, Harold, 292
COPD. *See* Chronic obstructive pulmonary disease
Copenhagen Prospective Study of Asthma, 127
Coral, investigating, functional gene arrays and, 61
Coronal sulcus of penis, bacterial communities present on, 197
Corynebacterium, in BV-negative women, 183
Coverage of a sample, 90–91
cpn60 target, vaginal microbiome and, 181
CRISPRS. *See* Clustered regularly interspaced short palindromic repeats
Crohn's disease, 21, 45, 250, 252, 265–266
  abnormal microbiota, younger age of surgery and, 110
  ascending colon of patient with, with adhesion of *Bacteroides* to epithelial surface and entrance of crypts, *235*
  bacterial inclusions in vacuoles of goblet cells in patient with, *236*
  Bifidobacterium longum and, 328
  description of, 105
  effect of susceptibility alleles on bacterial biodiversity in surgical specimens, *111*
  ileal, 107
  impact of host factors and enteric microbes on development/progression of, *113*
  loss of mutualistic interactions between host and microbe in, 114
  metagenomic tools for studying host-microbiota interactions in, *106*
  modeling dysbiosis and, 111–112, 114
  pathogenic microbial communities and, 108
  probiotics and, 318
  prolific adherent biofilm located between villi in ileum of patient with, *243*
  prolific bacterial biofilm covering colonic mucosa in patient with, *234*
  promoting, host genetics and enteric microbiota and, 105–115
  reported cases of, steady increase in, 238
  sigmoid colon of patient with, *240*
  single-nucleotide polymorphisms and increased risk of, 110*t*
  "skip" lesions and, 106
  as supraautoimmune disease, 114
  tobacco smoke detrimental in patients with, 242
Crop, in leech, 277
Cross-hybridization, array-based technology and, 82
Crypt abscesses, 229
Crypt cells, 244
Crypts of Lieberkuhn, bacterial concentrations found within, 241, *242*
Crystal cells, immune system and, 258
CSP. *See* Competence-stimulating peptide
Cultivation, 293–294, 295
  as foundation of microbiology, 63
  of human microbiota, 308–309
Cultivation techniques, development of, 291
Cultivation-independent techniques, 52–63
  community structure and composition using phylogenetic markers, 53–60
    diversity of genetic potential, 60–62
    transcriptomics, proteomics, and metabolomics, 62–63
Culture Collection from University of Goteborg, 8
Curved-rod organisms, bacterial vaginosis and, 187

CVD. *See* Cardiovascular disease
Cyanobacteria, in human gut, 256
Cyclic AMP, 63
Cystic fibrosis, 119, 121
  cause of, 121–122
  microbiome in, examining, 122–123
  next-generation sequencing approaches applied to CF airway specimens, 124–125
  patient airway microbiota
    PhyloChip and study of, 80
    phylogenetic tree displaying relationship between age and taxon relative abundance, *81*

DAC. *See* Data Access Committee
DACC. *See* Data Analysis and Coordination Center
Dactilosporangium, trap method and cultivation of, 308
Danio rerio (zebrafish), microbiome within, 257–258
Data Access Committee, 5, 10, 11
Data Analysis and Coordination Center, 5, 6, 7, 28
Data Analysis Working Group (DAWG), 7, 14, 28
DCs. *See* Dendritic cells
Demultiplexing data, 86
Denaturing gradient gel electrophoresis, 54, 55, 120, 177
Dendritic cells, 267
Denisovans, comparative genomics studies of, 48
Dental caries, 144, 145
Dental plaque, *146*
  quorum-sensing mechanisms and, 142
  streptococci and, 136
Desquamated epithelial cells, in saliva, *214*
Detergents, intestinal mucosa and, 240
Dextran sodium sulfate, 240
DGGE. *See* Denaturing gradient gel electrophoresis
Diabetes
  commensal bacteria and, 267–268
  gestational, periodontal disease and, 154
  incidence of, 267
  mouse animal models and learning about, 259
  oral communities and, 153–154
  type 1, 45, 153
  type 2, 153, 154
Diabetes mellitus, defined, 153
Dialister
  bacterial vaginosis and, 184, 185
  infected root canals and, 145
Dialister propionicifaciens, bacterial vaginosis and, 185
Diarrhea, 247
  acute, bacterial concentrations found in mucus of patients with, 237
  antibiotic-associated, probiotics and, 321
  in children, synbiotics and, 330
  idiopathic, hybridization silence of *Faecalibacterium prausnitzii* in patient with, *249*
Diet
  host, microbiota assembly within animal digestive tracts and, 281
  human evolutionary context and, 48
  microbiome dynamics throughout life and, 46
Diffusion chamber
  microbial species cultivation in, 304–306
  for *in situ* cultivation of microorganisms, *305*
Digestive tract, 3
Disease categories
  general, for non-HMP microbiome and disease association projects, 43*t*
  specific, for non-HMP microbiome and disease association projects, 44*t*
Disturbed microbiome
  subsequent disease and, 45
Diversity
  calculating, 57
Diverticulosis
  bacterial concentrations found in mucus of patients with, 237
DNA sequencing capabilities
  recent advancements in, 120
DNA sequencing of phylogenetic markers, 55–57
  analytic methods for 16S rRNA community sequence data, 57
  clone libraries and Sanger sequencing, 56
  next-generation sequencing, 56
DNA-DNA hybridization assays, 53
Döderlein, Albert, 172, 191
Döderlein's bacilli, 172
Drosophilia (fruit fly), 282
  microbial diversity in, analysis of, 256–257, *257*

microbial influences on epithelial tissues and, 283
as model for innate immunity, 258
dsDNA-binding dyes, 59
DSMZ. *See* German Collection of Microorganisms and Cell Cultures
DSS. *See* Dextran sodium sulfate
DSS1
in human gut, 256
Dual oxidase (DUOX)
homeostatic balance in fly gut and, 262
Duodenum
bacteria in, 217–218
Dysbiosis. *See also* Bacterial vaginosis
CD pathogenesis and, 108, *113*
enteric, 105
modeling Crohn's disease and, 111–112, 114

EAE. *See* Experimental autoimmune encephalomyelitis
ECM. *See* Extracellular matrix
Ectocervix, bacterial load of, 169
Ectopic pregnancy, 171
Eczema, 22–23
Eggerthella-like species, bacterial vaginosis and, 186
ELSI. *See* Ethical, legal, and societal implications
Emulsifiers, intestinal mucosa and, 241
Endocervical canal, 168
Endocervix, bacterial load of, 169
Endodontic infection, course of, *146*
Endometrium (uterus), 168, 170
Endoscopic retrograde cholangiopancreatography, 218
Enrichment techniques, development of, 291
Enteric microbes, development/progression of Crohn's disease and, *113*
Enteric microbial communities
Crohn's disease and, 108
healthy GI tract and, 112
Enteric microbiota
Crohn's disease and, 105
host genetics and shaping of, 110–111
IBD and, 107–108
pathogenic microbial communities, 108
specific microbial pathogens, 107–108
Enterobacter cloacae, 229
distal colon of mouse monoassociated with, *228*
proximal colon of mouse monoassociated with, *229*
Enterobacteriaceae, 231, *232*, *233*, 239, 244, 252
in fruit fly, 256
ventilator-associated pneumonia and, 125
Enterococci, 252
in BV-negative women, 183
endocarditis caused by, 222
Enterococcus faecalis
biofilm matrix and, 143
Crohn's disease and, 108
infective endocarditis and, 150
Enterococcus faecium PR88, 320
Environmental microbial ecologists, 54
Epithelial tissues, microbial influences on, 283
EPSS. *See* Extracellular polymeric substances
ERCP. *See* Endoscopic retrograde cholangiopancreatography
Errors
pyrosequencing and, 86
sequencing and, 56
Escherichia coli, 85, 277
Crohn's disease and, 107
PCR products and, 56
polyposis of the stomach and, 218
sepsis caused by, 222
Escherichia coli Nissle, protection from IBD and, 266
Escherichia coli Nissle 1917, 319, 322
Esophageal adenocarcinoma, 21
Estimation, prediction vs., 90
Ethical, legal, and societal implications
HMP, projects listed by project investigator and title or description, 27*t*
of microbiome research, 26–27
Eub338, 59
Eukarya, 291
Euprymna scolopes (Hawaiian bobtail squid), 275
European Commission, calls for studies on human metagenomics, 3
European Society for Paediatric Gastroenterology, Hepatology and Nutrition, 322
European Society of Paediatric Infectious Diseases, 322
Expected coverage of a sample, 90–91

Experimental autoimmune encephalomyelitis, 266, 267
Expression arrays, 76
Extracellular matrix, 149
Extracellular polymeric substances, 143
Extraction biases, 61
Extrapolation of environments
  practice for, 85–88
  theory for, 88–92
    expected coverage of a sample, 90–91
    paradigm shift, 89–90
    urn model, 88–89

FACS. *See* Fluorescence-activated cell sorting
Facultative pathogens
  increased exposure to, 239
  role of, 238–239
Faecalibacterium prausnitzii
  colonic biofermenter and, 222
  Crohn's disease and, 114, 237, 250
  in fecal flora, 231, *232*
  hybridization silence of, in patient with idiopathic diarrhea, *249*
  in large intestine, 220
  protection from IBD and, 266
  self-limiting colitis in ascending colon and, *238*
Fallopian tubes (oviduct), 168, 170
"Fatty streak," atherosclerotic lesion and, 151
Feacalibacteria, in human intestine, 256
Fecal cylinder
  from healthy person, alcian stain of, *231*
  of patient with ulcerative colitis, *251*
  seen with DAPI stain, *251*
Fecal microbiota, 230–234
  biostructure of, 230–231, 234
    in health, IBD, and other GI diseases, 245, 247
Feces, human, methanogen diversity in, 60
Fecomucous, 231
Female reproductive tract
  anatomical features of, with relevance to genitourinary microbes, *168*
  controversy over bacterial vaginosis, 172–176
  microbial ecosystems of and impact on reproduction, 167–172
  vaginal health in 2025, 188–196
  vaginal microbiome in age of high-throughput sequencing, 176–188
Fibrobacteres, in human gut, 256
Filofactor, infected root canals and, 145
Fingerprinting, 120
  airway microbiome studies and, 128
  microbiome of CF patients examined with, 122
Firmicutes
  Crohn's disease and, 108, 114
  within mammalian intestine, 256
  obesity and, 154, 155
  vaginal microbiome and, 178
FISH. *See* Fluorescence *in situ* hybridization
FISH analysis
  bacteria in stomach and duodenum and, 217
  of pancreatic tract microbiota, 218
Fish models, QTL mapping studies and, 282
FISH probes, 211, 212
FISH-SIMS method, 60
Flagella, 225
Flies, control of AMP levels by, 261
Flow cytometer, 59
Flow cytometry, 64, 292
Fluorescence *in situ* hybridization, 59–60
  Cesarean section, fetal membrane examination and, 171
  intestinal microbiota and use of, 211, 212
Fluorescence microscopy, 59
Fluorescence-activated cell sorting, 62
Fluorescently labeled DNA, production of, 61
Foreskin
  coronal sulcus of penis and, 197
  microbiota of penis and, 196–197
Formula feeding, 45
Fourier transform infrared spectroscopy, 295
Foxp3, functional homolog of, in zebrafish, 259
Fredricks, D., 310
French Longitudinal Study of Children, 47
Fructooligosaccharide (FOS) derivatives, 324, 325, 327
Fructosyltransferases, 143
Fruit fly *(Drosophilia melanogaster)*, 255, 274
  intestinal homeostasis maintained in, *261*
  low-complexity microbial communities and, *275*
  microbial and immune system diversity in models of commensal-host interactions, *257*

microbial diversity in, analysis of, 256–257, *257*
reactive oxygen species and, 262
symbiosis studies and, 278
FTFs. *See* Fructosyltransferases
Functional foods, 316–317
Functional gene arrays, 61
Functional gene PCR, cloning and sequencing, 60
Functional metagenomics, 61–62
Fungi, 291, 308
Funisitis, 171
Fusobacteria, 244
in human gut, 256
infected root canals and, 145
vaginal microbiome and, 178
Fusobacteriales species, bacterial vaginosis and, 186
Fusobacterium necrophorum, 237
Fusobacterium nucleatum
as "bridging" organism, 140
gum disease sites and, 136
periodontal disease, adverse pregnancy outcomes and, 155, 156

Galacto-oligosaccharides, 324, 325, 327
Gallstones, bacteria in, 219
Gambian Gut Microbiome Project, 4
Gardnerella vaginalis, 173, 175, 178, 189, 190, 196, 239, 244
in BV-positive and BV-negative women, 183, 184
male and female genital microbiota and, 198
male reproductive tract and, 197, 198
vaginal microbiome and, 182
Gas chromatography-mass spectrometry, 63
Gastric esophageal reflux disease, characteristics of, 21
Gastroenteritis
acute, probiotics and, 321–322
hybridization silence of *Faecalibacterium prausnitzii* in patient with, *249*
Gastrointestinal cancers, preventing, 285
Gastrointestinal (GI) tract. *See also* Upper gastrointestinal tract
diversity and density of bacteria in, 254
healthy, immune system and, 112
GBPs. *See* Glucan-binding proteins
GC-MS. *See* Gas chromatography-mass spectrometry
Gel microdroplets, 63, 64
Gemella
in BV-negative women, 183
oral cavity and, 136
GenBank, 9, 53
Genetic determinants of microbiota composition, host, identifying, 281
Genetic potential, diversity of, 60–62
functional gene arrays, 61
functional gene PCR, cloning, and sequencing, 60
functional metagenomics, 61–62
shotgun metagenomics, 60–61
single-cell genome amplification and sequencing, 62
Genitourinary tract. *See also* Female reproductive tract; Male reproductive tract
microbita of, 167–199
Genome Reference Consortium, 13
Genome sequences, 294
as foundation for microbial physiology, 291
number of, available per year, *290*
Genomic novelty, SSU gene sequences as predictors of, 54
GeoChip, 61
GERD. *See* Gastric esophageal reflux disease
Germ free animal models
microbial influences on epithelial tissues, 283
microbial influences on immune system, 283–284
German Collection of Microorganisms and Cell Cultures, 8
Germinal stocks, of colonic biofermenter, 245
Germinal zone
colonic bacteria in wildtype mouse and, *246*
of colonic biofermenter, 250
GI lymphoid tissues, 112
Gingivae, 126
Gingival tissues, diseased, population shift in, 145
Gingivitis, 148
Glucan-binding proteins, 143
Glucosyltransferases, 143
Glutens, intestinal mucosa and, 241–242
Glycoproteins, in salivary secretions, 138
GMDs. *See* Gel microdroplets
G908R, Crohn's disease and, 109

Goblet cells, viscosity of mucus secreted from, 230
Good-Turing estimator, 94
GOS. *See* Galacto-oligosaccharides
Gram-negative bacteria, periodontal disease and, 145
Gram-negative rods, in BV-positive women, 185
Gram-positive bacteria, periodontal disease and, 145
"Great plate count anomaly," 63, 76, 292, 303, 311
Greengenes, 53, 181
Growth of microorganisms, 293
Growth techniques, development of, 291
GTFs. *See* Glucosyltransferases
Gum disease sites, anaerobic bacteria and, 136
Gums, communities of bacteria in, *137*
Gut diseases, microbiome and, 21–22
Gut microbiome, as "cardinal microbiome," 46–47
Gut microbiota, obesity and, 154
Gypsy moth
  bacterial communities in midguts of, 277–278
  low-complexity microbial communities and, *275*

Haemophilus influenzae
  asthma and, 127
  COPD and, 127
  diffuse local infiltration of tonsil with, *215*
Hairpin binding, 59
Halitosis, 144
Hard palate, communities of bacteria on, *137*
Hawaiian bobtail squid *(Euprymna scolopes)*, 275
Hay fever, 45
Healthy Adult Cohort Study of Multiple Microbiomes, 10–17
  clinical phase, 10–11
  data processing and analysis phase, 13–17
  derivative datasets from, 28
  finalized datasets used by HMP DAWG for analysis of healthy cohort data, 16*t*
  percent human sequence reads in total sequences of whole-genome shotgun reads from, *15*
  range in DNA yield of samples collected from five major body sites in, 13*t*
  results from global analysis in, 14–17
  schematic of body sites samples for, *12*
  sequencing phase, 11–13
  subsites sampled within each body site, 11, *12*
Heatshock protein, *P. gingivalis* and, 152, 153
Helicobacter hepaticus, 266
Helicobacter pylori
  biofilms, data on, 217–218
  first description of, 239
  infection, 217
Heterorhabditiae, 276
High-density microarrays, 76
High-density phylogenetic array design, 77
High-performance chromatography-mass spectrometry, 63
High-throughput microbial cultivation, ichip for, *306*
High-throughput screening, vaginal microbiome in age of, 176–188
High-throughput sequencing, profiling of entire vaginal microbial communities and, 189
Hirudo verbana (leech), 277
HITChip, 58, 79
HIV
  bacterial vaginosis and, 176
  vaginal microbiota studies and, 180
HMP. *See* Human Microbiome Project
Homeostasis, intestinal, maintaining, 260–264, *261*
Hominid migrations, early, 48
Homo sapiens, comparative genomics studies of, 48
Homogeneous Poisson (point) process, 97–98, *98*
Homoserine lactones, 63
Horseradish peroxidase tag, 59
Host biology, function of resident microbial communities and, 282–284
Host development, bacteria and, 255
Host genetics
  Crohn's disease and, 105
  shaping of enteric microbiota and, 110–111
Host health
  role of commensal bacteria in, 264–268
    diabetes, 267–268
    inflammatory bowel disease, 265–266
    multiple sclerosis, 266–267
Host microbiota, disease and differences in, 121

Host-associated microbial communities
establishment of, 279–282
microbe-microbe interactions and, 281
HPLC-MS. *See* High-performance chromatography-mass spectrometry
HPP. *See* Homogeneous Poisson (point) process
HRP tag. *see* Horseradish peroxidase tag
HSP. *See* Heatshock protein
Human Genome Project, ELSI studies and, 26
Human genome sequence, publication of first drafts, 3
Human inflammatory bowel diseases, 106–107, 111. *See also* Inflammatory bowel disease
Human leukocyte antigen (HLA) class I genes, 48
Human microbiome, defined, 1
Human Microbiome Initiative (Canada), 4
Human Microbiome Project, 273
computational tools, 24–25
projects listed by project title and investigator, 25*t*
computational tools developed or modified for, 29*t*–*30*t
conceptual diagram of, *6*
data coordination and analysis
Data Analysis and Coordination Center, 7
HMP workshops, 7–8
defined, 1–2
demonstration projects of microbiome-disease associations, 17, 21–23
early results from, 17, 21
gut diseases and microbiome, 21–22
skin diseases and microbiome, 22–23
summary of, 18*t*–20*t*
urogenital diseases and microbiome, 22
distribution of reference sequence bacterial strains, by major body site, *9*
ELSI projects listed by project investigator and title or description, 27*t*
genesis of, 2–5
guiding principles and creation of a community resource project, 4–6
Healthy Adult Cohort Study of Multiple Microbiomes, 10–17
large-scale sequencing centers, 6–7
next-generation sequencing and, 56
NIH Common Fund-supported, launching of, 4
previously funded NIH microbiome-related RFAs and PAs not part of, 42*t*
products from, 27–28
computational tools for human microbiome research, 28
derivative datasets from Healthy Adult Cohort Study, 28
publications from, 28
PubMed publications list, 28
reference strain microbial genome sequences, 8–10
Research Network Consortium, 7
as scientific resource, 2
technology development, 23–24
projects listed by project title and investigator, 24*t*
web site for, 51
Working Groups, 6, 7–8
Human microbiome research
computational tools for, 28
future directions for, 45, 46–48
genesis of, 2–5
other NIH-supported, 41–42
Human microbiota
cultivation of, 308–309
"great plate count anomaly" in, 303–304
horizontal acquisition of, 279
Humans
animal model systems and understanding host-microbiota interactions relevant to, 275–279
diversity of commensal organisms of, vs. diversity in animal models, 256, *257*
as ecosystems, 273
innate and adaptive immunity in, 258
microbial and immune system diversity in models of commensal-host interactions, *257*
microbial diversity in, 255–258
Hybridization, 120, 128
Hybridization silence, *249*, 250
Hydra
low-complexity microbial communities and, *275*
symbiosis, evolutionary impact of microbe-dependent lifestyle and, 277
transgenic, 283
Hydrogen peroxide, *Lactobacillus* sp., protection against BV and, 191

Hydrolysis probe assays, 59
Hygiene hypothesis, 238

IBD. *See* Inflammatory bowel disease
IBD5 risk alleles, increased CD severity and, 110
ichip (isolation chip), 63, 306–307, 308, 309
  for high-throughput microbial cultivation, *306*
  miniature, for *in vivo* microbial cultivation in oral cavity, *310*
  new variant of, advantages with, *307*
ICR. *See* Ileocolic resection
IE. *See* Infective endocarditis
Ileal Crohn's disease, 107, 109
Ileocolic resection, 107
Ileum
  of healthy person, *221*
    DAPI fluorescence of all DNA structures at 400x magnification, *221*
  of interleukin 10 gene-deficient mouse and control IL10 gene-deficient mouse, *242*
  NOD2/CARD15 protein in, 109
Illumina GAII technologies, 13
Illumina sequencing, vaginal microbiome and use of, 177
Illumina/Solexa, 56
IL1, obesity and, 154
IL6, obesity and, 154
IL10-producing Tregs, protection from colitis and, 266
IL12B, inflammatory bowel disease and, 265
IL23R, inflammatory bowel disease and, 265
Immune dysfunction, immunomodulatory effects of microbiota members and, 285
Immune system
  development in animals: comparative immunity, 258–259
  diversity in animal models of commensal-host interactions, *257*
  microbial influences on, 283–284
Immunology field, revolution within, 3
in situ activity, 65
in situ hybridization, 53, 295
in situ PCR, 60
Infective endocarditis
  diagnosis, development and treatment of, 149
  oral communities and, *147*, 148–151
  promotion of, *139*
Infernal method, 88
Inflammation, atherosclerosis and, 151
Inflammatory bowel disease, 21, 45, 105, 111, *220*, 252, 265, *268*, 284, 318, 331
  biostrucure of fecal microbiota in, 245, 247
  break of mucus barrier in, 234, 236–238
  commensal bacteria and, 265–266
  crypt abscesses and, 229
  enteric microbiota and, 107–108
    pathogenic microbial communities, 108
    specific microbial pathogens, 107–108
  human genetic loci associated with, 108–109
  in humans, mouse animal models and learning about, 259
  prebiotics and, 326–327
  prevalence of, 265
  synbiotics and, 327–329
  Tregs and, 264
Inflammatory cytokines, obesity and, 154
Initiation mutants, *V. fischeri*, 280
Innate immune system, 3
Innate immunity, 258
INRA. *See* National Institute for Agricultural Research
Institut Pasteur, Biological Resource Center of, 9
Insulin, 153
Interbacterial interactions in oral cavity
  antagonistic, 141–142
  metabolic, 140–141
  physical, 140
Interbacterial relationships, understanding, 269
Interbacterial signaling, in oral cavity, 142–143
International Classification of Diseases 2010 (WHO), 42
International Human Microbiome, formation of, 3
International Human Microbiome Consortium, 51
Intestinal architecture, bacteria and, 255
Intestinal homeostasis
  maintenance of, 260–264, *261*
    microbial composition and, *268*
Intestinal inflammation, most prominent feature of, 234
Intestinal microbiota
  bacteria in upper GI tract, 212–219
    biliary tract, 218–219
    gallstones, 219

mouth, 212, *213*
pancreatic tract, 218
stomach and duodenum, 217–218
tonsils, 214, *215*, *216*, *217*
biostructure of fecal microbiota in health, IBD, and other GI diseases, 245–247
changes of colonic microbiota in disease, inflammatory bowel disease, 234–238
description of, 212
factors potentially constituting mucus barrier, 238
facultative pathogens, 238–239
fecal microbiota, 230–234
functional structure of, in health and disease, 211–253
large intestine, 220, 222
mucus barrier, 222, *223*, 224–225, 228–230
polymicrobial community of, 211
remodeling mucus barrier, possible strategies, 243–245
role of microbiota in colonic function, 222
site-dependent changes of colonic microbial biostructure, 247–251
germinal zone of colonic biofermenter, 250
mucus layer, 247–248
working area of colonic biofermenter, 248–250
small intestine, 219–220
substances reducing viscosity of mucus barrier, 239–243
bile acids, 241
detergents, 240
emulsifiers, 241
glutens, 241–242
smoking, 242
stress, 242
Intestinal tract, steady-state microbial communities of, 105
Intestine, function of, 211
Intrauterine contraceptive device, 170
Inulin, 327, 328
Irritable bowel syndrome, 21, 230
bacterial concentrations found in mucus of patients with, 237
cecal biopsy from patient with, *230*
increased mucus production in patients with, *248*
probiotics and, 320
Isogenic mice, comparative profiling of gastrointestinal microbiota from two groups of, 79
Isolation, 292–293, 295
Isolation chip. *See* ichip (isolation chip)
Isolation techniques, development of, 291
Isotopic signatures, 295
IUD. *See* Intrauterine contraceptive device
IV drug users, infective endocarditis and, 149

Jaccard index, 57
Jumpstart Human Microbiome Project (Australia), 4

Klenow fragment of DNA polymerase, 61
Koch, Robert, 292
Konopka, Alan, 292

Lachnospiraceae, 112
Lactic acid, competitive inhibition in vagina and, remaining questions about, 191–192
Lactin-V, 193–194
Lactitol, 331
Lactobacillales, among HIV-positive women with low Nugent scores, 180
Lactobacilli, 317
in healthy vagina, 183
postmenopausal women and, 194
probiotic, modulating of cytokine signaling and inflammation in reproductive tract and, 193
vaginal, primary goals of research on, 192–193
Lactobacillus, 189
early sexual debut and, 195
protection from IBD in mouse models and, 266
viscosity of mucus barrier and, 225
Lactobacillus acidophilus NCFM, 331
Lactobacillus bulgaricus, 317
Lactobacillus casei rhamnosus Lcr35, 322
Lactobacillus casei Shirota, 322
Lactobacillus crispatus, 180, 189, 191
Lactobacillus gasseri, 189, 191, 196
Lactobacillus GG, 322
Lactobacillus iners, 189, 190, 191
in BV-positive women, 184
male reproductive tract and, 197
vaginal microbiome and, 182
Lactobacillus jensenii, 189, 190
Lactobacillus johnsonii, 318
Lactobacillus plantarum, 257
Lactobacillus reuteri DSM 17938, 322

Lactobacillus rhamnosus GG, 318, 319, 320, 321, 322, 323
Lactobacillus salivarius, 319, 331
Lactobacillus sp.
vaginal, pregnancy and, 169
in vaginal bacterial community in women worldwide, 198
Lactulose, 327
Lamellocytes, 258
Large bowel, metabolic and physiologic processes mediated by bacteria growing in, *316*
Large intestine
bacteria in, 220, 222
as biofermenter, 222
function of, 212
prebiotics and health benefits for, 327
regulatory processes and complex microbiota of, 315
Large-scale sequencing centers, 6–7
Law of large numbers, 97
Lecythophora, modeling in zebrafish, 259
Leech *(Hirudo verbana)*
digestive tract symbioses and, 277
low-complexity microbial communities and, *275*
Lentzea, trap method and cultivation of, 308
Lepidopterans, 277–278
Leptotrichia, in BV-negative women, 183
Leptotrichia amnioniii, bacterial vaginosis and, 186
LF82 strain, Crohn's disease and, 107–108
L1007fs, Crohn's disease and, 109
LH-PCR, microbiome of CF patients examined with, 122
Lips, communities of bacteria on, *137*
Liquid serial dilution, 293
Long curly bacteria, viscosity of intestinal mucus and, 225, 228
Low birth weight, periodontal disease and, 155
Lungs
asthma and, 126–127
bronchopulmonary dysplasia and, 126
complex microbial communities in, 129
COPD and, 127–128
cystic fibrosis and, 121–125
mechanisms of innate immunity in, 119
microbiome studies, overview of, 123*t*
normal, 121
ventilator-associated pneumonia and, 125–126
Lymantria dispar, 278
Macrophage colony-stimulating factor, 151
MALDI-TOF, 110
Male reproductive tract
anatomy of, *196*
different microbial communities in different anatomically defined niches of, 196–197
Mammalian intestine, colonization of, 280, 281
MAP. *See Mycobacterium avium* subspecies *paratuberculosis*
Marine basalts, investigating, functional gene arrays and, 61
Marine environments, metatranscriptomics applied to, 62
Mathematical approaches for describing microbial populations, 85–101, *100*
average analysis of sample coverage, 92–97
building the taxonomy table, techniques, 86–88
assigning best taxonomy to OTUs once chosen, 88
demultiplexing the data, 86
level for clustering OTUs, 88
quality filtering, 86–87
conditional analysis of sample coverage, 97–100
embedding algorithm, 99–100
thinning property of Poisson point processes, 97–98
theory of extrapolation of environments, 88–92
expected coverage of a sample, 90–91
paradigm shift, 89–90
urn model, 88–89
Mating preferences, bacteria and, 255
Matrix-assisted laser desorption ionization-time of flight mass spectrometry. *See* MALDI-TOF
MBP. *See* Myelin basic protein
M-CSF. *See* Macrophage colony-stimulating factor
MDA. *See* Multiple displacement amplification
MDP. *See* Muramyldipeptide
Megasphaera
bacterial vaginosis and, 184, 185
in BV-negative women, 183
Mesalazine, 244, 318
Metabolic interbacterial interactions, in oral cavity, 140–141
Metabolic potential, 65

Metabolomics, 46, 63
Metagenomics, 3
Metagenomics of the Human Intestinal Tract (MetaHIT), 4
MetaHIT consortium, 61
MetaHIT project, 51
Metaomics studies, 295
Metaproteomics, 46, 62–63
Metatranscriptomics, 46, 62
Metchnikoff, Elie, 317
Methanobrevibacter smithii, in human intestine, 256
Methanogen diversity, in human feces, 60
Methotrexate, 244
Methylcoenzyme M reductase *(mcrA)* genes, 60
Metronidazole, for bacterial vaginosis, 173
MG1 (MUC5B), in saliva, 138
MG2 (MUC7), in saliva, 138
Mice. *See* Mouse (mice)
Microautoradiography (MAR)-FISH, 60
Microbacterium laevaniformans, 305
Microbe-microbe interactions, host-associated microbial communities and, 281
Microbial communities associated with human body
  methods for characterizing, 51–65, *52*
    cultivation, 63–64
    cultivation-independent techniques, 52–63
    insights into, and future projects, 64–65
Microbial community analysis, revolution in, 85
Microbial diversity
  in animal models of commensal-host interactions, *257*
  estimates of, changes in, 289
Microbial identification, developments in, 53
Microbial physiology, genome sequence as foundation for, 291
Microbial signaling molecules, 63
Microbiology, cultivation as foundation of, 63
Microbiome
  coining of term, 2
  function studies, future of, 48
  of human-associated microbial communities, 273
  profiling, 76
Microbiome research, ethical, legal, and societal implications of, 26–27
Microbiota
  of human-associated microbial communities, 273
  investigating, sample processing and molecular biology techniques, *120*
  wellbeing of multiple organisms and centrality of, 255
Microbiota composition, identifying host genetic determinants of, 281
Microbiota cultivation methodology, 304–308
  diffusion chamber, 304–306
  ichip, 306–307
  traps, 308
Microfluidics, 62, 292
MicroObes, 4
Microorganisms, 291
  with claims of probiotic properties in humans, 317, 317*t*
  diffusion chamber for *in situ* cultivation of, *305*
  growth of, 293
  ichip-grown, 307
  isolation of, 292–293
Minimum variance unbiased estimator of (1 – *M*), 95
Minitrap, *in situ* cultivation of oral microorganisms and use of, 308–309, *309*
Miscarriage, periodontal disease and, 155
MM probes, 77, 78, 82
Mobility of modern society, worldwide exchange of bacteria and, 239
Mobiluncus spp., 187, 196
MoBio Powersoil DNA extraction kit, 11
Molecular biology techniques, investigating the microbiota and, *120*
Monocultures, polymicrobial communities vs., 212
Moraxella catarrhalis
  asthma and, 127
  COPD and, 127
Morisita–Horn measurement, 57
mothur, 57, 88
Mouse diet, microbiota assembly within digestive tract and, 281
Mouse (mice), 255, 274
  distal colon of, monoassociated with *Enterobacter cloacae*, *228*
  gnotobiotic, function of human microbiota and studies of, 279

Mouse (mice) (*Continued*)
healthy, bacteria highly concentrated and distributed throughout proximal colon in, 226
high-complexity microbial communities and, *276*
innate and adaptive immunity in, 258
intestinal homeostasis in, *261*, 263–264
microbial and immune system diversity in models of commensal-host interactions, *257*
microbial influences on epithelial tissues and, 283
microbial influences on immune system and, 284
proximal colon in, *227*
Bacteroides, 227
monoassociated with *Enterobacter cloacae, 229*
mucus barrier in, 228, *229*
only *Bacteroides* within same microscopic field, *227*
Roseburia, 227
Mouse models
commensal bacteria and protection from IBD in, 265–266
of diabetes, 267
of multiple sclerosis, 266–267
QTL mapping studies and, 282
Mouth
anatomy of, 136–137
bacteria in, 212–213
massive bacterial biofilm attached to food remnants in, *213*
prebiotics and health benefits for, 327
Mouth, microbiota of, 135–156
from colonization to communities
antagonistic interbacterial interactions, 141–142
initial attachment, 138–140
metabolic interbacterial interactions, 140–141
physical interbacterial interactions, 140
communities in health and disease
disease, 145
health, 143–145
interbacterial signaling, 142–141
biofilm matrix, 143
oral communities and systemic disease
adverse pregnancy outcomes, 155–156
atherosclerosis, 151–153
bacteraemia, 146, 148
diabetes, 153–154
infective endocarditis, 148–151
obesity, 154–155
saliva, 137–138
Mouth epithelium, healthy, *213*
MS. *See* Multiple sclerosis
Mucins
oral cavity and, 138
in saliva, types of, 138
Mucociliary clearance, 119
Mucophob, 231
Mucophobic bacteria, 231, *233*
Mucotrop, 231
Mucotropic bacterial groups, *233*
Mucus barrier
prolonged compromise of, 252
remodeling, strategies for, 243–245
Mucus barrier, intestinal, 222–230
factors potentially constituting, 238
mucus surface layer of formed stools, 231
role of facultative pathogens, 238–239
substances reducing viscosity of, 239–243
bile acids, 241
detergents, 240
emulsifiers, 241
glutens, 241–242
smoking, 242
stress, 242
viscosity of, 224–225, 228–230
changes in, control of, 229–230
Mucus layer
in colonic biofermenter, 245
site-dependent changes of colonic microbial biostructure and, 247–248
Multiple displacement amplification, complications with, 62
Multiple sclerosis, 45, 265, *268*
commensal bacteria and, 266–267
mouse animal models and learning about, 259
Tregs and, 264
Muramyldipeptide, 109
Mutualism, 274
Mycobacterium, modeling in zebrafish, 259
Mycobacterium avium subspecies *paratuberculosis*, 107
Mycobacterium marinum, zebrafish model and study of, 258
Mycoplasma sp., in BV-positive women, 187, 188
Myd88, 283, 284
Myelin basic protein, 266

Myocardial infarction, 151
Myoplasma hominis, bacterial vaginosis and, 187–188, 196

Nasopharynx, microbiota of, 135
NAST, 88
National Academy of Sciences, 3
National Center for Biotechnology, 5, 181
National Children's Study, 47
National Collection of Type Cultures (UK), 8
National Human Genome Research Institute, 3, 26
National Institute for Agricultural Research (France), 3
National Institutes of Health, 1, 273. *See also* Human Microbiome Project
  Common Fund website, 5
  HIV Lung Microbiome Project, 4
Native valve infective endocarditis, 149
Natural killer (NK) cells, in zebrafish, 259
NCBI. *See* National Center for Biotechnology
NCBI Bioprojects pages, 5
NCS. *See* National Children's Study
NCTC. *See* National Collection of Type Cultures (UK)
Neanderthals, comparative genomics studies of, 48
Necrotizing enterocolitis (NEC), 21
Neisseria gonorrhoea, 171, 191
Neisseria spp., 136, 310
Nematode
  binary bacterial-animal mutualism and, 276–277
  binary host-microbe associations and, *274*
Next-generation sequencing technologies, 56, 57, 120, 273
NFkB proteins, intestinal homeostasis and, 260, 261, 262, 263
NHGRI. *See* National Human Genome Research Institute
NHGRI Sample Repository for Human Genetic Research, 11
NIH. *See* National Institutes of Health
NMR spectroscopy, 63
Nod2, 109
NOD2 alleles, 110, 112
NOD2 genotypes, Crohn's disease and, 111
NOD2/CARD15 loci, Crohn's disease and, 108, 109
NOD2/CARD15 risk alleles, younger age of surgery and, 110
Nonspecific vaginitis, 187
Nosocomial infective endocarditis, 149
Novel microbes, 64
NSV. *See* Nonspecific vaginitis
Nucleic acid sequencing technology, 53
Nugent scores, *179*, 182, 186, 191, 195, 198
  bacterial vaginosis and, 173, *174*, 175, 176
    ambiguity around intermediate scores, 175
  postmenopausal women and, 194
  vaginal bacterial community types and, 189–190
  vaginal microbiome studies and, 178
Nutrient media for microbial isolation, composition of, experimentation and, 310–311
Nutrition, bacteria and, 255. *See also* Diet

Obesity
  diabetes and, 153
  oral communities and, 154–155
Obesity epidemic, socioeconomic implications of, 154
Oligonucleotide probe sequence, 77
Olsenella, infected root canals and, 145
Omics studies, 295
Operational taxonomic units, 77, 86
  assigning best taxonomy to, 88
  comparisons of communities based on, 57
  diffusion chamber and, 305
  levels for clustering of, 88
  vaginal microbiota and, *179*, 181
Optical tweezers, 292
Oral bacteria, in health and disease, *144*
Oral cavity
  anatomy of, 136–137
  bacteria and stratified epithelium of, 213
  from colonization to communities, 138–143
    antagonistic interbacterial interactions, 141–142
    biofilm matrix, 143
    initial attachment, 138–140
    interbacterial signaling, 142–143
    metabolic interbacterial interactions, 140–141
    physical interbacterial interactions, 140
  communities in health and disease
    disease, *144*, 145
    health, 143–145, *144*

Oral cavity (*Continued*)
diagrammatic representation of, *137*
microbial colonization of, *139*
microbiota of, 135
miniature ichip for *in vivo* microbial cultivation in, *310*
minitrap used for *in situ* cultivation of microorganisms from, 308–309, *309*
mucosal surfaces of, 136
Oral communities and systemic disease, 146, 148–156
adverse pregnancy outcomes, 155–156
atherosclerosis, *147*, 151–153
bacteraemia, 146, 148
diabetes, 153–154
infective endocarditis, *147*, 148–151
obesity, 154–155
Oral mucosa, turnover rate for, 126
Oral sepsis, infective endocarditis and, 148
OTU-based clustering, advantage and disadvantages with, 86–87
OTUs. *See* Operational taxonomic units
Oviduct, 168, 170

Pancreatic duct, calcified, biofilm in, *218*
Pancreatic tract, bacteria in, 218
Paneth cells, 109, 111
Parotid glands, 137
Particulate matter, human airway and, 119
Parvimonas, infected root canals and, 145
Pathogenesis, 274
PCR. *See* Polymerase chain reaction
Pediatric patients, gut disease studies with, 21
Pellicle binding sites, pioneer colonizers competing for, 140
Pellicles, on enamel surface of teeth, 138
Pelvic inflammatory disease, bacterial vaginosis and, 171, 176
Penile microbiome, 22
Penis, microbiota of, 196–197, 198
Peptidoglycan (PG) levels, intestinal homeostasis and, 261
Peptidoglycan recognition proteins, deficiency in mice and disrupted intestinal homeostasis, 263
Peptoniphilus, in BV-negative women, 183
Peptoniphilus lacrimalis, in vagina, 185
Peptostreptococcus, in BV-negative women, 183
Peptostreptococcus anaerobius, in vagina, 185
Periodontal disease
adverse pregnancy outcomes and, 155
diabetes and, 153
gestational diabetes and, 154
oral microbiota and, 145
Periodontitis, 144, 148, 154
PerMANOVA, circumcision, alteration in penile microbiota and use of, 197
Persistence mutants, *V. fischeri*, 280
PGA, biofilm matrix and, 143
PGRPs. *See* Peptidoglycan recognition proteins
Photolithographic method, 77
Photorabdus, nematode and, 276
Phyloarrays, 75–82
PhyloChip, 58, 75, 79
COPD examination and, 127
microbiome of CF patients examined with, 123*t*, 124
probe set for each taxon detected housed by, 82
sample preparation for phylogenetic microarray profiling and, 78–79
ventilator-associated pneumonia and, 125, 126
Phylogenetic clustering algorithms, profiling of entire vaginal microbial communities and, 189
Phylogenetic markets
community structure and composition and use of, 53–60
community fingerprinting, 54–55
DNA sequencing, 55–57
FISH, 59–60
phylogenetic microarrays, 58
qPCR, 58–59
Phylogenetic microarrays, 58, 75
design, 76–77
sample preparation for profiling, 78–79
utility of, 79–82
Phylogenetic profiling tools, strengths and weaknesses of, 82
Physical interbacterial interactions, in oral cavity, 140
PID. *See* Pelvic inflammatory disease
Pili, 225
Planctomycetes, in human gut, 256
Plaque, dental, bacteria within, 148
Plaque biofilm development, insoluble glucans and, 143
Plate wash PCR, 64
PM probes, 77, 78, 82

Pneumonia, ventilator-associated, 125–126
Poisson point processes, thinning property of, 97–98
Polymerase chain reaction, 53, 76, 85
Polymicrobial communities, monocultures vs., 212
Polymicrobial diseases, formerly classified, entire microbiome and, 2
Porphyromonas gingivalis, 140, 141, 142, 143, 146
  atheromas and, 152, 153
  gum disease sites and, 136
  infected root canals and, 145
  periodontal disease, adverse pregnancy outcomes and, 155, 156
  periodontitis and, 153
Postmenopausal women, vaginal pH in, 194
Pouchitis, 22, 266, 319–320
Prebiotics, 285, *316*, 324–327, 331
  defining, 324
  health benefits associated with, 325, 327
  inflammatory bowel disease and, 326–327
Prediction, estimation vs., 90
Prednisolone, 244, 250
Preeclampsia, periodontal disease and, 155
Pregnancy
  adverse outcomes, oral communities and, 155–156
  vaginal microbes and, 169
Premature infants, bronchopulmonary dysplasia and, 126
Preterm birth
  bacterial vaginosis and, 176
  BV and, debate over, 173
  periodontal disease and, 155
Prevotella intermedia
  atheromas and, 152
  periodontitis and, 153
Prevotella timonensis, in BV-positive women, 184
Prevotellas spp.
  in BV-negative women, 183
  normal lung and, 121
Probiotics, 284, 285, *316*, 317–324, 331
  acute gastroenteritis and, 321–322
  aging and, 323–324
  allergy-related diseases and, 322–323
  antibiotic-associated diarrhea and, 321
  Crohn's disease and, 318
  defined, 317
  irritable bowel syndrome and, 320
  pouchitis and, 319–320
  ulcerative colitis and, 318–319
Probiotics research, current aims in, 193
Proinflammatory cytokines
  gestational diabetes and, 154
  vaginal microbes, conception and, 169
Proline-rich proteins, oral cavity and, 139
Propionate, physiologic effects of, produced by bacteria colonizing the large gut, 325, *326*
Propionibacterium freudenreichii subsp. *shermanii* JS, 320
Prosthetic valve infective endocarditis, 149
Proteins, in salivary secretions, 137–138
Proteobacteria
  asthma, COPD and, 121
  Crohn's disease and, 108
  in fruit fly, 256, 257
  in human gut, 256
  in vagina, 183
  vaginal microbiome and, 178
  zebrafish microbiota and, 258
Protists, 291
Providencia, in fruit fly, 256
PRPs. *See* Proline-rich proteins
PSA, 284
  antiinflammatory IL10 response and, 264
  protection from IBD and, 266
  protective in models of EAE, 267
Pseudomonadaceae, CF genotype, age and, 123*t*, 124
Pseudomonas aeruginosa
  biofilm matrix and, 143
  COPD and, 127
  cystic fibrosis and, 80, 122, 124
  ventilator-associated pneumonia and, 125, 126
  zebrafish model and study of, 257
Pseudoramibacter, infected root canals and, 145
Pseudorectum surgery, 22
PubMed, HMP publications in, 31*t*–41*t*
Putative peptide signals, 63
PWPCR. *See* Plate wash PCR
Pyrosequencing, errors and, 86

QIIME, 57, 88
qPCR, 58–59
Quality control, bioinformatic tools for, 56
Quality filtering, 86–87

Quantitative PCR. *See* qPCR
Quantitative trait loci (QTL) mapping strategy, 282

Raman transform infrared spectroscopy, 295
Raman-FISH method, 60
Random variable, average analysis of sample coverage and, 92–93
Rate measurements, 295
RDP
  classifier, 88
  in public databases, growth of, 53
Reactive oxygen species, homeostatic balance in fly gut and, 262
Real-time PCR, 58
Reference strain microbial genome sequence dataset, 5–6
Reflux esophagitis, 21
Relative abundance, accounting for, in community analyses, 57
Resident microbial communities, in host biology, function of, 282–284
Respiratory distress syndrome, 126
Respiratory tract microbiome studies, outlook for, 128–129
Resting zone, in colonic biofermenter, 245
Rheumatoid arthritis, 265
Ribosomal Database Project, 181
Ribosomal RNA Database, 54
Ribosomal RNA sequences, 53
Richness, calculating, 57
Rifaxamin, 318
Risk alleles, ileal CD and, 109
RNALater, 78
Robbins' estimator of $(1 - M)$, 93, 95, 97
Roche 454 FLX pyrosequencing platform, vaginal microbiome and use of, 177
Roche 454 sequencing technology, 13, 56
Root canals, infected, microbiota of, 145
ROS. *See* Reactive oxygen species
Roseburia spp., 237
  colonic biofermenter and, 222
  in fecal flora, 231, *232*
  in large intestine, 220
  in mouse with IL10 deficiency, *246*
  self-limiting colitis in ascending colon and, *238*
  between villi in ileum of patient with Crohn's disease, *243*
  viscosity of mucus barrier and, 224, 228
R702W, Crohn's disease and, 109

S. cristatus, 142
S. mutans, 143
Saccharomyces boulardii, 318, 321, 322
Saliva, 127–128
  bacteria in, 212
  isolated island of bacteria attached to desquamated epithelial cells in, *214*
  microbial colonization and, 136
Salivary glands, bacteria and epithelium of, 213
Salivary pellicle
  removing, 139
  salivary glycoproteins and, 137
Salmonella, 85, 277
Salmonella typhimurium, 109
Sample processing, investigating the microbiota and, *120*
Samples
  average analysis of sample coverage, 92–97
  conditional analysis of sample coverage, 97–100
    embedding algorithm, 99–100
    thinning property of Poisson point processes, 97–98
  expected coverage of, 90–91
SAMPs. *See* Symbiosis-associated molecular patterns
Sanger sequencing, 62
  of clone libraries, 56, 57
  microbiome of CF patients examined with, 123*t*, 124
  ventilator-associated pneumonia and, 126
SCFA. *See* Short-chain fatty acids
Secondary ion mass spectrometry, 295
Segmented filamentous bacteria, 256
  diabetes and, 267
  multiple sclerosis and, 266
  small intestine and, 219
Selenomonadales, BV-positive women and, 184
Selenomonas noxia, obesity and, 155
Separation process, 292
Sequence data, collecting, 120
Sequence read archive, 5, 28
Sequence-based approaches, 120
Sequencing biases, 61
Sequencing errors, 56, 86
Sequencing technology, 3
Serpulina, 237, 244
Serpulina biofilm adherent to mucosa, *239*

Serratia, in fruit fly, 256
Sexually transmitted diseases
female urogenital tract and, 171
in men, 198
Sexually transmitted infections, 22, 191
SFB. *See* Segmented filamentous bacteria
Shannon index, 57
Short bowel syndrome, 107
Short coccoid rods, viscosity of intestinal mucus and, 228
Short-chain fatty acids, 108, 317, 325
Shotgun metagenomics, 60–61, 62
Signal compounds, 292
"Silent infections," 239
SILVA web database, in public databases, growth of, 53
Simpson index, 57
Single-cell genome amplification and sequencing, 62
Single-cell techniques, 295
Single-nucleotide polymorphisms
Crohn's disease and, 108, 109
increased risk of, 110*t*
16S rRNA community sequence data, analytic methods for, 57
16S rRNA gene, 5
amplification, ventilator-associated pneumonia and, 126
oligonucleotides for, synthesizing, 77
ubiquity of, 85
vaginal microbiome, high-throughput screening and, 177
16S rRNA gene PCR, male reproductive tract and, 197
16S rRNA gene sequence databases, rapid expansion of, 76
16S rRNA PhyloChip, 75
cystic fibrosis patient airway microbiota study, 80
probe set for each taxon detected housed by, 82
version G2, 77
16S rRNA sequences
interpretation of, 8
in public databases, growth of, 53
Skin diseases, microbiome and, 22–23
"Skip" lesions, Crohn's disease and, 106
slp, 88
Small intestine
bacteria in, 219–220
function of, 212
rat, *221*
same microscope field in DAPI fluorescence, *221*
Smoking
COPD and, 127
mucus secretion and, 242
Sneathia, in BV-negative women, 183
Sneathia sanguinegens, bacterial vaginosis and, 186
Sneathia sequences, vaginal microbiome studies and, 178
SNP. *See* Single-nucleotide polymorphisms
Soft palate, communities of bacteria on, *137*
Soil, investigating, functional gene arrays and, 61
Sørenson index, 57
Species, definitions of in microbes, problematic nature of, 86
Sperm survival and motility, vaginal microbes and, 169
Spirochaetes, in human gut, 256
Squid
binary bacterial-animal mutualism and, 275–276
binary host-microbe associations and, *274*
Squid-*Vibrio* symbiosis, microbial influences on immune system and, 284
SRA. *See* Sequence read archive
SSU rRNA genes, 53, 54
Staley, James, 292
Staphylococcaceae, differentiating, 77
Staphylococcus, in BV-negative women, 183
Staphylococcus aureus, 23
cystic fibrosis and, 122, 124
infective endocarditis and, 149, 150
ventilator-associated pneumonia and, 125
zebrafish model and study of, 257
Staphylococcus spp., 121
Starr's estimator, 94, 95, 97
STAT3, inflammatory bowel disease and, 265
Statherin, oral cavity and, 139
Steinernematidiae, 276
Steptococcus mitus, oral surfaces and, 137
STIs. *See* Sexually transmitted infections
Stomach, bacteria in, 217–218
Strains Working Group, 9
Streptacidophilus, trap method and cultivation of, 308
Streptococci
dental plaque and, 136
healthy microbial communities in oral cavity and, 144

Streptococcus
in BV-negative women, 183
infected root canals and, 145
protection from IBD in mouse models and, 266
Streptococcus australis, 144
Streptococcus gordonii
infective endocarditis and, 150
in oral cavity, 140, 141, 142
oral surfaces and, 137
Streptococcus intermedius, subgingival oral surfaces and, 137
Streptococcus milleri group, CF airway and, 123
Streptococcus mitis, 144, 150, 151
Streptococcus mutans, infective endocarditis and, 150, 151
Streptococcus oralis, 144, 150
Streptococcus pneumoniae, 142
asthma and, 127
COPD and, 127
Streptococcus pyogenes, diffuse infiltration of tonsil with, *215*
Streptococcus salivarius, 136, 137, 145, 151
Streptococcus salivarius strain K12, intestinal homeostasis and, 263
Streptococcus sanguinis, 137, 140, 150, 151
Streptococcus spp., 310
Stress, viscosity of mucus and, 242
Structure, defined, 54
Sublingual salivary glands, 137
Submandibular salivary glands, 137
Supernatants, intestinal homeostasis and, 263
"Supershedder" mouse phenotype, 81
Supraautoimmune disease, Crohn's disease as, 114
Symbiosis, defined, 274
Symbiosis-associated molecular patterns, 264
Symbiosis-defective bacterial mutants, identifying, 280–281
Synbiotic 2000, 328, 329
Synbiotics, 327–331, 331
defined, 327
histopathology of rectal mucosa from UC patient presynbiotic and post-synbiotic treatment, *329*
IgE-associated disease and, 330
improved immune function in aging population and, 331
inflammatory bowel disease and, 327–329
specimen bowel habit diary from UC patient receiving *Bifidobacterium longum* and Synergy I during 4-week study period, 330*t*
Synergy I, 331
specimen bowel habit diary from UC patient receiving *Bifidobacterium longum* and, 330*t*

T helper cells, intestinal homeostasis and, four main populations of, 264
T lymphocytes, in zebrafish, 259
T regulatory cells (Tregs), 264, 266
Tannerella, infected root canals and, 145
Tannerella forsythia, periodontitis and, 153
TaqMan, 59
TaqMan allelic discrimination assays, 110
Target microbes, detecting growth of, 64
Taxonomic clustering, advantage and disadvantages with, 86
Taxonomy table
building, considerations in, 86–88
assigning best taxonomy to OTUs once chosen, 88
demultiplexing the data, 86
level at which to cluster the OTUs, 88
quality filtering, 86–87
Teeth
communities of bacteria on, *137*
microbial load of, 126
Temperature gradient gel electrophoresis, 54
Temporal temperature gradient gel electrophoresis
examining microbiota in clinical samples with, 120
microbiome of CF patients examined with, 122, 123*t*
Tenericutes, vaginal microbiome and, 178
Terminal restriction fragment length polymorphisms, 55, 120
bronchopulmonary dysplasia and, 126
microbiome of CF patients examined with, 122, 123, 123*t*
vaginal microbiome and, 177
Termites, *276*, 278–279
TGGE. *See* Temperature gradient gel electrophoresis
Th1 cells, 264
Th2 cells, 264
Th17 cells, 264, 284
Thermoactinomycetaceae, CF genotype, age and, 123*t*, 124

Thinning property of Poisson point processes, 97–98
TLR2. *See* Toll-like receptor 2
TNBS. *See* Trinitrobenzene sulfonic acid
TNFalpha
  obesity and, 154
  periodontal disease and, 153
Toll-like receptor 2, 264
Tongue, 126, *137*
Tonsillar epithelium, healthy, and free of bacteria, *215*
Tonsillar foci, infectious, composition of bacteria within, 214
Tonsillar lesions, local, occurrence of different bacterial groups within, such as fissures and diffuse infiltrates, 217*t*
Tonsillar surface
  superficial adherence of bacteria to, *216*
    DAPI stain of same microscopic field revealing DNA structures, *216*
Tonsillar tissue, macroabscess within, *217*
Tonsils
  bacteria in, 214
    communities of, *137*
  diffuse infiltration with *Streptococcus pyogenes, 215*
  diffuse local infiltration of, with *Haemophilus influenzae, 215*
  examples of microabcess, *216*
    DAPI stain of same microscopic field, *216*
  fissure filled with bacteria, *216*
    DAPI stain of same microscopic fields demonstrating bacteria surrounded by inflammatory cells, *216*
Tooth extraction, bactaeremia and, 148
Tooth pulp-root canal system, infections in, 145
Top-down approaches, 295
Transmembrane protein toll, *Drosophilia* and, 258
Trap method, *in situ* cultivation and, 308
Traps, 309
Treponema, infected root canals and, 145
Treponema denticola, 141, 152
T-RFLP. *See* Terminal restriction fragment length polymorphisms
Trinitrobenzene sulfonic acid, 259
TTGE. *See* Temporal temperature gradient gel electrophoresis
T300A polymorphism, ileal Crohn's disease and, 109
Twin Cohort Microbiome Diversity project (Korea), 4

UC. *See* Ulcerative colitis
uclust, 88
Ulcerative colitis, 106, 107, 248, 265–266
  fecal cylinder of patient with, *251*
  pathogenic microbial communities and, 108
  probiotics and, 318–319
  replacement of mucus layer by leukocytes in, 250
  reported cases of, steady increase in, 238
  showing cytopathogenic effects of bacterial adhesion, *235*
  sigmoid colon of patient with, *237*
  tobacco smoke beneficial in patiens with, 242
Unbiased estimator for (1—*M*), 93
Uncircumsized men, microbiota of circumcised men vs. microbiota of, 196–197
UniFrac, 57, 88
Universal target, vaginal microbiota studies and, 180
Upper gastrointestinal tract
  bacteria in, 212–219
    biliary tract, 218–219
    gallstones, 219
    mouth, 212–213
    pancreatic tract, 218
    stomach and duodenum, 217–218
    tonsils, 214, *215, 216, 217*
Ureaplasma sp., bronchopulmonary dysplasia and, 126
Urinary tract infection, Lactin-V and, 193
Urn model, 88–89
Urogenital diseases, microbiome and, 22
UT. *See* Universal target
Uterus, 168
UTI. *See* Urinary tract infection
Uvula, *137*

Vagina, microbial inhabitants of, common configuration for, 168–169
Vaginal atrophy, new findings on, 194
Vaginal bacterial community types, identification of, 189–190
Vaginal birth, 45
Vaginal dryness, absence of BV and, 194

Vaginal health in 2025, understanding the microbiome and future of women's healthcare, 188–196
Vaginal microbial community composition
  interpersonal variation in, 181
Vaginal microbiome, 22
  in age of high-throughput screening, perspectives on, 176–188
Vaginal microbiota
  classes of, 172
  longitudinal studies on, criticism of, 195
  molecular perspectives of, *179*
  women's lifelong health and, 194
Vaginal smears, Gram stain interpretation of, 173
Vaginitis, BV and, 194
VAP. *See* Ventilator-associated pneumonia
Vascular cell adhesion molecule 1 (VCAM1), 151
Veillonella spp.
  bacterial vaginosis and, 184, 185
  in BV-negative women, 183
  normal lung and, 121
  in oral cavity, 140, 141, 142
Ventilator-associated pneumonia, 125–126
Verrucomicrobia, in human gut, 256
Vertebrates, complex microbial communities harbored in, 275
Verucobacteriaceae, *233*
Vibrio fischeri, 275, 276, 280, 283–284
Viral identification, microarrays and, 58
Viridans streptococci, infective endocarditis and, 150
VSL3, 318
VSL3 cocktail, protection from IBD in mouse models and, 266

Washington University at St. Louis, 7
Washington University Sequenom Technology/Genotyping Core, 110
WGS. *See* Whole-genome shotgun sequencing
Whole-genome sequencing, 62
Whole-genome shotgun sequencing, for nucleic acid extracts, 5
Wildtype mouse
  distal colon of, universal Eub338 Cy3 probe, at 400', 226
  healthy, colonic bacteria in, *246*
Winterberg, Heinrich, 292
Wipple's disease, 237
Women who have sex with women, 196
Women's healthcare, vaginal health in 2015, understanding the microbiome and, 188–196
World Health Organization, International Classification of Diseases 2010, 42
WSW. *See* Women who have sex with women

Xanthomonadaceae, CF genotype, age and, 123*t*, 124
Xenorhabdus, nematode and, 276

Yersinia spp., 277
Yogurt, 317
Young Lives, 47

Zebrafish *(Danio rerio)*, 255, 274
  embryonic development studied in, 279
  high-complexity microbial communities and, *276*
  innate and adaptive immunity in, 258, 259
  intestinal homeostasis maintained in, *261*
  microbial and immune system diversity in models of commensal-host interactions, *257*
  microbial influences on epithelial tissues and, 283
  microbial influences on immune system and, 284
  microbiome within, 257–258
  NFkB pathway in, 262